International Symposium on Materials and Electrical Engineering (ISMEE 2019)

Advanced Materials for Strengthening Research in Electrical and Computer Engineering

IOP Conference Series: Materials Science and Engineering
Volume 850

Bandung, Indonesia
17 July 2019

Editors:

Budi Mulyanti
Ade Gafar Abdullah
Iwan Kustiawan
Didin Wahyudin

ISBN: 978-1-7138-3613-1
ISSN: 1757-8981

Printed from e-media with permission by:

Curran Associates, Inc.
57 Morehouse Lane
Red Hook, NY 12571

Some format issues inherent in the e-media version may also appear in this print version.

Institute of Physics
Dirac House, Temple Back
Bristol BS1 6BE UK

Phone: 44 1 17 929 7481
Fax: 44 1 17 920 0979

techtracking@iop.org

Additional copies of this publication are available from:

Curran Associates, Inc.
57 Morehouse Lane
Red Hook, NY 12571 USA
Phone: 845-758-0400
Fax: 845-758-2633
Email: curran@proceedings.com
Web: www.proceedings.com

TABLE OF CONTENTS

COMPUTER SCIENCE AND ENGINEERING

ELECTRICAL ENGINEERING

ELECTRONICS ENGINEERING

INFORMATICS

MATERIALS MANUFACTURING AND PROCESSING

MECHATRONICS

MATERIALS SCIENCE AND ENGINEERING

IOP Conf. Series: Materials Science and Engineering **850** (2020) 011001 doi:10.1088/1757-899X/850/1/011001

Preface

It is great honor and pleasure to bring you this collection of articles from the second International Symposium on Materials and Electrical Engineering (ISMEE). The symposium was hosted by the Department of Electrical Engineering Education, Universitas Pendidikan Indonesia and co-organized by Institute of Microengineering and Nanoelectronics (IMEN), Universiti Kebangsaan Malaysia, Japan Advanced Institute of Science and Technology (JAIST), Japan, and IEEE UPI Student Branch. It can be used as a scientific forum for accommodating discussion among researchers in the field of materials and electrical engineering, and engineering education. This symposium was held in the Grand Tjokro Hotel Bandung - Indonesia, July 17th, 2019.

We would like to express our gratitude to all keynote and invited speakers from overseas who travelled to our country to deliver and exchange their ideas. Our appreciation also goes to all the committee members who have worked hard to make this event possible.

The theme of this symposium focuses on "Advanced Materials for Strengthening Research in Electrical and Computer Engineering" which is important thing to accumulate lessons learned from practical uses of the technologies following the theories. So, the interaction between such theoretical and practical aspects of integrating the innovative and advanced technologies would construct new fundamentals of engineering field, which would be one direction ISMEE should pursue.

This volume of proceedings provides an opportunity for readers to engage with a selection of refereed papers that were presented during the ISMEE. We wish you enjoy and discover valuable your engagement with their ideas in sustaining your own professional development in the field of materials and electrical engineering. Thank you.

The Editors

Prof. Dr. Budi Mulyanti
Prof. Dr. Ade Gafar Abdullah
Iwan Kustiawan, Ph.D.
Didin Wahyudin, Ph.D

IOP Conf. Series: Materials Science and Engineering **850** (2020) 011001 doi:10.1088/1757-899X/850/1/011001

LIST OF COMMITTEE

Advisory Board

Prof. Hiroyuki Iida - Japan Advanced Institut of Science and Technology (JAIST), Japan
Prof. Dato' Burhanuddin Yeop Majlis - IMEN, Universiti Kebangsaan Malaysia (UKM), Malaysia
Prof. Dr. Bambang Riyanto, M.Eng. - Institut Teknologi Bandung
Assoc Prof. Shinobu Hasegawa - Japan Advanced Institut of Science and Technology (JAIST), Japan
Prof. Kuang-Hui Chi - National Yunlin University of Science and Technology, Taiwan
Prof. Dr. Khairurrijal - Institut Teknologi Bandung, Indonesia
Dr. Syarif Hidayat - Institut Teknologi Bandung, Indonesia
Dr. Suryono - Universitas Diponegoro, Indonesia
Dr. Yohandri - Universitas Negeri Padang, Indonesia
Dr. Mahardhika Prasetya Adji - Universitas Negeri Semarang, Indonesia
Dr. Eng. Khoeruddin - Universitas Negeri Yogyakarta, Indonesia
Dr. Eng. Tedi Kurniawan - Universiti Malaysia Pahang, Malaysia
Dr. Eng. Muhammad Aziz - Tokyo Institute of Technology, Japan
Dr. Eng. Farid Triawan - Sampoerna University, Indonesia
Dr. Eng. Topan Setiadipura - Badan Tenaga Atom Nasional (BATAN), Indonesia

Scientific Committee

Prof. Dr. Budi Mulyanti – Universitas Pendidikan Indonesia
Prof. Dr. Sumarto – Universitas Pendidikan Indonesia
Prof. Dr. Bachtiar Hasan – Universitas Pendidikan Indonesia
Prof. Dr. Mukhidin – Universitas Pendidikan Indonesia
Prof. Dr. Ade Gafar Abdullah – Universitas Pendidikan Indonesia
Dr. Enjang Akhmad Juanda – Universitas Pendidikan Indonesia
Dr. Tasma Sucita – Universitas Pendidikan Indonesia

Organizing Committee

Conference Chair:
Iwan Kustiawan, Ph.D.

Co-Conference Chair:
Dr. Siscka Elvyanti

Members:
Dandhi Kuswardhana, Ph.D.
Didin Wahyudin, Ph.D.
Dr. Maman Somantri
Dr. Dadang Lukman Hakim
Arjuni Budi P., M.T.
Dr. Hasbullah
Dr. Aip Saripudin
Erik Haritman, M.T.
Tommi Hariyadi, M.T.

IOP Conf. Series: Materials Science and Engineering **850** (2020) 011001 doi:10.1088/1757-899X/850/1/011001

CONFERENCE PHOTOGRAPH

IS 2019
MEE
International Symposium on Materials and Electrical Engineering

IOP Conf. Series: Materials Science and Engineering **850** (2020) 011001 doi:10.1088/1757-899X/850/1/011001

IOP Conf. Series: Materials Science and Engineering **850** (2020) 011001 doi:10.1088/1757-899X/850/1/011001

IOP Conf. Series: Materials Science and Engineering **850** (2020) 011002 doi:10.1088/1757-899X/850/1/011002

CONFERENCE PHOTOGRAPH

IOP Conf. Series: Materials Science and Engineering **850** (2020) 011002 doi:10.1088/1757-899X/850/1/011002

International Symposium on Materials and Electrical
"Advanced Materials for Strengthening Research in Electrical and
Grand Tjokro Hotel, Bandung, Indonesia,
Organized
by
JESTEC

Intern
"Advanced M

Keynote Speake

Peer review statement

All papers published in this volume of *IOP Conference Series: Materials Science and Engineering* have been peer reviewed through processes administered by the proceedings Editors. Reviews were conducted by expert referees to the professional and scientific standards expected of a proceedings journal published by IOP Publishing.

IOP Conf. Series: Materials Science and Engineering **850** (2020) 012001 doi:10.1088/1757-899X/850/1/012001

Utilization of augmented reality technology as a learning media for early childhood

M R Anggara*, N Azizah and I Kustiawan

Department of Electrical Engineering Education, Universitas Pendidikan Indonesia, Jl. Dr. Setiabudhi 207, Bandung, Indonesia

*muhamadrafianggara@student.upi.edu

Abstract. The purpose of this study is to describe the use of augmented reality technology as a learning medium for early childhood. This study used descriptive qualitative method. Data collection techniques using observation and interviews with 4 early childhood. The results of this study indicate that the use of augmented reality technology as a medium of learning in early childhood can provide an understanding of an object concretely, improvisation of sounds and images that support the learning atmosphere, and be able to attract the attention and focus of children in school. In addition, a lot of potential is developed when implementing augmented reality technology. It is hoped that the use of this technology can be applied to early childhood as a form of learning media innovation.

1. Introduction

Early childhood education is the most basic institution in shaping human resources. The nature of this education is to help children get the right stimulation to develop their potential to be able to enter the next level of education. Early childhood education helps stimulate child development through learning [1,2].

The media used by early childhood teachers generally use an educational game tool. However, the use of this tool as a learning medium has not been successful in providing a concrete picture of an object and has not been able to attract the attention of children in the learning process [3]. Early childhood is in the pre-operational stage. Pre-operational stage is the stage where children think concretely and cannot think abstractly. That is, it needs something concrete to explain an object to children. In addition, the child's focus point is also minimal, a maximum of only 15 minutes [4].

Early childhood institutions must innovate, especially in the use of instructional media. Bearing in mind, now entering the era of the industrial revolution 4.0 where technological developments are developing rapidly. Children are more interested in the use of gadgets as learning media compared to the use of educational game tool [5,6]. The Government of Indonesia issues regulations on the use of technology on ICT competency utilization maps, namely students can use applications to learn to recognize visual forms and images [7]. Therefore, through this research we designed augmented reality as a learning media in early childhood education that can provide real object images and can attract children's attention in the learning process.

2. Method

The methodology in this research is descriptive qualitative in which the natural setting as the direct source of data and the researcher is the key instrument. The point is the researcher as the main data collection tool. Qualitative descriptive research in this study is observational data about the activities of early childhood in utilizing augmented reality and described in descriptive form. The attitude shown from the beginning of the activity until the end of the utilization of augmented reality technology [8,9]. The subjects of this study were 3 early childhood.

Data collection techniques using observation and interviews. Data collection techniques in qualitative research can generally be grouped into two ways in which data collection techniques are interactive and non-interactive. In this study the authors used interactive techniques including observations and interviews. The observations referred to in this study are participant observation. Participant observation is an in-depth observation by blending into the middle of the research subject. So through observation, researchers learn about the behavior and meaning of the behavior. Observations were made to find out early childhood activities in using augmented reality as a learning medium.

Observations are carried out in an unstructured manner, then carried out repeatedly. We did the interview to conduct a preliminary study to find problems that must be investigated, and want to know things from respondents in more depth. Interviews as data collection techniques by way of question and answer are carried out systematically based on specific objectives. Interviews were conducted with 3 early children. Through interviews, efforts are made to dig deeper information. Interviews conducted in a structured and face to face.

Data obtained from the results of observations and interviews are outlined in writing and analyzed. At this stage the writer describes what is seen, heard, felt and asked. The author reduces all information that has been obtained by sorting data or choosing which data is interesting, important, useful and new. Data that is not used is removed. Based on these considerations, the data are then grouped into various categories determined as the focus of the study.

The next stage is the presentation of data, which presents information obtained from the data reduction process, then is organized and presented with a narrative text model. The next stage is the conclusion and verification of data that has been patterned, focused, and arranged systematically, then concluded, to further search for new data, as a test of tentative conclusions obtained.

The last step is data validation using triangulation/ combination which combines various existing data collection techniques and data sources [9]. The goal is to check and compare answers between participants and experience among fellow researchers conducting data collection.

3. Results and discussion

The application of augmented reality in normal children can increase children's focus point in learning, children get interactive learning media, can improve children's vocabulary and so on.Facing the industrial revolution 4.0 the education sector must change and keep up with rapid technological developments. Little by little the field of education began to develop, especially in utilizing learning media. Learning media can take advantage of technological developments such as augmented reality [10].

Augmented reality is a technology that combines 2-dimensional or 3-dimensional virtual objects into a real environment and then projects these virtual objects in real time. Augmented reality in education has a positive impact, which is interesting for multi-modal learning, increases the accessibility of educational content, increases student control of educational content, opens collaborative learning opportunities, motivates students to be actively involved, and turns abstracts into concrete [11].

Augmented reality-based learning for early childhood education has not been done much, although the potential use of augmented reality for early childhood education is very large and has unlimited opportunities. There are 3 principles of augmented reality, namely: (1) Augmented reality is an

IOP Conf. Series: Materials Science and Engineering **850** (2020) 012001 doi:10.1088/1757-899X/850/1/012001

amalgamation of the real and virtual worlds; (2) Augmented reality runs interactively and in real time; and (3) There is integration between 3-dimensional objects, namely virtual objects that are integrated in the real world [12].

Augmented reality systems have been developed in various applications, especially in the fields of education and entertainment. One of them is the Quiver Vision application. How to use it: (1) Download images provided by the Quiver Vision app; (2) Printing and coloring images; (3) Scanning image objects through the camera in a quiver vision application; and (4) Playing a projected 3-dimensional object.

The use of augmented reality in early childhood education can be innovated by making puzzles from images provided by quiver vision. The aim is to foster 21st century skills in early childhood such as critical thinking, collaboration, communication, and creativity [13]. The following are the stages in the use of augmented reality technology in early childhood education:

- The teacher downloads images from the quiver vision application. The pictures given to children are adapted to the theme and sub-themes of the day;
- The image has been downloaded, then edited to a number of puzzle pieces and prepare a base to be attached to the puzzle piece;
- Print the puzzle into A3 paper. Then cut the puzzle settings;
- The teacher prepares all materials and tools used in the use of augmented reality technology such as smartphones / tablets, quiver vision applications, colored pencils / crayons and so on;
- Children are directed to groups of 3-4 people. Every child is responsible for completing the puzzle picture arrangement;
- The child attaches the puzzle structure to the provided puzzle base;
- After sticking, children are directed to a perfectly arranged color image. Children are free to color using the colors they want, meaning that there is no compulsion in developing children's imagination;
- The next step, open the Quiver Vision application that has been downloaded on Google Play. Children are trained to patiently wait for their turn in the process of scanning colored augmented reality images;
- After successfully scanning and 3D images appear from each image that has been colored. The teacher explains about each picture. The teacher explains about dinosaur animals, starting from what is dinosaur, habitat, food, and everything about dinosaur animals. Children can also play the available games or quizzes.

After utilizing augmented reality as a learning media in early childhood, children's cognitive development is increasingly developing. Seen when children more easily recognize objects from the pictures they have colored, children can identify various animal shapes such as eyes, nose, mouth, feet, hands and various other body parts, children can know various colors, thinking skills critical since early childhood can be developed, children can find out animal sounds that are projected to be a vibrating vision application, children know and can operate the technology that developed in the era of the industrial revolution 4.0, namely augmented reality.

In addition, the potential gained by children when implementing augmented reality are: (1) children can use applications to learn to recognize objects concretely; (2) children are increasingly interested when using augmented reality technology as a learning medium; (3) Children can more easily recognize objects when using augmented reality than educational toys; (4) aspects of child development can be developed through the application of augmented reality; (5) 21st century skills in children can be developed; (6) Learning is more interesting and interactive.

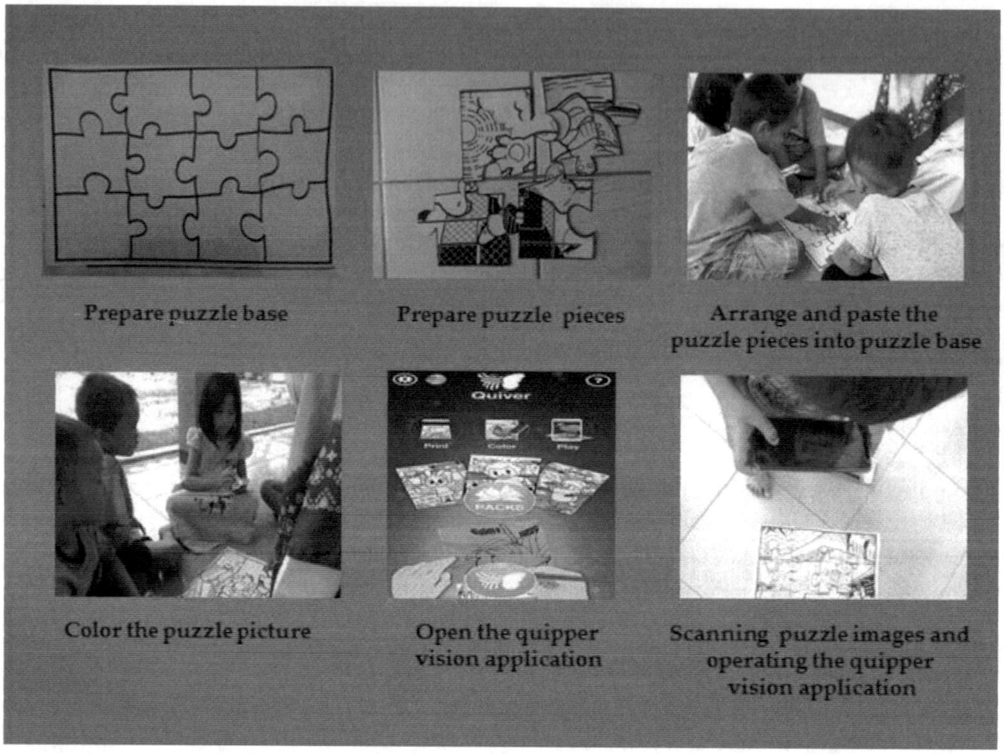

Figure 1. Steps for using augmented reality in early childhood.

4. Conclusion

The use of augmented reality technology as a learning medium in early childhood can provide an understanding of an object concretely, improvisation of sounds and images that support the learning atmosphere, and be able to attract the attention and focus of children in the learning process. In addition, a lot of potential is developed when implementing augmented reality technology. It is hoped that the use of augmented reality technology can be applied to early childhood as a form of learning media innovation in early childhood institutions.

References

[1] Currie J 2001 Early childhood education programs *Journal of Economic perspectives* **15**(2) 213-38

[2] Britto P R, Lye S J, Proulx K, Yousafzai A K, Matthews S G, Vaivada T, Perez-Escamilla R, Rao N, Ip P, Fernald L C and MacMillan H 2017 Nurturing care: promoting early childhood development *The Lancet* **389**(10064) 91-102

[3] Barr R 2008 Attention and learning from media during infancy and early childhood *Handbook on children, media, and development* 143-65

[4] Huitt W and Hummel J 2003 Piaget's theory of cognitive development *Educational psychology interactive* **3**(2) 1-5

[5] Paraskeva F, Mysirlaki S and Papagianni A 2010 Multiplayer online games as educational tools: Facing new challenges in learning *Computers & Education* **54**(2) 498-505

[6] Silawati E and Rachmania S 2016 Character building of early young learners through gadget as learning media in kindergarten *Advances in Social Science, Education and Humanities Research (ASSEHR)* **58** 152-156

[7] Syahid A A, Sunarya D T, Sudin A, Sunaengsih C and Karlina D A 2019 ICT competences of primary school teachers *Journal of Physics: Conference Series* **1318**(1) 012148

[8] Kim H, Sefcik J S and Bradway C 2017 Characteristics of qualitative descriptive studies: A systematic review *Research in nursing & health* **40**(1) 23-42
[9] Sousa D 2014 Validation in qualitative research: General aspects and specificities of the descriptive phenomenological method *Qualitative Research in Psychology* **11**(2) 211-27
[10] Creswell J W, Fetters M D and Ivankova N V 2004 Designing a mixed methods study in primary care *The Annals of Family Medicine* **2**(1) 7-12
[11] Wang X, Ong S K and Nee A Y 2016 Multi-modal augmented-reality assembly guidance based on bare-hand interface *Advanced Engineering Informatics* **30**(3) 406-21
[12] Azuma R, Baillot Y, Behringer R, Feiner S, Julier S and MacIntyre B 2001 Recent advances in augmented reality *IEEE computer graphics and applications* **21**(6) 34-47
[13] Soulé H and Warrick T 2015 Defining 21st century readiness for all students: What we know and how to get there *Psychology of Aesthetics, Creativity, and the Arts* **9**(2) 178

Designing smart greenhouse systems using SCADA based on IoT

D R Nurjannah[1,*], D Supriadi[1], A Sutiawan[1] and I Kustiawan[2]

[1] Electronics Engineering Study Program, Politeknik TEDC Bandung, Jl. Pesantren No.2, Cimahi 40513, Indonesia
[2] Department of Electrical Engineering Education, Universitas Pendidikan Indonesia, Jl. Dr. Setiabudhi No. 207, Bandung 40154, Indonesia

*deasy@poltektedc.ac.id

Abstract. The purpose of this study is to design a smart green home system to help the process of vegetable growth in urban areas by utilizing SCADA and IoT technology. This system can help planting vegetables in urban areas with natural conditions that are not possible and limited land. Using the hydroponic planting method will certainly be very appropriate because it does not need soil as its growing medium. The system that we designed can control and monitor parameters that affect the development of vegetables such as temperature, humidity, nutrition, water level, light, and water use in real time.

1. Introduction

The increasing lifestyle of people in urban areas and eating patterns that are classified as instant can make public health in urban areas less attention. The emergence of health problems may be due to unhealthy eating patterns and mostly consume foods that are instant and practical. This is where the need for a healthy lifestyle and self-awareness to maintain a balanced food intake [1,2].

With the start of a healthy lifestyle awareness in the community and awareness of healthy foods such as vegetables, it will increase the consumption needs of vegetables in the community so that the supply must always be maintained and always stay fresh [3]. Therefore, the supply must be balanced with a large and fast vegetable production. However, on the other hand it is almost impossible to plant vegetables in urban areas because in addition to natural conditions and land is an obstacle [4]. Therefore we need an artificial ecosystem that can resemble the real nature.

With artificial ecosystem technology, growing vegetables can now be applied on limited land such as in urban areas, so that it becomes more efficient. Green house technology can bridge the needs of vegetables in urban areas through a hydroponic planting system that does not require planting media such as soil and without pesticides. This hydroponic system will produce healthy vegetable products without chemicals such as pesticides and no seasonality in the planting process, so that people in urban areas will be able to get fresh vegetables for consumption every day [5-7]. Utilization of green houses in urban areas will be very efficient because it is not fixated on certain ecosystems. The availability of fresh and healthy vegetables on the market will continue to be maintained. To maintain the quality and the quality of the greenhouse production of this hydroponic system in its production requires precise data, is controlled quickly, and can be monitored directly so that the production process can produce good products. This requires a system that can control and monitor data directly on the green house [8].

The need for a SCADA (Supervisory Control And Data Acquisition) system in the application of green houses in urban areas with hydroponic planting is a major factor. We propose a system in such a way as to provide the needs for healthy, fresh, quality vegetables in urban areas and always maintain their availability on the market. Based on the aforementioned explanation, this study describes the design of a smart greenhouse system to be able to produce products with good performance, control the system, acquire data, and monitor the development data of vegetables grown on the green house system.

2. Methods

The design of the smart greenhouse system is carried out with several stages and several preparation processes, so that the system can be structured and made easier in the design process [9]. The initial stages of the design method are divided into 3 parts, namely RTU greenhouse A and B databases, data communication systems, and monitoring parameters for plant development of smart greenhouse systems.

In 2 units of greenhouse RTU, they communicate with SCADA CIMON with Modbus RTU RS-485 communication media which only uses 2 wire cables, namely A and B. In principle, the data communication system to these 2 RTU units communicates in sequence, starting from slaveID 1 then slaveID 2, where in this communication process each address variable sends or receives data sequentially starting from the 40001-40045 address. By using Team Viewer software, it is possible to communicate data with the smart greenhouse system remotely using a computer/laptop and can also communicate this data using a mobile phone.

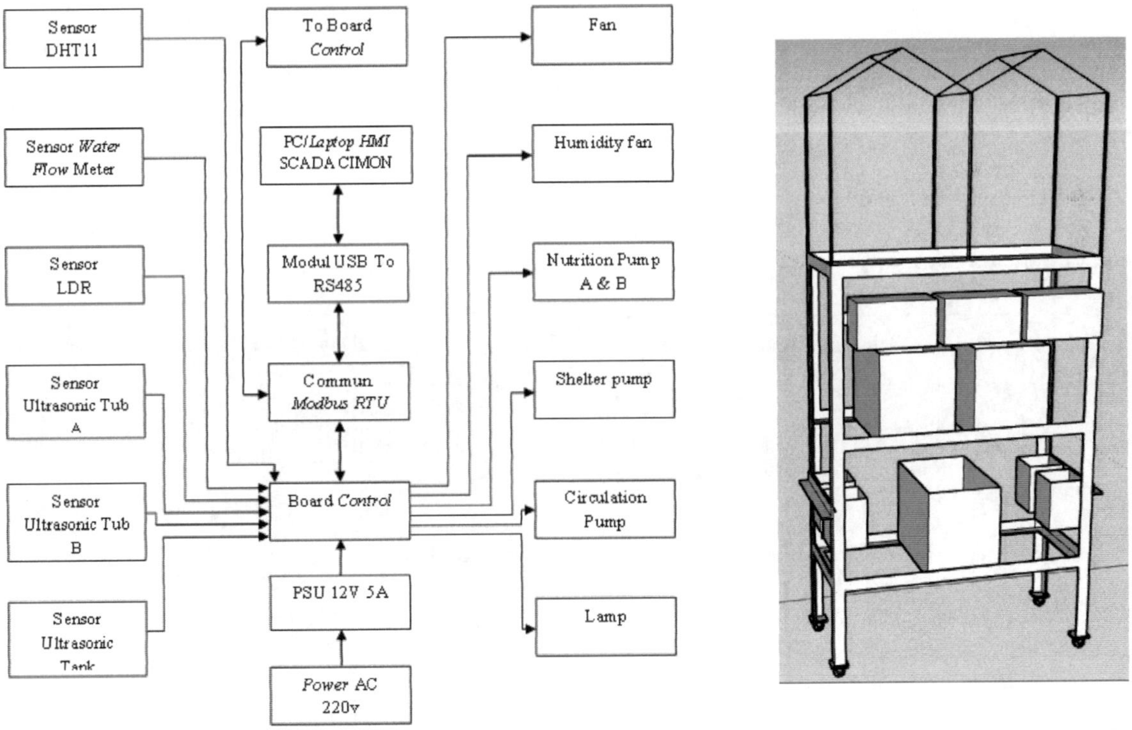

Figure 1. Diagram block and prototype of *Smart Green House.*

The design of smart greenhouse system framework as illustrated in Figure 1 is divided into 2 parts, namely 2 pieces of greenhouse room and the placement of supporting equipment such as storage tanks, 2 tanks, 4 tanks of nutrition, pipes, wiring, PSU, and Control Board.

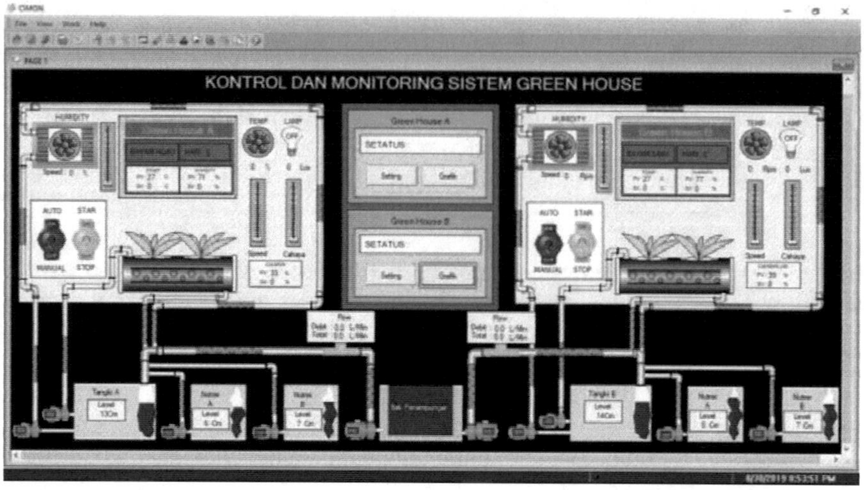

Figure 2. Display of HMI SCADA CIMON.

This greenhouse automation system uses the SCADA system, combined with an IoT-based control system. Some components that work automatically based on parameter data that will be inputted to SCADA, communicating with arduino nano as a controller are SCADA CIMON, Arduino nano microcontroller, RS485 module, DHT11 sensor, water flow sensor, ultrasonic sensor, LDR sensor, fan, pump, and lamp. SCADA software systems and hardware elements in Figure 2 illustrate functions to control processes locally, monitor, process real-time data, interact with devices such as sensors, valves, pumps, motors, and others through human-machine interface software, and record real-time data into log files.

3. Results and discussion

The system created is then tested based on a predetermined program. Data measured are in the form of temperature and humidity, light, water level, and water usage. The data obtained is as follows.

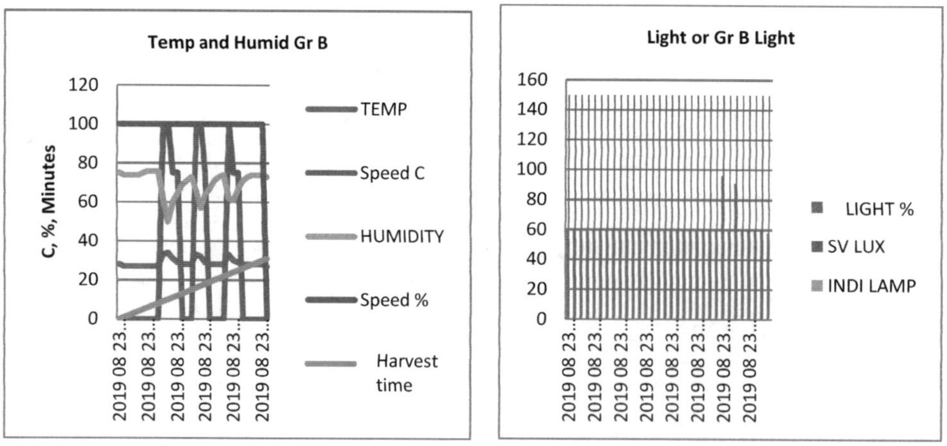

Figure 3. Graph of temp/humidity and light in greenhouse B.

From Figure 3 it can be explained that in 2 pieces of greenhouse A and B, obtained information on temperature, humidity, fan speed, and harvest time of each plant planted on this smart greenhouse system. There is a humidity parameter where the set point is not reached within 31 minutes, causing the fan speed to reach 100% from the start of the program start until the program stops. The humidity system in green house B is only able to reach 76%, which means that the fan humidity system is unable to reach the set humidity point of 80% according to the set point for 31 minutes.

The intensity of the light in the greenhouse room reaches 150 Lux maximal from the 12v LED after being calibrated using LUX meters. The LDR sensor set points outside the green house can work as desired. When the set point of light conditions at an intensity level of less than 80% on the LDR sensor readings that are located outside the green house, then the lights will turn on, otherwise the lights will go out.

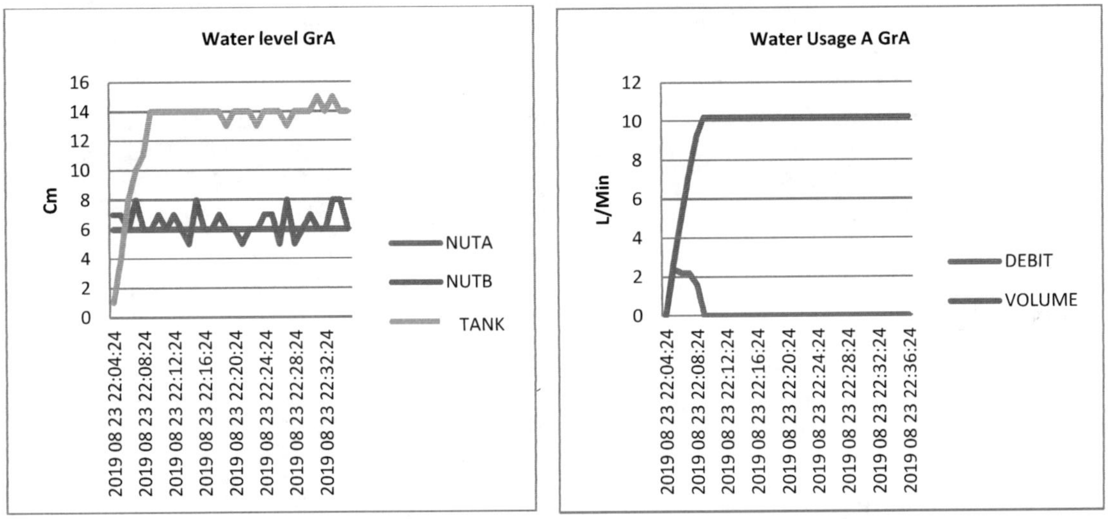

Figure 4. Graph of water level and water usage in greenhouse A.

Referring to the water level graph information with an ultrasonic sensor, it is found that the reading is not stable due to the low quality of the ultrasonic sensor used (Figure 4). We set the water level to 15cm, tank volume A 10.2 liter and Tank B 11.7 liter. 1.5 liter difference was found due to the unstable ultrasonic sensor reading.

4. Conclusion

Based on the results of the test equipment and data sampling acquired by SCADA, it can be concluded as follows:

- Greenhouse prototype that has been designed according to the needs of green spinach and mustard plants;
- The growth conditions parameters of the two types of plants work as expected; found an ultrasonic sensor reading that has not been stable;
- The temperature and humidity system is not maximal because it only uses a fan.

References

[1] Cullen F and Kingston H 2009 Analysis of rural and urban consumer behavior toward new food products using a food-related lifestyle instrument *Journal of Foodservice Business Research* **12**(1) 18-41

[2] Hu F B 2011 Globalization of diabetes: the role of diet, lifestyle, and genes *Diabetes care* **34**(6) 1249-1257

[3] Kempen E L, Muller H, Symington E and Van Eeden T 2012 A study of the relationship between health awareness, lifestyle behaviour and food label usage in Gauteng *South African Journal of Clinical Nutrition* **25**(1) 15-21

IOP Conf. Series: Materials Science and Engineering **850** (2020) 012002 doi:10.1088/1757-899X/850/1/012002

[4] Xiong T, Dumat C, Pierart A, Shahid M, Kang Y, Li N ... and Laplanche C 2016 Measurement of metal bioaccessibility in vegetables to improve human exposure assessments: field study of soil–plant–atmosphere transfers in urban areas, South China *Environmental geochemistry and health* **38**(6) 1283-1301
[5] Kumar K S, Tiwari K N and Jha M K 2009 Design and technology for greenhouse cooling in tropical and subtropical regions: A review *Energy and Buildings* **41**(12) 1269-1275
[6] Kumar A, Tiwari G N, Kumar S and Pandey M 2006 Role of greenhouse technology in agricultural engineering *International Journal of Agricultural Research* **1**(4) 364-372
[7] Grewal H S, Maheshwari B and Parks S E 2011 Water and nutrient use efficiency of a low-cost hydroponic greenhouse for a cucumber crop: An Australian case study *Agricultural Water Management* **98**(5) 841-846
[8] Ai W and Chen C 2011 Green house environment monitor technology implementation based on android mobile platform *2011 2nd International Conference on Artificial Intelligence, Management Science and Electronic Commerce (AIMSEC)* 5584-5587
[9] Albiñana J C and Vila C 2012 A framework for concurrent material and process selection during conceptual product design stages *Materials & Design* **41** 433-446

SafeDri: A mobile-based application for safety driving

H Maulid*, W Nurhidayat and S J Priyono

Department of Software Engineering, School of Applied Science, Telkom University, Jl. Telekomunikasi no 1, Terusan Buah Batu, Bandung 40257, Indonesia

*hariandimaulid@tass.telkomuniversity.ac.id

Abstract. Globally, around 1.3 million people die every year and more than 50 million people suffer permanent disability due to traffic accidents. This condition has worsened in many countries in the world in line with the high rate of ownership of motorized vehicles. The most common reasons for traffic accidents are distracted driving, over speeding, drunk driving and reckless driving. All those reasons are related with unsafe behaviour while driving. To cope with the problem, we developed a mobile based application that assists driver to drive more safely. The apps will collect information during a stipulated journey and recognize speeding facts, harsh acceleration, heavy breaking, and fuel consumption. While using this application, the drivers will learn how to drive safely, set goals and preferences, get alerts upon over limit violation, and receive feedback on their completed journeys. The results of a usability test on the application indicate that the app managed to identify misbehaviour while driving and rate driver behaviour based on journey history.

1. Introduction

Road safety is an issue that tends to emerge from year to year. Nowadays it has become a global problem and is not merely a matter of transportation but has become a social problem. According to World Health Organization (WHO) report, around 1.3 million people die every year globally and more than 50 million people suffer permanent disability due to traffic accidents [1,2]. The majority victims killed in traffic accidents are people between 15 and 44 years. This condition has worsened in many countries in the world in line with the high rate of ownership of motorized vehicles.

Some main contributors to traffic accident in South East Asian countries including unlicensed driver, traffic law violation, inadequate of comprehensive development of roads and other transport infrastructure, and rapid and unplanned urbanization. The most common reasons for traffic accidents are distracted driving, over speeding, drunk driving and reckless driving [3]. All those reasons are related with unsafe behaviour while driving. However, road traffic accidents are preventable. Some measures have been taken ranging from enforcing legislation on controlling speed and alcohol consumption, obligate the use of helmet and seatbelt, to designing safer roads and vehicles.

In the form of technological advancement, the development of in-vehicle sensing technology has contributed to measure driving style more objectively. Such system includes collision-avoidance, neglected and distracted driver monitoring system, automatic steering system, and blind spot warning [4]. However, despite the effectiveness of those systems in improving driver safety, the price was not quite affordable since they only found in a pricey vehicle. On the other hand, the advancement of mobile sensing platform on regular smartphone enables capturing and processing data from the real world in a cost-effective manner.

Some projects that incorporating mobile sensing for safety driving have been conducted such as SafeDrive [5] that automatically silent driver's smartphone when car speed reach the maximum limit. It also rewards the driver for not texting while driving. The drawback of that application is that it only available in Romania. Another mobile project was DriveWell [6], a behaviour-based program from Cambridge Mobile Telematics, that accurately measure the driving performance and vehicle dynamics to make better driving. Like SafeDrive, DriveWell also will only well suited when it is used in the country where it originates due to different geographical situation and road infrastructure.

Furthermore, there are two main focuses on the application of Intelligent Transportation System (ITS), efficiency and safety application. The latter formerly used for avoiding or reducing traffic incidents as well as their impact. With ITS, vehicles can communicate with each other through Vehicle-to-Vehicle communication or Vehicle-to-Infrastructure technologies [7]. Vehicular Ad hoc NETworks (VANET) is the example of vehicle-to-vehicle ad hoc network [8]. One vehicle can inform other vehicles about some specific information such as traffic information, traffic accident, road conditions, and driving navigation [9-12]. Some concerns in VANET are cyber-security challenge [7] and trust management [13].

Some ITS safety application had incorporated Internet of Things (IoT) technology. Goshi et al developed an Assistant System for Safe Driving by Informative Supervision and Training (ASSIST) based on IoT architecture [14]. Their system comprised of an on-board system inside a vehicle (microprocessor, GPS, camera) to get various data including position, speed, and fuel consumption. These data will be sent to supervision system to calculate the summary of driving. Aital and Singh in [15] proposed an IoT-based risk assessment and warning system that assist the driver in maintaining a safe distance between vehicles and avoiding unsafe conditions on roads. IoT also had been used to build safety application for detecting vehicle obstacle [16], ensuring safe driving [17], and tracking hand and head movement while driving using wearable magnetics [18].

In mobile platform, there has been an active research on leveraging smartphone sensing to assist driver. SmartLDWS was the first LDWS on smartphone that warns the driver when a car moves out of the lane [19]. On the other hand, SignalGuru detects and predicts traffic signals with the camera and advises the driver to adjust speed for fuel consumption efficiency [20]. Mobile sensing also has been used in various projects to monitor driving style, pattern, and behaviour. DriveSafe [4], CarSafe [21], and Driving coach [22] are the three examples. DriveSafe detects inattentive driving behaviour such as drowsiness and distraction using computer vision and pattern recognition techniques. Similarly, CarSafe uses dual camera feature in smartphone to detect drowsiness, inattentive, tailgating, drifting, and ignoring blind spot. The front-facing camera monitor driver condition while the rear-facing camera tracking road conditions. Unlike the two, driving coach uses smartphone's sensors and the state of the vehicle to detect driving pattern. It then will recommend new behaviours to the driver that supposedly to improve driving experience and fuel consumption efficiency. However, the most related work with our project on mobile safety driving application are SafeDrive [5] and DriveWell [6]. As aforementioned, the main drawback of those applications was they fit only specific geographical situation as well as road infrastructures where those applications originated from.

To address this issue, we argue for the possibility of utilizing a mobile based architecture on developing an application that assists the driver to drive more safely that well suited South-East Asian countries geographical and road infrastructures. The apps will collect information during a stipulated journey and recognize speeding facts, harsh acceleration, heavy-braking, and fuel consumption. While using this application, the drivers will learn how to drive safely, set goals and preferences, get alerts upon over limit violation, and receive feedback on their completed journeys. The goal of this work is to present the process of our mobile-based application development that targeted local users in order to gain user's specific needs and expectations toward the application.

2. Project design and implementation

We used User-Centered Design method in designing user needs and expectations as well as project implementation. Figure 1 depicts all stages of project design and development. It started with collecting data user needs and expectations, preliminary design, evaluation of design alternatives, prototypes, and final artifact.

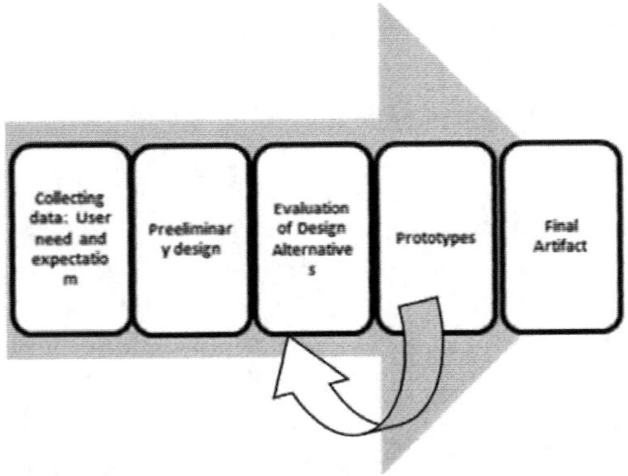

Figure 1. User-centered design stages.

2.1. Collecting data and preliminary design

The first stage is collecting user needs and expectations using various data collection techniques including interview and questionnaires. We recruited 10 (ten) respondents to participate in this stage. We decided to combine this stage with preliminary design (stage two) for two reasons: to give initial picture design of the system to the respondents and to save the time. In doing so, we divided respondents into two groups with equal number. While the first group doing interview, other group filling the questionnaires. For the group who do an interview, we also provided a preliminary design that consists of the main features of the system as shown by figure 2.

Figure 2. Preliminary design of the prototype.

The expected result of data collection is to determine the main goal of the system. All respondents agreed that the proposed system should manage to reduce traffic accident by minimizing its factors. Some actions identified as serious factors including over-speeding, drunk driving, distracted driving, and reckless driving. We also asked the respondents to comment on our proposed preliminary design whether it managed to achieve the goal or it required any improvement. At the end of these two stages we obtained user's goal and new improvement on our preliminary design as follows:

- Set your vehicle
  - Car – type and year
  - Motorcycle – type and year
- Start your journey
  - Detect over-speeding
    - Include time and location
  - Detect heavy breaking
    - Include time and location
- Stop your journey
  - Draw the polyline for visual journey history on Gmap
  - Display Journey History
    - Display distance covered
    - Display number of violation (over speeding and heavy-braking)

2.2. System implementation

The good thing of user-centered design is to involve users throughout all processes from design phase to implementation phase including its evaluation in an iterative manner until fit users' goal [23]. In this implementation phase, we did two iterations. The First iteration covers a sketch of low-fidelity and high-fidelity prototype. Prototype implementation and its evaluation conducted in the second iteration.

2.2.1. First iteration. In the first iteration, we discussed and refined low and high-fidelity prototype together with all respondents/users according to the result of preliminary design. We asked them to draw a sketch on the paper on expected interface. This low fidelity allowing respondents to create and modify interface straightforwardly. Upon agreed on a low-fidelity sketch, we designed its high-fidelity interface to be implemented at the next iteration. An example of low fidelity skecthes along with its high-fidelity interface are shown in figure 3 and 4.

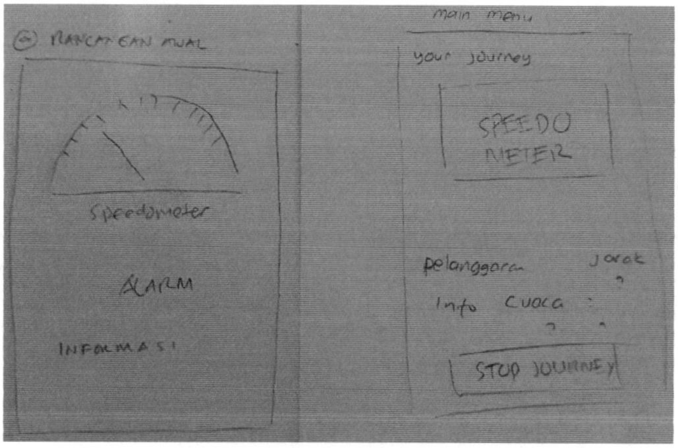

Figure 3. A low fidelity sketch.

Figure 4. High-fidelity prototype.

2.2.2. Second iteration. The second iteration consists of implementing low and high-fidelity interface into hard coding implementation.

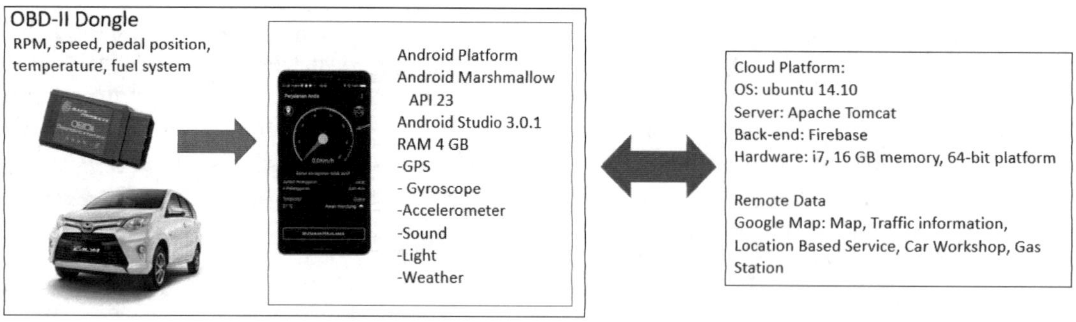

Figure 5. System architecture.

Beforehand, we have determined to use Android mobile platform and Firebase as a Backend as a Service (BaaS) as implementation environment of this system. We used several sources of raw data namely smartphone sensors (GPS, accelerometer, Gyroscope, sound, light, and weather), OBD-II Dongle with Bluetooth (RPM, speed, pedal position, temperature, and fuel system), and other third party data or remote data (map, traffic info, location, weather, car workshop, and gas station). The detailed system architecture can be found in figure 5.

IOP Conf. Series: Materials Science and Engineering **850** (2020) 012003 doi:10.1088/1757-899X/850/1/012003

The Android application (SafeDri) will be the heart of the system. It will collect and pre-process raw data that derived from smartphone sensors, OBD-II, and remote data to obtain meaningful information (data acquisition). To produce meaningful data, the SafeDri should create data fusion modules which acquire and process raw data and turn them into meaningful information. For example, to obtain over speeding data, a simple module recalls raw speed, RPM, and location from OBD-II and internal sensor before then combines them with speed limit rule. Another example of data fusion module is when the app requires location-based service (nearest gas station or nearest car workshop) it can query location service like Google Map or OpenStreetMap based on GPS current location sensor.

Based on first iteration result, we turned high-fidelity prototype into system prototype named SafeDri. This app will collect information during stipulated journey such as over speeding and heavy braking. After each ended trip, the app will calculate the number of violations relating to over speed and heavy braking. Moreover, not only the total of violation displayed on the history, but also time and location, when and where the user made miss-driving-behaviour as displayed by figure 6 to figure 8.

Figure 6. SafeDri main menu and features. Start the journey and display location.

Figure 7. Recent journey history and violation.

Figure 8. Detail of violation data (heavy braking).

3. Discussion

While developing heavy breaking detection, on the other hand, we encountered long debatable discussion. As heavy breaking could be categorized both for normal condition to avoid crash or abnormal condition (violation), however, any form of immediate breaking (extreme lower speeding until stopping) could lead to the series of collision. Imagine when a car speed reaches 60 km/hour and suddenly stops on avoiding a pedestrian, there is the possibility of other vehicles behind that car to hit and leading to the series of unintended accident.

To validate the system, we went through usability testing with System Usability Scale (SUS) to measure the usability aspect of the app [24]. SUS consists of 10 (ten) questions that give global subjective assessment view toward the UI/UX of the app. This test conducted during the second iteration phase. After the system being implemented, we asked 10 (ten) respondents to use and explore SafeDri app in one-week period and record the states.

Table 1. Result of SUS score.

Respondents	USERS Answer per-question	SUS Average per respondents	Final SUS Score
User 1	5, 1, 5, 2, 4, 2, 5, 2, 5, 2	87.5	
User 2	5, 2, 4, 1, 5, 1, 5, 2, 4, 3	85.5	
User 3	5, 2, 5, 2, 5, 2, 5, 1, 5, 1	92.5	
User 4	5, 1, 4, 2, 5, 1, 4, 1, 5, 2	90.0	
User 5	4, 2, 5, 1, 5, 2, 4, 1, 5, 2	87.5	
User 6	5, 2, 5, 2, 5, 2, 5, 2, 4, 1	87.5	**88.5**
User 7	4, 3, 4, 1, 5, 1, 5, 1, 4, 2	85.0	
User 8	5, 1, 5, 2, 5, 2, 5, 2, 5, 1	92.5	
User 9	5, 1, 3, 1, 4, 2, 5, 2, 5, 1	87.5	
User 10	5, 1, 4, 1, 5, 2, 5, 2, 4, 1	90.0	

Next, we asked them to fill out SUS test based on user experience towards application in previous week. We use 5-likert scale with 5 as strongly agree and down to 1 as strongly disagree as the answer score for every question. The calculation method is described as follows:

- For odd numbered questions (1, 3, 5, 7 and 9), subtract 1 from the score.
- For even numbered questions (2, 4, 6, 8 and 10), subtracts their value from 5.
- SUS Average per-respondent by adding up all total score then multiply this by 2.5
- Final SUS score calculated by totalling all SUS average per-respondent and then divided by the number of respondents.

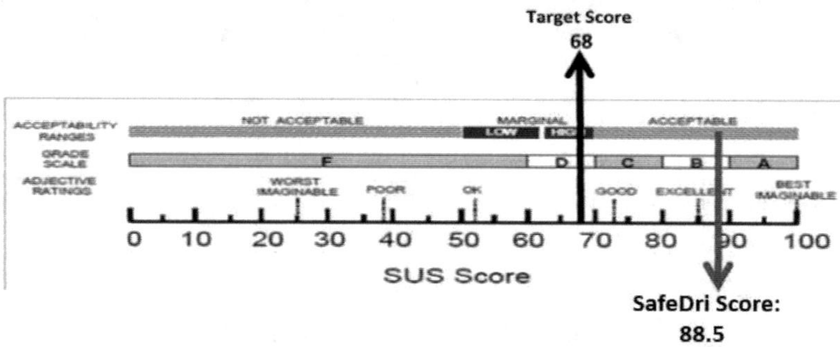

Figure 9. Acceptability of SAFEDRI app.

Table 1 shows the final average SUS score as users' feedback. It reached 88.5 beyond the minimal target which is 68. The test score expressed excellent acceptable range (B) as depicted by figure 9. However, since this project is just the first part (4 months) out of a three-year ambitious project, there are some possibilities to add some features and technologies advocated in the app. Some possible features including mood and fatigue detection, lane trajectory detection, collision avoidance, distracted driver detection, traffic congestion detection, motorway and hilly road detection, as well as city and countryside location determination. Consequently, some sensors and technologies should be considered such as Internet of Things (IoT), dual camera facility, not to mention machine learning and data science.

4. Conclusion

SafeDri takes advantage of the development of mobile sensing that provides an identification of misbehavior while driving and rate driver behaviour based on journey history especially when the driver tries to break the limit of speed or waste the fuel consumption and do a sudden stop. Although the SUS test score reached excellent, it still needs some improvement, especially in terms of features on various condition detection along with their technological advancement. We plan to implement some features that significantly affect the driver style and behaviour such as mood and fatigue, drowsiness, distraction, and surrounding environment detection.

References

[1] Peden M, Scurfield R, Sleet D, Mohan D, Hyder AA, Jarawan E, and Mathers C 2004 *World report on road traffic injury prevention* (Geneva: World Health Organization).
[2] Hu W, Hu X, Deng J, Zhu C, Fotopoulos G, Ngai E C H and Leung V C M 2014 *Proc. Int. on Middleware for Context-Aware Applications in the IoT (Bordeaux)* **1** pp 19 – 24.
[3] King E L 2017 available online: https://www.huffpost.com/entry/top-15-causes-of-car-accidents_b_11722196 last accessed 20 July 2019.
[4] Bergasa L M, Almería D, Almazán J, Yebes J J and Arroyo R 2014 *Drivesafe: An app for alerting inattentive drivers and scoring driving behaviors: Proc. Int. 2014 IEEE Intelligent Vehicles symposium proceedings (Miami)* pp 240-245.
[5] Cobalas T 2014 *SafeDrive app rewards your good behavior while driving* available online: https://devpost.com/software/safedrive-app last accessed 20 July 2019.
[6] Cambridge Mobile Telematics *DriveWell: A behavior-based program* available online: https://www.cmtelematics.com/drivewell/ last accessed 20 July 2019.
[7] Mukisa S S and Rashid A 2014 *Cyber-security challenges of agent technology in intelligent transportation systems.* In *Proceedings of the 1st International Workshop on Agents and CyberSecurity (Paris)* pp 9.
[8] Intelligent Transportation Systems Committee 2011 *IEEE Trial-Use Standard for Wireless Access in Vehicular Environments (WAVE)-Multichannel Operation. IEEE, Standard,* pp.1609-4.
[9] Choffnes D R and Bustamante F E 2005 *An integrated mobility and traffic model for vehicular wireless networks.* In *Proc. Int. of the 2nd ACM international workshop on Vehicular ad hoc networks (Cologne)* (New York: ACM) pp. 69-78.
[10] Härri J, Filali F, Bonnet C and Fiore M 2006 *VanetMobiSim: generating realistic mobility patterns for VANETs.* In *Proc. Int. of the 3rd international workshop on Vehicular ad hoc networks (Los Angeles)* (New York: ACM) pp. 96-97.
[11] Lochert C, Scheuermann B, Wewetzer C, Luebke A and Mauve M 2008 *Data aggregation and roadside unit placement for a vanet traffic information system.* In *Proc Int. of the fifth ACM international workshop on VehiculAr Inter-NETworking (San Francisco)* (New York: ACM) pp. 58-65.
[12] Cops M 2006 *Vehicle infrastructure integration (VII)-bringing vehicle adhoc network technology on the road.* In *the Third ACM International Workshop on Vehicular Ad Hoc Networks* (New York: ACM).
[13] Ma S, Wolfson O and Lin J 2011 *A survey on trust management for intelligent transportation System.* In *Pro. Int. of the 4th ACM SIGSPATIAL International Workshop on Computational Transportation Science (Chicago)* (New York: ACM) pp. 18-23.
[14] Goshi K, Hayahi M, Sumida Y and Matsunaga K 2016 *An Iot-based safe driving management System. In Proc. Int. on Mobile and Ubiquitous Systems: Computing, Networking andservices (Hiroshima)* vol 13 (New York: ACM) pp 280-281.
[15] Aithal A and Singh S 2018 *IoT Based Driving Risk Assessment and Warning System.* In *2018 International Conference on Smart Systems and Inventive Technology (ICSSIT) (Tirunelveli, India)* IEEE pp. 282-286.
[16] Umakirthika D, Pushparani P and Rajkumar M V 2018 *Internet of Things in Vehicle Safety—Obstacle Detection and Alert System. Int. J. Eng. Comput. Sci. 7* pp.23540-23551.
[17] Al Mamun M A, Puspo J A and Das A K. 2017 *An intelligent smartphone-based approach using IoT for ensuring safe driving.* In *2017 International Conference on Electrical Engineering and Computer Science (ICECOS)(Palembang)* IEEE pp. 217-223.

IOP Conf. Series: Materials Science and Engineering **850** (2020) 012003 doi:10.1088/1757-899X/850/1/012003

[18] Huang H, Chen H and Lin S 2019 *MagTrack: Enabling Safe Driving Monitoring with Wearable Magnetics.* In *Proc. of the 17th Annual International Conference on Mobile Systems, Applications, and Services (Seoul)* (New York: ACM) pp. 326-339.

[19] Lan M, Rofouei M, Soatto S and Sarrafzadeh M 2009 *SmartLDWS: A robust and scalable lane departure warning system for the smartphones.* In *2009 12th International IEEE Conference on Intelligent Transportation Systems (St. Louis)* IEEE pp. 1-6.

[20] Koukoumidis E, Peh L S and Martonosi M R 2011 *SignalGuru: leveraging mobile phones for collaborative traffic signal schedule advisory.* In *Proc of the 9th international conference on Mobile systems, applications, and services (Bathesda)* (New York: ACM) pp. 127-140.

[21] You C W, Lane N D, Chen F, Wang R, Chen Z, Bao T J, Montes-de-Oca M, Cheng Y, Lin M, Torresani L and Campbell AT 2013 *Carsafe app: Alerting drowsy and distracted drivers using dual cameras on smartphones.* In *Proc. of the 11th annual international conference on Mobile systems, applications, and services (Taipei)* (New York: ACM) pp. 13-26.

[22] Araújo R, Igreja Â, De Castro R and Araujo R E 2012 *Driving coach: A smartphone application to evaluate driving efficient patterns.* In *2012 IEEE Intelligent Vehicles Symposium (Alcala de Henares)* IEEE pp. 1005-1010.

[23] Maulid H, Azimi I and Fauzi A H 2018 *A user-centered design based collaborative system for Jum'a Preacher Scheduling. Int. Journal on Islamic Applications in Computer Science and Technology (IJASAT)* UK.

[24] Borsci S, Federici S and Lauriola M 2009 *On the dimensionality of the System Usability Scale: a test of alternative measurement models. Cognitive processing* vol 10 issue 3 (Springer: Berlin) pp.193-197.

Ability estimation in computerized adaptive test using Mamdani Fuzzy Inference System

W Ridwan*, I Wiranto and R D R Dako

Department of Electrical Engineering, State University of Gorontalo, Jl. Jendral Sudirman no.6, Kota Gorontalo, Indonesia 96128.

*wridwan@ung.ac.id

Abstract. Assessment is an activity to find out the learning outcomes of a course. Conventionally, assessment is done using a pencil and paper test. Along with the development of information and communication technology, evaluation can be done computerized, known as Computer Based Test (CBT). The Computerized Adaptive Test (CAT) is one form of CBT where the items given are chosen based on the students' abilities. This research aims to design a system that can estimate the ability of students based on the parameters of the questions and answers given. The estimation method uses the Mamdani Fuzzy Inference System (MFIS). The input MFIS is a level of difficulty and discrimination of the questions, the probability of students being able to answer correctly, and the student's answer, while the output is an ability estimated. Based on this fuzzy system output, the next item questions will be determined according to the ability of the students. 24 IF-THEN rules are used for fuzzy systems. CAT simulations are carried out for Linear Algebra Course with six topics, namely vector, matrices concepts, matrices operation, determinant, inverse, and matrices applications. The type of question given is multiple choice. Giving items will be stopped if the value of the estimated ability of the students has not changed. From the simulations carried out, for each topic, the ability of the student can be acquired with the number of questions as many as six questions. So that the CAT system can minimize the time of the exam, reduce the subjectivity of the assessment and can arrange for each student to get the questions according to his abilities.

1. Introduction

Assessment activity of learning outcomes in the cognitive/knowledge domain is usually done with an objective and descriptions test techniques with various types and carried out conventionally using paper or also called a pencil and paper test. Along with the development of information and communication technology, the assessment of student learning outcomes can be done computerized (known as Computer Based Test/CBT). The use of CBT in examinations can reduce the subjectivity of lecturers in conducting assessments, is not affected by the condition of the lecturers (tired, unfocused), and certainly can reduce the use of paper (following environmentally friendly principles). At CBT, it is possible to randomize questions between one student and another student. In the sense that every student is allowed to get a different sequence of questions. This can prevent students from collaborating in completing the exam. However, the form of an exam like this can cause inaccuracy in giving questions and inefficient time between one student and another.

Computerized Adaptive Testing (CAT) is a test method for test participants that is managed based on the level of participants' abilities [1]. Based on that, the score of the assessment of each examinee will not be much different, both those with more and moderate abilities. This is because the system used in The CAT can recognize the level of participants' abilities so that the questions given are a matter that can be done by the participants. The design of the CAT as an instrument for evaluating learning outcomes has advantages including (a) assisting lecturers in evaluating learning outcomes (reducing subjective elements, not affected by lecturers' condition/tired or out of focus); (b) in accordance with environmentally friendly principles because it reduces the use of paper in conventional examinations; (c) can be integrated with online learning (e-learning) so that it allows students to conduct self-assessment.

The primary key to the CAT program is the computer's ability to provide the right questions adaptively, according to the estimation of test-takers' abilities. In general, the CAT system consists of five main components, namely [2-4]:

- calibrated item bank,
- the starting point,
- item selection rule,
- scoring, and
- stopping rule.

Given 1 and 2, repeat 3 and 4 until 5 is satisfied.

At the stage of item selection, an algorithm is needed that can estimate the ability of the examinees so that the items to be issued are adjusted to the abilities of the participants. The classic method of knowing the level of participants' ability is to the Maximum Likelihood estimator (MLE) [5,6]. The Bayesian approach used in [7,8] to estimate participants' ability.

In this research, the estimation of participants' abilities will be carried out using fuzzy logic. Fuzzy logic estimates the ability of examinees based on the difficulty level of the question, the parameters of the question and the response/answers of the examinees. The estimation results are used to determine the next question that will be issued.

2. Method

In general, Fuzzy Logic Control (FLC) has four main parts, as shown in Figure 1. The four elements have the following functions [9,10]:

- Fuzzifier functions to transform input signals that are crisp to the fuzzy set using the fuzzifier operator.
- The Knowledge Base contains a database and basic rules that define fuzzy sets of input and output areas and arrange them in the control rules.
- Decision Making is at the core of FLC, which can be human in making decisions. Fuzzy set actions are inferred using fuzzy implications and fuzzy inference mechanisms.
- Defuzzification functions to transform conclusions about fuzzy set actions into an actual variable that is crisp by using defuzzifier operators.

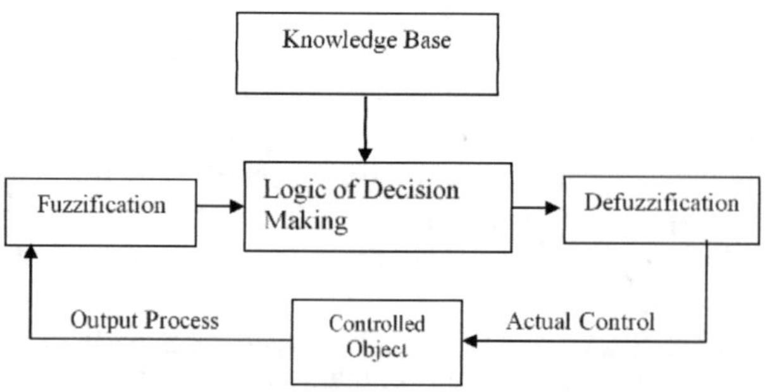

Figure 1. Fuzzy logic control.

In this research, the block diagram of the fuzzy logic system to estimate the ability of examinees, as shown in Figure 2. There are four inputs to the fuzzy system, namely:

- Item difficulty level (b_i), with three levels, namely easy, medium, and difficult.
- Question discrimination (a_i), with two levels, namely satisfactory and good.
- Probability of the examinees answer the question correctly (p_i), calculated using Equation 1 [6].

$$p_i(\theta) = c_i + (1 - c_i)\frac{e^{1.7a_i(\theta - b_i)}}{1 + e^{1.7a_i(\theta - b_i)}} \quad i = 1,2,3,\ldots,n \tag{1}$$

where,
$p_i(\theta)$: the probability that the student has the ability θ to answer the item i correctly
θ: student's ability to estimate
b_i: item difficulty level
a_i: item discrimination
c_i: guessing factor
n: the number of items
e: 2.718
The parameter c is worth 0.25 because there are only four possible answers to each question (multiple choice questions). This input is divided into two categories, namely minimum and maximum.

- Participant's response or answer (r_i), if the correct answer is worth 1, whereas if the wrong answer is 0.

Input 1 and input 2 are the question parameters.

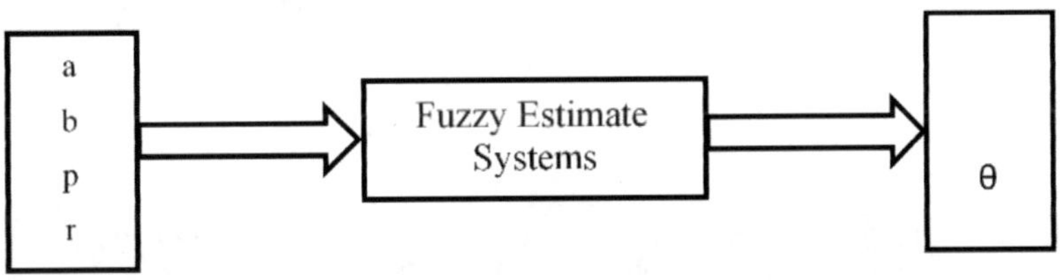

Figure 2. Block diagram of the fuzzy estimate system.

Fuzzy system output is the estimated value of the participant's ability (θ). This output is divided into five levels, namely Very Low (VL), Low (L), Average (Av), Great (G), and Excellent (Ex). These five output levels adjust to the values in the student scoring system, which is from the highest value A to the lowest value E. While the values B, C, and D are in the two values.

The membership functions of the four inputs and one output can be seen in Figure 3 to Figure 7.

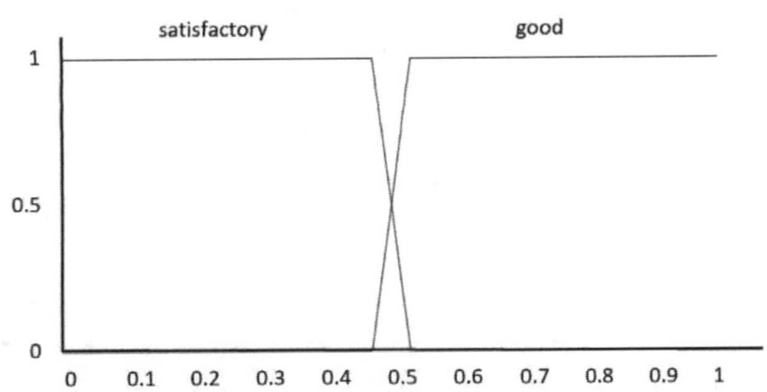

Figure 3. The membership function of the input variable "item discrimination".

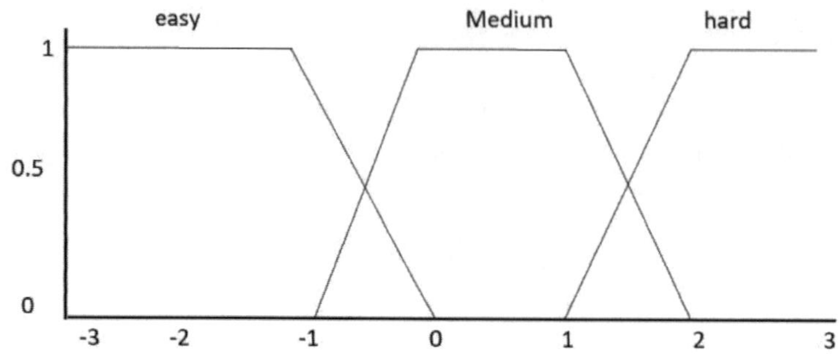

Figure 4. The membership function of the input variable "item difficulty level".

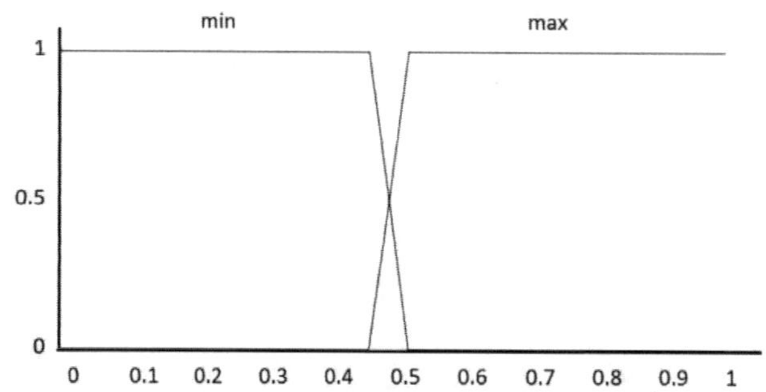

Figure 5. The membership function of the input variable "Probability".

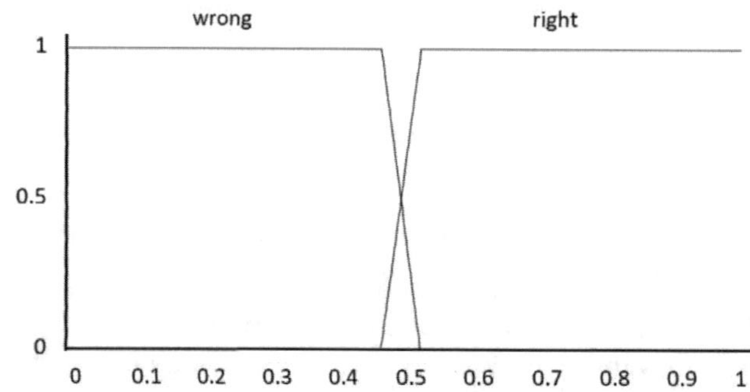

Figure 6. The membership function of the input variable "participants' response".

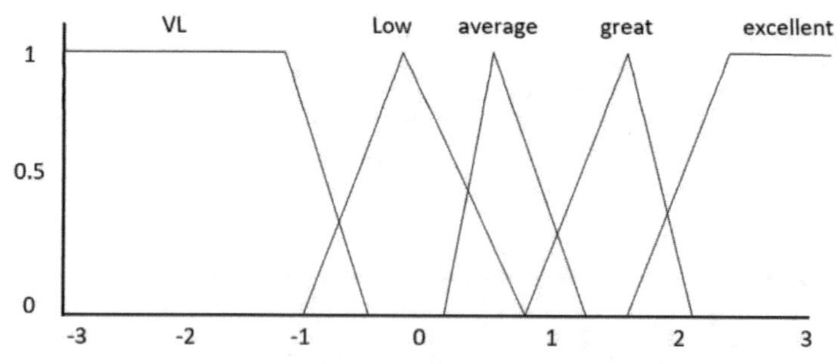

Figure 7. The membership function of output variable "participants' abilities level".

Based on the above input-output fuzzy system, fuzzy system rules are then compiled to estimate the ability of the examinees. There are 4 inputs namely item discrimination (divided into 2 levels), item difficulty level (divided into 3 levels), probability (divided into 2 levels) and participant's response (divided into 2 levels), so that the combination of these four inputs produces 24 rules (= 2 times 3 times 2 times 3). The 24 fuzzy IF-THEN rules summarized in Table 1. The value 1 in Table 1 shows input or output at that level. The explanation of Table 1 given below.

- IF discrimination is Satisfactory AND difficulty is Easy AND probability is Minimum AND response is Wrong THEN ability is Very Low (rule 1).
- IF discrimination is Satisfactory AND difficulty is Easy AND probability is Minimum AND response is Right THEN ability is Low (rule 2).
- IF discrimination is Satisfactory AND difficulty is Medium AND probability is Minimum AND response is Right THEN ability is Average (rule 6).
- IF discrimination is Good AND difficulty is Medium AND probability is Minimum AND response is Right THEN ability is Good (rule 18).
- IF discrimination is Good AND difficulty is Hard AND probability is Maximum AND response is Right THEN ability is Excellent (rule 24).

The logic that underlies rule number 1 is that if an exam participant is given an easy question, and he/she answers incorrectly, the level of ability is very low. Other fuzzy rules are built based on relevant logic.

Fuzzy system designed using Mamdani fuzzy model with the defuzzification process using center of gravity defuzzyfier.

Table 1. Fuzzy system rules.

Rule No.	Input									Output				
	Discrimination (a)		Difficulty (b)			Probability (p)		Response (r)		Ability (θ)				
	G	S	H	M	E	Max	Min	R	W	VL	L	Av	G	Ex
1		1			1		1		1	1				
2		1			1		1	1			1			
3		1			1	1			1	1				
4		1			1	1		1				1		
5		1		1			1		1		1			
6		1		1			1	1				1		
7		1		1		1			1		1			
8		1		1		1		1					1	
9		1	1				1		1			1		
10		1	1				1	1					1	
11		1	1			1			1			1		
12		1	1			1		1						1
13	1				1		1		1		1			
14	1				1		1	1				1		
15	1				1	1			1	1				
16	1				1	1		1				1		
17	1			1			1		1			1		
18	1			1			1	1					1	
19	1			1		1			1		1			
20	1			1		1		1						1
21	1		1				1		1			1		
22	1		1				1	1						1
23	1		1			1			1			1		
24	1		1			1		1						1

3. Result and discussion

The design of the fuzzy system to estimate the ability of examinees, then simulated using system simulation software. For simulation purposes, we use the item bank for Linear Algebra Course. There are 177 multiple choice questions for six topics, namely vector, matrices concepts, matrices operation, determinant, inverse, and matrices applications.

We can see the results of this simulation to what extent the fuzzy system that is built can produce the expected output. As the first question, a question with a moderate level (Medium) is chosen, and the probability of correct answer = 0.5. Table 2 to Table 4 shows the different scenarios of simulation for the Vector topic.

The first scenario is if the answer continues right (see Table 2). In this case, the difficulty level of the question will increase (medium to hard level) along with the increase in the estimation of participants' abilities (θ). The test for this subject will stop at the fourth question because, at that time, the estimated ability value (θ) has not changed anymore.

Table 2. Simulation results of vector topics if the response continues right.

Stage	Level	No. of Question	a	b	p	r	c	θ	$\Delta\theta$
1	M	16	0,8	-0,4055	0,5	1	0,25	1,874	
2	H	28	0,1	1,7346	0,6294	1	0,25	2,207	0,333
3	H	29	0,1	2,9444	0,6015	1	0,25	2,4589	0,2519
4	H	30	0,1	2,9444	0,6095	1	0,25	2,4589	0

The second scenario is if the participant's answer is wrong. In Table 3 can be seen that the level of difficulty of the question will go down (from medium to easy) along with the decrease in the estimation of the ability of the examinee (θ). The test will stop at the third question because the value of (θ) no longer changes.

Table 3. Simulation results of vector topics if the response continues wrong.

Stage	Level	No. of Question	a	b	p	r	c	θ	$\Delta\theta$
1	M	16	0,8	-0,4055	0,5	0	0,25	-1,0786	
2	E	12	0,5	-1,0986	0,6282	0	0,25	-1,8853	-0,8067
3	E	8	0,3	-1,7346	0,6105	0	0,25	-1,8853	0

The third scenario is if the participants' answers are right and wrong alternately (see Table 4). This results in the estimation of the ability (θ) to change continuously, or $\Delta\theta$ will not reach zero. Likewise, the level of difficulty of the problem will be fluctuating (medium-hard-medium). In this third case, the procedure for stopping the exam is if the number of questions has reached six items.

For other topics (namely matrices concepts, matrices operation, determinant, inverse, and matrices applications), we obtain the simulation result that tends to be the same as the results on the vector topic.

Table 4. Simulation results of vector topics if response right and wrong alternately.

Stage	Level	No. of Question	a	b	p	r	c	θ	$\Delta\theta$
1	M	16	0,8	-0,4055	0,5	1	0,25	2,0819	
2	H	28	0,1	1,7346	0,6360	0	0,25	0,2873	-1,7946
3	M	21	0,8	0,4055	0,5949	1	0,25	2,4589	2,1716
4	H	29	0,1	2,9444	0,6095	0	0,25	0,6657	-1,7932
5	M	26	0,3	0,619	0,6294	1	0,25	1,5	0,8343
6	H	30	0,1	2,9444	0,5791	0	0,25	0,6657	-0,8343

4. Conclusion

It has been successfully designed a Computerized Adaptive Test (CAT) with an estimation of participants' abilities using the Mamdani Fuzzy Inference System. The simulation process was carried out for Linear Algebra Courses with six topics, namely vector, matrices concepts, matrices operation, determinant, inverse, and matrices applications. The simulation results show that for each topic, at most 6 questions are delivered, and the system has been able to know the participants' abilities.

Acknowledgments

Authors are grateful to the Research Institute and Community Development of the State University of Gorontalo and Directorate of Research and Community Development – Ministry of Research, Technology and Higher Education Republic of Indonesia for providing financial support to complete this research.

References

[1] Balas-Timar D V and Balas V E 2009 Ability estimation in CAT with Fuzzy logic *Proc. Int. Symp. Comput. Intell. Intell. Informatics* pp 55–62
[2] Cisar S M, Radosav D, Pinter R and Cisar P 2012 Computer adaptive tests: a comparative study *IEEE 10th Jubil. Int. Symp. Intell. Syst. Informatics* pp 499–504
[3] Eggen T J H M 2012 Computerized Adaptive Testing Item Selection in Computerized Adaptive Learning Systems *Psychom. Pract* pp 14-25
[4] Van Der Linden W J and Pashley P J 2010 Elements of Adaptive Testing
[5] Hambleton R K, Swaminathan H and Rogers H J 1991 *Fundamentals of Item Response Theory* (California: SAGE Publications)
[6] Baker F B 2001 *The Basic of Item Response Theory* (ERIC)
[7] Van der Linden W J 1998 Bayesian item selection criteria for adaptive testing *Psychometrika* Vol 63 (2) pp 201-216
[8] Veldkamp B P 2010 Bayesian item selestion in constrained adaptive testing using shadow tests *Psicologica* Vol 31 pp 149-169
[9] Passino K M and Yurkovich S 1997 Fuzzy Control (Boston: Addison Wesley Logman)
[10] Jang J S R, Sun C T and Mizutani E 1997 Neuro-Fuzzy Fuzzy and Soft Computing: A Computational Approach to Learning and Machine Intelligence (New Jersey: Prentice Hall International)

IOP Conf. Series: Materials Science and Engineering **850** (2020) 012005 doi:10.1088/1757-899X/850/1/012005

Design of Graphical User Interface (GUI) for a remote laboratory on programmable logic controller experiments

M I Hambali, E Haritman* and A B Pantjawati

Department of Electrical Engineering Education, Universitas Pendidikan Indonesia, Jl. Dr. Setiabudhi 207, Bandung 40154, Indonesia

*erikharitman@upi.edu

Abstract. Remote Laboratory is the use of telecommunications technology to conduct remote experiments. In this study, remote lab operates through the Graphical User Interface (GUI) display, where interactions are carried out online through computer devices. This GUI is designed to support experiments on Programmable Logic Controller (PLC) and must be able to facilitate the user and provide feedback directly to confirm that an action has been taken. This GUI design includes color, image, icon, label, and text navigation to present information to users when needed. This research was carried out through several stages, namely the design of the GUI by using Microsoft Visual Studio software, coding and testing of appearance and function. The results showed that the GUI design met the desired specifications, both in appearance and function. From the users testing, the GUI design get positive responses from all users.

1. Introduction

The 4.0 Industrial Revolution Era has influenced all aspects of human life, including education, especially in the learning process. One of the implications of the industrial revolution on the educational aspect is the development of remote laboratories. Remote laboratory is the use of telecommunications technology to conduct real-distance experiments. Unlike the virtual lab, the remote lab is designed to enable students or researchers to carry out real experiments [1]. By applying the concept of lab management or Digital Laboratory Information System (DLIS), students can access remote lab effectively and efficiently [2].

In this study, the remote lab was designed to run a remote industrial automation laboratory, in this case controlling a programmable Logic Control (PLC). PLC is a special computer that is used to control and operate the manufacturing and machining processes in sequence [3].

Remote lab is a laboratory system that is integrated with remote operation. To make it easier for users to access the object to be controlled, it needs a display interface that connects users and computers. Graphical User Interface (GUI) can meet these needs, which is an interface on a computer operating system that uses a graphical menu. The purpose of using a graphical menu is to provide a display that can make the operating system more responsive, providing convenience and comfort for users in operating the computer. Computer users more easily understand the instructions in the form of images rather than a long description [4]. The main idea of the GUI is to accommodate the user interface with the application intuitively and simply [5]. Software that can be used to create a GUI is Microsoft Visual Studio with Visual Basic, which is software for developing and building applications on the .NET Framework system [6].

IOP Conf. Series: Materials Science and Engineering **850** (2020) 012005 doi:10.1088/1757-899X/850/1/012005

In this study, the GUI at a remote laboratory for industrial automation labs was designed using Microsoft Visual Studio as the main software and Visual Basic. NET is used as a programming language that is equipped with design features to create the desired interface.

2. Research methods

The GUI design was conducted using the waterfall method that is described as follows:

2.1. Need analysis

At this stage a needs analysis is done that underlies the determination of the GUI specifications to be designed.

2.2. System design

At this stage the GUI is designed with the following features:

- Splash Screen panel
- Program Menu. In this menu the users can choose which software will be used for doing experiments
- Camera menu. In this menu the users can open the attached camera to see the output from the PLC
- Lamp menu. In this menu the users can turn lights on and off
- Instruction Menu. In this menu the user can see the steps in conducting experiments with this remote lab
- About Menu. In this menu the users can see a brief explanation of the remote laboratory
- Timer
- Notifications

Figure 1 shows the remote laboratory system flow chart.

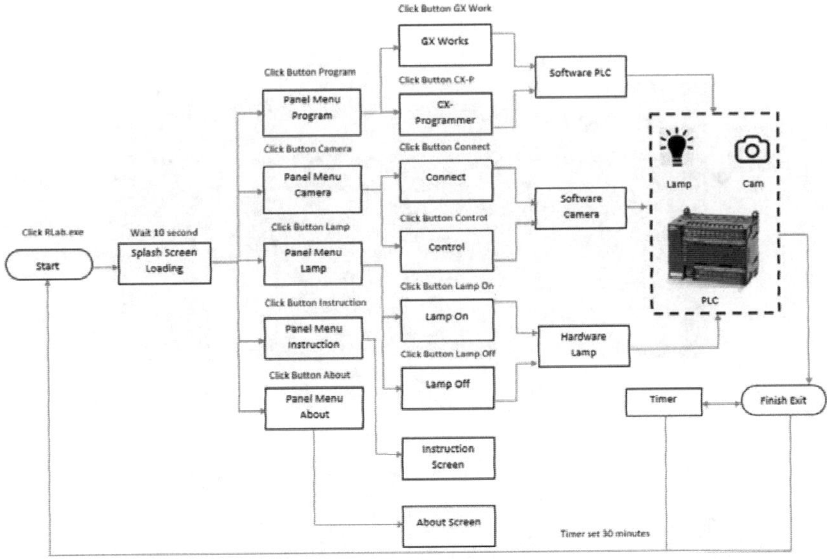

Figure 1. The remote laboratory system flow chart.

2.3. Coding and testing

This is the most important stage in working on the GUI design, where the system design that has been created will be coded. The results of the coding are tested first to detect and correct errors that occur in the system. Then the GUI is tested on the user.

2.4. GUI testing

If the design stage is complete and the GUI meets the desired specifications, the GUI is then tested on a number of users to get a response to the function, appearance and features. Illustration of user testing shown in figure 2.

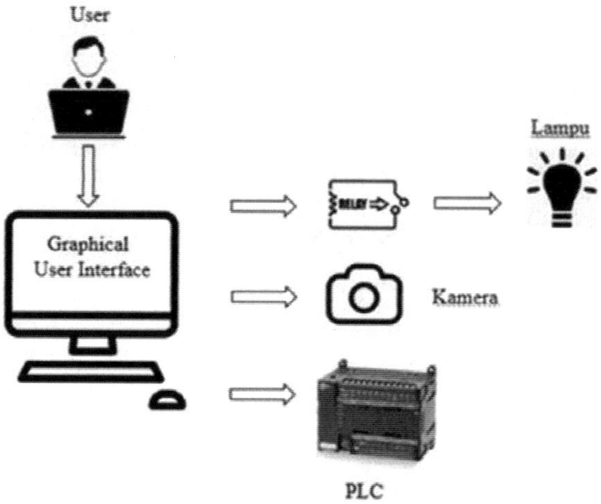

Figure 2. GUI system users testing chart.

2.5. Evaluation

After all testing is done, the last step is to evaluate and analyse the GUI system that was designed.

3. Results and discussion

Graphical User Interface (GUI) design results can be accessed by users using window-based applications on PC or smartphone devices. Figure 3 and 4 show the Splash Screen and the Program menu panels respectively.

Figure 3. Splash screen panel.

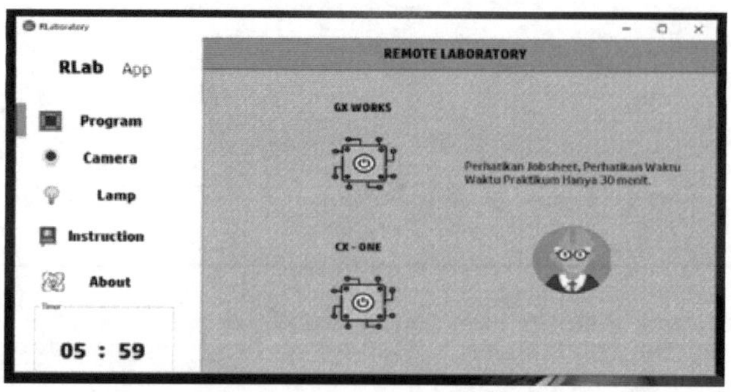

Figure 4. Program menu panel.

IOP Conf. Series: Materials Science and Engineering **850** (2020) 012005 doi:10.1088/1757-899X/850/1/012005

When the user clicks on RLab.exe, the splash screen panel, as shown in figure 3, will appear for 10 seconds. After the Splash Screen panel disappears, the Program menu panel, as shown in figure 4, will appear and the timer will automatically countdown. In this program menu the user can choose the software that will be used to carry out the desired experiment.

When the user clicks on the Camera menu, a display will appear as shown in figure 5. The user can click on the connect icon to connect to the camera and the control icon to connect to camera controls. On the Lights menu panel, as shown in figure 6, users can click on the lights on the icon to turn lights on or off. While the combo box is used to choose which port is connected to the PC.

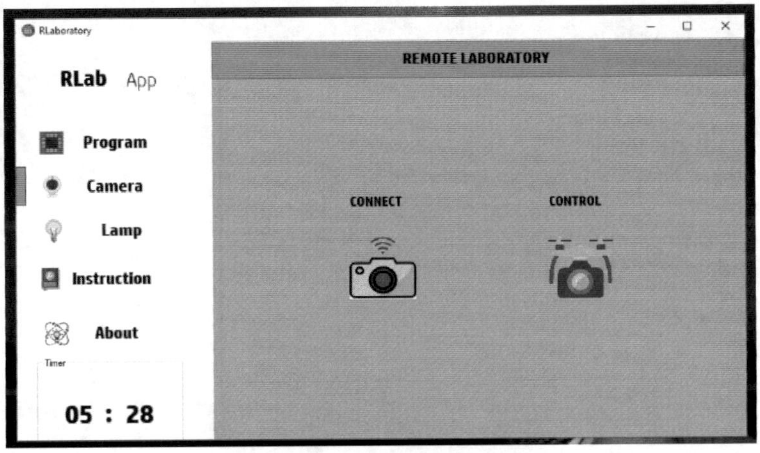

Figure 5. Camera menu panel.

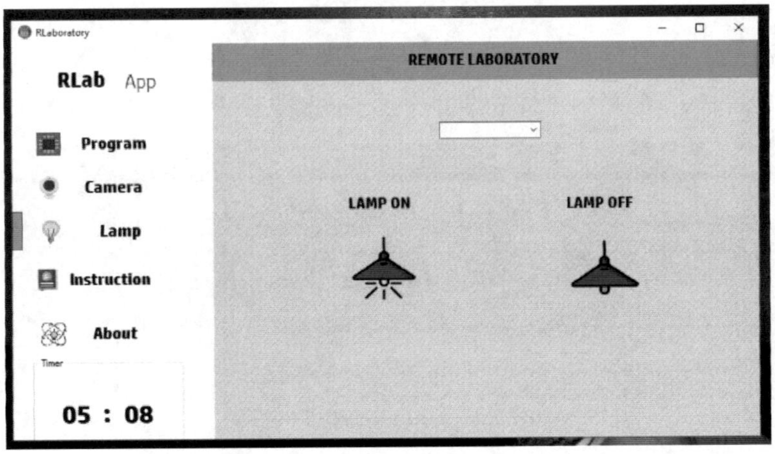

Figure 6. Lamp menu panel.

Users can see the steps of the experiment in the Instruction menu panel, as shown in Figure 7, and see a brief description of the Remote Laboratory, in the About menu panel, as shown in Figure 8.

If there are only 5 minutes left to conduct the experiment, a notification will appear on the screen as shown in figure 9.

IOP Conf. Series: Materials Science and Engineering **850** (2020) 012005 doi:10.1088/1757-899X/850/1/012005

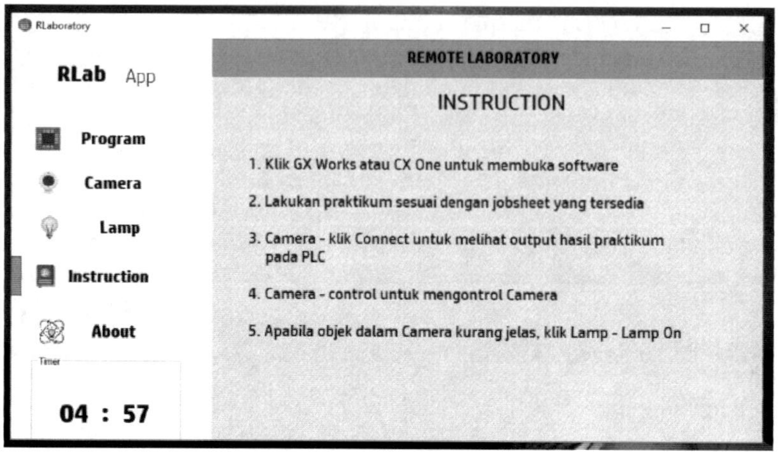

Figure 7. Instruction menu panel.

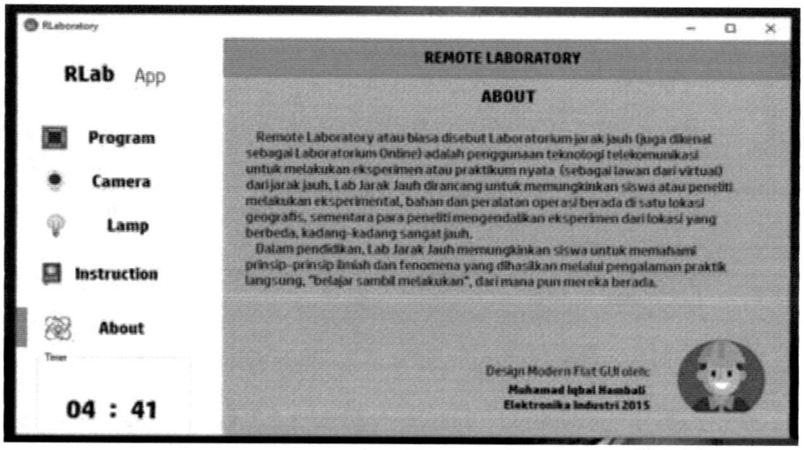

Figure 8. About menu panel.

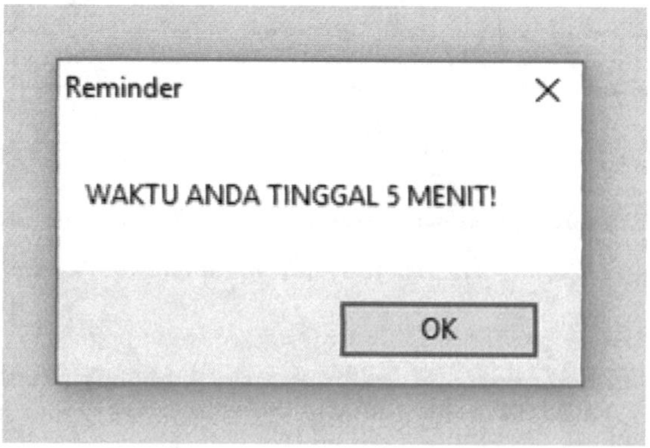

Figure 9. Time left notification.

In the final stage, 5 users were asked to test the appearance and function of the GUI. Table 1 shows the feedback given by the users in the trial.

Table 1. Users responses.

No	Feature name	Display that appears when buttons clicked	Yes/No answers				
			U1	U2	U3	U4	U5
1	Program menu	Program panel display, contain GX works button, CX programmer button and reminder	Yes	Yes	Yes	Yes	Yes
2	GX works button	GX works display	Yes	Yes	Yes	Yes	Yes
3	CX programmer button	CX programmer display	Yes	Yes	Yes	Yes	Yes
4	Camera menu	Camera panel display, contain connect and control button	Yes	Yes	Yes	Yes	Yes
5	Connect button	Logitech webcam camera display	Yes	Yes	Yes	Yes	Yes
6	Control button	Logitech webcam control display	Yes	Yes	Yes	Yes	Yes
7	Lamp menu	Lamp panel display, contains combo box, Lamp on button and lamp off button	Yes	Yes	Yes	Yes	Yes
8	Lamp on button	Turn on the light	Yes	Yes	Yes	Yes	Yes
9	Lamp off button	Turn off the light	Yes	Yes	Yes	Yes	Yes
10	Combo box	Choose the port to use	Yes	Yes	Yes	Yes	Yes
11	Instruction menu	Instruction display, contains experiment steps	Yes	Yes	Yes	Yes	Yes
12	About menu	About display, contains a brief description of the remote lab	Yes	Yes	Yes	Yes	Yes

Table 1 shows that the five users stated that all GUI features function properly. Users are also satisfied with the colors and modern designs of the designed GUI, so it can be concluded that the color aspect has an important role in the appearance design. This is consistent with the results of research which states that color can affect human psychology [7]. The success of the display design, is also determined by how to enter supporting elements such as icons, images, labels, text panels and colors, so that the interface display design can create an attractive and interactive impression [8].

4. Conclusion

The conclusion of this study is that the Graphical User Interface (GUI) for Remote Laboratories in PLC experiments has been successfully created and meets the desired specifications. Users respond very well to the appearance and features provided by the GUI.

References

[1] Callaghan M J, Harkin J, McGinnity T M and Maguire L P 2007 Paradigms in Remote Experimentation *International Journal of Online Engineering* **3**(4)
[2] Fushshilat I, Rahmat A, Somantri Y and Haritman E 2018 Laboratory management: digital laboratory information system (DLIS) concept *IOP Conference Series: Materials Science and Engineering* **434** 1 012286
[3] Petruzella F D 2005 *Programmable logic controllers* (McGraw-Hill Education)
[4] Avila J L and Amaya D 2014 Use of Remote Laboratories for Education in Industrial Processes and Automation *Proceedings of The 2014 IAJC-ISAM International Conference* 978-1-60643-379-9
[5] Ashvins Group 2005 *Designing Resourceful Graphical User Interfaces (GUIs) for Healthcare Applications* 3-5
[6] Gunarto H 2006 *Introduction to Visual Basic .NET Programming* (Singapore: Tech Pub.)
[7] Birren F 1961 *Colour Psychology and Colour Therapy* (New York : University Books Inc.)
[8] Pile J F 1995 *Interior Design* (New York : Harry N. Abrams Inc.)

IOP Conf. Series: Materials Science and Engineering **850** (2020) 012006 doi:10.1088/1757-899X/850/1/012006

The use of smartphone on learning activities: Systematic review

R O Firmansyah[1,*], R A Hamdani[2] and D Kuswardhana[3]

[1] Study Program of Technological and Vocational Education, Universitas Pendidikan Indonesia, Bandung, Indonesia
[2] Department of Mechanical Engineering Education, Universitas Pendidikan Indonesia, Bandung, Indonesia
[3] Department of Electrical Engineering Education, Universitas Pendidikan Indonesia, Bandung, Indonesia

*rianoktariana@gmail.com

Abstract. This article aims to analyze the use of smartphone technology in education field, with a focus on exploring the effects of smartphone and its software application use in learning activities. This study was conducted using the method of systematic literature review through 3 stages, namely: searching related journals through an online database of several publishers, analyzing the search results journals and dividing them into 3 groups according to the research objectives, and describing the results of the analysis in the form of scientific and logical arguments. From the results of the study it can be concluded that smartphone is significantly useful in student-centered learning, can be used in various disciplines, and at all levels of education, and there are various smartphone applications that can be used for learning needs.

1. Introduction

Smartphone is a mobile device that people use to communicate with each other which is capable of carrying out various activities, such as internet access, social media applications, and document processing with good screen resolution [1,2]. Nowadays smartphones are used as learning facilities in almost all fields, such as health, languages, engineering and education [3,4]. Indonesia is a developing country with a very large number of smartphone users and occupies the 2nd largest social network user in the world [5].

The development of smartphone technology is very fast. Almost all activities can be carried out and accessed using a smartphone [6]. In the education field, for example, almost all students have smartphones so they can use them to support the learning process. Several studies reveal the importance of integrating smartphone technology into the learning environment [1]. Smartphones can facilitate students in finding learning resources and use learning applications at any time and anywhere [7,8]. In this case, the achievement of learning objectives is inseparable from the teacher's role in determining the right strategy in utilizing smartphones in the learning process [1]. The negative impact of smartphone usage, such as game addiction, must be anticipated [8].

This article aims to analyze the development of learning using smartphone technology, with a focus on exploring the use of types of application software on smartphones and their influence in learning activities at various levels of the education unit.

2. Theoretical framework of smartphone for learning

In mobile learning, it is assumed that students can use smartphone devices, with various applications in them to support learning activities, wherever they are [9].

2.1. Conceptualization of mobile learning

The concept of mobile learning is a learning process that is centered on students in learning various materials so that they gain experience when they learn through mobile devices. Mobile learning is part of the new conception of society. Research and reflection on mobile learning must stimulate multidisciplinary and interdisciplinary thinking and methods in education [10].

2.2. Mobile device application on knowledge management

Implementation of mobile applications must be able to create a different learning environment, and increase student motivation. A teacher must know the experiences of students while learning and find strategies to improve the learning process [11]. Mobile learning can encourage students to acquire, store, share, apply, and create knowledge [12].

2.3. Mobile device application on language learning

Language is the main thing in social life, because people who are proficient in language will be able to communicate and socialize well. Smartphones with adequate internet network support can be good language learning media [13,14].

2.4. Existing learning environment

The current learning environment aims to integrate learning activities on the spot so students can interact appropriately with each other and the environment [15]. The use of mobile technology continues to grow, including in the field of education. Smartphones can facilitate and enhance learning interactions by accessing, discussing and sharing related information through social networks at anytime and anywhere [16].

3. Methods

The focus of this article review is the use of smartphones in learning. Studies related to this article are e-learning and e-module learning which are currently developing into m-learning [17]. Election focuses on studies based on the rapid development of smartphone technology that resembles computer capabilities, with prices that are relatively cheap and can be utilized in learning [18]. Taking this theme is considered interesting because almost all individuals have a smartphone, and in this article we will analyze how smartphones are used in learning, in terms of application usage, the benefits provided and the population of learning media use at various levels of education.

The method used in writing this article is a systematic literature review. This method was chosen because it can identify, evaluate, and interpret research data and results in accordance with research questions [19]. In addition, a systematic literature review can be used to minimize bias, disseminate the literature search framework, and allow for comprehensive analysis [20].

The systematic literature review in this article aims to explore the benefits and effects of smartphone implementation in learning, the type of application software used, and its application at the education unit level. This method of systematic literature review is carried out through 3 stages [21], namely searching and filtering literature, analyzing search results in accordance with the purpose of writing articles, and pouring the results of analysis in the form of scientific and logical arguments.

The literature search process is carried out from various online sources. Some of the literature databases used are Google Scholar, Scopus, and Thomson. The publishing houses used are science direct, springer and taylor and francis [22-24]. The keywords used for the literature search process are smartphones, smartphones for lerarning, m-learning, study on smartphones, learning using smartphones, smartphone apps for learning and smartphone utilities for learning. In addition to keywords, there are criteria for filtering the literature traced, i.e. published articles in the last five years, in the form of journals, and are research articles. Each article is reviewed by researchers with reference to the use of smartphones in learning in terms of the benefits provided, the type of application software used and the application of learning media at various levels of education unit. The last stage of the literature review is to provide a review of argumentation theory and logical argument writing.

4. Results

The research topics identified were grouped into three groups, namely: a) smartphone utilization in learning, b) types of software applications used for learning, and c) the application of learning media at various level of educational units. Data analysis uses the literature review which is divided into three stages with the aim of getting relevant journals according to the needs of the article.

The first stage is searching, filtering and collecting initial data from predefined databases and publishing houses, which obtained 100 journals based on the keywords: smartphone for lerarning, m-learning, study on smartphones, learning using smartphones, smartphone app for learning and smartphone utilities for learning. The second stage is analyzing the search results in accordance with the focus of article writing, which is carried out by the translation and analysis of each journal. The analysis is outlined in a tabular form containing the title, author, participant (to find out the level of education of the respondent), context (knowing the activities and applications used in research), type of data and basic finding (knowing the results or usefulness of the smartphone in learning), as shown in table 1.

Table 1. Decomposition of each journal at the data analysis stage.

No	Sub	Description
1	Tittle	Explore patterns of use of mobile applications for education for undergraduate students
	Author	Iris Shuk Han Wai, Sara Sze Yan Ng and Dickson K. W. Chiu Kevin K. W. Ho Patrick Lo
	Year	2016
	Participant	150 undergraduate students in business, education and engineering
	Context	this study aims to provide an up-to-date description of the use of cellular applications in higher education.
	Type Of Data	Quantitative and qualitative methods were used to collect data from 150 students
	Basic Finding	The results show undergraduate students often use mobile applications to engage in learning activities related to their academic studies,
2	Tittle	Use of smartphones and smartphone addiction in middle school students in Korea: Prevalence, social networking services, and use of games
	Author	Seong-Soo Cha and Bo-Kyung Seo
	Year	January-June 2018: 1 –15
	Participant	1,824 middle school students using smartphones.
	Context	This study aims to examine smartphone usage patterns, smartphone addiction characteristics, and smartphone addiction prediction factors in middle school students in South Korea.
	Type Of Data	Face to face interview
	Basic Finding	According to the Smartphone Addiction Scale score, 563 (30.9%) were classified as risk groups for smartphone addiction and 1261 (69.1%) were identified as normal user groups. Teenagers use cellular messenger for a long time, followed by surfing the internet, playing games, and using social networking services.
3	Tittle	iMidwife: smartphone use by students, midwifery technology as a mediated educational tool in the clinical environment
	Author	Annemarie DeLeo and Sadie Geraghty
	Year	(Received 9 January 2017; accepted 5 December 2017)
	Participant	76 midwifery students
	Context	survey of midwifery students using a questionnaire that facilitates both qualitative and quantitative responses
	Type Of Data	mixed methodology,
	Basic Finding	The participants were asked whether they intended to work in rural/remote areas or in the metropolitan area after graduation; 85% (n = 25) of participants intended to work in metropolitan hospitals at graduation and 15% (n = 4) said they intend to work in rural/remote areas of Western Australia.
4	Tittle	Factors that influence teacher decisions to use smartphone clicker applications to improve teaching interaction in university classrooms in Saudi Arabia
	Author	Abdulaziz Aljaloud, William Billingsley & Paul Kwan
	Year	Received 16 February 2017; Accepted 26 March 2018
	Participant	33 teachers from the computer science faculty
	Context	33 teachers and 14 were interviewed
	Type Of Data	Descriptive of the interview
	Basic Finding	The main implication of this finding is that smartphone clicker application developers and user training coordinators must consider the teacher's perceptions of the suitability of technology and their desire to design learning assignments to facilitate students' participation and involvement.

After describing each journal, the next step is to select and group journals in accordance with the focus of writing, with the scheme as shown in figure 1.

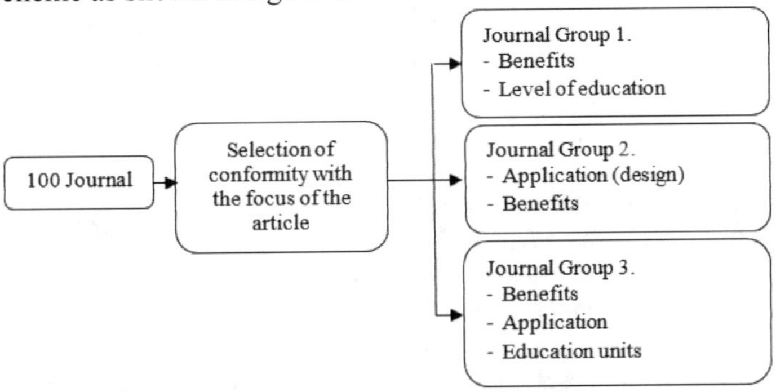

Figure 1. The journal grouping scheme in accordance with the focus of the discussion.

Data analysis aims to assess the suitability of the journal with the focus of the discussion in this article. The results of the data analysis found that group 1, 33 journals, contained aspects of benefits and education units, group 2, 19 journals, contained aspects of benefits and application software, and group 3, 48 journals, contained aspects of benefits, application software and applications to the education unit, as shown in figure 2.

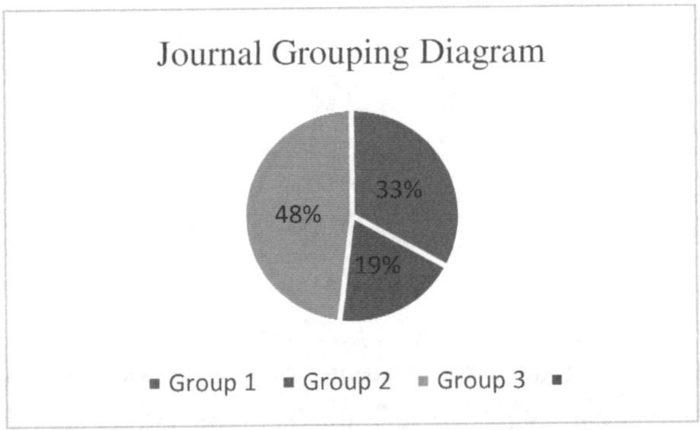

Figure 2. Grouping the results of data analysis.

5. Discussion

The last stage is to provide an overview of argumentation theory and logical arguments on the results of the analysis in the second stage. The overview focused on the benefits of smartphones in learning, the type of application software used and the application of instructional media in educational units.

5.1. Use of smartphones in learning

The application of smartphone technology in the learning process analyzed from relevant journals shows that the use of technology in learning can produce significant benefits if properly managed by the teacher [25,26]. Learning uses a smartphone with a student center strategy, is facilitated by teachers to achieve learning goals in the classroom [27].

Smartphone technology is applied not to complicate the learning process but rather to facilitate the mobile learning process. To create innovative learning opportunities, their usefulness and rationality need to be considered [28].

There are two converging functions that are influenced by cellular technology: (1) convergence between cellular technology as a learning media and learning design, and smartphone marketing that combines communication and computer functions in one device, (2) convergence between new personal and mobile technologies and a new conception of learning as a privately managed lifelong activity [29].

5.2. Smartphone applications used in learning
Smartphone application plays an important role in supporting learning activities, because it will determine the learning patterns and stages. Based on the results of analysis on relevant journals, it shows that there are several applications in smartphone that have been used in various disciplines, e.g. effective questionnaire applications [30], effective lesson schedulers [31], social networks for learning [32], language learning [26], and information in the health field [33].

5.3. Learning implementation using a smartphone
Learning by using a smartphone has been widely applied in various fields and levels of education. In this article smartphone users are grouped into 6 groups, namely kindergartens, elementary schools, junior high schools, senior high schools, colleges and the public consisting of various communities and informal institutions.

As shown in figure 3, smartphones are used at all levels of the education unit, with the most users in universities, which is 40%.

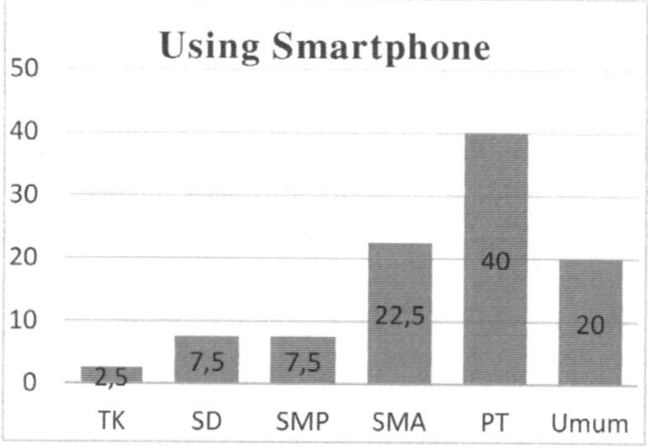

Figure 3. Smartphone usage at various education unit level.

6. Conclusion
From this study it can be concluded that the use of smartphones in learning facilitates significantly the achievement of student learning goals with student-centered learning models in various disciplines. Teachers can utilize various applications on smartphones for learning needs, such as applications for effective questionnaires, lesson schedulers, social networking, language learning and practice guidance applications. Smartphone applications in learning activities are used at all levels of education from elementary school to college.

References
[1] Jeng Y L, Wu T T, Huang Y M, Tan Q and Yang S J 2010 The add-on impact of mobile applications in learning strategies: A review study *Educational Technology & Society* **13**(3) 3-11

[2] Yu Chen L, Chuang Yu and Wen Tzy T 2018 Why not be "smarter"? Examining the factors that influence the behavioral intentions of non-*smartphone* users **20**(3) 1-20

[3] Tossell C C, Kortum P, Shepard C, Rahmati A and Zhong L 2015 You can lead a horse to water but you cannot make him learn: *Smartphone* use in higher education *British Journal of Educational Technology* **46**(4) 713-724

[4] Buck J L, McInnis E and Randolph C 2013 The new frontier of education: The impact of *smartphone* technology in the classroom *American Society for Engineering Education* **1** 1 1-11

[5] Poushter J 2016 *Smartphone* ownership and internet usage continues to climb in emerging economies *Pew Research Center* **22** 1-44

[6] Chen H R and Huang H L 2010 User acceptance of mobile knowledge management learning system: Design and analysis *Educational Technology & Society* **13**(3) 70-77

[7] Lan Y J, Sung Y T, Tan N C, Lin C P and Chang K E 2010 Mobile-device-supported problem-based computational estimation instruction for elementary school students *Journal of Educational Technology & Society* **13**(3)
[8] Seong-Soo Cha and Bo-Kyung Seo 2018 *Smartphone* use and *smartphone* addiction in middle school students in Korea: Prevalence, social networking service, and game use. 1, 1-15.
[9] El-Hussein M, Osman M and Cronje J C 2010 Defining mobile learning in the higher education landscape *Journal of Educational Technology & Society* **13**(3)
[10] Traxler J 2007 Defining, Discussing and Evaluating Mobile Learning: The moving finger writes and having writ.... *The International Review of Research in Open and Distributed Learning* **8**(2)
[11] Yau J Y K and Joy M 2010 A context-aware personalized m-learning application based on m-learning preferences *2010 6th IEEE International Conference on Wireless, Mobile, and Ubiquitous Technologies in Education* (pp. 11-18) IEEE
[12] Chen H R and Huang H L 2010 User acceptance of mobile knowledge management learning system: Design and analysis *Educational Technology and Society* **13**(3) 70-77
[13] Huang Y M, Hwang W Y and Chang K E 2010 Guest Editorial-Innovations in Designing Mobile Learning Applications *Educational Technology and Society* **13**(3) 1-2
[14] De Jong T, Specht M and Koper R 2010 A study of contextualised mobile information delivery for language learning *Journal of Educational Technology and Society* **13**(3) 110
[15] Patten B, Sánchez I A and Tangney B 2006. Designing collaborative, constructionist and contextual applications for handheld devices *Computers and education* **46**(3) 294-308
[16] Martin F and Ertzberger J 2013 Here and now mobile learning: An experimental study on the use of mobile technology *Computers and Education* **68** 76-85
[17] Kim Y and Shon J G 2011 A Study on Design of K-12 e-learning system for utilization *smartphone Journal of Internet Computing and Services* **12**(4) 135-143
[18] Kafyulilo A 2014 Access, use and perceptions of teachers and students towards mobile phones as a tool for teaching and learning in Tanzania *Education and Information Technologies* **19**(1) 115-127
[19] L García-Borgoñón 2014 Software process modeling languages: A systematic literature review *Inf. Softw. Technol.*, **56** 2 103–116
[20] J K Roehrich 2014 Are public-private partnerships a healthy option? A systematic literature review *Soc. Sci. Med.* **113** pp. 110–119
[21] Crossan M M and Apaydin M 2010 A multi-dimensional framework of organizational innovation: A systematic review of the literature *Journal of management studies* **47**(6) 1154-1191
[22] Jacsó P 2005 Google Scholar: the pros and the cons *Online information review* **29**(2) 208-214
[23] Mongeon P and Paul-Hus A 2016 The journal coverage of Web of Science and Scopus: a comparative analysis *Scientometrics* **106**(1) 213-228
[24] Bernstein B 2018 On the classification and framing of educational knowledge. In *Knowledge, education, and cultural change* (pp. 365-392). Routledge.
[25] Chan N N, Walker, C, and Gleaves A 2015 An exploration of students' lived experiences of using *smartphones* in diverse learning contexts using a hermeneutic phenomenological approach *Computers and Education* **82** 96-106
[26] Aljaloud A, Billingsley W, and Kwan P 2018 Factors that influence teachers' decisions to use *smartphone* clicker apps to enhance teacher-student interactions in university classrooms in Saudi Arabia *Learning: Research and Practice* 1-20.
[27] Sandholtz J H 1997 *Teaching with technology: Creating student-centered classrooms* Teachers College Press, Teachers College, Columbia University, 1234 Amsterdam Ave., New York, NY 10027
[28] Jeng Y L, Wu T T, Huang Y M, Tan Q and Yang S J 2010 The add-on impact of mobile applications in learning strategies: A review study. *Educational Technology and Society* **13**(3) 3-11

[29] Antoun C, Katz J, Argueta J and Wang L 2017 Design Heuristics for Effective *Smartphone* Questionnaires *Social Science Computer Review* 0894439317727072.

[30] Yau J Y K, Joy M and Dickert S 2010 A Mobile Context-aware Framework for Managing Learning Schedules-Data Analysis from a Diary Study *Journal of Educational Technology and Society* **13**(3)

[31] Huang J J, Yang S J, Yueh-Min H and Hsiao I Y 2010 Social learning networks: Build mobile learning networks based on collaborative services *Journal of Educational Technology and Society* **13**(3) 78

[32] De Jong T, Specht M and Koper R 2010 A study of contextualised mobile information delivery for language learning *Journal of Educational Technology and Society* **13**(3) 110

[33] DeLeo A and Geraghty S 2017 iMidwife: midwifery students' use of *smartphone* technology as a mediated educational tool in clinical environments *Contemporary nurse* 1-10

IOP Conf. Series: Materials Science and Engineering **850** (2020) 012007 doi:10.1088/1757-899X/850/1/012007

Exhaust temperature and peltier element optimization of thermoelectric generator output

I Setyowidodo*, S Sutanto, A Mufarrih and I M Sholehah

Mechanical Engineering, Universitas Nusantara PGRI Kediri, Indonesia

*irwan.setyowidodo@gmail.com

Abstract. The purpose of this study are to obtain (1) the optimum exhaust temperature for the highest voltage output, (2) the optimum number of peltier elements for the highest voltage output, (3) the optimum peltier circuit type for the highest voltage output. The analysis were done to obtain the optimum Seebeck coefficient as a representation of the thermoelectric generator performance. The design of this study is an L20 orthogonal matrix with three replications using a factorial design. The independent variables varied were exhaust temperature and number of peltier elements, while data analysis using ANOVA Two Way. Based on the results of the research data it can be concluded that the furnace wall temperature affects the TEG current output. The optimal wall temperature was 480 °C that produces a TEG current output of 168 mA. This indicates an increase in the temperature of the used metal smelting furnace walls cause the increase in the TEG current output. Meanwhile, the number of peltier elements influenced the TEG current output. The optimum values combination of exhaust temperature and peltier's number which cause the highest TEG output voltage are 480 °C and 6 (six) peltier element in series circuit, respectively.

1. Introduction

The development of advanced technology requires a lot of equipment in everyday life using ferro and non-ferrous metals. This cause metal waste is spread throughout the archipelago. The problem becomes an opportunity for small and medium industries to process ferro and non-ferro waste. The metal smelting industry requires high temperature to melt used metal and requires long melting time depending on the type of scrap metal, for example, the melting point of aluminium is 660°C or 1220 °F. High temperature and old metal smelting processes cause a lot of heat energy to be wasted during the metal smelting process. At present, the rapid grow of metal smelting industry causes a lot of heat energy wasted, whereas at this time energy must be processed efficiently to overcome the electricity needs.

Thermoelectric generator (TEG) is a technology that can directly change the temperature difference into electrical voltage. The use of waste heat from the melting furnace can add a list of alternative energy that can be used to overcome the energy crisis at this time. Some of the advantages of this TEG include, it is very reliable (usually over 100,000 hours of operating conditions), without sound when operated because it has no moving mechanical parts, does not require more maintenance, is simple, compact and safe, has a very small size and very light, capable of operating at high temperatures, capable of operating for small scale and remote locations, environmentally friendly, and a flexible energy source. In addition to the advantages, this TEG has disadvantages, namely at low efficiency, but if TEG is used to convert waste heat into electrical energy, low efficiency in TEG can be ignored [1].

IOP Conf. Series: Materials Science and Engineering **850** (2020) 012007 doi:10.1088/1757-899X/850/1/012007

The low efficiency of TEG will only produce a small TEG output while the electrical energy requirements for daily life are quite large. The value of the TEG current output is influenced by several factors including the number of peltier elements used and the hot side temperature of the peltier. Assume increasing the number of peltier elements and providing a high temperature on the peltier heat side to affect the TEG current output [2]. The purpose of this study are to obtain (1) the optimum exhaust temperature for the highest voltage output, (2) the optimum number of peltier elements for the highest voltage output, (3) the optimum peltier circuit type for the highest voltage output, and (4) optimum point of the Seebeck coefficient on the performance chart. The advantages of the research that is expected to be able to utilize thermal energy from the exhaust channel into a form of energy that can be more utilized so that thermal energy is able to have a better use value. Previous research on the effectiveness of the Tec-12706 peltier module as a generator utilizing heat energy from the Peltier Tec-12706 module [3]. Based on the studies it can be concluded that the amount of peltier current is influenced by several factors including the type of peltier, number of elements, type of electrical circuit, and temperature difference. The more the number of peltier elements the greater the TEG current output, the greater the temperature difference the greater the TEG current output and the type of electrical circuit that can increase the acquisition of TEG current output is a parallel circuit. Therefore, further research is done on how exhaust temperature optimization, the number of peltier elements to produce optimum voltage in a metal melting furnace, so that the lost heat energy can be reused in the form of electrical energy.

2. Material and methods

2.1. Experimental procedure

The flow diagram of this thermoelectric generator research is shown in figure 1.

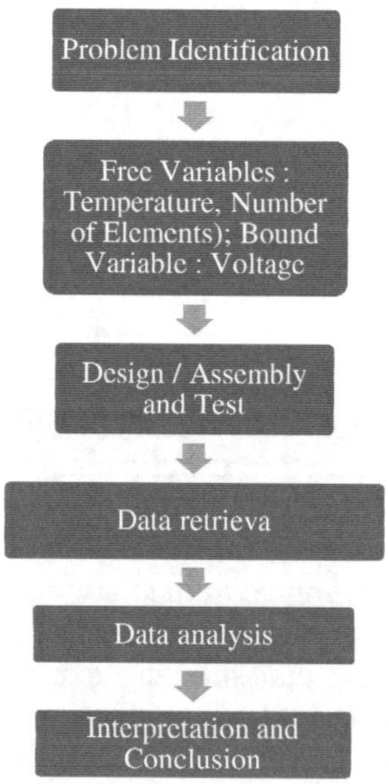

Figure 1. Research flow chart.

2.2. Research instrument

- Peltier module used is TEG type SP 1848 27145 SA are shown in figure 2.

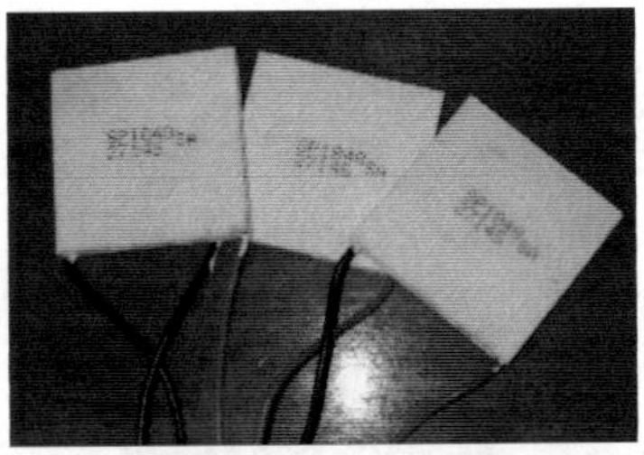

Figure 2. Peltier module *Thermometer Infrared.*

- Cable for compiling parallel electrical circuits are shown in figure 3.

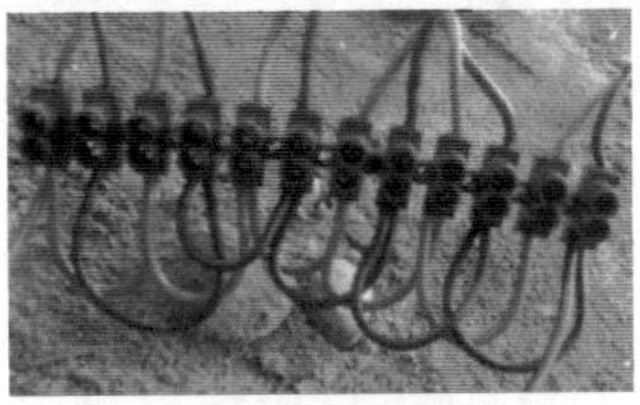

Figure 3. Parallel sequence.

- Digital multitester.
- Heatsink.
- Nuts and bolts.
- Thermal paste.
- Expenses.

2.3. Data collection procedures
Data retrieval is done repeatedly 3 times. The steps of data collection in this study are shown in figure 4-7.

- Measuring the temperature of the metal melting furnace wall with a predetermined temperature variation.

Figure 4. Measurement of furnace wall temperature.

- Measuring the temperature of the heat heatsink attached to the wall of the furnace insulator.

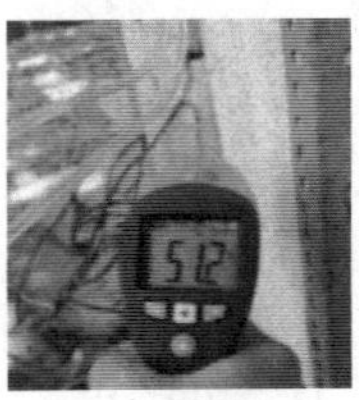

Figure 5. Heat heatsink temperature measurement.

- Measuring the temperature of the cold heatsink.

Figure 6. Cold heatsink temperature measurement.

- Measuring the TEG current output using a digital multitester connected with an assembled peltier element electrical cable.

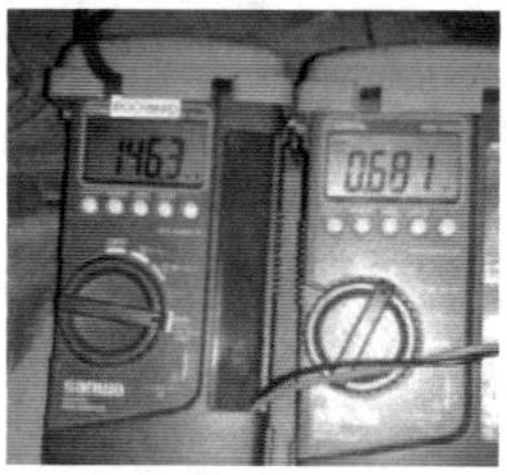

Figure 7. Record the results of the research data in the design data table of the research results.

2.4. Data analysis
Analysis of this research data is quantitative analysis, therefore the data analysis uses statistical analysis of variance (ANOVA) method with a significant level of 5%.

3. Results
Retrieval of research data that has been done with replication 3 times the data obtained as shown in table 1.

Table 1. TEG output data.

Total of Peltier	Exhaust's Temperature (^{0}C)	Current (ma)		
		1	2	3
4	400	135.5	144.5	141.5
	420	139.5	148.5	146.8
	440	145.3	153.5	150
	460	148.5	154.4	157
	480	150	156	161
6	400	132	141.4	140
	420	147.3	146	146
	440	152.6	149	148.2
	460	156	160	149
	480	160.3	168	153.6

Based on table 1, it can be seen that the largest current that can be produced is 168 mA which is located at a temperature of 480 °C and the number of elements 6. The results of the test data analyzed by Two Way ANOVA obtained P-value and F count, which were then used as guidelines to test the hypothesis with the following results:

IOP Conf. Series: Materials Science and Engineering **850** (2020) 012007 doi:10.1088/1757-899X/850/1/012007

3.1. Comparison of P-value with α

The second hypothesis testing is to compare the value of P-Value of each independent variable with a significant level, namely $\alpha = 0.05$ with the provision that if the value of P-Value $<\alpha$ then there is the effect of independent variables on the TEG current output and if the value of P-Value$> \alpha$ then there is no effect of independent variables on TEG current output as shown in table 2.

Table 2. P-value comparison.

No.	Research Factors	*P- Value*		*α*
1	T.Exhaust	0.000	<	0.05
2	Total of Peltier	0.491	>	0.05

3.2. Comparison of $F_{0,05}$ with F_{value}

After being analyzed with ANOVA, it was obtained that F_{value} of each independent variable was compared with $F_{0,05}$ with the provision that if the value of $F_{value} < F_{0,05}$ then H_0 is accepted and if the value of $F_{value} > F_{0,05}$ then H_0 is rejected as shown in table 3.

Table 3. F-value comparison.

No.	Research Factors	F_{value}		$F_{0,05}$
1	T.Exhaust	15.85	>	2.76
2	Total of Peltier	0.49	<	4.20

3.3. Test the hypothesis

Based on the second hypothesis testing, it can be seen in table 4.

Table 4. Test hypotesis.

No.	Research Factors	Conclusion
1	T.Exhaust	accepted
2	Total of Peltier	rejected

3.4. Main effect plot

Data analysis with the main effect plot was used to support the research conclusions, can be seen in figure 8.

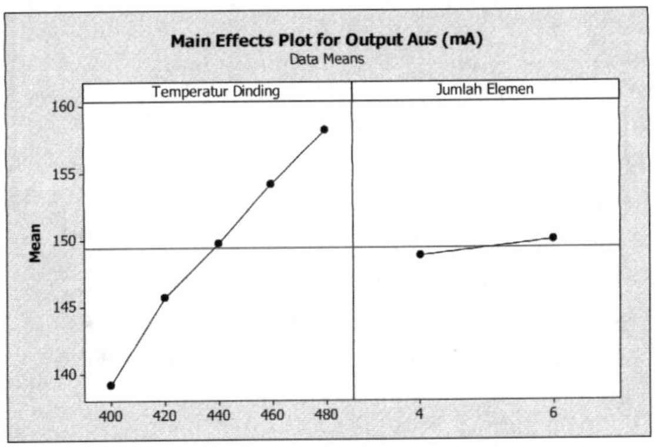

Figure 8. Main effect plot.

The Main Effect Plot image shows that:

- Furnace wall temperature has a significant effect on TEG current output.
- The number of peltier elements has an influence on the TEG current output but is not too significant even though the hypothesis test shows no influence on TEG current output.

4. Discussion

Based on the experiment's results, the optimum furnace temperature is 480ºC that produce 166 mA of current. The higher the temperature of the furnace wall the greater the current. This due to the increase in the heatsink temperature (while the cold heatsink is kept constant <100ºC) causes a large temperature difference between the two heatsinks. When the heat increase, the electric ion mobility will rise in thermoelement, which eventually enlarge the current. The most optimum difference in temperature to obtain a large current output is the highest temperature difference in the independent variable of the researcher and the largest arrangement [4,5].

The second independent variable is the number of peltier elements that have 2 levels, among them 4 peltier elements and 6 peltier elements. Based on the results of hypothesis testing there is no effect of the number of peltier elements on the TEG current but in showing a large increase in the current as the number of elements increases but does not have a large effect because the unbalanced voltage has a direct effect on the current, because the unbalanced voltage has an angle not the same and can be broken down into 3 sequences namely positive, negative and zero [6]. At balanced voltages have the same angle and only produce a symmetrical sequence without zero and negative, this applies the same as the current. When arranged in parallel, the voltage will be divided equally and the voltage is not balanced because the negative and zero prices appear causing the outgoing voltage to be disturbed and the value is small, this affects the current produced. The peltier voltage inequality of parallel circuits, peltier which has a smaller voltage will be a burden but not fully and causes the current produced by peltier with the highest voltage to pass through another peltier which has a lower voltage. Based on research on the value of Vp (parallel voltage) the current TEG can be positive or negative [7]. A positive value current means an open circuit that is no external load applied, whereas if the current is negative it means that the element module absorbs the current during the heating process.

5. Conclusion

Based on the study, it can be concluded that the variations of wall temperature effect TEG output voltage. The optimum value of wall temperature is 480ºC resulting in TEG current output of 168 mA. The greater the temperature the greater the TEG current output. From hypothesis test statistical analysis, the increase in the number of peltier elements does not significant affect TEG current output. The optimum values combination of exhaust temperature and peltier's number which cause the highest TEG output voltage are 480 ºC and 6 (six) peltier element in series circuit, respectively.

References

[1] Ismail B I and Ahmed W H 2009 Thermoelectric power generation using waste-heat energy as an alternative green technology *Recent Patents on Electrical & Electronic Engineering (Formerly Recent Patents on Electrical Engineering)* **2**(1) 27-39

[2] Fikri H A A 2016 *Efektifitas Modul Peltier TEC-12706 Sebagai Generator Dengan Memanfaatkan Eenergi Panas Dari Modul Peltier TEC-12706* (Doctoral dissertation, Surakarta: Universitas Muhammadiyah Surakarta)

[3] Sugiyanto S and Siswantoro S 2013 Perbandingan Penggunaan Thermoelectric Generator Tipe TEG127-40A dengan TEG126-40A sebagai Media Konversi Panas Menjadi Listrik pada Kompor Gas LPG dengan Pendinginan Alami *Rekayasa Teknologi Industri dan Informasi* 60-65

[4] Putra N, Koestoer R A, Adhitya M, Roekettino A and Trianto B 2009 Potensi Pembangkit Daya Termoelektrik untuk Kendaraan Hibrid *Makara Journal of Technology* **13**(2) 53-58

[5] Setyowidodo I, Fadhila A T and Mufarri A 2018 Analysis Effect of Exhaust's Temperature, Total of Peltier's Element and Circuit Type for Thermoelectric Generator Output Voltage at Exhaust Internal Combustion Vehicle *IPTEK Journal of Proceedings Series* **1**(1) 3-8
[6] Sarimun W 2012 *Proteksi Sistem Distribusi Tenaga Listrik* (Jakarta: Garamond)
[7] Montecucco A, Siviter J and Knox A R 2014 The effect of temperature mismatch on thermoelectric generators electrically connected in series and parallel *Applied Energy* **123** 47-54

IOP Conf. Series: Materials Science and Engineering **850** (2020) 012008 doi:10.1088/1757-899X/850/1/012008

Seawater distillation using vacuum pumps and ultraviolet lights to improve desalination process and freshwater quality

M S Rosyad*, S Ramadhani, I Riski, H Sutanto, Y Wahyono and W Indrawan

Diponegoro University, Jl. H. Prof. Soedarto, SH, Tembalang, Semarang, Central Java-Indonesia

*s4bil.54@gmail.com

Abstract. Lack of clean water supply quantitatively because of 97% water in earth is seawater with 35% salinity. One effort that can be done to overcome the clean water crisis is through desalination of seawater to produce water with low salinity. Some of the desalination methods that ever existed were MSF (Multi Stage Flash Distillation) and Reverse Osmosis. However, both of these methods have high investment cost. Therefore, a cheaper and easier method to apply such as distillation is required. In this research, distillation system is made of stainless steel and aluminium combined with vacuum pumps and ultraviolet lights. The use vacuum pump can speed up the evaporation process of seawater. Ultraviolet lights function to degrade heavy metal pollutants and eliminate harmful bacteria. Tests carried out include evaporation rate and water quality. This technology can increase water productivity up to 2 times and water quality shows a decrease in some pollutant parameters including 99.7% TDS, 99.93% salinity, 100% total coliform, and heavy metals Fe of 96,1%, Cr of 90.33% and Zn of 98,9%, and pH of seawater being neutral from 7.5 to 6.9. The technology has the potential to be applied to the community.

1. Introduction

Clean water is the most important requirement for human activities. Quantitative lack of clean water is due to 97% of the water on earth being seawater. Seawater has an average salt content of 35%. This means that for every one litter of seawater there are 35 grams of salt dissolved in it [1]. This high salinity level makes seawater cannot be consumed directly, so the processing needs to be done first. In addition, fresh water content is also declining due to sustainable development without regard to the environment to minimize the area of infiltration of rainwater. The freshwater content in the soil is dwindling as it is taken continuously so that more seawater is absorbed into the soil replacing the freshwater position.

In the world's the average sea water oceans has a salinity of 3.5%, which means that for every litter of sea water there is 35 grams of salt dissolved in it. The main salt content found in seawater include chloride (55%), sodium (31%), sulphate (8%), magnesium (4%), calcium (1%), potassium (1%), and the rest (less than 1%) consisting of bicarbonate, bromide, borax acid, strontium, and fluoride. The presence of these salts affects the physical properties of seawater such as density, compressibility and freezing. Water with salinity is certainly not consumable.

Seawater with a high salinity level can actually be used as one solution to meet the needs of clean water by separating the content of mineral salts dissolved with water through desalination.

Desalination technologies have been used rapidly over the past few decades throughout the globe to produce clean drinking water from seawater to improve the quality of already existing supplies of fresh water for human consumption. Several methods of water desalination ever existed were MSF (Multi Stage Flash Distillation) and Reverse Osmosis. However, the disadvantage of these two methods is to have a high investment cost. Therefore, a cheaper, and easier to apply method such as distillation is required.

One of the desalination methods available is distillation. Distillation is a method of separating seawater with salt by heating seawater by utilizing sunlight to produce water vapour, which is then condensed to produce clean water. Distillation can separate mineral salts such as sodium chloride (NaCl), calcium carbonate, magnesium sulphate, and magnesium bromide from water compounds (H_2O) so that clean water is produced with a total value of dissolved solids or small TDS (Total Dissolved Solid). Distillation also has the ability to reduce the concentration of various kinds of pollutants including suspended solids, heavy metals, organic matter, ions and radionuclides. However, in its application, the conventional distillation method takes a considerable amount of time and much water falls back into the evaporative container during the condensation process [2].

Increasing the temperature difference between the water cover is the focus when trying to improve the rate of condensation, and consequently the productivity of the still is increased. This can be done by either decreasing the cover temperature or increasing the water temperature. Continuous supply of water film is fed over the cover in order to reduce the temperature. However, this method requires raising the water container to the level of the highest part of the still or using pump water. Therefore, the using vacuum pumps on distillation system is right solution. Water pumps are machines or mechanical equipment used to raise fluids from lowlands to highlands or to drain fluids from low-pressure areas to high-pressure areas. The water pump can be a booster flow rate on a piping network system so that the water speed is faster. By exploiting the nature of this vacuum pumps placed in the distillation system it can speed up the increase of water temperature and water evaporate faster.

Ultraviolet lights have attracted much attention because of their application in the decolourization of bacteria and harmful pollutants such as dyes, chemicals, and toxic gases. UV light can replace sunlight in sterilizing the results of evaporation of seawater against bacteria that remain after the evaporation process. The use of UV with medium pressure mercury lamps can produce ultraviolet radiation output that is greater than lethal mercury low pressure lamps for protozoan microorganisms, viruses and algae [3].

The goal of this study is to develop a optimize the freshwater productivity by distillation method using vacuum pumps and ultraviolet lights as accelerator and degradation with good visible-light driven activity and effectively solve desalination by distillation method problem [4]. By using vacuum pumps and ultraviolet lights, the seawater can evaporate faster and improve freshwater productivity and quality.

2. Methodology

This technology is made from stainless steel and aluminium, the use of aluminium aims to hold the heat received so that the system heat faster and as a good conductor to receive heat from an electric stove. With these materials, the heat can be channelled well to seawater so that the desalination process can take place faster.

Vacuum pumps are used for accelerators and accelerate the process of evaporation of seawater. The vacuum pump used is a pump type vacuum pump rocker 300 oil. This vacuum pump is placed in the upper corner of the beam-shaped evaporated water container. The vacuum pump can suck and drain the water vapour received through the water in the evaporation process of seawater so that waters from the desalination process is received more and the desalination process can take place faster [5].

Ultraviolet light is used to produce ultraviolet light. Ultraviolet lights used are ultraviolet lamps with a sterilization rate of 180 gpm (medium pressure). This ultraviolet lamp is placed in the evaporation water container. Ultraviolet light can sterilize the results of seawater evaporation of bacteria and protozoa microorganisms, viruses and algae that remain after the evaporation process [6]. So that we get the results of clean water that comes out through the tap to the reservoir of clean water.

The distillation process is carried out on a laboratory scale. 5 L of seawater is put into a beaker and then covered on top with a box that is given ultraviolet light and combined with a vacuum pump. The technology tests carried out include the speed test of the desalination process and the water quality test of desalination results including heavy metal Atomic Absorption Spectrophotometry (AAS) test, electrical conductivity adapter and multimeter test, concentration pH meter test, Salinity Meter salinity test, TDS tester turbidity and test MPN (Most Probably Number) of total coliform and E-coli bacteria.

3. Results and discussion

In this section, we discuss how to format the title, authors and affiliations. Please follow these instructions as carefully as possible so all articles within a conference have the same style to the title page. This paragraph follows a section title so it should not be indented.

3.1. Rate evaporation testing of water production speed

To determine the speed of production of water, in this technology testing is carried out on the effect of using a vacuum pump as a fixed parameter that continues to run during the desalination process. Water productivity testing is carried out by using the initial capacity of 5 litters of seawater.

Table 1 show effects of using vacuum pumps on water capacity. It can be seen that in the first 30 minutes after distillation heating without using a vacuum pump, the water produced was 200 mL, compared to distillation using a vacuum pump, the water produced was 400 mL, the use of this vacuum pump could increase the productivity of the resulting water twice as much distillation. This happens because the use of a vacuum pump can focus the direction of the water vapour produced [5]. It is different from without the use of a water pump where the direction of water vapour spreads to several parts such as the walls of the evaporation container so that it slows down the water production process.

Table 1. Effects of using vacuum pumps on water capacity.

Desalination time	Volume without pump	Volume with pump
30 minute	200 mL	400 mL
60 minute	450 mL	950 mL
90 minute	650 mL	1400 mL
120 minute	900 Ml	1950 mL

3.2. Atomic Absorption Spectrophotometry (AAS) analysis

Atomic Absorption Spectrophotometry (AAS) (Figure 1) is one type of spectropometric analysis where the basis of its measurement is the measurement of absorption of a ray by an atom, a light that is not absorbed, is transmitted and converted into a measured electrical signal. Welsh (Australia) first introduced AAS in 1955. AAS is a popular method for metal analysis, because besides being simple, it is also sensitive and selective [7].

Figure 1. Atomic Absorption Spectrophotometry (AAS).

Table 2 can be explained that the test of this chemical parameter includes testing of heavy metals Pb, Cd, Cu, and Fe which were tested using the AAS (Atomic Absorption Spectrophotometry) testing method. From the results of testing the AAS heavy metals obtained results include, Fe decreased by 96.1%, Cr by 90.33%, and Zn by 98,9%.

Table 2. Heavy metal Atomic Absorption Spectrophotometry (AAS) test.

Characteristic test	Unit	Quality standard	Sea water	Distillation	Ultraviolet and distillation
Fe	ppm		7,4285	1,4491	0,2896
Cr	Ppm	0,3	0,208	0,207	0,293
Zn	Ppm	0	< 0,0001	< 0,0001	< 0,0001

3.3. Salinity meter analysis

Figure 2. Salinity meter analysis.

Table 3. Data result of salinity testing.

Sample	Salinity	Unit	Method
Sea Water	29.460	ppm	Salinity meter instrument
Distillation Water	165	ppm	
Distillation and UV Water	20,7	ppm	

The more ions in water the greater the current value. Conversely, the fewer ions in the water, the smaller the current value. By decreasing the value of electrical conductivity, it also indicates that the water becomes tasteless because the ionic salt content has been reduced to solids so that it cannot deliver electricity anymore [8]. This is also proven by the reduced salinity of seawater [9]. It can be seen that the treatment using ultraviolet lamps can reduce salinity with an efficiency of 99.93%.

3.4. pH meter analysis

Figure 3. pH meter analysis from seawater.

Figure 4. pH meter analysis from distillation water.

Figure 5. pH meter analysis from distillation and UV water.

The test results of the water pH before processing were 7.5 while the water after treatment was 6.9. A decrease in pH value to pH 7 (neutral). The deposition of impurities that make water acidic results in an increase in pH [10]. Moreover, in the presence of ultraviolet light it triggers a decrease in pH near pH 7 (neutral).

Table 4. Data result of pH concentrate testing.

Sample	Value pH	Method
Sea Water	7,5	pH meter digital instrument
Distillation Water	6,5	
Distillation and UV Water	6,9	

3.5. Total Dissolved Solid (TDS) tester analysis

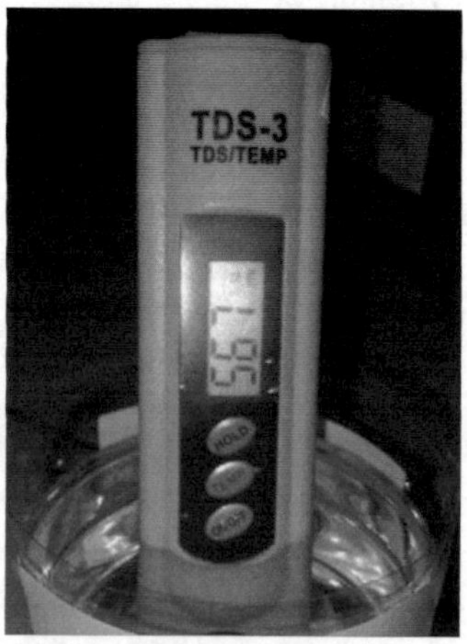

Figure 6. Total dissolved solid tester analysis from seawater.

Figure 7. Total dissolved solid tester analysis from distillation.

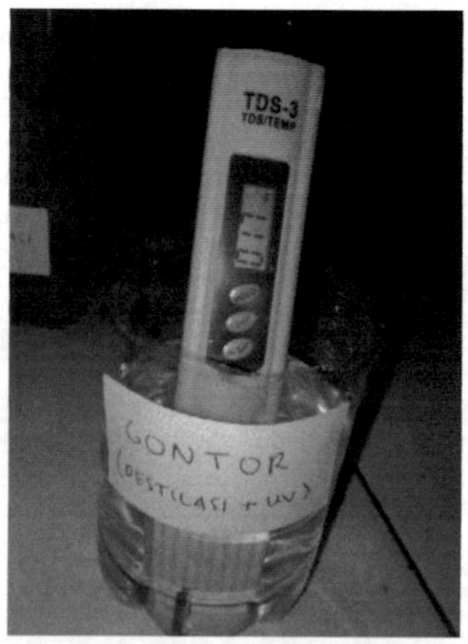

Figure 8. Total dissolved solid tester analysis from distillation and UV.

The total test results for dissolved solids before processing were 5670 ppm while the water after treatment was 17 mg / L. There was a decrease in the total value of dissolved solids 99.7%. The distillation function can reduce the value of dissolved solids. Dissolved solids are retained with mineral salts during the evaporation process. Impurity mass greater than the mass of evaporating water particles causes impurity particles to be lifted along with water vapour [11].

Table 5. Data result of dissolved solid testing.

Sample	Value TDS	Unit	Method
Sea Water	5670	mg/L	TDS meter digital instrument
Distillation Water	22	mg/L	
Distillation and UV Water	17	mg/L	

3.6. Most Probably Number (MPN) analysis

Figure 9. Total coliform bacteria test from seawater.

Figure 10. Total coliform bacteria test from distillation.

Figure 11. Total coliform bacteria test from distillation and UV.

The results of the bacterial test for total coliform of water before processing were 8400 MPN / 100 mL while the water after processing 0 MPN / 100 mL. There was a decrease in the total coliform bacterial value of 100%. During the desalination process, seawater is heated to cause some bacteria found in seawater to die due to the influence of high temperatures. In addition, the presence of ultraviolet lights can also reduce the amount of bacteria present in water. Ultraviolet light has a photon energy of 3.2 eV which can kill bacteria in water [12].

Table 6. Data result of most probably number total coliform bacterial testing.

Sample	Quantity	Unit	Method
Sea Water	8400	MPN/100 mL	0,399
Distillation Water	3000	MPN/100 mL	0,121
Distillation and UV Water	0	MPN/100 mL	7,4285

4. Conclusion

Seawater distillation using a vacuum pump and ultraviolet light has been successfully carried out. A systematic study of the effect of vacuum pumps and Ultraviolet lamp content on changes in seawater temperature and water productivity during distillation and ultraviolet activity in degrading harmful pollutant pollutants has been carried out. From testing the vacuum pump, the changes in water productivity increase, it can be concluded that the use of vacuum pumps affects water in increasing water productivity. From the AAS data, it can be seen that ultraviolet lights can reduce heavy metal content. From the results of bacterial tests, it can be seen that the bacteria contained decreased and died. Quality analysis shows that ultraviolet lamps are placed on the roof surface, degrades bacterial bacteria and harmful pollutant pollutants through the process of ultraviolet radiation. Test result of water quality show that seawater distillation with ultraviolet activity is better than conventional seawater distillation method.

Acknowledgments
The authors would like to thank The Ministry of Research, Technology and Higher Education, the Republic of Indonesia for funding this work through 2018's Student Creativity Program.

References
[1] Hidayat R R 2011 *Design of Salt and Freshwater Separation Equipment Using Solar Energy in the Department of Marine Science and Technology* (Bogor: Bogor Agricultural Institute) pp 166-168
[2] Cammack R 2006 *Oxford Dictionary of Biochemistry and Molecular Biology* (New York: Oxford University Press)
[3] Tose H Q, Ma S Y, Khranovskyy Z M, Gao J M and Zhu H B 2014 Microstructure and Raman scattering of ultraviolet lights deposited on buff er layers *J. Cryst. Growth* **394** 132–136
[4] Vignesh K, Suganthi A, Rajarajan M and Sara S A 2012 Ultraviolet activity under visible light irradiation *Powder technology* **224** 331-337
[5] Zhang V, Ekblad T, Yakimova R and Hultman L 2012 Surface morphology effects on the light-controlled wettability of ultraviolet nanostructures *Applied Surface Science* **258** 8146-8152
[6] Wang Y, Huang F, Pan D, Li B, Chen D, Lin W ... and Lin Z 2009 Ultraviolet-light-induced bactericidal mechanism on single crystals *Chem. Commun* **44** 6783–6785
[7] Liu K, Yang B F, Yan H, Fu Z, Wen M, Chen Y and Zuo J 2008 Strong room-temperature ultraviolet emission from nanocrystalline atomic absorption spectrophotometry ultrasonic spray pyrolysis *Applied Surface Science* **255** 2052-2056
[8] Rif'an M, Pramono S H, Shidiq M, Yuwono R, Suyono H and Suhartati F 2012 *Optimization of Utilization of Solar Energy Electricity in the Department of Electrical Engineering* (Malang: Brawijaya University) pp 44-48
[9] Banedic S A 2016 Enhanced solar still performance using a vacuum pumps on the distillation *Nineteenth International Water Technology Conference* **32** 124-133
[10] Haros M H and Sheibani R 2013 Nanostructure ultraviolet lights coating on box prepared by lamp lights deposition process: Application for removal of mercaptan *J. Industrial Chem.* **19** 161-165
[11] Sin J C, Abdullah A Z, Liu K and Mohamed A R 2012 Degradation of wastewaters containing organic dyes Ultraviolet Lights oxide: a review *Desalination and Water Treatment* **41** 131-169
[12] Watanabe, Sun R D, Lam S M, Fujishima A T and Hashimoto K 2011 Photoinduced Surface Wettability Conversion of Ultraviolet Lights *J. Phys. Chem. B* **105** 1984-1990

IOP Conf. Series: Materials Science and Engineering **850** (2020) 012009 doi:10.1088/1757-899X/850/1/012009

Reactive power injection at the North Makassar distribution system with D-STATCOM

Z Muslimin*, A Suyuti, E Palantei, I Indrabayu and I C Gunadin

Department of Electrical Engineering, Hasanuddin University, Indonesia

*zaenabandy@gmail.com

Abstract. The need for electrical energy is increasing in line with the growth of population. As consequences of high demand, electrical power network is also extended massively. A wider distance between the power generator to far load centers also consider as drawback in performances due to power loss along the transmission. The problem of power loss and voltage drop in the distribution system can be solved by installing DSTATCOM. This paper aims at finding the effect of DSTATCOM installation on real power losses and reactive power, as well as its effect on voltage profiles on the North Makassar Distribution System. The simulation results show that by installing DSTATCOM on 92 Bus North Makassar Distribution System on several buses, increasing of performances can be achieved. The average active power loss decreased by 2.5%, declining of the reactive power loss averaged 26.7% and the average voltage profile improved by 0.12% (2.5 kV).

1. Introduction

Electrical energy is one of the needs that cannot be separated from people's lives in the era of globalization. The dependence of electrical energy is very large in various activities of modern human life can be found in many sectors such as in the world of transportation (two-wheeled and four-wheeled vehicles such as motorcycles and cars) in the service sector and industry, government sector, and housing sector. The use of electrical energy increases as the demand for new pairs increases. The electrical energy produced by power plant centres is generally located far from load centres. The generated electricity will be channelled to load centres through transmission and distribution networks. A considerable distance in the delivery of power from the plant to the consumer will cause a decrease in voltage which results in a power loss. Voltage drop is one of the problems of electric power quality. There are many ways to reduce voltage drops in transmission and distribution systems. At present, various controllers are very flexible, utilizing power electronic components, providing fast and reliable control of transmission parameters such as voltage, channel impedance and phase angle between the sender and receiver ends known as FACTS.

FACTS is an electronic-based equipment that maintains the quality of electric power by maintaining better power flow by changing the parameters of voltage, impedance, and phase angle. On the other hand, special power equipment is used to improve power quality in the distribution system. This special power equipment functions are similar to FACTS, which includes special power equipment is D-SATCOM, UPQC, DVR including D-STATCOM which is most often used because it provides a low cost in compensating for reactive power [1]. D-STATCOM can handle a variety of functions, namely minimizing active and reactive power losses, harmonics, and their size is relatively small and inexpensive [2,3].

The passive control of the STATCOM-level converter integrated with the distribution transformer has been carried out resulting in good performance integrated with D-STATCOM with reactive power compensation and financing functions [4,5]. D-STATCOM analysis and control under unbalanced voltage conditions, contribute when the voltage is not balanced between phases [6,7]. Then the D-STATCOM design uses high power based on the dual converter topology in an isolated system that has been carried out by Sreenivasarao et al. which aims to optimize the voltage using a converter. Gupta and friends have developed a variety of load models to determine the D-STATCOM allocation on the distribution network [8]. The optimal placement of D-STATCOM using sensitivity in the distribution system that is affected by varying load growth [9-10]. The reactive power compensation using DSTATCOM, giving results that conceptually are similar to STATCOM on transmission [11]. The control scheme must be such that it provides better compensation for reactive power and profile stress so as to improve power quality at the end of the distribution [12].

Much research has been done before to reduce power losses and increase voltage profiles in distribution systems, besides using D-STATCOM also uses shunt capacitors, synchronous condensers. However, the results of the study show that shunt capacitors and synchronous condensers have disadvantages, that is, for shunt capacitors cannot supply variable reactive power while synchronous condensers are expensive and require routine maintenance [13]. For the northern Makassar distribution system, for now, it still uses shunt capacitors and has a network topology that is long enough to allow for considerable voltage drop and power losses when there is a dynamic load. Therefore, this is very interesting to study which aims to improve the efficiency and quality of utilities on the distribution network.

In this paper, D-STATCOM is used to increase the voltage profile and reduce power losses, with the following arrangement: in the first part the background is explained, the second part is the distribution system, the third part is the explanation of D-STATCOM, 4 simulation results and fifth conclusion.

2. Distribution systems

Distribution System of North Makassar (DSNM) is shown in Figure 1 which is used to analyse the system load especially the Mandai substation, having an average voltage value of 0.964 pu (graph of voltage profile shown in Figure 2.), total power losses of 195.94 kW and 68.36 kVar. This amount is before installation of DSTATCOM, by installing DSTATCOM, it will increase the voltage value and reduce the number of power losses.

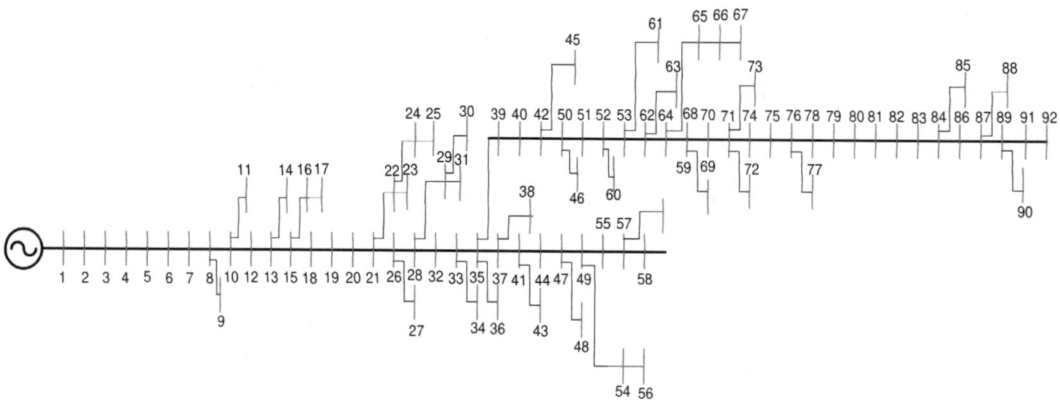

Figure 1. Distribution system of North Makassar.

To install DSTATCOM on the Distribution system of North Makassar in improving the voltage profile and reducing losses based on the load flow system using steps as shown in Figure 3.

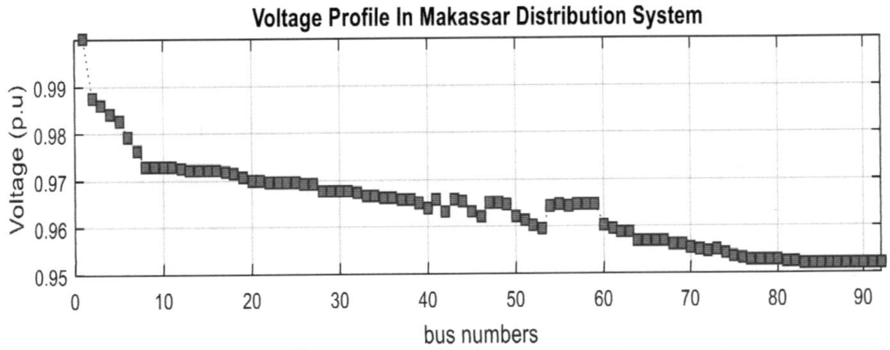

Figure 2. Voltage profile before D-STATCOM installation.

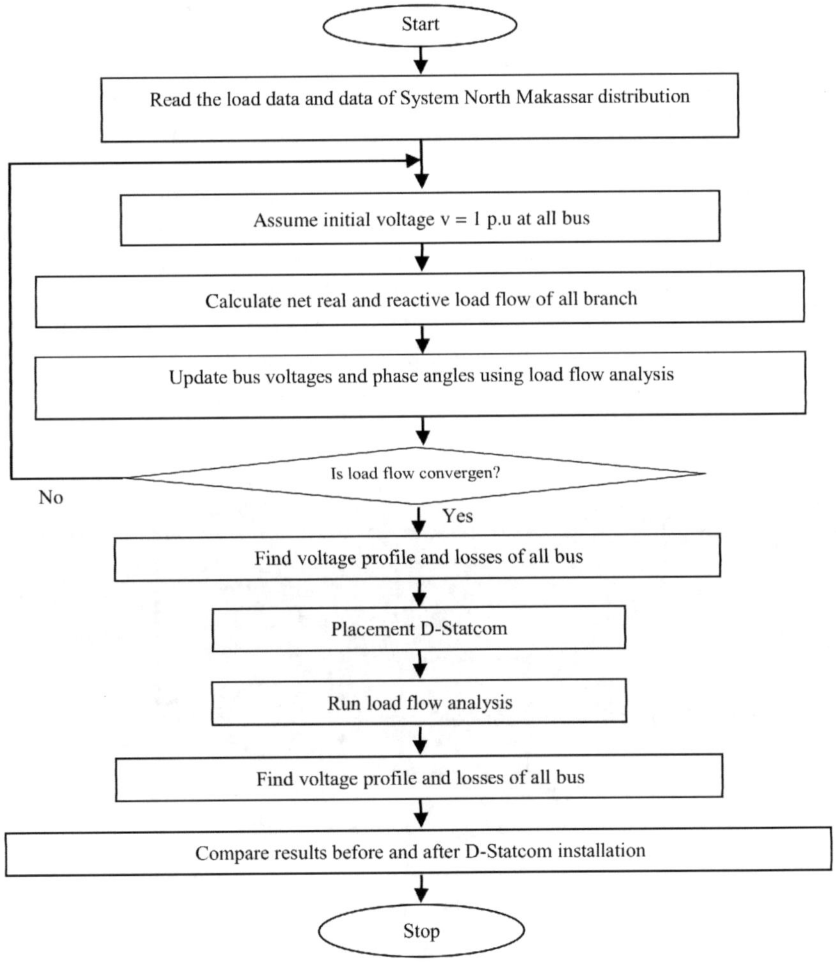

Figure 3. Flow chart to determine the voltage and amount of losses.

3. Distribution static compensator (D-STATCOM)

To analyze the Distribution System of North Makassar, the equivalent circuit of the distribution system is modeled in Figure 4, which is a simplification of Figure 1.

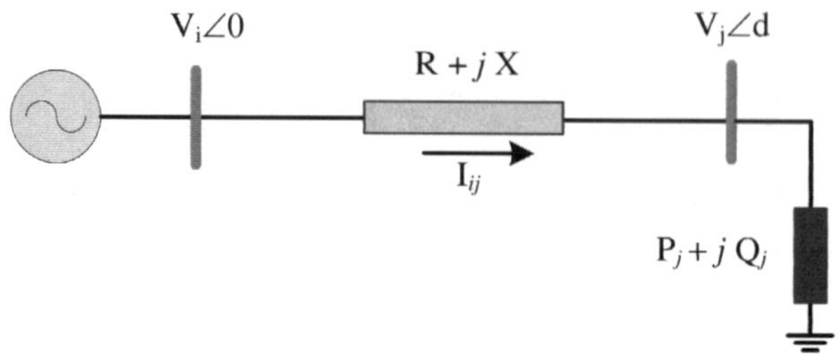

Figure 4. Equivalent circuit model of Distribution System of North Makassar.

D-STATCOM is an inverter-based voltage source which is a special power device, connected in parallel to the power system and can supply reactive power to increase the voltage profile and reduce power losses on the channel [14]. The main components of D-STATCOM include VSC (Voltage Source Converter), energy storage equipment, transformers, and a controller. For more details, the basic structure of D-STATCOM can be seen in Figure 5. The basic principle of D-STATCOM installed in a distribution system will produce an ac voltage source that can be controlled by a voltage source converter (VSC). The output voltage of D-STATCOM can control the exchange of active and reactive power between D-STATCOM and the distribution network system. Very flexible output voltage control from VSC is utilizing high frequency switching with pulse width modulation (PWM) technique at constant dc voltage source, then the average of the output voltage waveform is obtained to get the fundamental voltage component that can be adjusted by its magnet The PWM technique provides advantages in which low-order harmonics are reduced so that it will reduce the number of harmonics and harmonic filters. The higher the frequency ratio of transmission to the fundamental frequency, the lower the low-order harmonics that appear, but this can cause additional loss.

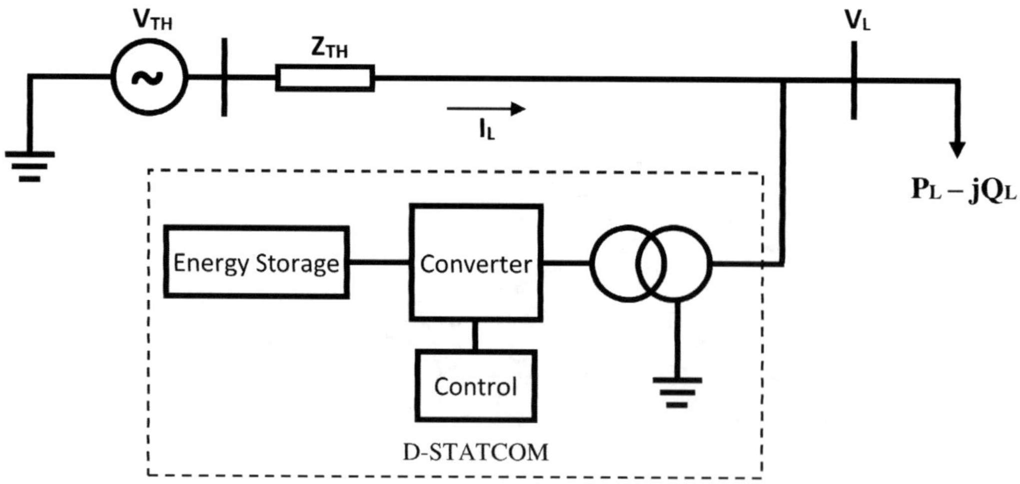

Figure 5. D-STATCOM Basic structure in distribution system.

4. Simulation results

The distribution system of north Makassar consisting of 92 buses has been modeled using the following scenario: case 1 is the condition before the installation of D-STATCOM, cases 2, 3,4 and 5 are to add D-STATCOM to each bus on bus 20, bus 35, bus 51 and bus 81.

The simulation results from the North Makassar Distribution System with several scenarios result in a significant increase in voltage and a reduction in losses shown in Table 1 and Table 2. Table 1 is the voltage on the bus and Table 2 is the total active and reactive power losses of the simulation results by installing D- STATCOM is 250 KV.

Table 1. Voltage on the bus.

Scenario	Voltage (pu)			
	Bus 20	Bus 35	Bus 51	Bus 81
Case 1	0.97000	0.96616	0.96132	0.95249
Case 2	0.97101	0.96718	0.96233	0.95351
Case 3	0.97101	0.96742	0.96258	0.95375
Case 4	0.97101	0.96742	0.96301	0.95418
Case 5	0.97101	0.96742	0.96301	0.95548

Table 2. Total active and reactive power losses.

Total Losses	Case 1	Case 2	Case 3	Case 4	Case 5
Aktif Power (kW)	195.94	192.49	191.44	190.79	189.60
Reaktif Power (kVar)	-68.36	-53.68	-51.82	-49.50	-45.38

Tables 1 and 2 show that the installation of D-STATCOM on a bus will increase the voltage profile and reduce active power losses and reactive power losses. Of the five scenarios, the most effective D-STATCOM installation is in case 5, namely bus 81, resulting in an increase in voltage profile of 3.24% and reduction in a reactive power loss of 33.6% and active power loss of 3.2%.

The graph of the comparison of active power losses and reactive power losses for various scenarios on the 92 buses system is shown in Figure 6., Figure 7., and Figure 8.

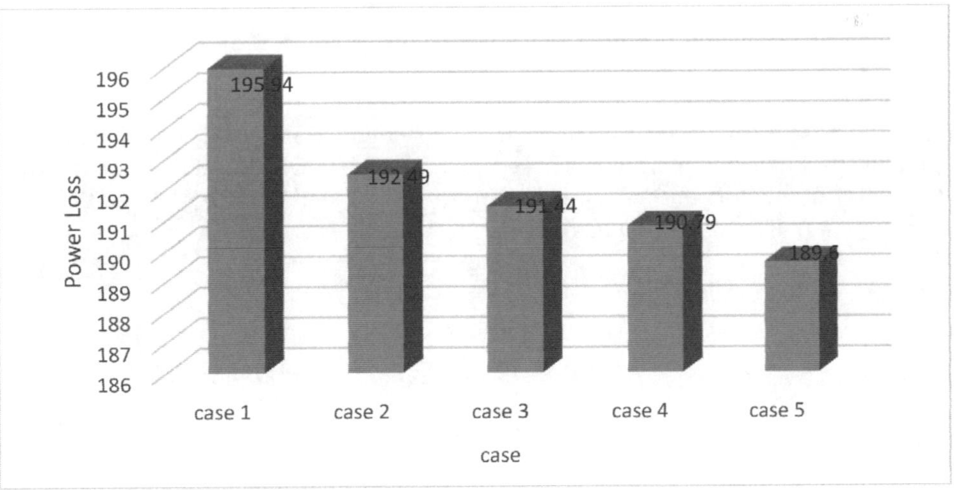

Figure 6. Comparison of Active power loss for different cases DSNM in 92 bus.

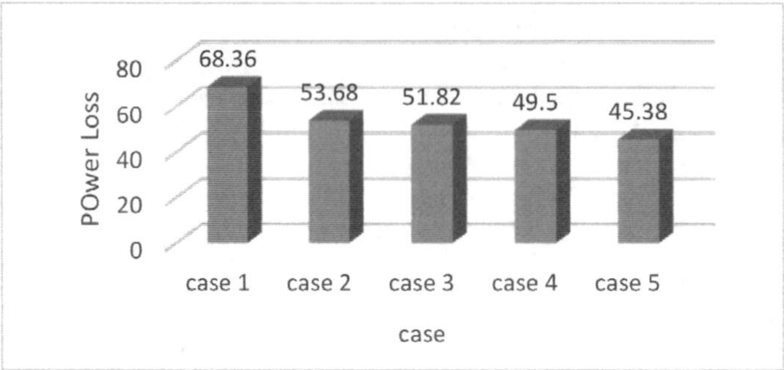

Figure 7. Comparison of Reactive power loss for different cases DSNM in 92 bus.

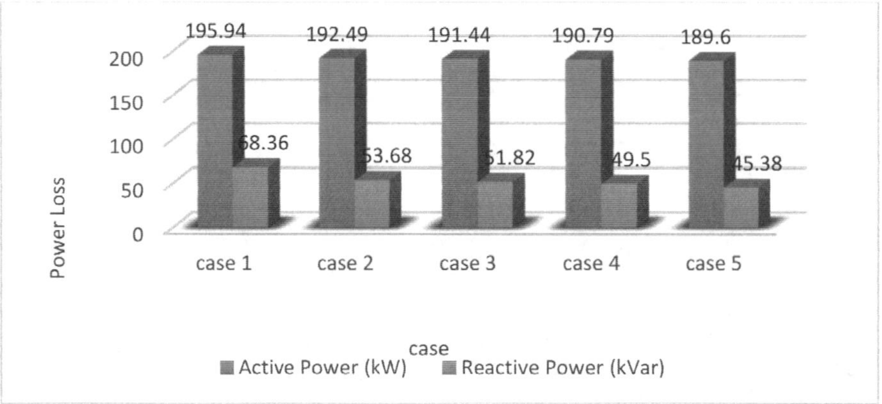

Figure 8. Comparison of Active power and Reactive power loss for different cases DSNM in 92 bus.

For voltage profiles on each bus on the Distribution System of North Makassar 92 buses are shown in Figure 9, Figure 10, Figure11, and Figure12. Figure 9, D-STATCOM installation on bus 20, in Figure 10, D-STATCOM installation on bus 35, in Figure 11 is an installation D-STATCOM on bus 51 and Figure 12 mounting D-STATCOM on bus 81.

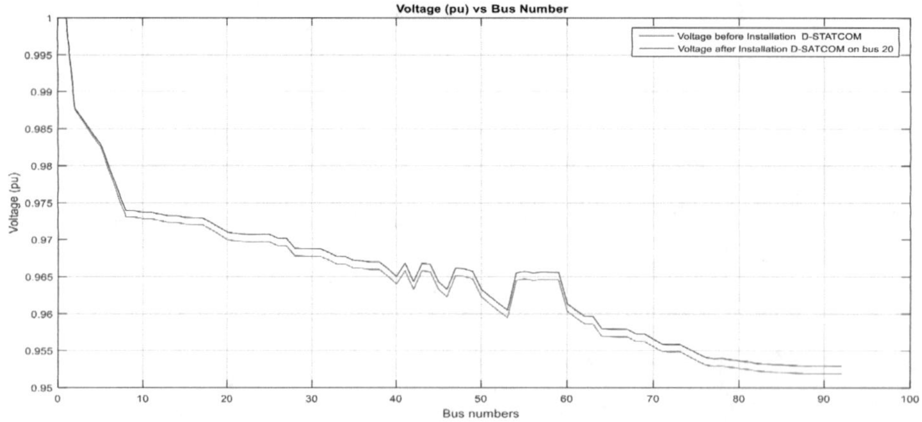

Figure 9. Installation of D-STATCOM in bus 20.

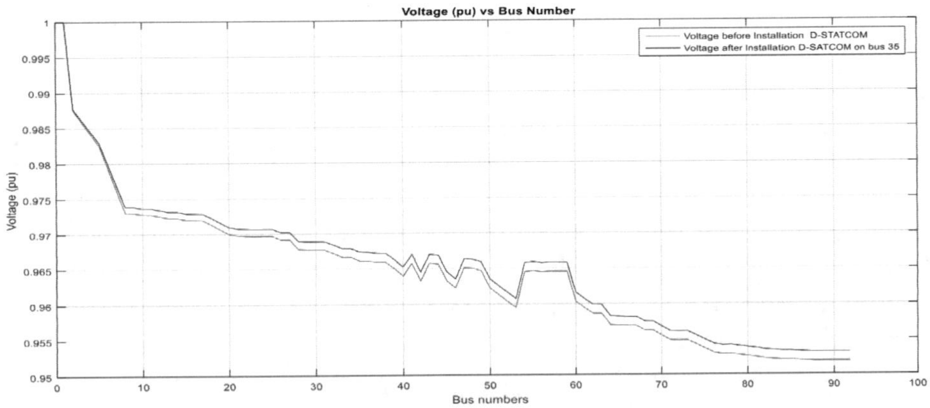

Figure 10. Installation of D-STATCOM in bus 35.

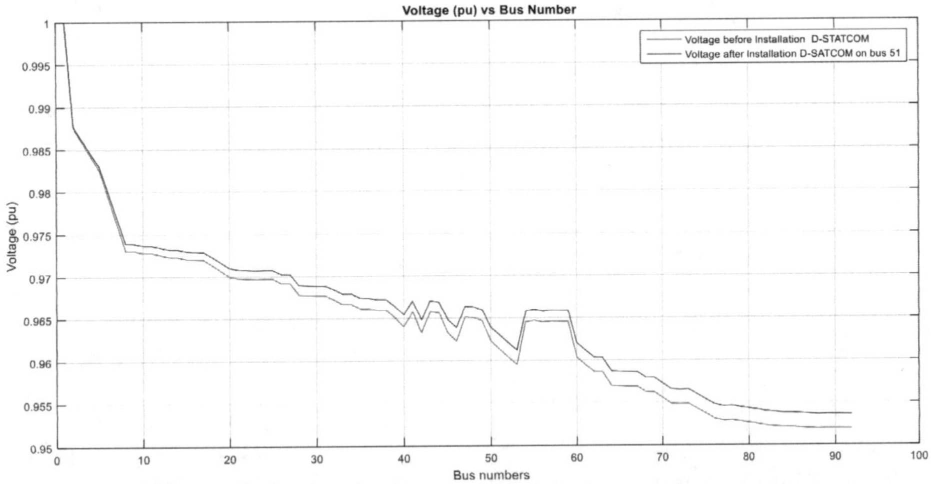

Figure 11. Installation of D-STATCOM on bus 51.

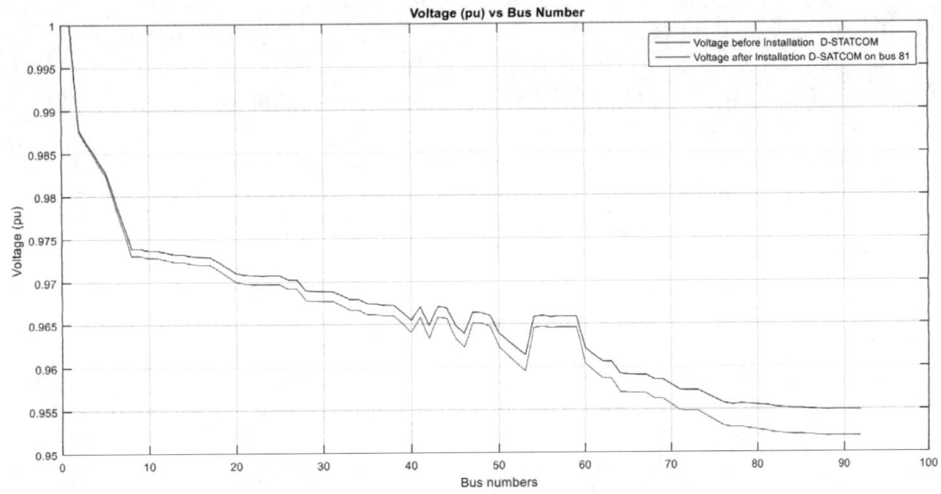

Figure 12. Installation of D-STATCOM on bus 81

The graph in Figure 13 illustrates the comparison of the stress profile of each bus in five scenarios.

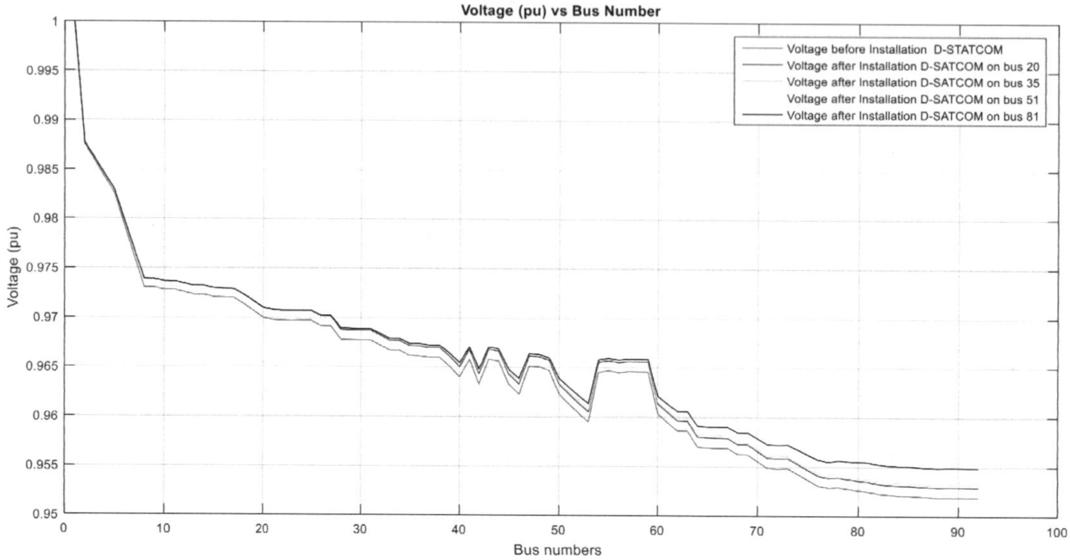

Figure 13. Comparison of voltage profiles in five scenarios on DSNM 92 bus.

5. Conclusion

Simulations are carried out on Distribution System of North Makassar 92 buses, especially Mandai Substation with several scenarios. With the installation of D-STATCOM in several scenarios, the simulation results show an increase in voltage profile and active and reactive power losses. Case 1 (without D-STATCOM) shows a voltage value of 0.970pu on bus 20, bus 35: 0.966pu, bus 51: 0.961pu and bus 81: 0.952pu while the total active power loss is 195.94kW and total reactive power loss is 68.36 kVar. Case 2 (installation of D-STATCOM on bus 20) shows a voltage value of 0.971pu on bus 20, bus 35: 0.967pu, bus 51: 0.962pu and bus 81: 0.953pu while total active power loss is 192.49kW and total reactive power loss is 53.68 kVar. Case 3 (installation of D-STATCOM on bus 35) shows a voltage value of 0.971pu on bus 20, bus 35: 0.967pu, bus 51: 0.963pu and bus 81: 0.953pu while total active power loss is 191.44kW and total reactive power loss is 51.82 kVar. Case 4 (installation of D-STATCOM on bus 51) shows a voltage value of 0.971pu on bus 20, bus 35: 0.967pu, bus 51: 0.963pu and bus 81: 0.954pu while total active power loss is 192.49kW and total reactive power loss is 53.68 kVar. Case 5 (installation of D-STATCOM on bus 81) shows a voltage value of 0.971pu on bus 20, bus 35: 0.967pu, bus 51: 0.963pu and bus 81: 0.955pu while total active power loss is 189.60kW and total reactive power loss is 45.38 kVar.

The simulation results show that the conclusion of this paper with the installation of D-STATCOM on the Makassar North 92 Distribution System can increase the average profile of 0.12% (2.5 kV) and the reduction in active power losses by an average of 2.5% and reactive power losses 26.7 %. The most effective installation of the five scenarios is case 5, namely on bus 81 with an increase in voltage profile of 0.16% (3.2kV) and a reduction in an active power loss of 3.24% and reactive power loss of 33.61%.

References

[1] Pinapatruni K K and Krishna Mohan L 2012 DQ based Control of DSTATCOM for Power Quality Improvemen *VSRD International Journal of Electrical, Electronics & Communication Engineering* **2**(5) 207-227

[2] Awasth V M and Huchche V A 2012 Reactive Power Compensation using D-STATCOM *IEEE* **8** 583-585

[3] Das P 2015 Optimal Sizing And Placement of Distributed Generation in A Radial Distribution System Using Loss Sensitivity Factor And Firefly Algorithm *Int. J. Sci. Res. Manage (IJSRM)* **3**(4) 2611–2618
[4] Iqbal F, Khan M T and Siddiqui A S 2017 Optimal Placement of DG and D-Statcom for Loss Reduction and Voltage Profile Improvement *Alexandria Engineering Journal* **57**(2) 755-765
[5] Chen Y, Wen M, Lei E, Yin X, Lai J and Wang Z 2018 Passivity-Based Control of Cascaded Multilevel Converter Based D-Statcom Integrated with Distribution Transformer *Electric Power Systems Research* **154** 1-12
[6] Choudhury S, Bhowmik P and Rout P K 2018 Economic load sharing in a D-STATCOM Integrated Islanded Microgrid based on Fuzzy Logic and Seeker Optimization Approach *Sustainable Cities and Society* **37** 57-69
[7] Wang Y, Tang J and Qiu X 2011 Analysis and Control of D-Statcom under Unbalanced Voltage Condition *International Conference on Mechatronic Science Electric Engineering and Computer (MEC)* **1** 1623-1625
[8] Sreenivasarao D, Agarwal P and Das B 2013 Performance Enhancement of A Reduced Rating Hybrid D-Statcom for Three-Phase, Four-Wire System *Electric Power Systems Research* **97** 158-171
[9] Mehouachi I, Abbes M and Chebbi S 2019 Design of A High Power D-Statcom Based on The Isolated Dual-Converter Topology *International Journal of Electrical Power & Energy Systems* **106** 401-410
[10] Gupta A R and Kumar A 2018 Optimal Placement of D-Statcom Using Sensitivity Approaches in Mesh Distribution System with Time Variant Load Models under Load Growtt *Ain Shams Engineering Journal* **9**(4) 783-799
[11] Reddy S S K, Babu A V, Suman S and Manasa K D 2011 Reactive Power Compensation Using D-Statcom *International Journal of Engineering Research and Applications (IJERA)* **2** 1-6
[12] Gupta A R and Kumar A 2016 Energy Saving Using D-STATCOM Placement in Radial Distribution System under Reconfigured Network *Energy Procedia* **90** 124-136
[13] Sreejith S, Bose U, Vachana K M D S and Jyothi V 2014 Application of D-STATCOM as Load Compensator for Power Factor Correction *International Conference on Control, Instrumentation, Communication and Computational Technologies (ICCICCT)* **2** 595-600
[14] Gunadin I C, Muslimin Z and Siswanto A 2017 Transient Stability Improvement Using Allocation Power Generation Methods Based on Moment Inertia *International Conference on Electrical Engineering and Informatics (ICELTICs)* **5** 147-152

IOP Conf. Series: Materials Science and Engineering **850** (2020) 012010 doi:10.1088/1757-899X/850/1/012010

Optimal distributed generation placement to reduce power loss using Particle Swarm Optimization method

M G Firmansyah*, **Y Mulyadi, H Hasbullah and A Saripudin**

Department of Eectrical Engineering Education, Universitas Pendidikan Indonesia, Jl. Dr. Setiabudi No. 229, Bandung 40154, Jawa Barat, Indonesia

*giri@student.upi.edu

Abstract. In electrical power systems, distribution networks have considerable power losses compared with the other parts of electrical power systems. This caused by the distance of power plant to the load center. The high value of power losses causing disadvantage for the electricity provider company. This study aims to reduce power losses in the MLBC feeder distribution network by erecting the distributed generation. The optimal location for erecting distributed generation sought by using the Particle Swarm Optimization (PSO) method. Before distributed generation erected in MLBC feeder distribution network, the power losses' value is 128.1 kW. Locating the most optimum location for erecting the distributed generation is using the particle swarm optimization method, bus 13 on the MLBC feeder distribution network was the optimum location for distributed generation erection. Hence, the loss value from load flow simulation is 67.5 kW or 60.6 kW for power loss reduction after distributed generation installed on bus 13.

1. Introduction

As technology advances, electrical energy became a primary need in people's lives. Electrical energy generated on a large scale, centralized and located far from the load in general. It requires a transmission and distribution network to distribute electricity. Transmission and distribution network have a long structure that has impact on power losses and voltage drops. The power losses cause the efficiency of a network reduced, while voltage drops cause damage to consumer's electrical equipment. Therefore, installing a Distributed Generation will anticipate the effect.

Distributed Generation is a small-scale generator located near to the load center. Installation of distributed generation has positive impact. Reliability of the network, voltage profile and power quality will increase, while power losses can be reduced by installing distributed generation. Erecting distributed generation to the worst voltage profile in a system can cause positive impact to the system [1].

In this research, Particle Swarm Optimization method is used to obtain the smallest power losses value. Particle Swarm Optimization is a population-based stochastic algorithm for optimization which is based on social-psychological principles [2]. The simulation is done using ETAP and MATLAB software, so the optimal location of distributed generation can be obtained and the power losses can be reduced.

2. Methods

In 1995, James Kennedy and Russell Eberhart put forward a new method for completing nonlinear functions. Particle Swarm Optimization is similar to a Genetic Algorithms, in that the system is initialized with a population of random solution [3]. The basic of this algorithm inspired by social animals. This social relationship is a natural process for communicating with each other in groups that seek food and migrate together to a group such as groups of birds, bees and so on. When one group member gets the desired path, the other members will follow quickly [4].

In multi-variable optimization such as particle swarm optimization, a herd is considered to have a certain size and a random position for each particle in a multidimensional space. Each particle in the particle swarm optimization is considered to have two characters, that is position and speed. Each of these particles moves in a certain multidimensional space and each particle will remember the best position that the previous particle has passed. Each particle contained in a multi-dimensional space of the particle swarm optimization will transmit its best position to the other particles so that each particle will adjust its position and speed according to the information received regarding that position. Particle swarm optimization is an algorithm for finding minimum or maximum function values based on new populations. Particle swarm optimization has advantages in finding complex non-linear optimization values. This algorithm is proven to be able to find the best solution for optimization problems.

In this particle swarm optimization algorithm, it will use a several numbers that will used as a candidate location for connecting a distributed generation and realized into a data. A general flow of data analysis carried out in this research can be seen in Figure 1. This research begins with conducting literature studies and field studies to obtain the knowledge base needed in research. After that, a problem was raised to become a topic that would become the object of research. The minimum loss value after the installation of distributed generation is the purpose of this study. Furthermore, observations and data collection were carried out to support the research. After the data is collected, the data is processed using ETAP 16.0 and MATLAB r2017a software to get the desired results.

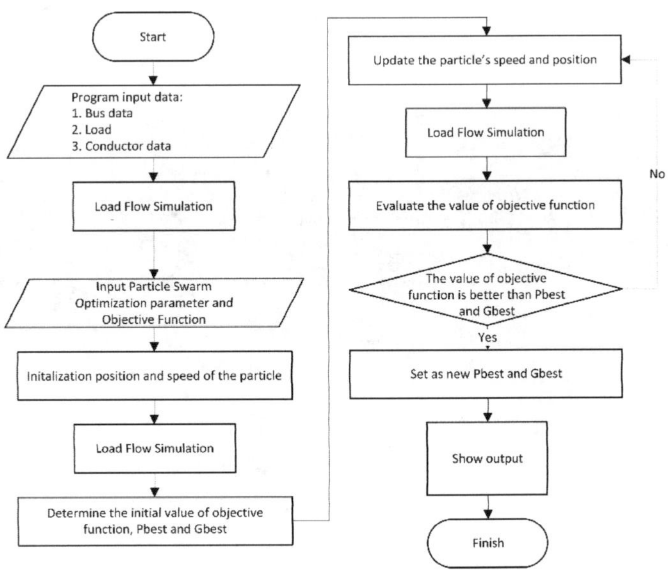

Figure 1. Research flowchart.

The stages of the Particle Swarm Optimization program are as follows.

- Determine the swarm size, value, and particle speed randomly.
- Evaluate objective function from each particle.
- Determine Pbest and Gbest value.

- Calculate the particle speed from previous iteration by using this equation.

$$V_j=\theta V_j(i\text{-}1)+c_1 r_1\left[P_{bestj}\text{-}X_j(i\text{-}1)\right]+c_2 r_2[G_{bestj}\text{-}X_j(i\text{-}1)] \quad (1)$$

where i is iteration, j is integer (1, 2, 3, 4, ..., N), and r_1 and r_2 are random values.
- Determine particle position and next iteration by using this equation.

$$X_j(i)=X_j(i\text{-}1)+V_j(i) \quad (2)$$

where X_j is next iteration particle's position and V_j is next iteration particle's speed.
- Evaluate again objective function from next iteration.
- Renew the value of Pbest and Gbest.
- Check whether the solution for optimization has been optimal or not. If it is optimal, the algorithm process will stop, but if not, the algorithm will return to step number 4.

3. Results

3.1. Load flow simulation before installation of distributed generation

Malangbong-Coklat feeder using a 20 kV distribution voltage that supplied from Malangbong substation. One-line diagram of Malangbong-Coklat feeder is shown in Figure 2. Afterwards, a power-flow simulation was performed using Newton-Raphson method using MATLAB r2017a software.

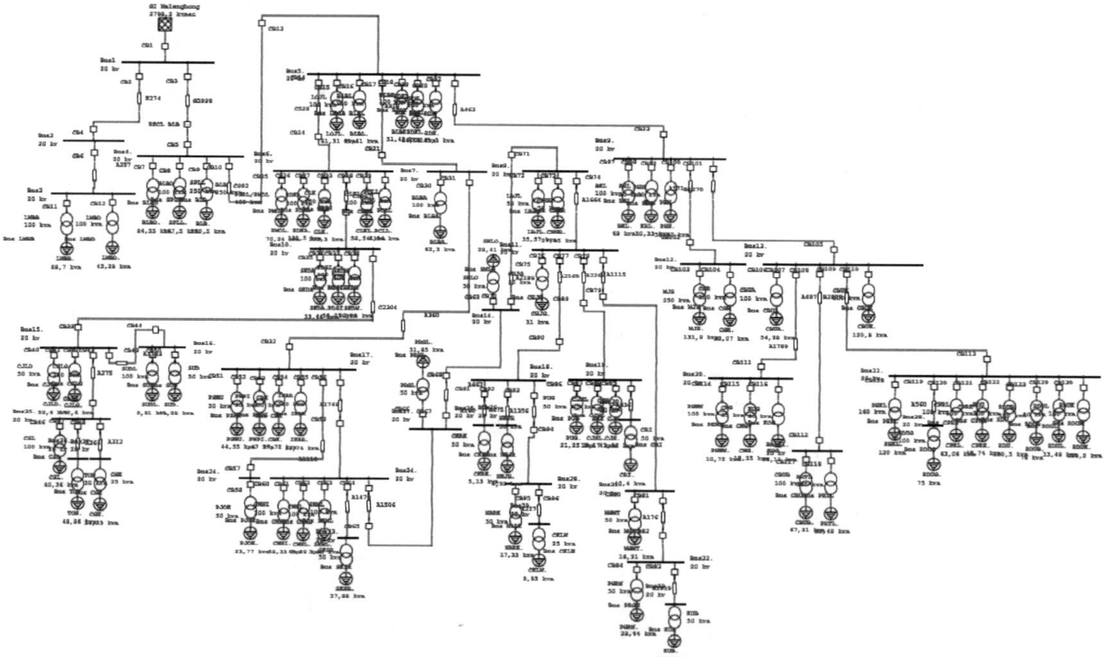

Figure 2. One-line diagram of Malangbong-Coklat feeder's.

Table 1. Load flow simulation results before installation of distributed generation.

No	Power	kW	Information
1	Total Power Injected	3243.5	Malangbong Substation
2	Power Loss	128.1	-

Malangbong-Coklat has 38 buses that connected by 37 lines, there also 71 distribution transformator. The result load flow simulation is shown at Table 1. The total power supplied by Malangbong Substation in Malangbong-Coklat feeder is worth 3,243.5 kW, and the power loss value of the Malangbong-Coklat feeder is worth 128.1 kW.

3.2. Capacity of distributed generation

The capacity of the distributed generation to be installed is obtained from the total load that's connected to the Malangbong-Coklat feeder. Based on Table 1, the total of load connected on Malangbong-Coklat feeder is 3243.5 kW.

The distributed generation that will be installed in this simulation has power capacity of 4.7 MVA with a power factor of 0.85. To find the active power to be supplied can be calculated by Equation (3).

$$P = S\cos\varphi = 4.7 \times 10^6 \times 0.85 = 4.0\ \mathrm{MW} \tag{3}$$

The real power that will be supplied to the distribution network by a distributed generation is equal to 4.0 MW. This value is applied so the capacity of the distributed generation is not fully used.

3.3. Locating the optimal placement of distributed generation

Distributed Generation will be installed on one of the 38 buses on Malangbong-Coklat feeders. The optimal location of distributed generation installation will be sought using the Particle Swarm Optimization (PSO) method.

Table 2 displays bus data on Malangbong-Coklat distribution network that will be processed into input data in the MATLAB application. Meanwhile, Table 3 shows the line data of Malangbong-Coklat distribution network. It consists of from-to bus, line name, length, and the line impedance.

Table 2. Bus data on Malangbong-Coklat distribution network.

Bus Number	Transformator	kW	kVAR	Total kW	Total kVAR
Bus 2	-	-	-	-	-
Bus 3	LMBD	38.4	23.9	113.8	70.7
	LMBB	75.4	46.8		
Bus 4	BLBD	71.9	44.6	418,6	259.5
	SPLL	159.3	98.8		
	BLB	187.4	116.1		
...	...	...	...	...	...
Bus 36	TON	41.6	25.8	41.6	25.8
Bus 37	CGE	29.2	18.1	29.2	18.1
Bus 38	RDUG	63.8	39.6	63.8	39.6
	Total			**3,441.8**	**2,133.8**

Table 3. Line data on Malangbong-Coklat distribution network.

From Bus	To Bus	Line	Length (km)	R	Xj
1	2	E274	0.274	0.07252	0.02611
1	4	G5998		1.58750	2.48317
2	3	A257	0.285	0.28587	0.30620
...	...	...	...	...	...
35	36	A264		0.29366	0.03146
35	37	A312		0.34705	0.03718

After the bus and line data are processed, simulation of locating the optimal location of distributed generation can be done, the graph of power losses value after installation distributed generation, the location and capacity of installation of distributed generation that installed.

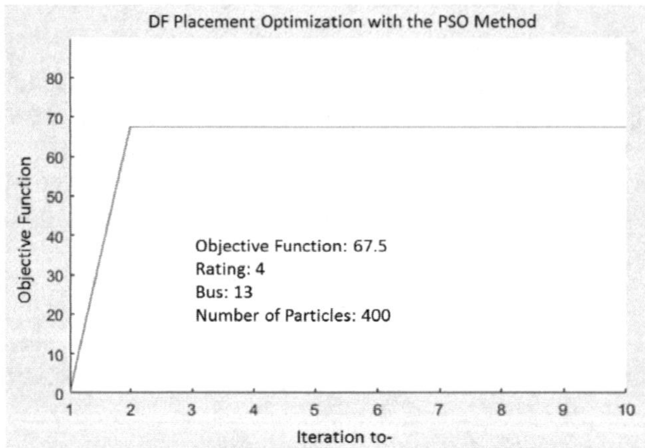

Figure 3. Objective function graph to iteration in locating the optimal location of distributed generation simulation.

After 10 times of iterations, it was found that the optimal location for installing distributed generation is on bus 13. The objective function is the value of power loss after the installation of a distributed generation on bus 13 that is equal to 67.5 kW. Rating is the capacity of a distributed generation installed on bus 13 which is equal to 4 MW. While the number of particles is the number of particles used in the particle swarm optimization algorithm.

3.4. Load flow simulation after installation of distributed generation

Load flow simulation after installation of a distributed generation on bus 13 still using the Newton-Raphson method. After simulating the installation of distributed generation, the load flow simulation results obtained as shown in Table 4.

Table 4. Load flow simulation after installation of distributed generation.

No	Power	kW	Information
1	Total Power Injected	1881	Malangbong Substation
		1450,746	*Distributed Generation*
2	Power Loss	67,5	-

4. Conclusions

The value of power losses in Malangbong-Coklat distribution network is equal to 128.1 kW. This power loss can be reduced by the installation of a distributed generation on the Malangbong-Coklat distribution network. After the load flow simulation and locating the optimum location for the installation of distributed generation using particle swarm optimization method, it was found that the most optimum location for distributed generation installation was bus 13 located in the Cibiuk area. Installation of distributed generation on bus 13 results in reduced power loss from before the installation of distributed generation. The power loss after installation of distributed generation is 67.5 kW, reduced by 60.6 kW or 47.3% from before the installation of distributed generation.

IOP Conf. Series: Materials Science and Engineering **850** (2020) 012010 doi:10.1088/1757-899X/850/1/012010

References

[1] Jaganathan R K and Saha T K 2004 Voltage stability analysis of grid connected embedded generators *Proceedings of the Australasian Universities Power Engineering Conference–AUPEC* 1-6

[2] Kennedy J and Eberhart R 1995 Particle swarm optimization *Proceedings of ICNN'95-International Conference on Neural Networks* **4** 1942-1948

[3] Satria D A 2015 *Penempatan DG Menggunakan Metode PSO pada Jaringan Distribusi Mikrogrid untuk Meminimalisasi Rugi Daya* (Doctoral dissertation, Surabaya: Institut Technology Sepuluh Nopember)

[4] Abugri J B and Karam M 2015 Particle Swarm Optimization for the Minimization of Power Losses in Distribution Networks *2015 12th International Conference on Information Technology-New Generations* 73-78

IOP Conf. Series: Materials Science and Engineering **850** (2020) 012011 doi:10.1088/1757-899X/850/1/012011

Optimization of capacitor capacity to reduce electric power losses in distribution networks in Jampang Kulon Sukabumi

J Nurrizkiani*, H Hasbullah and A Saripudin

Department of Electrical Engineering Education, Universitas Pendidikan Indonesia, Jl. Dr. Setiabudhi 207, Bandung 40154, Indonesia

*julianurrizkiani0407@gmail.com

Abstract. The electric power system consists of three main parts, one of which is the distribution system. The distribution system has the role of channeling electrical energy from the substation to the consumer. In the distribution of electric power, high reliability is needed so that the quality of electric power is maintained well. There are many factors that can reduce power quality, including power losses. One effort to maintain the quality of electrical power by reducing electrical power losses is to install capacitors on distribution channels. The capacitor is installed as a reactive power compensator so that the power factor approaches 1. In the Jampang Kulon feeder, there are considerable power losses, namely on bus 85-16-1 at 61.6 kW and bus 86-14-1 at 16.5 kW. By determining the capacitor capacity using ETAP, a 3 x 900 kVAR capacitor is mounted on bus 85-16-1 and a 3 x 300 kVAR capacitor on bus 86-14-1. The highest reduction in power losses occurred on buses 85-16-1 and buses 82-17-1. On bus 85-16-1 the power losses have decreased from 61.6 kW to 50 kW (reduced by 11, 6 kW), while on bus 82-17-1, the power losses have decreased from 59.4 kW to 49.2 kW (reduced by 10.2 kW).

1. Introduction

There are three main parts to the electric power system, namely the generation system, transmission system, and distribution system. The distribution system is one part of the electric power system which is located closest to the consumer. The distribution system has a very important role because it is directly related to the distribution of electrical energy from the substation to the consumer [1].

To meet the electricity needs, the electric power system must have a high level of reliability so that electricity with good quality power is available. There are several important parameters that must be considered so that the quality of electric power is maintained, including harmonic problems, voltage fluctuations, frequencies, power factors, power losses and voltage drops. From these parameters, low power factor and power losses are common problems [1].

One way to improve the power factor is to reduce reactive power in the distribution channel. If the reactive current component can be reduced, the total reactive current will be reduced so that the power factor will be greater as a result of reduced reactive power. Installation of capacitors in the distribution channel will reduce reactive power because the capacitors supply reactive power to the load [2,3]. The value of capacitor capacity was found using ETAP [4,5].

2. Methods

This research is a quantitative research. Determination of the capacitor capacity on the distribution network was done by descriptive analysis methods, namely by collecting data, analysing data, and drawing conclusions. It was necessary to determine the capacitor capacity to be installed on the distribution network.

In Figure 1, a research flow diagram was presented. In this study, the data collected were power losses, power factors, and active power distributed in the Jampang Kulon feeder distribution channel. Therefore, calculate the value of the capacitor capacity to be used to correct power losses.

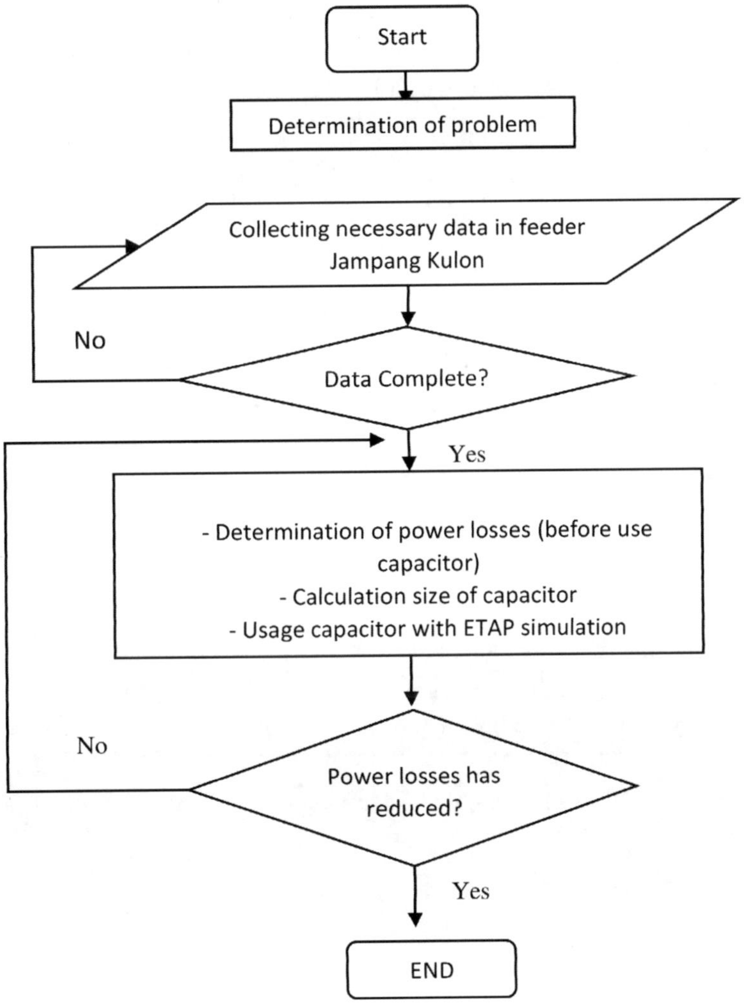

Figure 1. Research flow diagram.

The determination of the capacitor capacity value is done using the following equation.

$$Qc = P \times \left[\left(\sqrt{\frac{1}{pf_{awal}^2} - 1}\right) - \left(\sqrt{\frac{1}{pf_{baru}^2} - 1}\right)\right] \quad (1)$$

where Qc is the capacitor capacity (kVAR), P is Active power (kW), the initial Pf is the original power factor, and the new Pf is the desired power factor.

After getting the results of the calculation of the capacitor capacity to be installed, an estimate of the reduction in power losses can be calculated. Calculation of reducing power losses that occur in the Jampang Kulon channel was done using the ETAP 12.6.0 simulator. When the capacitor value was known, the capacitor was placed in a position in the distribution channel using ETAP.

Preliminary data obtained from the source, namely data on power losses that occurred before the installation of capacitors were listed in Table 1, while the power factor data on feeder buses were in Table 2. It can be seen that the power losses in the Jampang Kulon feeder distribution channel were sufficient and the power factor was less than 1. To overcome this, a capacitor have been installed. Using this data, the capacitor capacity needed to approach the power factor of 1 have been determined or calculated.

Table 1. Power losses in Jampang Kulon feeder.

No.	Device ID	Losses (kW)
1	bus 538-2-1	13,8
2	bus 51-17-1	27,6
3	bus 39-15-1	27,6
4	bus 62-17-4	10,7
5	bus 62-17-1	10,4
6	bus 207	19,2
7	bus 62-17-3	10,5
8	bus 65-16-1	26,7
9	bus 66-15-1	19
10	bus 85-16-1	61,6
11	bus 82-17-1	59,4
12	bus 86-14-1	16,5
13	bus 83-16-1	33,1
14	bus 84-18-1	23,6
15	bus 84-18-3	30,6

Table 2. Power factor data on Jampang Kulon feeder buses.

No	Device ID	%PF	Cos φ
1	bus 538-2-1	81,5	0,815
2	bus 51-17-1	81,7	0,817
3	bus 39-15-1	81,6	0,816
4	bus 62-17-4	81,5	0,815
5	bus 62-17-1	85	0,85
6	bus 207	81,6	0,816
7	bus 62-17-3	81,6	0,816
8	bus 65-16-1	81,7	0,817
9	bus 66-15-1	85	0,85
10	bus 85-16-1	81,6	0,816
11	bus 82-17-1	85	0,85
12	bus 86-14-1	82,5	0,825

Table 2. Cont.

No	Device ID	%PF	Cos φ
13	bus 83-16-1	82,5	0,825
14	bus 84-18-1	82,1	0,821
15	bus 84-18-3	81,9	0,819

3. Results and discussion

3.1. Bus data on distribution channels

Data on measurements of power losses, power factor, and active power have been taken from PT. PLN (Persero) Rayon Pelabuhan Ratu, Sukabumi Regency. Distribution Channels used as the object of research were the channels on Jampang Kulon feeders. There are 314 buses. However, in this study, only 15 buses were used as the study sample, namely buses with high power losses.

Table 3. Bus data on Jampang Kulon Feeder in PT. PLN (Persero) Rayon Pelabuhan Ratu.

Device ID	Power Losses (kW)	%PF	P (MW)
bus 538-2-1	13.8	81.5	5147
bus 51-17-1	27.6	81.7	5078
bus 39-15-1	27.6	81.6	5098
bus 62-17-4	10.7	81.5	4576
bus 62-17-1	10.4	85.0	4465
bus 207	19.2	81.6	4397
bus 62-17-3	10.5	81.6	4515
bus 65-16-1	26.7	81.7	4246
bus 66-15-1	19.0	85.0	4360
bus 85-16-1	61.6	81.6	3979
bus 82-17-1	59.4	85.0	3788
bus 86-14-1	16.5	82.5	1518
bus 83-16-1	33.1	82.5	3827
bus 84-18-1	23.6	82.1	3912
bus 84-18-3	30.6	81.9	3862

3.2. Calculation of capacitor capacity

At this stage, a capacitor have been installed in a 20 kV distribution system on the Jampang Kulon feeder. Installation of capacitors aims to minimize electrical power losses. The bus that should be installed on the Jampang Kulon feeder capacitors can be seen in Table 4.

Table 4. Bus data that should be installed by capacitor.

No.	Device ID	*Pf*	P (MW)
1	bus 85-16-1	0,816	3979
2	bus 86-14-1	0,825	1518

IOP Conf. Series: Materials Science and Engineering **850** (2020) 012011 doi:10.1088/1757-899X/850/1/012011

Using Equation (1), the capacitor capacity values to be installed in the Jampang Kulon feeder are presented in Table 5. After getting the capacitor capacity needed, the capacitors were installed on the buses that have been determined using the ETAP simulation application 12.6.0.

Table 5. Capacitor capacity that should be in Jampang Kulon Feeder.

No.	Location	Capacitor Capacity (kVAR)
1	bus 85-16-1	3 x 900
2	bus 86-14-1	3 x 300

From the results of calculations and simulations with ETAP, we got the results of changes in new power losses as presented in Table 6. Comparison of power losses that occur in the Jampang Kulon feeder distribution channel by placing capacitors can be seen in Table 7.

Table 6. Power loses after installed a new capacitor.

No.	Device ID	Power Loses (kW)
1	bus 538-2-1	10.3
2	bus 51-17-1	20.6
3	bus 39-15-1	20.6
4	bus 62-17-4	8.2
5	bus 62-17-1	8.0
6	bus 207	14.9
7	bus 62-17-3	8.1
8	bus 65-16-1	21
9	bus 66-15-1	14.8
10	bus 85-16-1	50.0
11	bus 82-17-1	49.2
12	bus 86-14-1	13.6
13	bus 83-16-1	27.4
14	bus 84-18-1	19.6
15	bus 84-18-3	25.4

Table 7. Comparison of power losses before and after installed a new capacitor.

No	Device ID	Power Losses (kW)		Power Losses Decreasing (kW)	Power Losses Decreasing (%)
		Before	after		
1	bus 538-2-1	13.8	10.3	3.5	33.98
2	bus 51-17-1	27.6	20.6	7.0	33.98
3	bus 39-15-1	27.6	20.6	7.0	33.98
4	bus 62-17-4	10.7	8.2	2.5	30.49
5	bus 62-17-1	10.4	8.0	2.4	30.00
6	bus 207	19.2	14.9	4.3	28.86
7	bus 62-17-3	10.5	8.1	2.4	29.63
8	bus 65-16-1	26.7	21.0	5.7	27.14

Table 7. Cont.

No	Device ID	Power Losses (kW)		Power Losses Decreasing (kW)	Power Losses Decreasing (%)
		Before	after		
9	bus 66-15-1	19.0	14.8	4.2	28.38
10	bus 85-16-1	61.6	50.0	11.6	23.20
11	bus 82-17-1	59.4	49.2	10.2	20.73
12	bus 86-14-1	16.5	13.6	2.9	21.32
13	bus 83-16-1	33.1	27,4	5.7	20.80
14	bus 84-18-1	23.6	19.6	4.0	20.41
15	bus 84-18-3	30.6	25.4	5.2	20.47

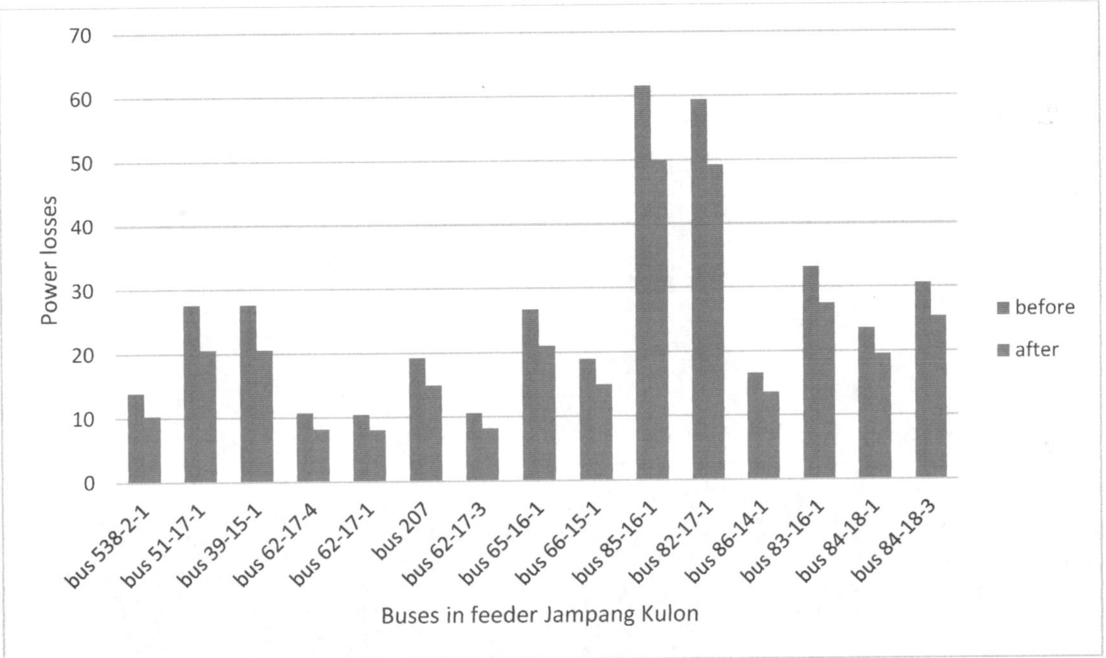

Figure 2. Chart of comparison of power losses before and after installed a new capacitor.

4. Conclusion

Power losses in the distribution channel have been decreased by installing a new capacitor. Installing capacitors on buses 85-16-1 and buses 86-14-1 reduced the value of power losses of 2.4 kW to 11.6 kW. The highest reduction in power losses occurred on buses 85-16-1 and buses 82-17-1. Power losses have been reduced by increasing the value of the power factor on the distribution channels.

References

[1] Schavemaker P and Van der Sluis L 2017 *Electrical power system essentials* (US: John Wiley & Sons)

[2] Mahdavian M, Kafi M H, Movahedi A and Janghorbani M 2017 Improve performance in electrical power distribution system by optimal capacitor placement using genetic algorithm *2017 14th International Conference on Electrical Engineering/Electronics, Computer, Telecommunications and Information Technology (ECTI-CON)* 749-752

[3] Askarzadeh A 2016 Capacitor placement in distribution systems for power loss reduction and voltage improvement: a new methodology *IET Generation, Transmission & Distribution* **10**(14) 3631-3638

[4] Kulkarni S and Sontakke S 2015 Power system analysis of a microgrid using ETAP *International Journal of Innovative Science and Modern Engineering (IJISME)* **3**(5)

[5] Ullah N, Khan F, Ali S M, Syed S, Alam Z and Assad M 2018 Electric Supply Substation Simulation using ETAP Software *University of Sindh Journal of Information and Communication Technology* **2**(4) 176-179

IOP Conf. Series: Materials Science and Engineering **850** (2020) 012012 doi:10.1088/1757-899X/850/1/012012

Reliability impact and assesement of distributed generation integration

R D N Sidik*, Y Mulyadi and B Trisno

Department of Electrical Engineering Education, Universitas Pendidikan Indonesia, Jl. Dr. Setiabudhi No.207 Bandung, Indonesia 40154

*rizaldaffa@student.upi.edu

Abstract. Distribution network is a part of an electric power system that has more intensity of failure than other parts of the power system. This is due to the close distribution network at the load which results in more factors that can be the cause of the failure. The more interference, the worse the reliability index. One effort to improve the feeder reliability index is by connecting Distributed Generation (DG) to feeders. In this study the effect of DG on the reliability index of Malangbong Chocolate (MLBC) feeders will be observed. MLBC is a power secondary substation in the district of Garut in West Java province, Indonesia. The method used in this study is to simulate the reliability of MLBC feeders before and after connecting with DG using the ETAP Power System 16.0 software. The results of the two simulations are then compared to see the effect of DG on the system reliability index. The results show that the SAIFI values before and after using DG connected to bus 4 and bus 5, are 3.8865 and 1.1802 per customer respectively. Whereas the value of SAIDI are 4.3079 and 1.1079 hours per customer in one year respectively. So it can be concluded, the use of DG on feeders can improve the reliability index of the system.

1. Introduction

The distribution network is the final part of the electric power system that carries electric power from the transmission line to each user. The distribution network has more interference intensity than other parts of the power system [1]. This is due to the proximity of the distribution network to the load, thus providing more factors causing disruption [2]. The more interference, the worse the system reliability index.

To determine the level of reliability of the distribution feeder, the reliability index parameter is used. Reliability indices that are commonly used in distribution networks are SAIFI (System Mean Disruption Frequency Index), SAIDI (System Mean Disruption Duration Index), and CAIDI (Average Customer Disruption Duration Index) [3].

One solution to improve the distribution network reliability index is Distributed Generation (DG) [4]. DG is a generator that is connected directly to the distribution network so that no new transmission line construction is needed [5]. In several previous studies it was stated that the application of DG is one of the most economical solutions to overcome the problem of electricity load growth, increased line power loss, and increased voltage drop [6-8].

This study focused on the impact of DG on the reliability index on Malangbong Chocolate feeders, which are power secondary substations in the district of Garut in West Java province, Indonesia. This research uses data from Indonesia Electricity Company in the district of Garut. The impact of DG to the reliability index is simulated using ETAP 16.0 software. The most commonly used reliability index to show the reliability of electric power, and used in this study, are SAIFI and SAIDI [9].

2. Reliabilty index

2.1. SAIFI (System Average Interruption Frequency Index)

SAIFI is a reliability index that related to how many times interruption happens in one period of time. SAIFI can be defined by this equation [9].

$$SAIFI = \frac{\sum_i \lambda_i N_i}{N_t} \tag{1}$$

Notes:
λ = Failure rate (interruption per year)
N_i = Number of customers that have black out.
N_t = Number of customers that connected to feeder.

2.2. SAIDI (System Average Interruption Duration Index)

SAIDI is a reliability impact related to how long the blackout happens in one period of time. SAIDI can be defined by this equation [9]:

$$SAIDI = \frac{U_i . N_i}{N_t} \tag{2}$$

Notes:
U_i = Interruption duration (Hours)
N_i = Number of customers that have black out.
N_t = Number of customers that connected to feeder.

3. Methodology

This research was conducted using secondary data from PLN in Garut district. The data used are one-line diagrams from MLBC distribution feeder, conductor specifications, and interrupt recapitulation. The data is then used to make a one-line diagram of the MLBC feeder in ETAP 16.0 software and run a reliability assessment to obtain a reliability index.

3.1. Data input

This research is used data one-line diagram MLBC feeder, interruption recapitulation during 2018, and the number of loads that connected to MLBC feeders. Table 1 shown interruption data from 2018.

Table 1. Interruption data.

No	Date	Rec-ssogh	Duration (minutes)
1	01/06/2018	S.BBTS	60
2	01/06/2018	S.BBTS	84
3	01/11/2018	S.BBTS	6
4	19/2/2018	S.BBTS	47
5	17/4/2018	S.BBTS	160
6	20/4/2018	GH.WBK	202
7	27/4/2018	S.BBTS	19
8	05/04/2018	S.BGP	94

Table 1. Cont.

No	Date	Rec-ssogh	Duration (minutes)
9	05/09/2018	R.BLB	46
10	05/12/2018	S.BBTS	15
11	13/5/2018	S.BBTS	34
12	14/5/2018	S.CWNI	64
13	25/7/2018	S.CWNI	5
14	26/7/2018	S.NBK	29
15	08/03/2018	S.NBK	95
16	08/03/2018	R.BLB	101
17	25/8/2018	R.BLB	88
18	09/03/2018	S.CWNI	35
19	20/9/2018	S.BBTS	26
20	25/9/2018	S.BBTS	20
21	10/08/2018	S.BBTS	51
22	10/10/2018	R.BLB	13
23	23/10/2018	R.BLB	18
24	23/10/2018	GH.WBK	116
25	24/10/2018	S.BBTS	13
26	25/10/2018	R.BLB	0
27	11/06/2018	S.BBTS	62
Total interuption			1503

3.2. Simulation unsing ETAP 16.0

The simulation is done by entering the following data: the number of loads connected to the feeder, the number of customers, and the conductor specifications. Feeder configurations made are the same as radial one-line diagram configurations. Figure 1 shows the MLBC feeder in the ETAP software.

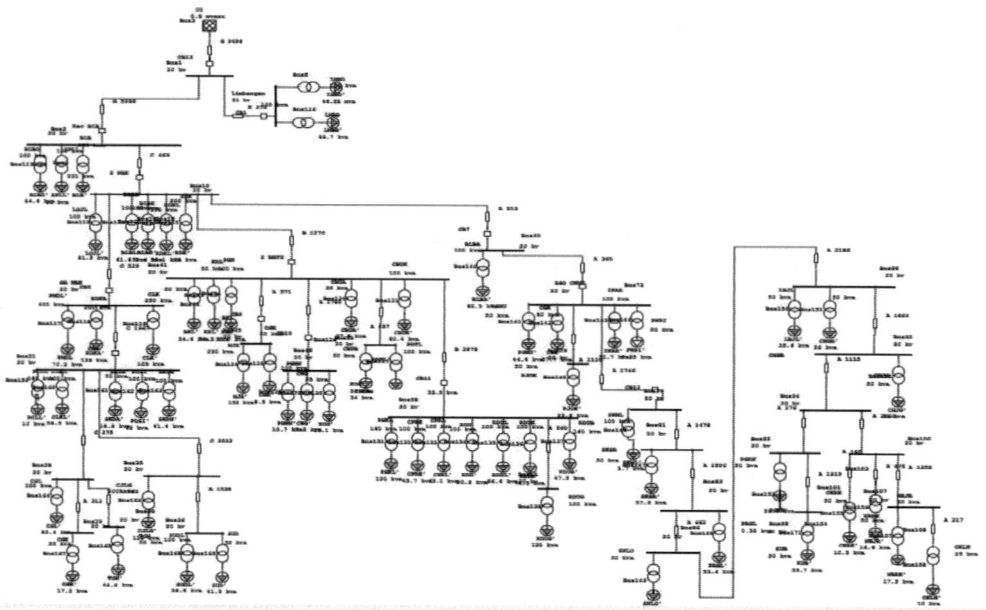

Figure 1. MLBC one-line diagram in ETAP 16.0.

MLBC distribution feeders supply electricity to 72 distribution transformers with a total load and number of customers 2.8 MW and 30,862 customers respectively. Figure 2 shows a simple model of one-line diagram of MLBC feeders. MLBC feeders have 5 main busses each connected to different loads. Table 2 shows the number of loads and customers connected to each bus.

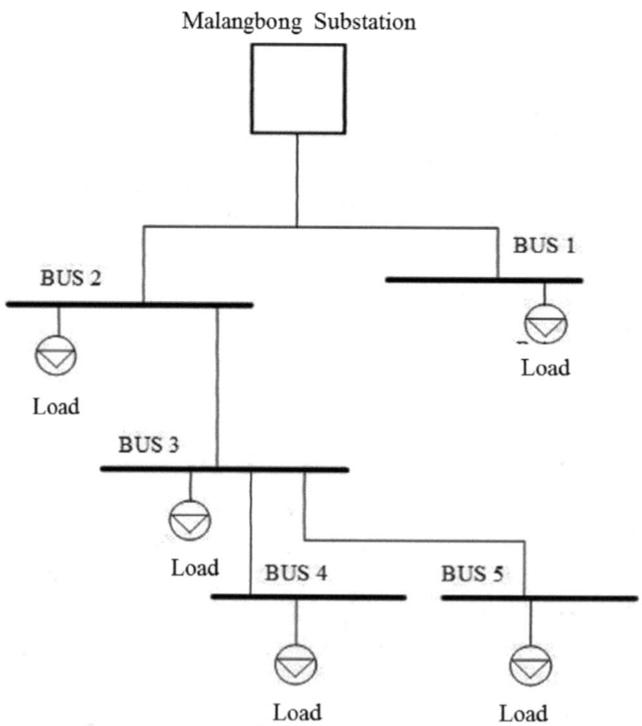

Figure 2. Simple MLBC one-line diagram.

IOP Conf. Series: Materials Science and Engineering **850** (2020) 012012 doi:10.1088/1757-899X/850/1/012012

Table 2. Load ad customers data.

NO	BUS	Load (kW)	Customers
1	BUS 1	114.73	631
2	BUS 2	2738.91	11640
3	BUS 3	2402.65	11314
4	BUS 4	894.12	4736
5	BUS 5	448.41	2541
TOTAL		2853.64	30862

Based on table 2, the total load connected to MLBC feeders is 2.8 MW, therefore the DG capacity that used in this research is 4 MVA.

In a study conducted by Ngaopitakkul et al, it was concluded that DG capacity and placement influence the reliability of feed distribution [5](Ngaopitakkul et al., 2013). The focus of this research is the impact of DG placement on the reliability of MLBC feeders.

DG placement is shown in figure 2 where the MLBC feeder has 5 main buses connected to the load. To determine the DG installation point, there are several things to consider, namely:

- Interruption location.
- Feeders configuration.
- Number of loads and customer connected.

After considering those points, the DG installation will be simulated with three different scenarios:

- One unit DG connected to one bus
- One unit DG connected between two bus.
- Two unit DG connected to two different bus.

4. Results and discussions

The simulation results are displayed in table number 3-5. Table 3 for the condition of one DG unit connected to one bus, table number 4 for the condition of one DG unit connected between two buses, and table number 5 for the condition of two DG units connected to two buses.

This study uses the SPLN 68-2 standard on reliability indexes for spindle configuration with Recloser. Based on these standards, the system is said to be reliable if the SAIFI value is less than 2.4 interruptions per year and the SAIDI value is less than 12.8 hours per year.

Table 3. One unit DG connected to one bus results.

No	BUS	SAIFI(f/cs.yr)	SAIDI(hr/cs.yr)	CAIDI (hr/f)
	without DG	3.8865	4.3079	1.108
1	1	2.5817	2.1388	0.828
2	2	1.9866	1.7215	0.867
3	3	1.5171	1.3804	0.910
4	4	1.4428	1.3140	0.911
5	5	1.2545	1.1743	0.936

The simulation of MLBC feeders reliability index before connected to DG shows that SAIFI 3.8865 interruption per customer in one year and SAIDI 4.3079 hours per cutomer in one year.

DG simulation with the first scenario shows that the smallest reliability index is when DG is connected to bus number 5. In this condition, the reliability index decreases a lot with a SAIFI value of 1.2545 per customer in one year and SAIDI of 1.1743 hours per customer in one year. This happens because most disruptions occur near the MLBC substation, so that when the disturbance occurs, the feeder has a backup supply from DG that is connected to the bus.

Table 4. One unit DG connected between bus results.

No	BUS	SAIFI (f/cs.yr)	SAIDI (hr/c.yr)	CAIDI (hr/f)
	without DG	3.8865	4.3079	1.108
1	1 and 2	3.2162	3.8304	1.191
2	3 and 4	2.6598	3.3397	1.256
3	3 and 5	2.8446	3.4415	1.210
4	4 and 5	2.2423	2.5935	1.157

Table 4 shows the simulation results of DG installation between two buses. It appears that the installation of DG affects the reduction in SAIFI and SAIDI values of the MLBC feeder. The biggest decrease is when DG is connected between bus 4 and bus 5 with SAIFI 2,224 interruptions per year and SAIDI 2.5935 hours per interruption. This is because on bus 4 there are BBO SSO (Sectionalizer) buses and CWNI SSO which have a high failure rate in 2018. BBTS SSO has a failure rate of 1,083 failures per year and CWNI SSO has a failure rate of 0.25 failures per year. So, if there is a disruption to the SSO BBTS and SSO CWNI, the electricity supply will not be interrupted because it gets electricity supply through the connected DG.

Table 5. One unit DG connected between bus results.

No	BUS	SAIFI (f/cs.yr)	SAIDI (hr/cs.yr)	CAIDI (hr/f)
	without DG	3.8865	4.3079	1.108
1	1 and 2	1.9319	1.6775	0.868
2	1 and 3	1.4623	1.3363	0.914
3	1 and 4	1.3881	1.2700	0.915
4	1 and 5	1.1997	1.1303	0.942
5	2 and 3	1.5171	1.3804	0.910
6	2 and 4	1.4428	1.3140	0.911
7	2 and 5	1.2545	1.1743	0.936
8	3 and 4	1.4428	1.3140	0.911
9	3 and 5	1.2545	1.1743	0.936
10	4 and 5	1.1802	1.1079	0.939

Table 5 is the simulation result of DG installation between two buses. It appears that the installation of DG affects the reduction in the value of SAIFI and SAIDI for MLBC feeders. The biggest decrease is when DG is connected to bus 4 and bus 5 with a value of SAIFI 1.1802 disruptions per year and SAIDI 1,1079 hours per interruption. This condition occurs because in bus 4 there is a SSO BBTS (Sectionalizer) which has a high failure rate of 1,083 failures per year. Whereas on bus 5 there is a CWNI SSO which has a failure rate of 0.25 failures per year. So if there are disturbances in the SSO BBTS and

SSO CWNI, the electricity supply will not be interrupted because it receives an electricity supply from DG that is connected to each bus.

5. Conclusions

From the results of research and analysis can be concluded as follows:

- Distributed Generation installations on the MLBC feeder bus affect the feeder reliability index.
- Installing DG on an MLBC feeder can significantly reduce the system's reliability index.
- The scenario of installing two DG units on different buses gets the biggest decrease in the reliability index compared to the other two scenarios.
- The optimal DG position is on buses 4 and 5, where the reliability index value gives the best results.

References

[1] Marsudi D 2006 *Operasi Sistem Tenaga Listrik* (Yogyakarta: Graha Ilmu)

[2] Hlatshwayo M, Chowdhury S, Chowdhury S P and Awodele K O 2010 Impacts of DG penetration in the reliability of Distribution Systems *2010 International Conference on Power System Technology* 1-8

[3] Kurniawan R and Pane Z 2015 Studi Keandalan Jaringan Distribusi 20 kV yang Terinterkoneksi Dengan Distributed Generation (Studi Kasus: Penyulang PM. 6 Photo Gardu Induk Pematangsiantar) *Singuda ENSIKOM* **11**(31) 100-105

[4] Kaur P, Kaur S and Khanna R 2016 Optimal placement and sizing of DG comparison of different techniques of DG placement *2016 IEEE 1st International Conference on Power Electronics, Intelligent Control and Energy Systems (ICPEICES)* 1-4

[5] Ngaopitakkul A, Pothisarn C, Bunjongjit S, Suechoey B, Thammart C and Nawikavatan A 2013 A reliability impact and assessment of distributed generation integration to distribution system *Energy and Power Engineering* **5**(04) 1043

[6] Yousefian R and Monsef H 2011 DG-allocation based on reliability indices by means of Monte Carlo simulation and AHP *2011 10th International Conference on Environment and Electrical Engineering* 1-4

[7] Sami T, Mahaei S M, Namarvar M H and Iravani H 2011 Optimal placement of DGs for reliability and loss evaluation using DIgSILENT software *2011 10th International Conference on Environment and Electrical Engineering* 1-5

[8] Zhu D, Broadwater R P, Tam K S, Seguin R and Asgeirsson H 2006 Impact of DG placement on reliability and efficiency with time-varying loads *IEEE Transactions on Power Systems* **21**(1) 419-427

[9] Short T A 2018 *Electric power distribution equipment and systems* (US: CRC press)

IOP Conf. Series: Materials Science and Engineering **850** (2020) 012013 doi:10.1088/1757-899X/850/1/012013

Analysis of use and need of sectoral electrical power and energy in West Java Region

T Sucita* **and Y Mulyadi**

Department of Electrical Engineering Education, Universitas Pendidikan Indonesia Jl. Dr. Setiabudhi No. 229, Isola, Sukasari, Bandung 40154

*tasmasucita@upi.edu

Abstract. The current and future usage and power needs of electricity are important in electricity utilities. This study aims to predict the amount of power and electricity usage and demand from various sectors in order to be used as a reference and an overview of the amount of sectoral load and electrical energy needs in the West Java region in the future. The method for determining the estimated electrical load and energy requirements can be calculated by a combined method (analytical, econometric and trend) and a simple linear regression method with a sectoral approach by grouping customers into five sectors (household, business, industrial, general and social). The findings of the study using the combined method of the percentage of the amount of power connected by 5.46% with an error of 4.42%, and the amount of energy sold at 4.78% with an error of 4.49%. Whereas by using the regression method the percentage of total connected power is 5.27% with an error of 3.54%, and the amount of energy sold is 4.24% with an error of 3.98%. By doing these forecasts it is hoped that the future electricity needs can be met.

1. Introduction

Electricity changes every day due to modern civilization and technological advancements [1]. Where along with increasing population growth in an area each year, making the need for electrical energy has increased in accordance with population growth and regional economic growth. The growth of these two conditions will lead to an increase in demand for electricity and the need for electricity reserves for the next few years. This is to anticipate power cuts, especially at peak load times. Thus, it requires development in the field of electrical energy to balance the demands and needs of a region.

Estimates of electricity consumption and demand for electricity in the next decade or two are important in electricity utilities [2,3]. Forecast in the future is needed to determine and estimate the amount of load needed so that it can prepare the electricity generation capacity needed to anticipate the growing demand for electrical energy demand in the future [1,4]. In addition, load forecasts help electricity utilities to make important decisions, including decisions on purchasing and generating electricity, the development of transmission and distribution infrastructure [5].

Factors that influence the use of loads are such as population, weather forecasts, and others. This is done by using the value of the demand for electricity in the future by mostly estimating past load consumption and considering other factors that influence the amount of electricity load used in an area [6]. Load forecasting is a method of estimating the electrical load required by consumers [7].

The aim is to balance the electricity supply from the demand for electricity. Medium-term load forecast is a load forecast for a period of one month to one year, which is used to determine the planning of both the transmission and distribution network expansion planning and the addition of a new electricity center. And the long-term load forecast is the estimated load over a period of over one year, which is used for planning and forecasting the condition of the electricity grid in the future and determining energy needs in the future [8,9]. Short-term, medium-term and long-term forecasts in the demand forecast load requirements, are indispensable for the correct operation of the electric utility and are also needed for proper scheduling activities, such as scheduling generator operation, maintenance scheduling and scheduling of electricity investment in the future [10]. In this study, we predict the amount of power and electricity usage and demand from various sectors as a reference and an overview of the amount of sectoral load and electrical energy needs in the West Java region in the future.

2. Methods

2.1. Research design

Research "Forecast Analysis of Long-Term Electricity Load Needs for Various Sectors in PT. West Java Regional PLN" is using a quantitative approach, which is to determine how large the number of customers, connected power, and energy sold per sectoral from 2018 to 2022. In this study, the method used is to compare the results of the number of customers, connected power, and energy sold by sector from 2018 to 2022 using the combined method with the regression method.

There are several data findings that become the main data for data processing. The data will be input for calculations using Microsoft Office Excel software. Data obtained in the form of data on the population of West Java last 5 years, West Java GRDP data of the last 5 years contained in West Java BPS data and data on the number of customers, connected power data, energy data sold per sectoral contained in the data ESDM electricity system. The following are data obtained from the West Java BPS data and the EMR Electricity System in 2013 to 2017 showed as Table 1, Table 2, and Table 3.

Table 1. Number of customers, connected power and electricity energy sold.

Sector	Year 2013			Year 2017		
	Number of Customers	Connected Power (MVa)	Sold Energy (MWh)	Number of Customers	Connected Power (MVa)	Sold Energy (MWh)
Household	9.698.695	8.189,72	14.486.340	12.388.399	10.858,71	17.555.200
Business	310.838	2.847,01	3.398.560	506.173	3.591,92	5.231.900
Industry	12.471	6.246,44	19.879.770	14.590	7.948,31	22.956.680
Public / Public	50.675	301,65	587.030	79.087	406,59	730.510
Social	203.449	413,99	529.960	265.362	661,48	951.150
Total	10.276.128	17.998,81	38.881.660	13.253.611	23.467,00	47.425.440

Table 2. Data of growth rate of gross regional domestic product of ADHK 2010.

Year	Total	Growth (%)
2013	1.093.543,55	-
2014	1.149.216,06	5,09
2015	1.207.232,34	5,05
2016	1.276.627,64	5,75
2017	1.342.953,38	5,20
Growth Rate		5,27

Table 3. Data on population numbers.

Year	Total Population	Average Population Growth
2013	45.340.799	1,56
2014	46.029.668	1,52
2015	46.709.569	1,48
2016	47.379.389	1,43
2017	48.037.827	1,39
Average population growth		1,476

2.2. Forecast phase with combined method

In calculating the estimated electrical energy requirements for each sector, it is determined by data that has been obtained previously, while the variables used are as follows: Total Population. Electricity customers, Connected Power, and Energy Consumption [11].

Estimates of total energy consumption (ET) are obtained from the sum of energy consumption in the household (RT), industry (I), business (B), general (U), and social (S) sectors expressed by the equation (1) as in [11], that is,

$$ET_t = ERT_t + EI_t + EB_t + EU_t + ES_t \tag{1}$$

2.3. Forecasting phase with a simple linear regression method

Simple linear regression is a method used to test the extent of the relationship between the cause variable (*X*) and the effect variable (*Y*). Simple linear regression (Simple Linear Regression) is a statistical method that is often used in production both for predicting or estimating the characteristics of quality and quantity [12].

The linear regression equation model is like the equation (2):

$$Y = A + BX \tag{2}$$

The cause variable used is the period of year (*X*) that will be predicted and the consequent variable is the household expense (*Y*), industry (*Y*), business (*Y*), general (*Y*) and social (*Y*), on electricity customers, connected power, and energy is sold [12].

2.4. Accuracy value forecast

The accuracy of the forecast is a very important thing in the forecast where, the data that already exists with the forecast data is measured accordingly. There are several calculations that can be used to calculate the estimated error rate of the total. There are three most well-known calculations in calculating forecast errors including the mean absolute deviation (MAD), mean square error (MSE), and the mean absolute percent error (MAPE) [13].

$$MAD = \sum_{t=1}^{n} \frac{|Xt - Ft|}{n} \tag{3}$$

$$MSE = \sum_{t=1}^{n} \frac{(Xt - Ft)^2}{n} \tag{4}$$

$$MSE = \left(\frac{100\%}{n}\right) \sum_{t=1}^{n} \frac{|Xt - Ft|}{n} \tag{5}$$

3. Results and discussion

Based on the results of the calculation of the number of residents and the need for electrical energy loads for the number of customers, connected power and energy sold sectorally from the West Java BPS data and ESDM electricity system data. From the results of the study obtained data as follows:

Table 4. Forecast of population of West Java in 2018 – 2022.

Year	Population	Average Population Growth
2018	48.746.865	1,48
2019	49.466.369	1,48
2020	50.196.493	1,48
2021	50.937.393	1,48
2022	51.689.229	1,48

In Table 4. the population of West Java in 2018 - 2022 increases in income every year with an average increase of 1.48% and the estimated population in 2022 is 51,689,229.

Table 5. Forecast results of West Java electric energy load needs for the period 2018-2022 using the combined method and regression method.

Information	Combined			Regression		
	Year		Growth Rate	Year		Growth Rate
	2018	2022		2018	2022	
Total population	48.746.865	51.689.229	1,48	48.746.865	51.689.229	1,48
Number of Customers	13.704.888	17.010.507	5,12	14.026.605	17.023.702	5,14
Household	13.041.362	16.015.996	5,27	13.114.364	15.823.584	5,02
Business	561.594	850.972	10,95	532.355	727.771	7,54
Industry	15.077	17.193	3,34	15.087	17.211	3,36
Public / Public	86.855	126.346	9,82	85.237	113.451	7,48
Social	280.562	350.582	5,73	279.562	341.685	5,19
Connected Power (MVA)	24.132,35	30.596,45	5,46	24.853,73	30.335,33	5,27
Household	11.416,58	14.098,69	5,36	11.497,67	14.188,82	5,5
Business	4.058,94	6.497,46	12,59	3.671,97	4.317,35	3,75
Industry	8.207,57	9.334,07	3,27	8.543,02	10.337,08	5,4
Public / Public	449,27	666,23	10,38	428,39	533,84	5,6
Social	695,73	853,53	5,23	712,68	958,25	7,7
Sold Energy (MWh)	48.841.539	59.896.476	4,78	49.951.682	58.366.380	4,24
Household	18.428.240	22.376.708	4,97	18.750.836	21.832.434	4,46
Business	5.834.228	9.021.511	11,51	5.803.051	7.618.985	7,82
Industry	23.807.470	27.537.840	3,71	23.559.755	26.532.289	2,94
Public / Public	771.601	960.417	5,63	763.187	903.093	4,33
Social	1.104.961	2.012.514	16,17	1.074.854	1.479.580	9,26

Figure 1. shows that the results of forecasts using the regression method of population increases each year are proportional to the increase in the need for electrical energy load each year which covers the number of customers, connected power and energy sold. However, for the amount of energy sold in excess of the population of West Java this is because each year the need for energy is sold is increasing where the industrial sector has the greatest need for energy needs sold each year. Table 5 also shows that the average total growth rate of all sectors for the number of customers is 5.14%, while the average total growth of all sectors for the total connected power is 5.27%, and the average growth rate for all sectors for all sectors the amount of energy sold was 4.24%.

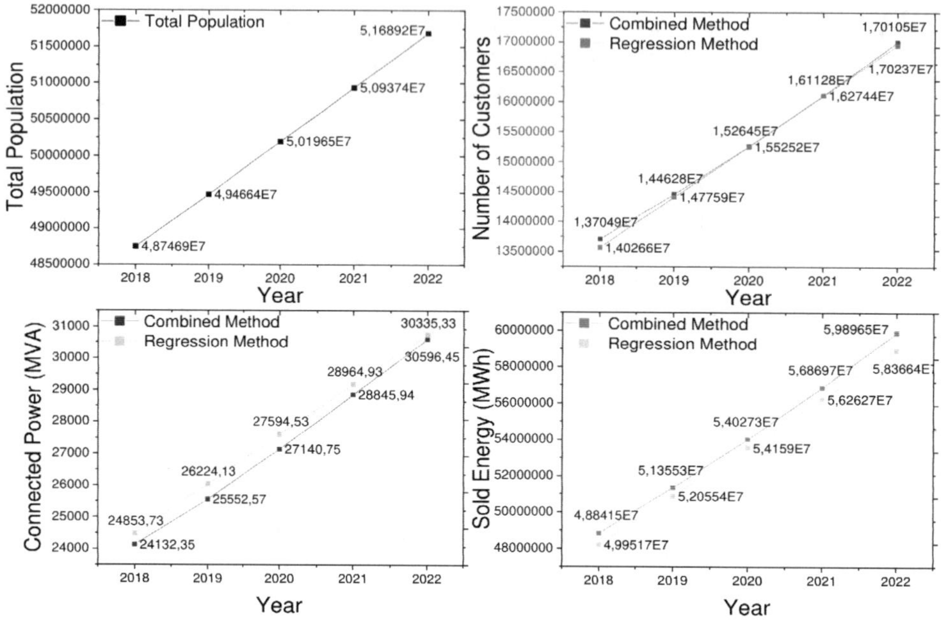

Figure 1. Graph of increase in population by increasing electricity load needs for 2018-2022 by the combined and regression methods.

Table 6. Comparison of the accuracy of forecasting the combined method with the regression method.

No	Information	Error	Combined	Regression
1	Number of Customers	MAD	13.671,60	36.800,80
		MSE	435.747.560,80	3.605.602.610
		MAPE	2,59	3,38
2	Connected Power	MAD	150,596	138,04
		MSE	38.165,76	36.566
		MAPE	4,42	3,54
3	Sold Energy	MAD	308.070,20	353.641,80
		MSE	240.580.808.055	268.353.459.910
		MAPE	4,49	3,98

Table 6. illustrates that:

- The estimated error rate of all customers in all sectors using the combined method has a relatively smaller error rate compared to the estimated number of customers using the regression method. Where if seen from the error level of the mape model the estimated number of customers with the combined method has an error value of 2.59% while the regression method has an error value of 3.38%. These results show that the accuracy of the estimated number of sectoral customers with the combined method is more accurate than the regression method.
- The error rate for power connected to all sectors by the combined method has a relatively greater error rate compared to the power forecast connected to the regression method. Where if seen from the error level of the mape model the power forecast connected to the combined method has an error value of 4.42% while the regression method has an error value of 3.54%. From these results, it is shown that the accuracy of the sectoral connected power forecast by the regression method is more accurate than the combined method.
- The estimated error rate of energy sold by all sectors using the combined method has a relatively higher error rate compared to the estimated energy sold by the regression method. Where if seen

from the error level of the MAPE model the estimated energy sold by the combined method has an error value of 4.49% while the regression method has an error value of 3.98%. From these results, it shows that the accuracy of the energy forecast sold by the sectoral regression method is more accurate than the combined method.

4. Conclusions

Based on the results of a research analysis of the forecast needs of the West Java regional electrical energy load the authors conclude the following matters. The estimated electricity demand for the next 5 years using the combined method is for the number of sectoral customers of 17,023,702, the sectoral connected power of 30,335.33 MVA, and the energy sold by the sector at 58,366,380 MWh. The results of the calculation of the forecast needs of the West Java regional electrical energy load from 2018 to 2022 obtained the results of the regression method for the average growth rate of all sectors, namely the number of customers by 5.14% with an estimated error value of the MAPE model of 3.38% , the amount of connected power is 5.27% with the estimated MAPE model error value of 3.54%, and the amount of energy sold is 4.24% with the estimated MAPE model error value of 3.98%. The increasing number of populations each year is proportional to the increase in electricity energy needs each year which includes the number of customers, connected power, and energy sold per sectoral (household, business, industrial, general and social sectors). The estimates of the connected power and energy sold the sectoral method of regression is more accurate than using the combined method.

References

[1] Ali D, Yohanna M, Puwu M I and Garkida B M 2016 Long-term load forecast modelling using a fuzzy logic approach *Pacific Sci. Rev. A Nat. Sci. Eng.* **18** 123–7

[2] Mohammed N A 2018 Modelling of unsuppressed electrical demand forecasting in Iraq for long term *Energy* **162** 354–63

[3] Sangrody H, Zhou N, Tutun S, Khorramdel B, Motalleb M and Sarailoo M 2018 Long term forecasting using machine learning methods *2018 IEEE Power and Energy Conference at Illinois (PECI)* (IEEE) pp 1–5

[4] Carvallo J P, Larsen P H, Sanstad A H and Goldman C A 2018 Long term load forecasting accuracy in electric utility integrated resource planning *Energy Policy* **119** 410–22

[5] He W 2017 Load forecasting via deep neural networks *Procedia Comput. Sci.* **122** 308–14

[6] Samuel I A, Chihurumanya F-N, Adewale A A and Awelewa A A 2014 Medium-term load forecasting of covenant university using the regression analysis methods *J. Energy Technol. Policy* **4**

[7] Adepoju G A, Ogunjuyigbe S O A and Alawode K O 2007 Application of neural network to load forecasting in Nigerian electrical power system *Pacific J. Sci. Technol.* **8** 68–72

[8] Chen T 2012 A collaborative fuzzy-neural approach for long-term load forecasting in Taiwan *Comput. Ind. Eng.* **63** 663–70

[9] Zhao K, Gan L, Wang Z and Xiong Y 2012 Application of combination forecast model in the medium and long term power load forecast *Advanced Materials Research* vol 490 (Trans Tech Publ) pp 1362–6

[10] Xia C, Wang J and McMenemy K 2010 Short, medium and long term load forecasting model and virtual load forecaster based on radial basis function neural networks *Int. J. Electr. Power Energy Syst.* **32** 743–50

[11] Hidayati D N and Agus Supardi S T 2018 Perkiraan Kebutuhan Konsumsi Energi Listrik Di Kabupaten Pati Pada Tahun 2026 Dengan Menggunakan Metode Gabungan

[12] Septyawan R 2018 Analisis Peramalan Kebutuhan Energi Listrik Pln Area Batam Menggunakan Metode Regresi Linear

[13] Margi K and Pendawa S 2015 Analisa Dan Penerapan Metode Single Exponential Smoothing Untuk Prediksi Penjualan Pada Periode Tertentu (Studi Kasus: Pt. Media Cemara Kreasi) *Pros. SNATIF* **2**

IOP Conf. Series: Materials Science and Engineering **850** (2020) 012014 doi:10.1088/1757-899X/850/1/012014

IoT scheme for surveillance system and laboratory security access

I Fushshilat[1,*] and Y Yogasmana[2]

[1] Department of Electrical Engineering Education, Universitas Pendidikan Indonesia, Jalan Dr. Setiabudhi No.229, Bandung, Indonesia
[2] Department of Mechanical Engineering Education, Universitas Pendidikan Indonesia, Jalan Dr. Setiabudhi No.229, Bandung, Indonesia

*imanf@upi.edu

Abstract. This paper proposes to design a modern surveillance system and laboratory access using digital technology, through the concept of the Internet of Things (IoT) which is becoming one of the trending topics in the development of world technology nowadays. IoT makes several electronic devices such as actuators, biometric sensors, RFID Reader, servers, and IP cameras are connected to an Internet network. Thus, those devices can be accessed anytime anywhere outside laboratory. identification of visitors are using RFID cards or biometrics sign such as fingerprints or faces that have been recognized by the system and every time visitors come to laboratory, they are recorded directly into the server automatically, IP camera allows displaying in real-time condition of the Laboratory, The system is utilizing the indoor sound system as a device for conducting teleconferences and appealing for information to visitors to the laboratory remotely via the internet, all of these are very useful for establishing a better security and surveillance system and facilitating work of laboratory administrators.

1. Introduction

As facility for education, and research. Laboratory has a lot of data, component, tools, device, valuable instrument that needs security [1]. Nowadays, most of Surveillance and security system of laboratory is still done in the conventional way handled manually by a laboratory staff / admin, whereas these opportunities can be optimized through a technology system.

Internet becomes technology that has developed very rapidly in this century, internet support devices such as smartphones, tablets, and computer. Internet is accessible by someone, anywhere and anytime [2]. In its application, internet does not only act as a provider of information such as news, knowledge, email, shopping, entertainment, etc. but also the concept of the Internet of Things (IoT) can be used as a media to connect / control a device in real time, anytime and anywhere.

IoT is a communication paradigm that refers to the idea of connecting daily life objects to Internet, where the object is combined with a microcontroller. In the other side, transceivers are able to communicate and configured with certain protocols so that it can interact between an object and others to achieve certain goals without human intervention [3]. briefly IoT is a technology that facilitates interconnection between systems so that they can be connected to each other [4].

This paper proposes to design the laboratory security access and supervision using IoT technology to improve the security system in the laboratory. This system integrates several electronic devices such as IP cameras, Motor DC, solenoid, RFID, controller in a network so each other can communicate and be controlled remotely by the administrator / laboratory staff, it can improve the laboratory security access and laboratory supervision.

2. Related work

Many studies have been conducted on electronic-based security access. For instances, the door access security system that can be controlled via Bluetooth [5]. The digital smart door security system integrated through several components such as Solenoid Actuators, RFID and Zigbee [6]. Wireless communication modules, Security system Non-contact based access has been by utilizing PSTN telephone tones that commonly known as DTMF (Dual Tone Multi Frequency) using Arduino as Controller and Solenoid as actuator which is mounted on the door as access security [7].

The use of biometrics as the key to access to a security system has also been studied, one of them by [8] was discussed about Fingerprint-based automatic door lock system to identify entry person access, then data from the fingerprint reading is sent to central communications and then forwarded to users via the telephone network, so user can find out who is entering the room. Meanwhile, a smart house model uses biometrics face detection as an identification feature of entry access, camera taking a photo of the visitor's face then through the face recognition method, smart home detects and identifies the faces of visitors so that foreigners cannot enter the house, only faces that have been registered in the system database can be entered into the house [9]. The development of other access security systems based on electronic was developed by utilizing GSM modules for receiving phone calls from users, then they can detect Caller ID which is used as the authorization key to access the system [10].

Based on the comparison with some relevant study results, we proposed an innovation design laboratory security access and surveillance system that adopted IoT technology by combining several electronic components with each other connected to an internet network, to improve security and to facilitate supervision in the laboratory, and to facilitate the work of Laboratory Staff.

3. System work scenarios

Laboratory users who will access the laboratory devices must go through the Laboratory door which has an electronic lock that can only be opened through RFID, or uses biometric sign such as fingerprints and faces that have been registered in the database server computer, Laboratory users must do a tapping identity card that has been equipped with an RFID chip (RFID Card) that has been previously registered, so that it can be recognized by the system, besides that it can also use Biometrics features such as faces or fingerprints that have previously been registered to the database server so can be recognized by the system, so everyone who enters the Laboratory will be recorded in database on server computer. For certain cases, if someone does not have authentication id to enter the laboratory and laboratory officers are not there, then user can contact the laboratory officer or administrator to request access into the laboratory by contacting the admin through taking a picture by scanning their face on face recognition camera. And then, the face data are stored on the server in order to be verified by the admin. If he approved, then user is able to enter the laboratory. Laboratory administrators can provide access by opening doors through electronic devices such as computers, hand phone or tablet that is connected to the internet, thus the administrator can provide access anytime anywhere even though outside the laboratory.

The IP Camera installed in the Laboratory allows Laboratory administrators to conduct remote monitoring through computer media, hand phone, tablet or electronic gadget that are connected to the Internet. Indoor Sound System can be used to provide advice or information remotely via the internet to laboratory users through administrator electronic gadgets or other users who have access to authorization and authentication to the Server.

4. System architecture

The architecture of the supervision system and laboratory access security proposed in this paper consists of several components including Controlled Device, Node Controller, Accesses Identifier, Router, Server, IP Camera, Indoor Sound System as shown in Figure 1.

Controlled device is a series of devices installed on objects that will be controlled by security access, for example such as doors, gates, windows, cabinets, lockers, and so on. Controlled Device consists of several components such as an electric motor and an electric solenoid as an electronic actuator, an electric motor functions as a door, gate or device driver which will be controlled remotely, Electronic solenoid as a key that will be attached to the door, gates and other objects where the lock and release systems are carried out electronically, then the components of limit switch and proximity sensor as a position sensor for the door or other objects that will be controlled movement, Mechanical Structure is a mechanical framework composed by several metals assembled in such a way that serves to convert the rotation of an electric motor into a motion that can be used to open and close the door or gate of the laboratory.

The Node Controller component responds to the data coming from a Sensor on a controlled device, access identifier and admin command then process it to make decisions to controlling the actuator on the controlled device, on this architecture a component of node controller handles a packet of controlled devices, this is because a node controller has limited General Purpose Input Output (GPIO) which is to receiving data and performing controls. Access Identifier consists of RFID Reader and fingerprint and face scanner, RFID Reader functions to read data from RFID cards that are owned by visitors, as well as fingerprint and face scanner both function to read data from biometric features, namely fingerprints and faces that are owned by visitors. User data as a result of verification and identification of Access Identifiers are then sent to the Node Controller to be identified and verified according to the data on the server.

The router functions to regulate data traffic from the Node Controller, IP Cam to Server and or vice versa in a network. In this communication the Router functions as an Access Point where the Node Controller and IP Cam communicate with the Server via Wireless Router. Beside it, Router also functions to manage data traffic from the Server to Admin or User and vice versa through the Internet network. The server is the center of the system with activities including: database storage, provider of User Interface Graphs, giving the action command to the node controller. IP Cam serves to perform live monitoring of the conditions of the laboratory, and record it in videos stored on the server, admin can see the condition of the laboratory live through a GUI that can be opened in electronic gadgets outside laboratory whenever and wherever, via Internet, Indoor Sound System can be used as an audio tool when conducting a teleconference can also be used by the admin to convey appeals or information to laboratory visitors, and appeal can be carried out remotely from outside the laboratory via gadget connected to the Internet.

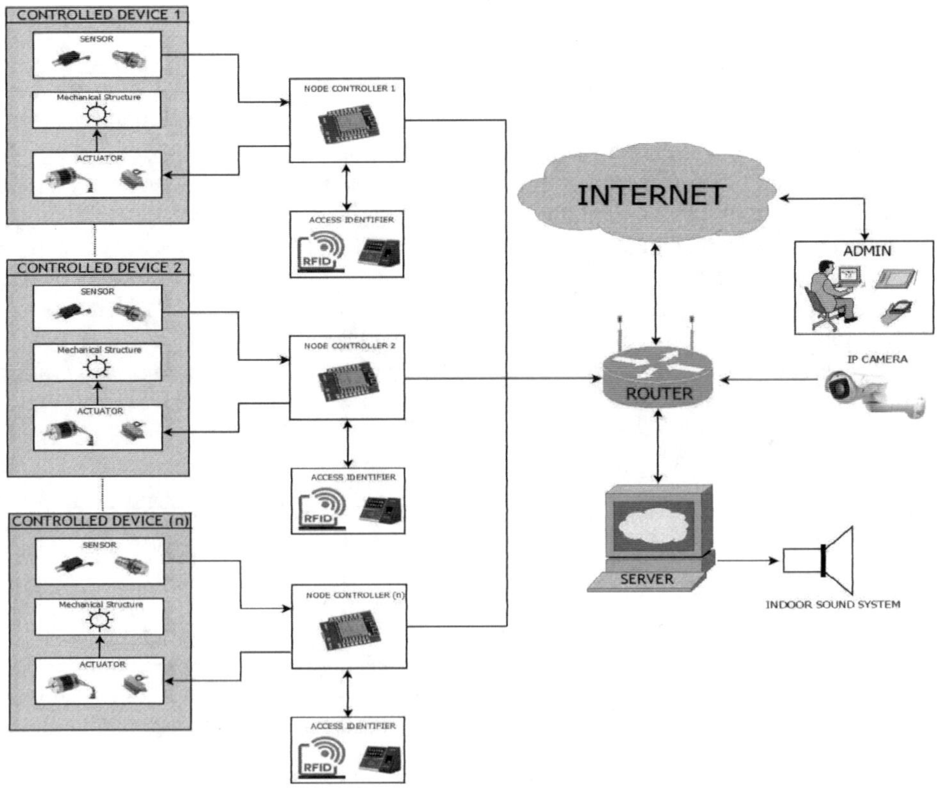

Figure 1. IoT scheme for laboratory access control and security systems.

5. Blok diagram system

Work details of the proposed System are shown in Figure 2. The Node Controller controls the actuators that is DC motor and Solenoid DC that are connected to the GPIO Pin through a power gain module, controlling of actuator is done digitally by issuing logic 1 or 0 to the GPIO to turn on and turn off the actuator, Sensor input, Limit Switch and Proximity Sensors send digital signals 1 or 0 directly to Node Controller via the GPIO pin which states the position of the Controlled Device, the input signal is read and processed by the Node Controller to determine the action, different from actuator and sensor components, Components in access identifier are RFID Reader and Biometrics Sensor use the communication protocol in its Digital communication with Node Controller, therefore a special pin, Serial Communication Pin (UART) is used which consists of Pin RXD and TXD for access identifier that uses the UART communication protocol, whereas for Access Identifier is using Serial Peripheral Interface (SPI) communication protocols, special pin pins such as MISO, MOSI, SCK, and RST are used for Node Controllers to communicate.

Access Identifier transfers data read from the user in form of a unique RFID Card ID, and digital data from the scanning of biometric sign to the Node Controller then process data which is matched with the database contained on the server then decide whether the visitor / user has access rights or not to enter the laboratory.

Node Controller has been programmed to do all actions needed by system, using programming languages that are compatible with Arduino on the basis of languages C and C ++. Node Controller is Embedded System which has a control base ESP8266 chip that already have features to communicate via Wi-Fi, technically it can be a WeMos module or NodeMCU. Node Controller communicates with the server to transfer the data and commands via wireless router where Node Controller has a unique Internet Protocol (IP) that can be recognized locally by the server or other device in a local network. Node Controller can access the database on the server, then server can give node controller commands to take action. IP Camera components are connected to the server via wireless router, digital data from the IP camera visualization are stored on the hard disk server, both video recording and live video that show the real-time condition of laboratory. Indoor Sound System connects to the server computer through wired or wireless connection, server can directly access to the Indoor Sound System when they want to appeal or provide information to laboratory visitors.

Laboratory admin can perform action commands to Controlled Devices with services provided by web server engine found on server computer, through Graphics User Interface (GUI) on electronic gadget, Computers or tablets that are connected to internet. GUI provided by the server computer when accessed by the laboratory admin via gadget, allow admin to control the controlled device, see the condition of live streaming laboratories and make audio call announcement to laboratory visitors. For technical requirements, setting and troubleshooting, admin laboratory can directly control the server computer through remote desktop applications located on the server computer remotely via the internet.

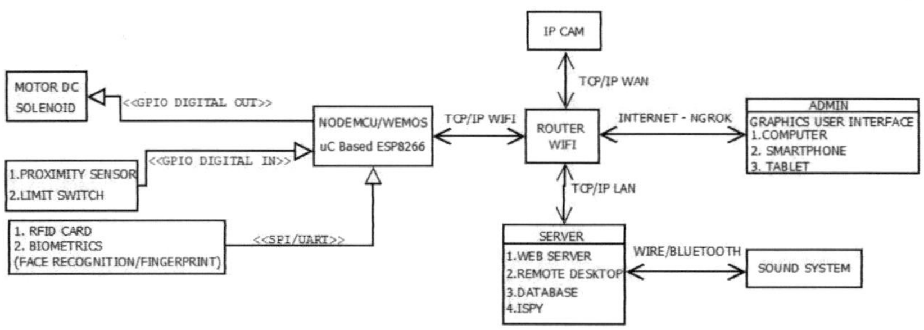

Figure 2. Blok diagram system.

6. Detail component

To form the proposed system, requires several components both hardware and software, table 1 shows the list of hardware utilization and its specifications, including controlled devices (actuators), sensors, microcontrollers, camera, and internet peripherals.

Table 1. Hardware requirements

No	Hardware	Specification
1	Motor	DC 12 V, Torque: 20 Kg.Cm
2	Solenoid Lock	DC12V, 2A
3	Limit Switch	SPDT Roller Level ARM
4	Proximity Switch	Photoelektric Difuse reflection
5	Node Controller	NodeMCU/ WeMos D1
6	Wireless Router	2,4 Ghz, 10/100 WAN LAN, IEEE 802.11n, IEEE 802.11g, IEEE 802.11b
7	RFID Reader	13,56 Mhz, USB Connection
8	IP Camera	1280 effective Pixel, 30 fps
9	Sound System	Surround, Bluetooth Support
10	Biometric Attendance Machine	Finger Print Scan, Face Scan, USB interface, DC supply, SDK
11	Internet Connection	2 Mbps

Table 2 shows the list of software utilization with its specifications, including software that works for website services, database provider, remote desktop to remote computer server, camera viewer software (option because it can be embedded directly on web pages), and software for programming microcontrollers.

IOP Conf. Series: Materials Science and Engineering **850** (2020) 012014 doi:10.1088/1757-899X/850/1/012014

Table 2. Software requirements

No	Software	Specification /Type
1	Web Server	Apache
2	Database	MySQL
3	Remote Desktop	Team Viewer
4	IP Cam Viewer	iSpy
5	μController IDE	Arduino

7. Conclusions and further work

In this paper, a digital surveillance and laboratory access security system are proposed using the IoT concept that integrates several electronic components in an Internet network so that each other is connected and accessible anywhere at any time outside the laboratory. We have explained system work scenario, system architecture, system block diagram and detail of the components needed to make the system. We hope it can improve security and facilitate supervision in the laboratory, and can facilitate the work of administrators / Laboratory staff.

In the future we will try to design the proposed system and will implement it in our work environment at the Industrial Electronics Laboratory, Faculty of Technology and Vocational Education, Universitas Pendidikan Indonesia.

References

[1] Fushshilat I, Rahmat A, Somantri Y and Haritman E 2018 Laboratory management: digital laboratory information system (DLIS) concept *IOP Conference Series: Materials Science and Engineering* **434** 1 012286

[2] Bermúdez-Ortega J, Besada-Portas E, López-Orozco J A and Jesús M 2017 A new open-source and smart-device accessible remote control laboratory *2017 4th Experiment@ International Conference (exp. at'17)* 143-144 IEEE

[3] Poongothai M, Subramanian P M and Rajeswari A 2018 Design and implementation of IoT based smart laboratory *2018 5th International Conference on Industrial Engineering and Applications (ICIEA)* 169-173 IEEE

[4] Al-Fuqaha A, Guizani M, Mohammadi M, Aledhari M and Ayyash M 2015 Internet of things: A survey on enabling technologies, protocols, and applications *IEEE communications surveys & tutorials* **17**(4) 2347-2376

[5] Hadis M S, Palantei E, Ilham A A and Hendra A 2018 Design of smart lock system for doors with special features using bluetooth technology *2018 International Conference on Information and Communications Technology (ICOIACT)* 396-400 IEEE

[6] Park Y T, Sthapit P and Pyun J Y 2009 Smart digital door lock for the home automation *TENCON 2009-2009 IEEE Region 10 Conference* 1-6 IEEE

[7] Rangkuti H A and Simatupang J W 2015 Security lock with DTMF polyphonic tone sensor *2015 International Conference on Automation, Cognitive Science, Optics, Micro Electro-Mechanical System, and Information Technology (ICACOMIT)* 119-122 IEEE

[8] Ping W, Guichu W, Wenbin X, Jianguo L and Peng L 2010 Remote Monitoring Intelligent System Based on Fingerprint Door Lock *2010 International Conference on Intelligent Computation Technology and Automation* **2** 1012-1014 IEEE

[9] Sahani M, Nanda C, Sahu A K and Pattnaik B 2015 Web-based online embedded door access control and home security system based on face recognition *2015 International Conference on Circuits, Power and Computing Technologies [ICCPCT-2015]* 1-6 IEEE

[10] Raju N G, Vikas J, Appaji S V and Hanuman A S 2018 Smart Lock Controlled using Voice Call *2018 International Conference on Smart Systems and Inventive Technology (ICSSIT)* 97-103 IEEE

Design and fabrication of double layer bowtie antenna with modification of wing edge as UHF sensor for partial discharge detection in air insulation

J P Uwiringiyimana* and U Khayam

School of Electrical Engineering and Informatics, Institut Teknologi Bandung, Bandung, Indonesia

*upeter1@yahoo.fr

Abstract. Partial discharge (PD) has been used as a reliable diagnostic technique for High voltage equipment. UHF methods have been widely applied for the detection of electromagnetic waves emitted from PD source in High Voltage equipment. EM wave propagation produced by PD can be detected by using UHF sensors. This paper discusses the new design of double layer bowtie antenna with modification of the wing edge of the antenna as a UHF sensor for PD detection in air insulation. The double layer bowtie antenna was designed by using FR4-epoxy substrate type and later on printed on PCB. The 3D-HFSS software was used to design the bowtie antenna. Based on simulation results of double layer bowtie antenna, the optimum design for PD measurement was obtained with a minimum return loss (RL) of -34.7 dB, the bandwidth of 560 MHz, and Voltage Standing Wave Ratio (VSWR) of 1.03. Based on the measurement results for the designed bowtie antenna by using the Vector Network Analyser (VNA), the bowtie antenna has a return loss of -14 dB, the bandwidth of 555 MHz, and VSWR of 1.05. The designed UHF double layer bowtie antenna is capable of detecting PD in UHF range since PD signal have a frequency range of 300 MHz - 3 GHz.

1. Introduction

High voltage insulation system plays a vital role in high voltage equipment used in an electric power system. Electrical insulation has the main role to withstand the high electric field between phase conductors or between phase and neutral conductors [1]. In excessive high electric field due to the appearance of field enhancement sites like void or protrusion, Partial Discharge (PD) may occur. The presence of PD in the electrical insulations may indicate the insulation aging and in the long term, this may further reduce the integrity of the insulation leading to the failure of the equipment. Diagnosis of PD in the early stage is needed to prevent the breakdown of the high voltage equipment. The failure of high voltage equipment may significantly contribute to the failure of the power system. Several key high voltage types of equipment such as the generator, power transformer, GIS, power cables and insulators play an important role in determining the reliability of the power system. The high voltage equipment in general contains gas, liquid or solid insulation [1-2]. Defects and a particular condition may lead to an extremely high electric field. Particularly at interfaces between gas and solid and liquid insulations, the excessive electric field may cause the breakdown of the gas while the liquid and solid insulations are still able to withstand the electric field. This phenomenon is referred as Partial Discharge (PD). Partial discharges in high voltage equipment for long time may degrade the quality of the insulation and at the same time, the appearance of discharges can be used as a diagnostic signal for assessing the condition of the insulation inside the high voltage equipment. Partial discharge has been an important subject investigated by many researchers and discussion in many international conferences as well as international journals or transactions. Diagnosis of PD is considered to highly represent the actual condition of the equipment [1-10]. PD detection is the first step in the power apparatus diagnosis. In order to maintain the reliability and availability of the power system, it is needed to design the online PD measurement system. PD occurring in power apparatus releases energy in various forms such as light emission, acoustic wave, heat, and the electromagnetic wave which can be detected by the suitable sensors [2]. The UHF method is one of the methods used to detect the PD signal. The UHF antennas are used to detect the PD signal in high voltage equipment [2-11]. They can be used as external sensors to detect PD signal online by placing the antennas near the high voltage equipment [2]. Different types of antennas have been developed in previous many kinds of research to detect PD in high voltage equipment [2-9]. PD signal has a frequency range of 300 MHz - 3 GHz [5-11]. Bowtie antenna is one example of the antennas types that have been developed to detect PD in air insulation [5-10]. In this paper, the UHF double layer bowtie antenna with modified wing edge has been designed and fabricated to detect partial discharge in air insulation. The designed double layer bowtie antenna has a minimum return loss value of -34.7 dB and a bandwidth of 560 MHz based on the simulation results. Based on the measurement results by using a vector network analyser, the designed double layer bowtie antenna has a return loss value of -14.7 dB and a bandwidth of 555 MHz.

2. Design and simulation of bowtie antenna

2.1. *Previously designed double layer bowtie antennas*

The new design of double layer bowtie antenna is based on the previous researches on bowtie antennas [8-9]. However, the previous design of bowtie antenna has a smaller bandwidth and greater return loss. The first design was double layer bowtie antenna whose bandwidth is 345 MHz and a return loss of -13.17dB while the second design of bowtie antenna in the previous research was double layer bowtie antenna with edge modification whose bandwidth is 370 MHz and a return loss of -24.7 dB [8-9]. The new design of double layer bowtie antenna with modification of wing edge was designed to have the lower minimum return loss and a wider bandwidth compared to two designs that were previously designed. The designed antenna is the double layer bowtie antenna with a flare angle of 60° and a wing radius of 36 mm. This double layer bowtie antenna has been designed based on the modification of the edge of the antenna. Figure 1 represents the design of the bowtie antenna in the previous research. The design of the bowtie antenna was simulated in a 3D-HFSS software and fabricated on a double-layer PCB with FR4-epoxy substrate material whose relative permittivity 4.4 and the substrate thickness of 1.6 mm.

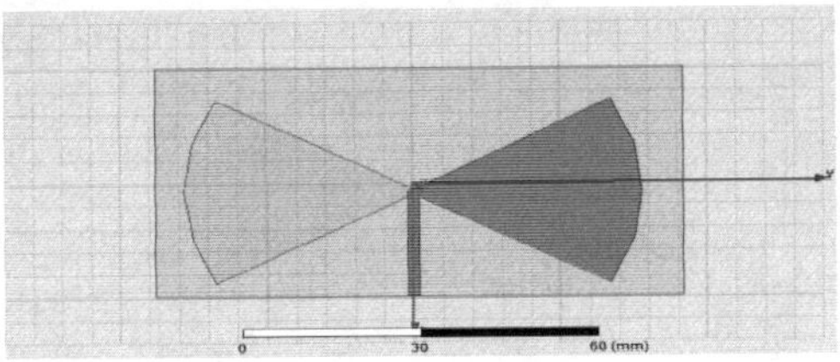

Figure 1. The initial design of double layer bowtie antenna in previous research [8].

Figure 2. Design of bowtie antenna with edge modification in previous research [9].

The initial design of double layer bow-tie antenna in the previous research has the following specification parameters: Wing radius = 40 mm, gap distance = 2.0 mm, flare angle = 60°, substrate dielectric = 4.4, stub = 23 mm.

2.2. Newly designed double layer bowtie antenna with modification of wing edge

The design of double layer bowtie antenna with modified wing edges has been designed based on the initial shape of the bow-tie antenna design in the previous research. The new design of double layer bowtie antenna was simulated with FR4-epoxy substrate material whose thickness is 1.6 mm. The optimal designed double layer bowtie antenna has a flare angle of 60° with wing radius of 36 mm, a rectangular plane substrate of 70 mm by 40 mm and with a stub length of 20 mm. In this study, the modifications were made on the edge of the double layer bowtie antenna by cutting the part of the antenna wing with low current destiny in order to increase current distribution on the surface of the antenna. During the modification of the wing of the bowtie antenna, the edge of the antenna was curved to the extent that the patches of bowtie antenna have a concave form based on the radius of curvature created in the wing of the antenna. The optimal design by modification of the wing edge of the bowtie was achieved by varying the radius of the curvature of the antenna wing. Figure 3 shows the modification of the wing edge of the double layer bowtie antenna based on the initial design of the bowtie antenna in the previous research.

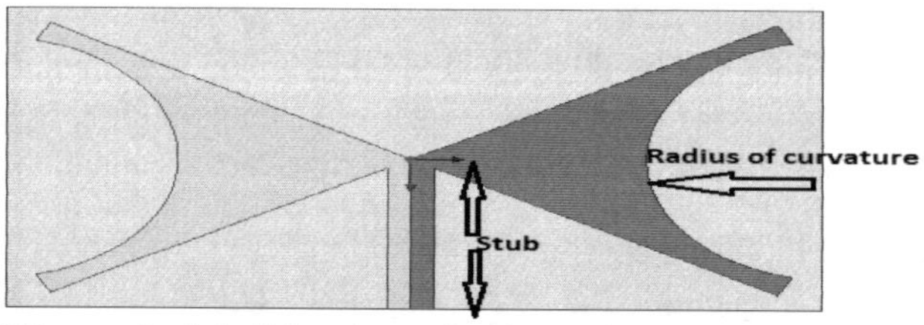

Figure 3. Modification of wing edges of the double layer bowtie antenna.

2.2.1. Optimization of wing radius for double layer bowtie antenna. The designed double layer bow-tie antenna with wing edge modification was designed by optimizing the wing radius where the optimal design has a wing radius of 36 mm instead of 40 mm of the initial design in the previous research [8]. The variation of wing radius has influenced return loss and bandwidth of the bowtie antenna. Figure 4 represents the optimization of wing radius based on the variations from 32 mm, 34 mm, 36 mm and 40 mm against return loss value, bandwidth and resonance frequency for each wing radius value used in the simulation. The effect of changes in wing radius is seen in the width of the bandwidth produced as well as in return loss of the antenna. The optimal design of double layer bowtie antenna with a modified wing edge is achieved by using a wing radius of 36 mm long. This wing radius of 36 mm gives a return loss (RL) value of -34.7 dB, the bandwidth of 560 MHz at a resonance frequency of 2.4 GHz. Table 1 shows the RL, bandwidth and resonance frequency of four samples of wing radius used in the simulation.

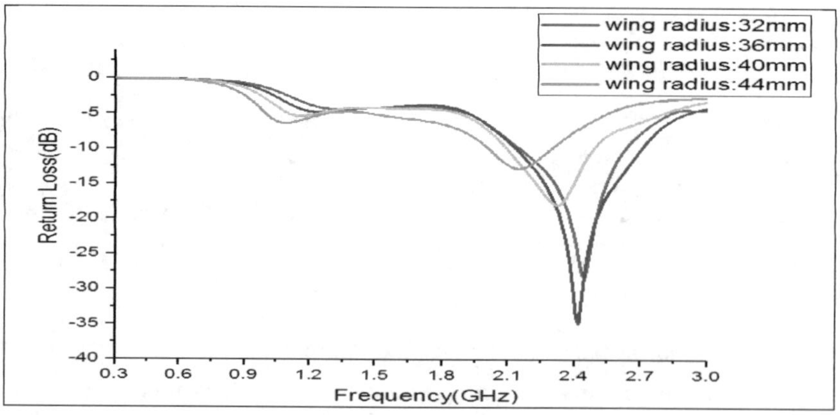

Figure 4. Wing radius against return loss (RL) value.

Table 1. Simulation results based on variation of wing radius.

Wing radius	***Bandwidth(MHz)***	***Return Loss(dB)***	**Resonance Frequency (GHz)**
32 mm	459	-28.2	2.4
36 mm	560	-34.7	2.4
40 mm	378	-17.9	2.3
44 mm	270	-12.7	2.14

2.2.2. Optimization of stub length for double layer bowtie antenna. After optimization of the wing radius, the next step was to find the optimal design by variations of stub lengths. Figure 5 shows the variations of stub length from 20 mm, 23 mm, 25 mm, and 27 mm. The stub length of 20 mm gives optimal result compared to other stub lengths used in the simulation. As it can be seen from the simulation results based on the variation of stub length of the antenna in Figure 5, the optimal design of double layer bowtie antenna with modification of antenna tip was obtained by using the stub length of 20 mm due to its minimum return loss (RL) of -34.7 dB with a corresponding bandwidth of 560 MHz. Table 2 shows the RL, bandwidth and resonance frequency of four samples of stub length used in the simulation.

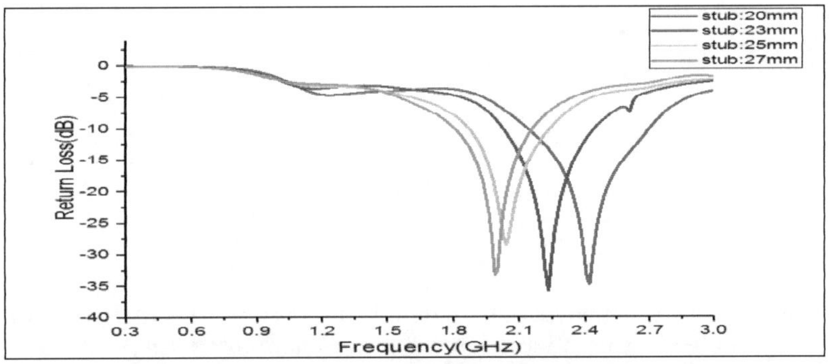

Figure 5. Stub length against return loss (RL) value.

Table 2. Simulation results based on variation of stub length.

Stub length	***Bandwidth (MHz)***	***Return Loss (dB)***	**Resonance Frequency (GHz)**
20 mm	560	-34.7	2.4
23 mm	410	-35.9	2.23
25 mm	360	-28.6	2.04
27 mm	350	-34.2	2.0

2.2.3. Variation of the radius of curvature for double layer bowtie antenna. The modification of double layer bowtie antenna was conducted based on the variation of radius of curvature of the patch of the antenna. The radius of curvature used in the simulation of double layer bowtie antenna is of four samples namely, 16 mm, 16.5 mm, 17 mm, and 17.5 mm. The simulation results for double layer bowtie antenna based on the modification of the wing edges are shown in Figure 6. The radius of curvature of 17.5 mm has demonstrated a significant optimal design where the associated minimum RL is -34.7 dB, the bandwidth of 560 MHz and resonance frequency of 2.4 GHz. Table 3 shows the RL, bandwidth and resonance frequency of four samples of the radius of curvature used in the simulation.

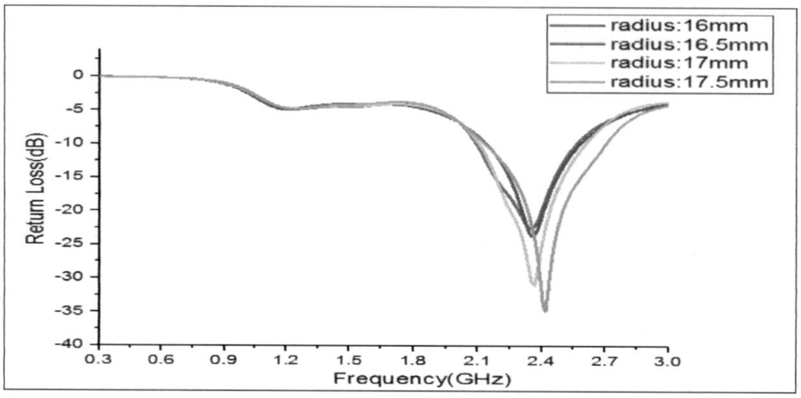

Figure 6. Radius of curvature against return loss (RL).

Table 3. Simulation results based on variation of radius of curvature.

Radius of curvature	*Bandwidth(MHz)*	*Return Loss(dB)*	Resonance Frequency (GHz)
16mm	450	-22.5	2.36
16.5mm	420	-23.7	2.36
17mm	490	-31	2.35
17.5mm	560	-34.7	2.4

2.2.4. Current distribution for double layer bowtie antenna. The modification of the bowtie antenna was performed based on the analysis of the current distribution on the antenna surface. The edge of the bowtie antenna had the lowest current density. The bowtie antenna has been modified by cutting the edge of antenna wing which was having a low current distribution. This was done in such a way that the antenna patch has a concave form in order to have a uniform current density distribution on the surface of the antenna. To make a concave form in the antenna wing, a radius of curvature was created in the wing of the antenna and four variations of the radius of curvature have been used in simulation as it is indicated in Figure 7. The radius of curvature of 17.5 mm produces high current distribution compared to radius of curvatures of 17 mm, 16.5 mm, and 16mm.

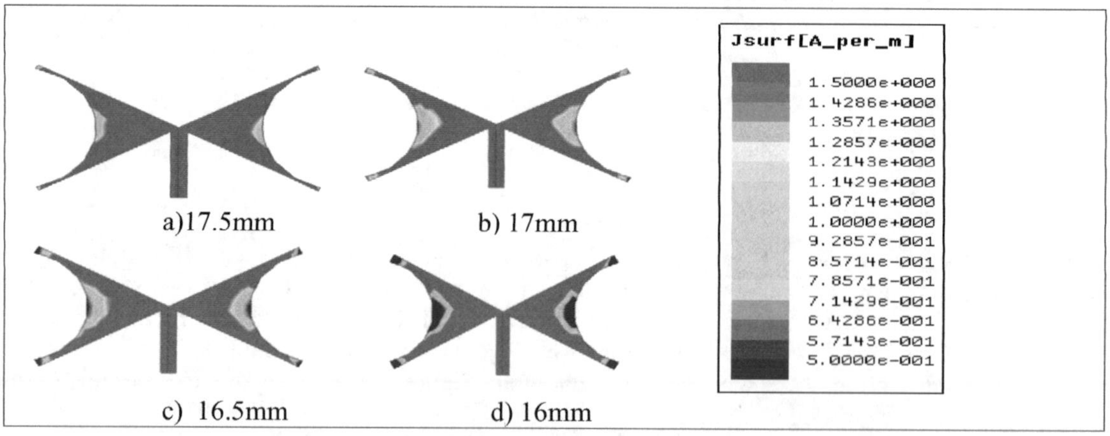

Figure 7. Current distribution for the bowtie antenna with a variation of radius of curvatures.

Based on the simulation results for the double layer bowtie antenna with modification of the wing edge of the antenna with optimization of wing radius, stub length and radius of curvature, the following specifications parameters for the double layer bowtie antenna were confirmed for the optimal design as follows: Wing radius = 36 mm, gap distance = 2 mm, flare angle = 60°, dielectric substrate = 4.4, stub length = 20 mm, radius of curvature =17.5 mm, substrate material: FR4 epoxy, substrate thickness = 1.6 mm, substrate form = Rectangular, Substrate dimensions = 70 mm x 40 mm.

3. Fabrication of double layer bowtie antenna with wing edge modification

After design and simulation of the double layer bowtie antenna with wing edge modification, the next stage was to fabricate the bowtie antenna. The double layer bowtie antenna was fabricated on the double layer printed circuit board (PCB) with FR4-epoxy substrate. Figure 8 shows the fabricated UHF double layer bowtie antenna for PD detection.

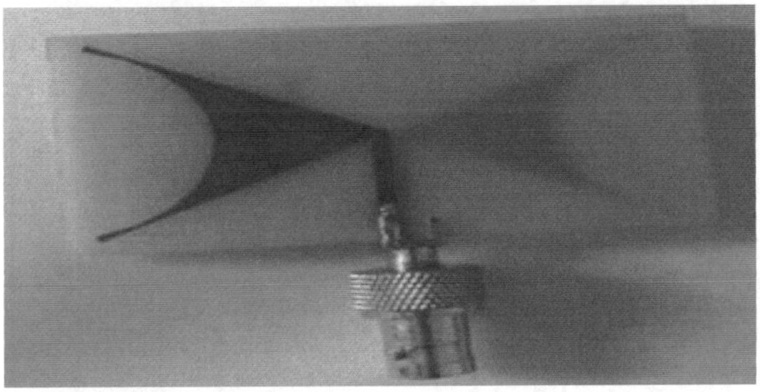

Figure 8. Fabricated double layer bowtie antenna.

4. Testing the designed bowtie antenna by using Vector Network Analyser

After fabrication of double layer bowtie antenna with modified wing edges, the next stage was to test the designed bowtie antenna by using the vector network analyser (VNA). The purpose of this testing was to measure some of the antenna parameters such as return loss (RL), bandwidth, vector standing wave ratio(VSWR) and resonance frequency. Figure 9 shows the measurement setup with the VNA. Table 4 shows the comparison between simulation results by using HFSS and the measurement results obtained by using the VNA. Figure 10 shows the comparison between simulation and measurement results of return loss for the designed bowtie antenna while Figure 11 shows the comparison between simulation and measurement results of VSWR for the designed bowtie antenna.

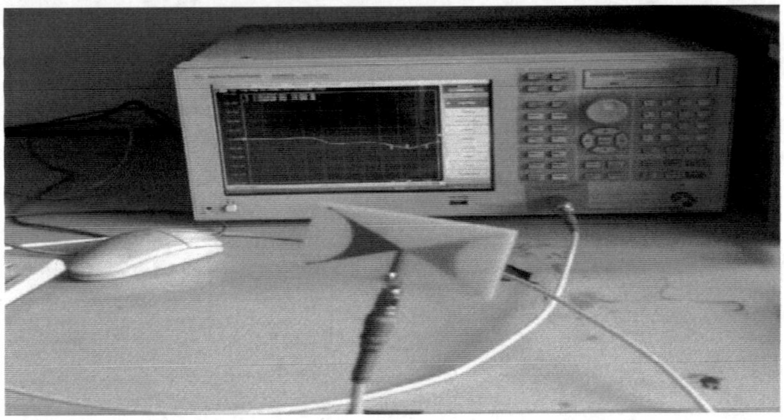

Figure 9. Measurement setup for the antenna with VNA.

Table 4. Comparison between simulation results and measurement results for the designed bowtie antenna.

Results	*Bandwidth(MHz)*	*Return Loss(dB)*	VSWR	Resonance Frequency (GHz)
Simulation	560	-34.7	1.03	2.4
Measurement	555	-14	1.5	2.5

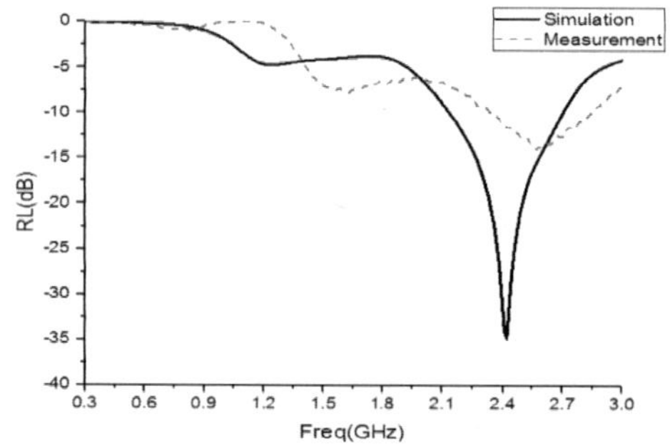

Figure 10. Comparison of simulation and measurement of RL for a designed bowtie antenna.

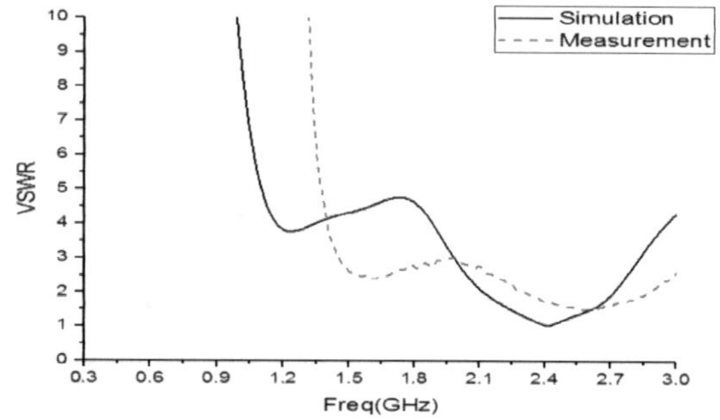

Figure 11. Comparison of simulation and measurement of VSWR for a designed bowtie antenna.

5. Partial discharge detection in air insulation by using the designed double layer bowtie antenna

After testing the designed double layer bowtie antenna by using the VNA, the antenna was implemented in detecting the PD in air insulation. The needle-plate electrode model was used as PD source to generate the electromagnetic waves captured by the designed double layer bowtie antenna placed near the PD source. The experiment for PD measurement in air insulation was performed at three different applied voltage levels namely, 5 kV, 6 kV and 7 kV. Figure 12 shows the PD measurement circuit used to conduct the experiment.

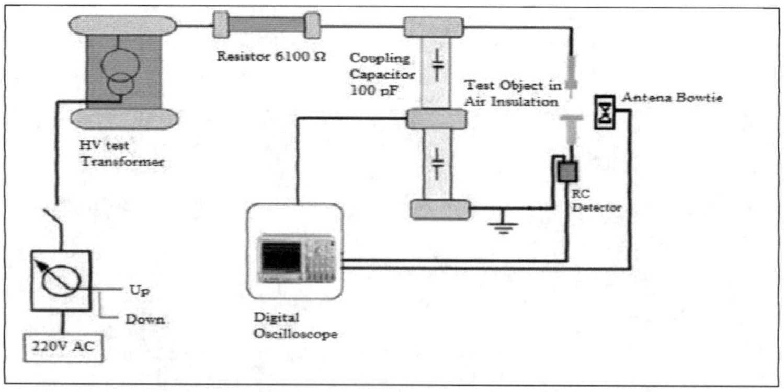

Figure 12. PD measurement circuit.

5.1. Background noise

Background noise is the amount of noise present in the laboratory before changing gradually the applied voltage. The measurement of background noise is conducted in order to obtain accurate data during the PD measurement experimental. There are two kinds of background noise, namely background noise OFF (BGN OFF) which is observed on the oscilloscope when the power source is still in OFF mode and background ON that is found on the oscilloscope when the power source is switched ON. The background noise is the number of noise signals that can originate from external systems or from the internal system itself of the measurement circuit. The background noise observed on the oscilloscope comes from the environment around the experimental circuit, for example, from the car engines or due to from electromagnetic interference of some equipment in the laboratory. The background noise measurement is conducted before the PD experiment to distinguish the noise signal from PD signal [3]. Figure 13 and 14 shows the background noise ON for the designed antenna and for the RC detector (comparison sensor). The peak-to-peak voltage (Vpp) magnitude for BGN ON detected designed double layer bowtie antenna and RC detector are 24 mV and 12 mV respectively.

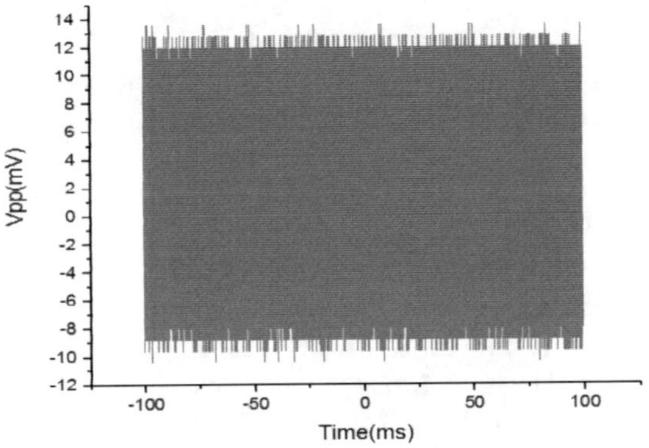

Figure 13. BGN ON for bowtie antenna.

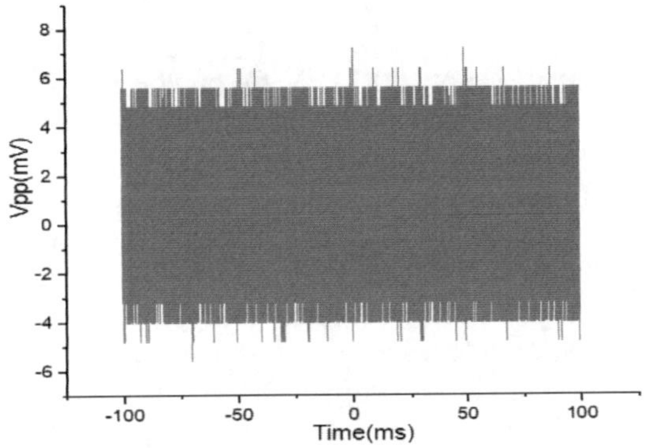

Figure 14. BGN ON for RC detector.

5.2. Partial discharge inception voltage

The partial discharge inception voltage (PDIV) is the possible minimum voltage needed to make the first PD signal to appear. Figure 15 and 16 show the negative PDIV (first PD which occurs in a negative half cycle) while the Figure 17 and 18 show the positive PDIV (first PD signal which occurs in a positive half cycle) for both antenna and RC. The Vpp magnitude for negative PDIV detected by double layer bowtie antenna and RC detector are 15.2 mV and 33.3 mV respectively while the Vpp magnitude for positive PDIV detected by double layer bowtie antenna and RC detector are 36.35 mV and 188.3 mV respectively. Alternatively, the needed applied voltage for negative PDIV detected by double layer bowtie antenna and RC detector are 3.36 mV and 2.84 mV respectively and the applied voltage for positive PDIV are 4.4 mV and 4.0 mV respectively.

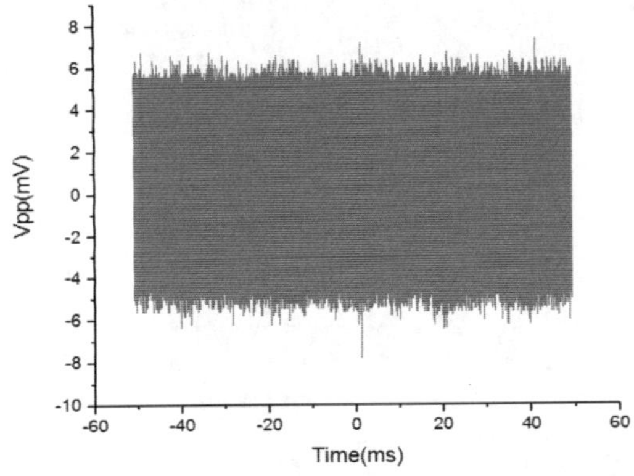

Figure 15. Negative PDIV for bowtie antenna.

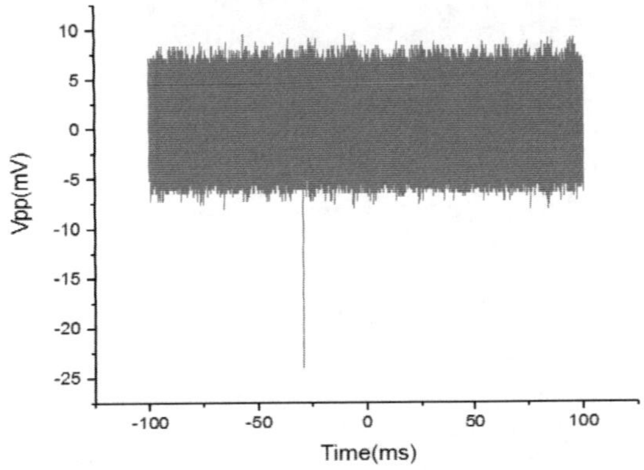

Figure 16. Negative PDIV for RC detector.

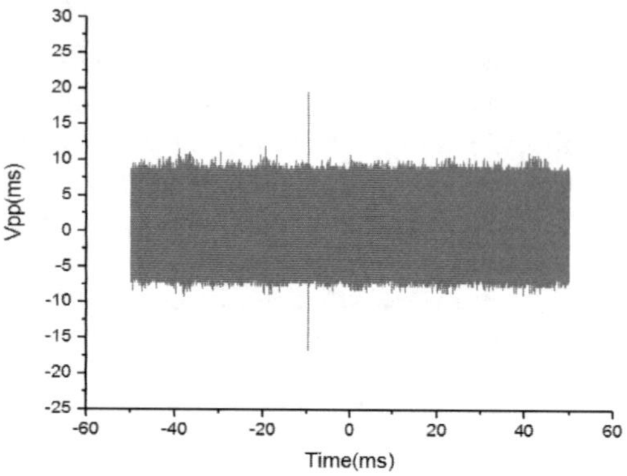

Figure 17. Positive PDIV for bowtie antenna.

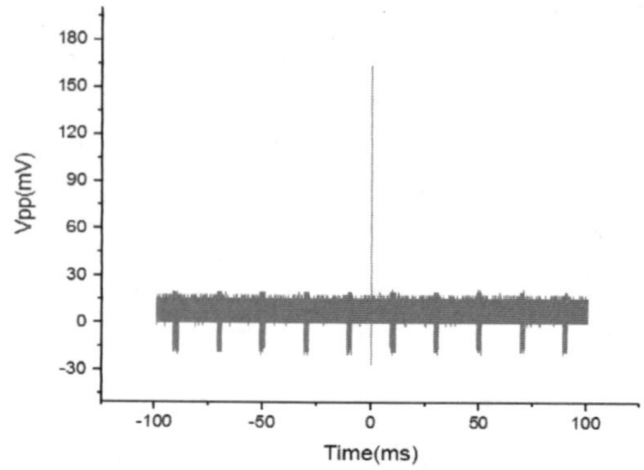

Figure 18. Positive PDIV for RC detector.

5.3. *Partial discharge waveform*

The PD waveform is another PD parameter that was measured during the experiment. Figure 19 and 20 show the shape of negative PD waveform while Figure 21 and 22 show the positive PD waveform detected by the designed antenna and RC detector. The PD measurement data was performed at 5 kV. As it can be seen in figures below, the Vpp magnitude for negative PD waveform detected by the designed double layer bowtie antenna and RC detector are 10 mV and 27.4 mV respectively while the Vpp magnitude for positive PD waveform detected by the designed double layer bowtie antenna and RC detector are 32 mV and 437.2 mV respectively.

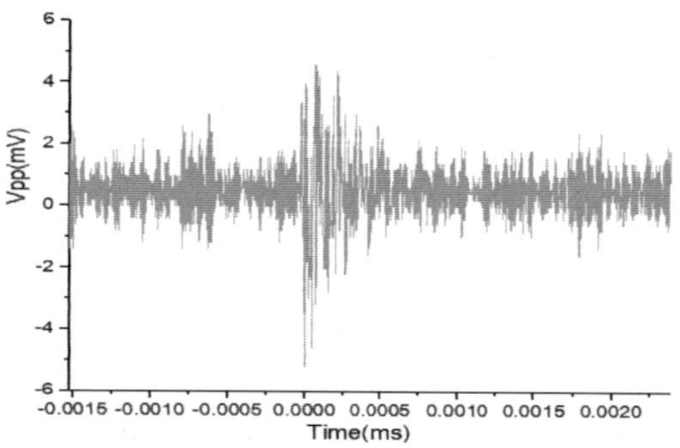

Figure 19. Negative PD waveform detected bowtie antenna .

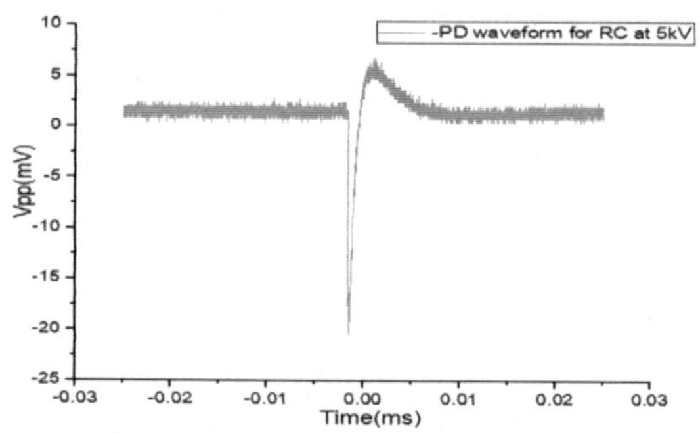

Figure 20. Negative PD waveform by detected RC detector.

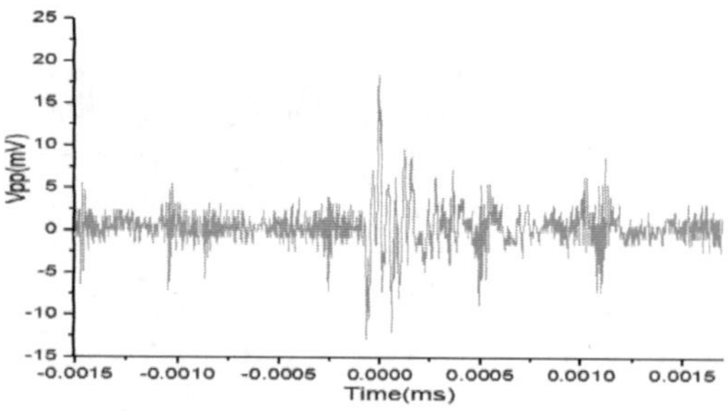

Figure 21. Positive PD waveform detected bowtie antenna.

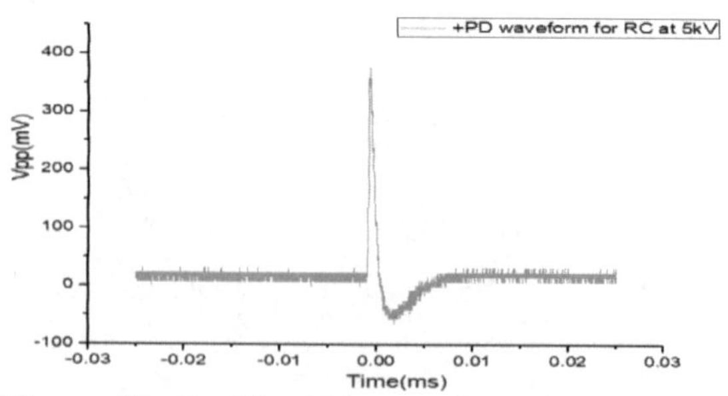

Figure 22. Positive PD waveform by detected RC detector.

5.4. *Partial discharge patterns*

The PD patterns is another PD characteristic which indicates the PD charge magnitude, PD phase and PD number. In other words, it is a graph ϕ - q - n that shows the appearance of the phase location

IOP Conf. Series: Materials Science and Engineering **850** (2020) 012015 doi:10.1088/1757-899X/850/1/012015

where PD occurs and the charge magnitude of the PD occurrence. Figure 23 and 24 show the PD patterns detected by the designed bowtie antenna and RC detector at 5 kV.

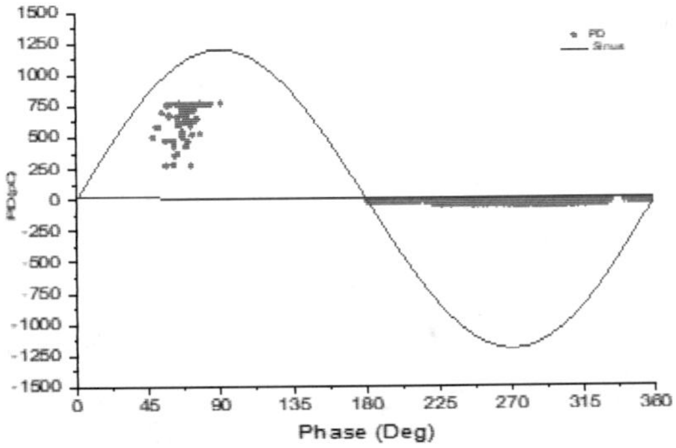

Figure 23. PD patterns by bowtie antenna.

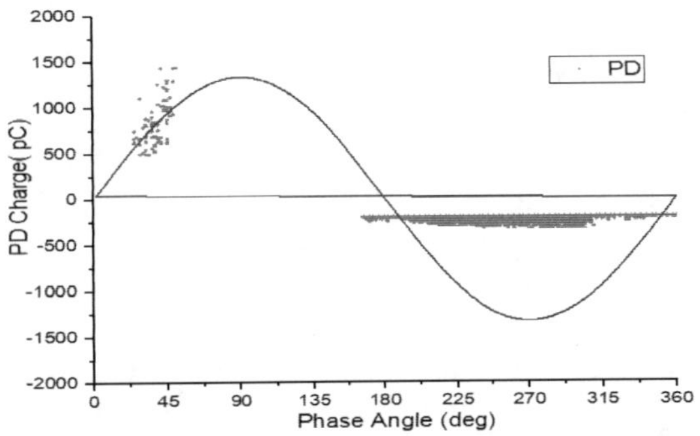

Figure 24. PD patterns by RC detector.

6. Discussion and analysis

The designed double layer bowtie antenna with modified wing edges was designed based on the initial design in the previous research [8-9]. The optimal design for double layer bowtie antenna has a wider bandwidth of 560 MHz, a minimum return loss of -34.7 dB and a resonance frequency of 2.4 GHz based on the simulation results while in the previous research, the double layer bowtie antenna that was designed had only a bandwidth of 345 MHz and a return loss of -13.17 dB, and a resonance frequency of 1.8 GHz. It is clearly seen that the new design of double layer bowtie antenna has been improved to have better sensitivity for PD detection in air insulation in the ultra-high frequency range, 300 MHz - 3 GHz [11]. After design and simulation for the new design of the bowtie antenna, the antenna was fabricated on double layer PCB with FR4-epoxy substrate. After fabrication of the double layer bowtie antenna with modification of wing edges, the antenna was tested by using the vector network analyser (VNA) to measure some antenna parameters such as, return loss (RL), bandwidth, VSWR and resonance frequency of the antenna. The antenna measurement by using VNA as shown that the designed antenna has a bandwidth of 555 MHz, return loss of -14 dB and a resonance frequency of 2.5 GHz. There has been a little bit difference between simulation results by using HFSS and measurement results by using VNA due to some errors that might have happened during the fabrication of the antenna. The fabricated antenna was applied in detecting PD in air insulation by using RC detector as a comparison sensor. The designed double layer bowtie antenna was used to measure BGN, PDIV, PD waveform and PD phase patterns. According to the PD phase patterns, the designed double layer bowtie antenna detect PD signals in the same patterns as the RC detector. According to the PDIV measurement, it can be said that the RC detector detects the first PD signal earlier than the designed double layer bowtie antenna because RC detector needs lower applied voltage for both negative PDIV and positive PDIV compared to the double layer bowtie antenna which needs a little bit higher voltage to detect negative PDIV. This indicates that the RC detector is more sensitive than the double layer bowtie antenna because RC detector is directly connected to the PD measurement circuit and it uses the electrical method while the bowtie antenna is not connected to the circuit, it is only placed near the PD source to detect the induced EM waves for PD source. So, the two sensors use a different method in detecting PD. Bowtie antenna uses a UHF method (non-electric) while the RC use an electric method. This is one of the reasons that makes RC detector more sensitive than an antenna.

7. Conclusion

The designed double layer bowtie antenna with modification of wing edges has proven to detect PD in air insulation. The radius of curvature created in the wing of the bowtie antenna has influenced the antenna characteristics, namely return loss (RL), bandwidth and vector standing wave ratio (VSWR). During the PD measurement, the designed double layer bowtie antenna has proven to detect PD in air

IOP Conf. Series: Materials Science and Engineering **850** (2020) 012015 doi:10.1088/1757-899X/850/1/012015

insulation in the same PD patterns as the RC detector except for a little bit different in the sensitivity. The RC detector has higher sensitivity compared to the designed double layer bowtie antenna since RC detector uses an electrical method in detecting PD while Antenna uses UHF method.

References

[1] Suwarno 2016 *Int. J.of Electrical Engineering and Informatics* **8** 147163
[2] Khayam U 2017 *4thInt.Conf.on Electrical Engineering,Computer Science and Informatics* (Yogyakarta: IEEE)
[3] Uwiringiyimana J P and Khayam U 2017 *Int. Conf. on High Voltage Engineering and Power Systems* (Bali: IEEE) pp. 35–39
[4] Hanalde A and Khayam U 2013 *Int. Conf. on Information Technology and Electrical Engineering* (Yogyakarta: IEEE) pp. 365-369
[5] Joko M, Suwarno, Khayam U and Masayuki H 2013 *Int. Conf. on Electrical Engineering and Informatics* (Malaysia: IEEE)
[6] Joko M, Suwarno, Khayam U and Masayuki H 2013 *48th Universities' Power Engineering Conference* (Dublin: IEEE)
[7] Khusnul K, Khayam U, Suwarno, Yuki T, Masahiro K and Masayuki H 2015 *5th Int. Conf.on Electrical Engineering and Informatics* (Bali: IEEE)
[8] Farrah V, Suwarno and Khayam U 2014 *Int.Conf.on Sensor, Sensor System, and Actuat*or (Bnadung: IEEE)
[9] Asep A and Khayam U 2014 Int.Conf.on Smart Green Technology in Electrical and Information Systems (Bali: IEEE) pp. 98-101
[10] Inu S, Khayam U, Suwarno, Kiichi N, Mohamad K, Masahiro K and Masayuki H 2014 *Int. Symp on Electrical Insulating Material* (Japan: IEEE) pp. 330-333.
[11] Judd, M D,Yang, L, Hunter and I 2005 *IEEE Electr. Insul. Mag*. pp 5–14.

GEMONG: A blind drinking aid

D Rustandi*, R M Fauzi, L A Purwanto and Y Somantri

Faculty of Technology and Vocational Education, Universitas Pendidikan Indonesia, Jl. Dr. Setiabudhi 229 Bandung, West Java, Indonesia

*dickrust30@upi.edu

Abstract. The purpose of this study was to design the GEMONG (A Blind Drinking Aid) tool. This device provides information in the form of sound to the blind when measuring water into the glass, starting to fill the water up to the size of the glass by making a sound on the loudspeaker programmed through the Arduino microcontroller. This glass is designed portable and can be washed as desired. The material used to make glass comes from stainless steel with PLA as a mechanical manufacture for the system using a 3D printing machine.

1. Introduction

Blind people are those who have no vision at all and those who still have residual vision but cannot use their vision to read 12 normal points of writing under normal light conditions even with the aid of glasses [1]. In daily life, blind people have difficulty in carrying out their activities. They rely on their senses of touch and hearing, or depend on their lives for the people around them to carry out their daily activities.

One of the problems faced by blind people is difficulty in eating or drinking [2]. Especially for drinking purposes, now there are many dispensers that help blind people to distinguish between cold and hot water so they can avoid the danger of hot water coming from the dispenser. However, this dispenser also still has a problem, which is not able to know whether or not the water is full in the glass. So if it is too full then water can certainly be spilled.

With the help of technology, these problems can be overcome so that blind people can take drinking water comfortably. Previously there had been technological innovations for the visually impaired, namely the modification of drinking water dispensers (minerals) that could greatly facilitate its users, especially for blind people [3]. Other examples namely a device for assisting the blind who are guided via RFID [4], and how to gather information, tips and indications on interaction with a touch-screen by blind users [5].

In this study we designed the glass as the latest technological innovation specifically for the blind so that it is easier and able to drink independently. This technology can tell by human voice if the water in the glass is in the maximum limit. The tool is called GEMONG (A Blind Drinking Aid). With this tool, it is hoped that blind people will not worry about spilling if they fill water into a glass. This tool is considered useful and practical use for drinking needs for the blind.

2. Methods

Some components used in the design of GEMONG, including Arduino Nano as a microcontroller, Dfplayer as an audio player module that transmits audio files from SD cards to Arduino Nano, speakers as audible outputs, load cels as sensors to detect the weight and amount of water. The design of the implementation of this research refers to the stages of the research and development approach. This approach is used to produce certain products, and test the effectiveness of these products [6]. Figure 1 shows the implementation method.

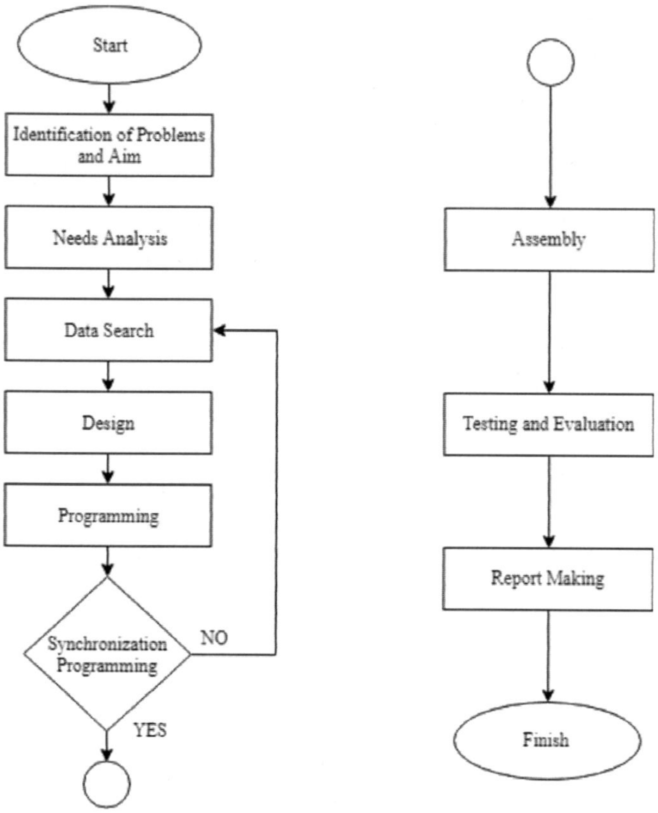

Figure 1. Implementation method flowchart.

The first thing to do is identify the problem and then analyse the needs. In designing this tool, several components are needed to make tools such as microcontrollers, sensors, polylactic acid, etc. Next is the search for data sheets for the needs of the appropriate specifications to be used in this tool. Tool design consists of mechanical design and hardware design. Glass is designed to be able to accommodate a set of hardware and glasses that can be stored and removed from a mechanical system. Figure 2 is the mechanical design of GEMONG.

Figure 2. GEMONG design.

In the process of designing hardware, electrical circuit design is required. The hardware design process includes the breadboard design. This breadboard design will be carried out as a work test of the device or component that has been prepared before installation directly to the tool.

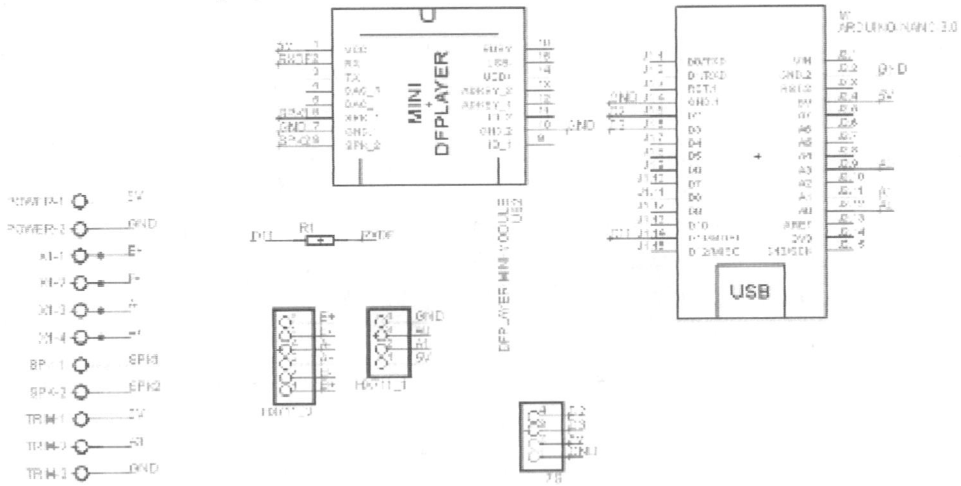

Figure 3. Hardware schematic.

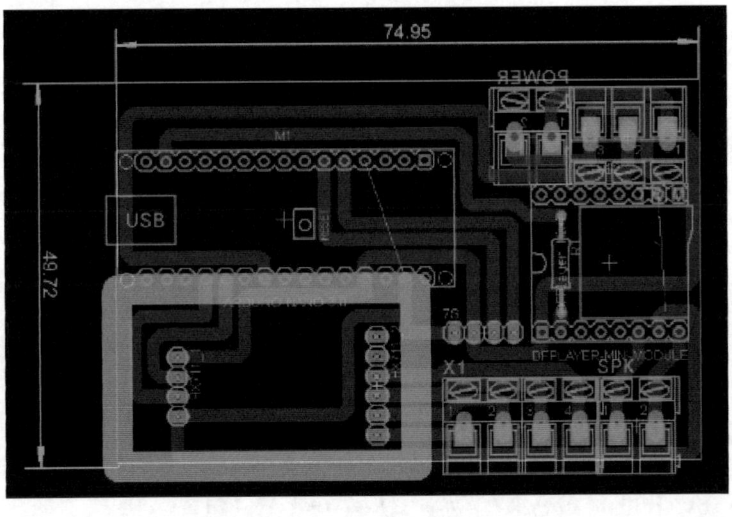

Figure 4. Hardware layout.

Use of this program aims to provide programs or commands on the device where the making of this program is done using Arduino software. The device can function if the software and hardware can run properly and synchronously. If there are still errors, it will be checked again in software and hardware to get the desired results and make sure the tools used are functioning.

Next is assembly. Tool assembly is done after the PCB layout and tool component assembly with the aim of completing the manufacture of the tool as planned. Meanwhile, test and evaluate the device by looking at the effectiveness and performance of the use that this tool is functioning properly. Shortcomings that are still found in this system will be an issue for evaluation.

3. Results and discussion

Tests were carried out on GEMONG performance. Testing is done by testing the suitability of performance between design and results. The following is a table of the results of tests conducted. Some conditions used as experiments include testing the Turn ON button, Start filling water, the water is half a glass, the water is full, Glass is lifted and water is drunk, and Glass saved. In general, the test results show that the performance of the tool can meet the specifications designed.

Table 1. Experimental result.

No	Condition	Trial									
		1	**2**	**3**	**4**	**5**	**6**	**7**	**8**	**9**	**10**
1	Turn Button ON	√	√	√	√	√	√	√	√	√	√
2	Start filling water	√	√	√	√	√	√	√	√	√	√
3	The water is half a glass	X	X	√	√	√	√	√	√	√	√
4	The water is full	X	X	X	√	√	√	√	√	√	√
5	Glass is lifted and water is drunk	X	X	X	√	√	√	√	√	√	√
6	Glass saved	X	X	X	√	√	√	√	√	√	√

4. Conclusion

GEMONG provides information in the form of sound to blind people when measuring water into glass from starting to fill up to the size of the glass there will be sound coming out through the speakers. GEMONG can also be carried everywhere, refilled with a charger, and washed as desired.

References

[1] Hallan D P, Kauffman J M and Pullen P C 2009 *Exceptional Learners (11th ed)* (USA: Pearson Education, Inc.)

[2] Pertuni 2014 *Persatuan Tunanetra Indonesia* [Online] Retrieved from: http://pertuni.idp-europe.org/. 21 September2018 (16:00)

[3] Scholl G T 1986 *What Does it Mean to be Blind? Definitions, Terminology, and Prevalence. InG.T. Scholl (Ed.), Foundations of Education for Blind and Visually Handicapped Children andYouth: Theory and Practice (pp. 23-33)* (New York: AFB Press)

[4] Tanveer M S, Hashem M M and Hossain M K 2015 Android assistant EyeMate for blind and blind tracker *2015 18th international conference on computer and information technology (ICCIT)* IEEE

[5] Chiti S and Leporini B 2012 Accessibility of android-based mobile devices: a prototype to investigate interaction with blind users *International Conference on Computers for Handicapped Persons* (pp. 607-614) (Berlin, Heidelberg: Springer)

[6] Haryati S 2012 Research and Development (R&D) sebagai salah satu model penelitian dalam bidang pendidikan *Majalah Ilmiah Dinamika* **37**(1) 15

IOP Conf. Series: Materials Science and Engineering **850** (2020) 012017 doi:10.1088/1757-899X/850/1/012017

Development of remote laboratory based on HTML5

A T Nugraha and E Haritman*

Department of Electrical Engineering Education, Faculty of Technology and Vocational Education, Universitas Pendidikan Indonesia, Jalan Dr. Setiabudhi No.229 Bandung 40154

*erikharitman@upi.edu

Abstract. The Laboratory is an important of means for engineering students and the researchers to do various kinds of experiments. Amount of equipment and time to be able to access the laboratory is limited. therefore, in this research developed of remote labs that can be accessed via web anytime and anywhere as long as available internet network. This is in line with the development of Industry 4.0 which introduces *cyber physical* technology based on internet or wireless network. This technology is not only used for communication but also includes controls and remote control. The results of this research showing the remote lab can be accessed by web via internet and via localhost and compatible with web browser as *google chrome, Mozilla, opera* etc.

1. Introduction

Remote Lab (Rlab) is a future of laboratory, because it provides opportunities for effective and holistic learning for students and researchers to provide the flexibility and additional time needed to complete an experiment or exciting breakthrough [1].

Rlab is also one of online learning, in this case practical learning can access physical laboratories via the internet (*Cyber Physical*). This learning in the line with development of industry 4.0 who introduce a system *cyber-physical* based on internet or *wireless* network, this technology is not only for communication but includes controls and remote control [2]. Rlab also one of the open educational resources, interactive. The main purpose of this Rlab is to provide a real physical laboratory from afar.

Universities such as the University of Deusto made the Rlab Called WEBLAB DEUSTO, Sydney University of Technology made LABSHARE (SAHARA) and the Massachusetts Institute of Technology create iLAB (ISA) accessible to everyone in the world. Some Rlab had been cross-platform means that can be accessed in different devices such as PC, Laptops even Smartphones and most client server communications are HTTP-based.

Universitas Pendidikan Indonesia (UPI) has an Rlab which is placed at the Industrial Electronics Laboratory (Rlab Elind) which can be used for industrial automation practice remotely. The Rlab model used is still application-based, where the user must install the third-party application, TeamViewer. Events where a PC or computer device that exists in a particular location, can be accessed and used from a different place called the Remote Desktop [3]. TeamViewer is one of the Remote Desktop applications that is compatible to remote a PC. As stated earlier that in order to be able to remotely, the user must install TeamViewer first. So, Rlab Elind becomes less flexible in terms of access. To resolve these problems, researchers will develop Remote labs based on HTML5.

IOP Conf. Series: Materials Science and Engineering **850** (2020) 012017 doi:10.1088/1757-899X/850/1/012017

2. Methodology

The study was conducted using the waterfall model which has saveral sequential steps, I.e : *requirement design system, coding* and *Testing*.

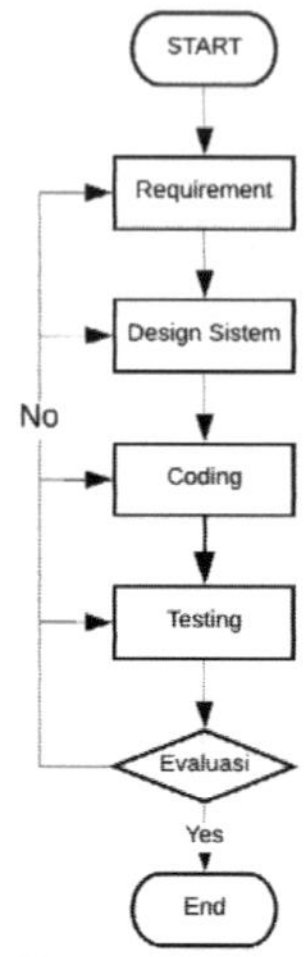

Figure 1. Research methodology.

Description on figure 1:

- Requirement, conducting literature study and identification of problems related to the development of Lab based on HTML5.
- Design system, make the Rlab architecture that will be developed.
- Coding, modify from the thinVNC a remote desktop based on HTML5.
- Testing, do a trial by user from Rlab that has been made and do the testing using Apache Jmeter™.

2.1. Architecture of remote lab based on HTML5

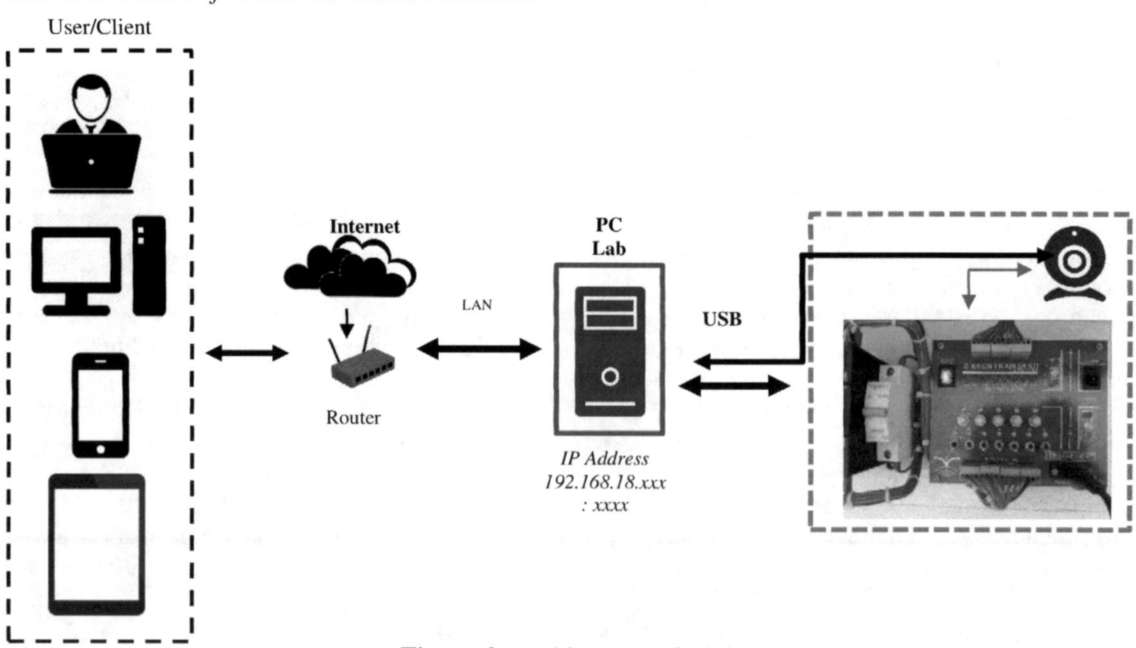

Figure 2. Architecture of Rlab.

IOP Conf. Series: Materials Science and Engineering **850** (2020) 012017 doi:10.1088/1757-899X/850/1/012017

The general architecture of an Rlab depends on the experiments carried out with a structure consisting of a computer and hardware connected to the computer such as a webcam, microphone and feedback from the result of the experiment back to the computer [4]. Figure 2, shows the architecture of RLAB that is being developed based on HTML5, where *users* can use Laptop, PC, Netbook or smartphone to remote the PC in lab then the webcam is used to display changes in conditions from the PLC.

Medium of long distance communication between the *user* and PC lab use via the internet while connecting the two platforms is needed a *Remote Desktop* (RD). from figure 2 the PC Lab connected to the router so that it has an IP Address : 192.168.18.xxx : xxxx (port), and then the PLC used in the Rlab is Omron CP1L and the webcam used to see changes in PLC conditions after the program has been loaded by the user.

2.2. Remote desktop

ThinVNC is a remote desktop based on HTML5 by taking full control or by sharing their windows desktop. This application supports of Internet *Explorer 9, Firefox, Chrome, Safari,* and *web browser* other HTML5 abilities. *IE8* and previous versions can be upgraded with HTML5 features by adding the Frame Chrome plug-in [5]. *Remote Desktop* (RD) used in this research is thinVNC. Through the application RD of thinVNC, Rlab Industrial Electronics can be accessed by users with real-time anytime and anywhere without the distance and time limit as long as the user is connected to the internet through electronic gadgets such as: *smartphone*, tablet, *notebook,* or the other electronic gadgets that support and have web browser applications [6,7].

3. Results and discussion

In this section will discuss the testing Remote Desktop is accessed through the web. There are several tests that are carried out, namely comparing access via the Internet to localhost, testing based on the number of users who access the web, testing based on network traffic and check browser compatibility. The tests uses *software* Apache JMeter™, JMeter is a Java-based *open source* application that can be used for performance test. JMeter usually used to perform *load/stress testing Web Application, FTP Application dan Database server test*. The following are the results of testing using Apache JMeter™:

3.1. Testing number of users

Testing the number of users is done in stages where the first test is by 1 user, both 2 users, 3 users and so on. The user here is a virtual user from *software* Apache JMeter™, The following is the result of testing the number of user:

Table 1. Access 1 *user* via internet and localhost.

Number of Users	Time	Sample Time (ms)	Status	Bytes	Sent Bytes	Latency	Connect Time (ms)
Via Internet							
1	09 : 19	4879	ok	5426	124	4879	3556
Localhost							
1	09 : 19	22	ok	5426	118	9	3

Table 2. Access 2 *user* via internet and localhost.

Number of Users	Jam	Sample Time (ms)	Status	Bytes	Sent Bytes	Latency	Connect Time (ms)
Via Internet							
1	09 : 21	787	ok	5426	124	787	259
2	09 : 21	826	ok	5426	124	826	277
Localhost							
1	09 : 21	28	ok	5426	118	7	2
2	09 : 21	9	ok	5426	118	7	2

Table 3. Access 3 *user* via internet and localhost.

Number of Users	Jam	Sample Time (ms)	Status	Bytes	Sent Bytes	Latency	Connect Time (ms)
Via Internet							
1	09 : 30	1316	ok	5426	124	1316	436
2	09 : 30	1093	ok	5426	124	1091	387
3	09 : 30	4103	ok	5426	124	4092	3303
Localhost							
1	09 : 30	878	ok	5426	118	812	12
2	09 : 30	447	ok	5426	118	447	6
3	09 : 30	419	ok	5426	118	419	5

Table 4. Access 5 user via internet and localhost.

Number of Users	Jam	Sample Time (ms)	Status	Bytes	Sent Bytes	Latency	Connect Time (ms)
Via Internet							
1	13 : 41	926	ok	5426	124	926	336
2	13 : 41	868	ok	5426	124	868	303
3	13 : 41	788	ok	5426	124	788	262
4	13 : 41	929	ok	5426	124	929	278
5	13 : 41	3842	ok	5426	124	3842	3296
Localhost							
1	13 : 41	12	ok	5426	118	9	5
2	13 : 41	13	ok	5426	118	10	4
3	13 : 41	12	ok	5426	118	12	5
4	13 : 41	12	ok	5426	124	9	5
5	13 : 41	42	ok	5426	124	42	13

3.2. Access via the internet and localhost by time

Testing internet access using Ngrok, a proxy server to create or open a private network through NAT or a firewall to connect localhost with a secure tunnel. So when testing using JMeter then, the address from Ngrok will be entered into *server name or IP* in JMeter. Whereas, testing on localhost only needs to enter the IP Address from the server computer (PC Lab). Than, the scenario created during the test is the number of users as many as 20 people every 1 second (20/20), will sending 1 *request* to *server* then the total number of samples = 20 (20x1). The scenario of *performance test* can be changed in value as needed. The following table test results between access via internet and localhost, testing performed 10 times at different hours, taken value - average of latency, sample time (ms) and the connect time (ms):

Table 5. Access via internet (Ngrok).

Time	Sample Time (ms)	Status	Bytes	Sent Bytes	Latency	Connect Time (ms)
12 : 48	840.4	ok	5426	124	840.2	280.65
13 : 34	908.9	ok	5426	124	908.8	363.25
14 : 13	855.9	ok	5426	124	855.75	314.85
15 : 04	857.8	ok	5426	124	857.5	319.5
15 : 30	928.85	ok	5426	124	928.8	391.5
20 : 43	1665.85	ok	5426	124	1658.65	553.9
21 : 21	1972.9	ok	5426	124	1756.6	850
22 : 06	1423.85	ok	5426	124	1375.5	557.35
22 : 30	1536.15	ok	5426	124	1303.15	420.15
8 : 22	1345.65	ok	5426	124	1345.45	576.25

Table 6. Access via localhost.

Time	Sample Time (ms)	Status	Bytes	Sent Bytes	Latency	Connect Time (ms)
14 : 13	46.9	ok	5426	118	39.75	21.85
15 : 04	213.95	ok	5426	118	213.85	177.45
15 : 30	26.2	ok	5426	118	24.6	8.65
8 : 24	435.45	ok	5426	118	372.3	312.2
8 : 58	52.75	ok	5426	118	44.2	24.6
9 : 07	378.45	ok	5426	118	317.45	196.1
9 : 09	213.2	ok	5426	118	210.35	173.7
9 : 10	53.3	ok	5426	118	49.55	29.6
9 : 12	659.3	ok	5426	118	649.75	609.2
9 :14	2846.2	ok	5426	118	2059.9	1604.65

Latency (delay) is the time needed for data to travel the distance from origin to destination. Delay can be influenced by distance, physical media, congestion or long processing time [6]. From the results of testing the average value generated via the internet latency is higher than the latency of localhost, the higher of latency value, the higher of access failure.

3.3. Browser compatibility

As explained in the previous chapter, HTML5 can be accessed through popular web browsers such as *google chrome, Mozilla firefox, Opera* etc. Tests carried out at this stage are any browser that can access this web-based Remote Desktop, along with the results of testing browser compatibility:

Table 7. Browser compatibility.

Browser	**Status**
Google Chrome	Connect
Mozilla Firefox	Connect
Opera	Connect
Maxthrone	Connect
Internet Explorer	Not connect
Safari	connect

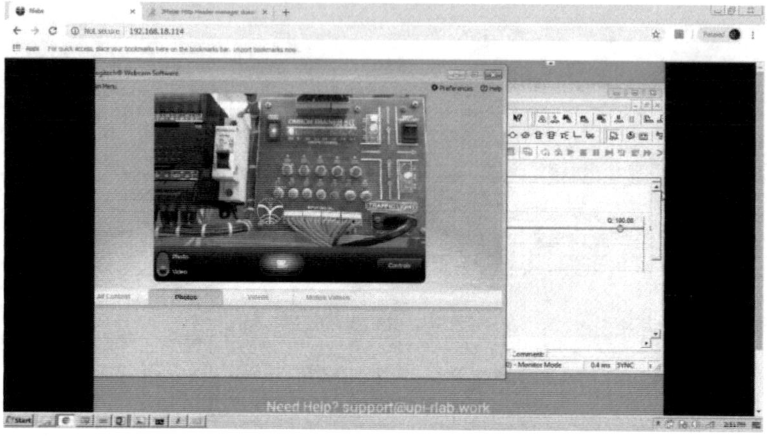

Figure 3. Learning in Rlab.

4. Conclusion

The conclusion that can be drawn from the results of developed Remote Lab based on HTML5 are, Rlab is successfully accessed via the web in two ways, namely via the internet and localhost using a web browser such as: *google chrome, Mozilla Firefox, Opera* and *Maxthrone*. Based on the results of testing using Apache JMeter™ software are as follows:

- The value of latency via the internet is greater than localhost, because there are several factors, namely the quality of the network between the user and server, time, distance and the software used is still free, the features are also limited.
- The value of latency via localhost is smaller than via the internet, because localhost is only one network between user and server but if many platforms enter the same network, the latency value will increase so that when in the remote process it will feel slowly.
- Response time when remote is still slower compared to Teamviewer's remote desktop, but its high flexibility can be accessed anywhere and anytime provided there is a good internet network available.

References

[1] Pruthvi P, Jackson D, Hegde S R, Hiremath P S and Kumar S A 2015 A distinctive approach to enhance the utility of laboratories in Indian academia *Proceedings of 2015 12th International Conference on Remote Engineering and Virtual Instrumentation (REV)* 238-241

[2] Kagermann H, Wahlster W and Helbig J 2013 Recommendations for implementing the strategic nitiative Industrie 4.0: Final report of the Industrie 4.*0 Working Group* **8**

[3] Pardede J, Hermana A N and Swarghani G 2017 Perbandingan Metode Breadth First Search dan Backlink pada Web Crawler *MIND Journal* **2**(2) 61-69

[4] Gowripeddi V V, Ram B K, Pavan J, Devi C Y and Sivakumar B 2018 Role of Wi-Fi data loggers in remote labs ecosystem *Online Engineering & Internet of Things* 235-249

[5] Wang V, Salim F and Moskovits P 2013 *The definitive guide to HTML5 WebSocket* Vol 1 (New York: Apress)

[6] Wulandari R 2016 Analisis Qos (Quality Of Service) Pada Jaringan Internet (Studi Kasus: Upt Loka Uji Teknik Penambangan Jampang Kulon Â€"LIPI) *Jurnal teknik informatika dan sistem informasi* **2**(2) 170

[7] Fushshilat I, Rahmat A, Somantri Y and Haritman E 2018 Laboratory management: digital laboratory information system (DLIS) concept *IOP Conference Series: Materials Science and Engineering* **434**(1) 012286

IOP Conf. Series: Materials Science and Engineering **850** (2020) 012018 doi:10.1088/1757-899X/850/1/012018

Design and development of telemedicine based heartbeat and body temperature monitoring tools

I P C Gunawan*, D H Andayani, T Triwiyanto, E Yulianto, T Rahmawati, L Soetjiatie and S D Musvika

Department of Electromedical Engineering, Poltekkes Kemenkes, Surabaya, Jl. Pucang Jajar Timur No. 10, Surabaya, 60245, Indonesia

*ccahya.ggunawan@gmail.com

Abstract. Body temperature and heart rate monitoring based on telemedicine is an electromedical device served to determine the condition of the heart rate per minute (BPM) and the patient's body temperature in realtime remotely. The purpose of this study is to design the heart rate and body temperature monitor based on telemedicine uses the Internet of Things (IoT). The system allows unlimited distance monitoring by utilizing the internet as a medium for sending the data. The body temperature and heart rate monitoring system uses DS18B20, Arduino, and Raspberry Pi temperature sensors and finger sensors as data processing and transmission microprocessors, LCD Character displays the results of heart rate readings and body temperature per minute. The results of heart rate measurements and body temperature are sent to the database via microservice and displayed on a web page in the form of numerical data and responsive plot graphics on desktops and mobile phones. Based on the measurement and comparison to standard apparatus, the average error of BPM measurement is 0.72%, with a maximum permissible tolerance of ± 5%. While in the body temperature parameter the average difference is 0.3 ° C with a maximum permitted tolerance of ± 1 °C.

1. Introduction

Monitoring devices in the medical world can monitor a variety of vital signs of patients, including the electrical signals of the heart and body temperature. For heart parameters, it is usually displayed on ECG charts and beats per minute called beats per minute (BPM). Adults' normal heart rates range from 60 - 100 bpm. After exercise, the range is from 200 to 220 bpm [1,2]. For body temperature parameters are usually displayed in units of Celsius/Fahrenheit. Normal body temperature usually ranges from 37°C. Measurement of body temperature can be done on the tympanic membrane, rectal, oral, and axillary [3,4]. Monitoring for the parameters above is very important because if the symptoms occur as above, action must be taken so that the patient's condition does not deteriorate. The patient's condition can worsen anytime and anywhere [5]. For patients who are outpatient at home (home care) who do rehabilitation of heart disease, of course, not always accompanied by a nurse/personal doctor [6]. For that, we need a system that can inform the results of monitoring BPM records and temperature to the doctor so that an evaluation of the patient's development can be carried out and take action if the patient shows abnormal heart rate or body temperature [7-9].

Rizal et al. designed of monitoring body temperature parameters, PPG and ECG with the wireless monitoring method. These parameters are only focused on the local area network [10]. Pierleoni et al.

designed a heart rate monitoring system for elderly people and patients with heart disease with a cellular connection (phone/SMS) and internet connection to convey information to the doctor. The drawback of this research is that the display output is only limited to Android smartphones, which have limitations on the display screen resolution [11]. Kakria et al. designed the monitoring the heart rate on an Android smartphone and then sending information to doctors with an internet connection. The weakness of this study is still using smartphones as interfaces that are limited in display screen resolution [12]. Gunawan et al. designed the monitoring system of heart rate and body temperature with interfaces on android smartphones and telemedicine communication using the SMS gateway. The weakness in this study lies in the limited data transmission distance because it only utilizes a Bluetooth connection. Wardhana et al. designed a system of monitoring heart rate and body temperature with interfaces on the web through an intermediary PC (Personal Computer). In this study, the weakness of the data sent is not real data; only image data that is not possible to edit.

Based on the results of the identification of the chronological problems above, the objective of this study is to design and development of telemedicine-based heartbeat and body temperature monitoring tools by using the internet and webpages as a medium for sending and displaying data to overcome the shortcomings of previous studies [13,14].

2. Methods

2.1. Experimental setup

This study was applied to normal human subjects aged over 17 years. Data retrieval is done in realtime with a measurement time of 1 minute for parameters of heart rate and body temperature. Sampling data is taken six times for each parameter.

2.2. Materials and tool

This study uses the Nellcor brand finger sensor on the parameters of heart rate and DS18B20 temperature sensor for body temperature parameters. The finger sensor is attached to the patient's finger, and the DS18B20 sensor is attached to the patient's axilla. The component used uses Arduino Nano as a Microcontroller for the acquisition of heart rate and body temperature data, the Raspberry Pi 3B is connected via the serial port to the Arduino module while functioning for the transmitter to display on the character LCD and web page interface.

2.3. Experiment

In this study, the researcher compiled a module conditioning the heartbeat signal from the finger sensor. The results of processing heart rate data and body temperature are displayed on LCD on apparatus display. Normally in 1 minute apparatus will send data to the server and displayed in the webpage interface. If one of the parameters has abnormal value, the apparatus will send data to the server automatically by interval 10 seconds. The researcher conducted several tests, including:

- Tests for processing heart rate data using an auto reference system against the results of Beat calculations per Minutes
- Test lost data delivery with a method of comparing the number of module shipment data with the amount of data displayed on the web display
- Delivery time testing with a method of calculating the process of sending data from the module until the data is displayed
- Comparison of the measurement results of the module with a comparison device for body temperature parameter using the thermometer, also for heart rate parameter using a pulse oximeter.

IOP Conf. Series: Materials Science and Engineering **850** (2020) 012018 doi:10.1088/1757-899X/850/1/012018

2.4. Telemedicine system

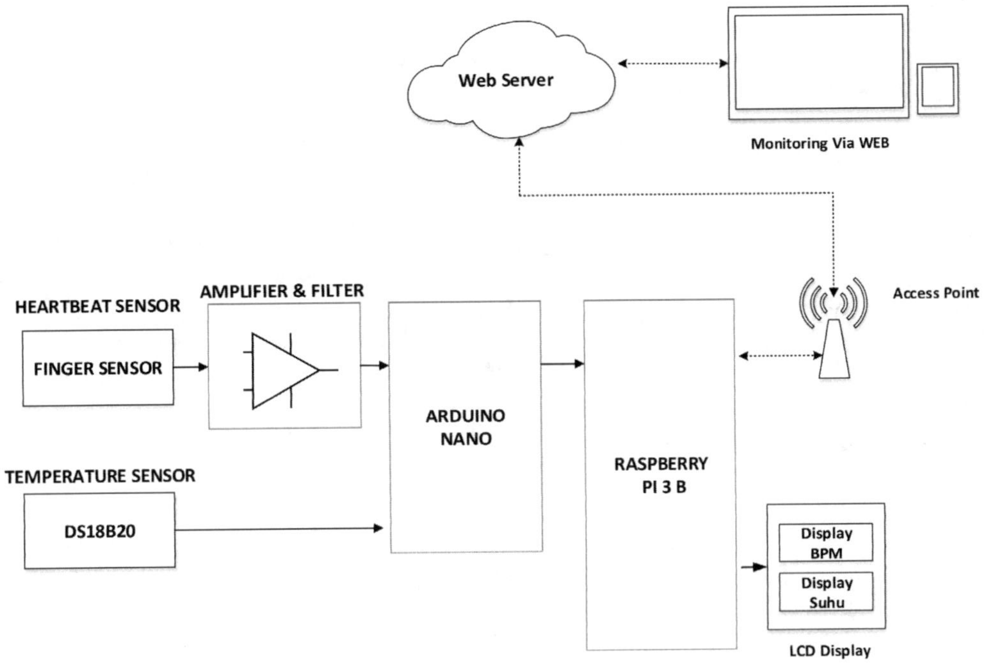

Figure 1. The diagram block of design and development of telemedicine based heartbeat and body temperature monitoring tools.

The circuit consists of a finger scanner to sensing the patient finger. Filter circuits and amplifiers function to filter unneeded signals and amplify the voltage divider resulting in the sensor circuit. Bandpass filter circuit with a cut-off frequency below 0.48 Hz and a cut-off frequency of 4.19 Hz and a gain of 101 times. The minimum system consists of two parts, namely the Arduino microcontroller as a device for processing data and the Raspberry Pi 3 B as a device for sending data to the database and arranging displays on 16 x 2-character LCDs. In this study, comparison of data results with predetermined comparison tools, measurement and analysis of delivery times, and calculation and analysis of lost shipping data were conducted.

As shown in Figure 1, the flow of reading the temperature value starts from reading the digital data generated by the DS18B20 sensor stored in a variable and sent to raspberry pi via serial communication. The flow of the heart rate reading starts with reading the ADC as a result of processing analogue signals on the finger sensor. The ADC signal is processed and calculated to get a heart rate per minute then sent via serial communication to the raspberry pi. The flow of data transmission starts from reading serial data; then, the results are displayed on the LCD. When the timer reaches 1 minute, the data is sent to the database via the data upload service. The flow of data appearance on the web page interface starts from retrieving data from the database with microscope access then displayed on plotting labels and graphics.

3. Results and discussion

3.1. Heartbeat calculation using auto reference result

The data processing program listing on Arduino starts from reading the ADC through pin A0 and converting ADC data to voltage. In the above program, listings use an auto reference system that uses the highest percentage of signal voltage to calculate heart rate.

Table 1. Result of auto reference testing.

Threshold setting	Apparatus	Average	Error
10%	Design	87.2	1%
	Standard	87.2	
20%	Design	92.2	1%
	Standard	91.7	
30%	Design	93.8	1%
	Standard	93.5	
40%	Design	71.3	25%
	Standard	94.5	
50%	Design	49.5	47%
	Standard	94.2	
60%	Design	42.0	55%
	Standard	92.7	
70%	Design	28.8	68%
	Standard	91.5	
80%	Design	0.0	100%
	Standard	89.8	
90%	Design	0.0	100%
	Standard	90.2	

From the results of testing that has been done with the retrieval of heart rate data carried out on a fixed subject and the auto reference value set with a range of 10%, the lowest error is obtained at 10%, and the highest error is at 90% as shown in Table 1.

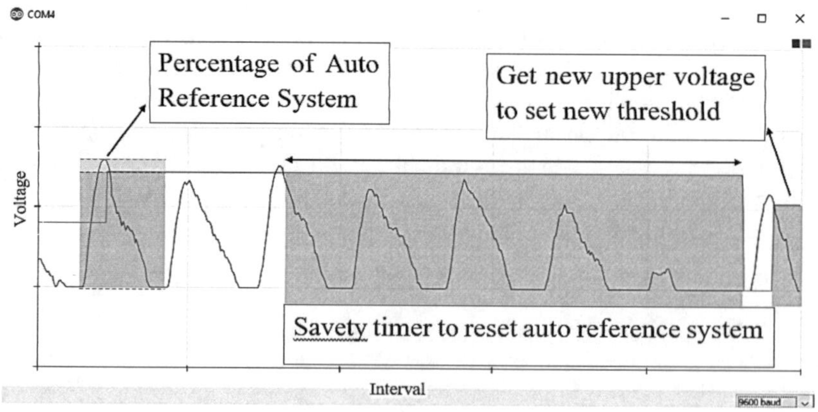

Figure 2. Auto reference system.

Analysis of the results of testing auto references that are applied to human heart rate calculations, as shown in Figure 2. Can be stated that the auto reference application works more effectively against a stable input pulse. At input pulse with a low level of stability, auto reference works effectively with settings below 30%.

3.2. Analysis and result lost data measurement

Table 2. Measurement lost data.

No	Time(s)	Amount of Transmitted Data	Amount of Received Data
1	100	10	10
2	200	20	20
3	300	30	30
4	400	40	40
5	500	50	50

The result data shown in lost data processing shows an error of 0%, which shows the amount of data sent is equal to the amount of data received at each measurement point 10, 20, 30, 40, and 50 sample data sent. Data testing is done by comparing the amount of data sent with the amount of data received as shown in Table 2.

3.3. Analysis and result interval of transmission data

Retrieval of data when sending data is done by taking the time when the data is sent via raspberry pi and retrieving data when the microservice program displays the data. The results between data are calculated as the time difference. In the above data retrieval carried out at internet download speeds of 36 Mb / s, the average shipping speed is 1 s per data transmission.

3.4. Result of the heartrate and temperature measurement

A comparison of data retrieval, as shown in Table 3, is made by comparing the module with a comparison device (Pulse Oximeter and Body Thermometer) carried out on five different subjects for temperature and bpm parameters. In the BPM parameter, the highest error calculated is 1.31%, and the highest temperature parameter error is 1.49%.

Table 3. Measurement error between module and comparison tools.

No	Subject	Error Heartrate (%)	Error Temperature (%)
1	P1	1.09	1.09
2	P2	1.23	0.68
3	P3	0.76	0.45
4	P4	1.31	1.40
5	P5	1.28	1.49

After comparing the results between the module and the measuring instrument, the average error in the parameters of the BPM is 1.14%, which is still within the tolerance range of the BPM value of 5%. For temperature parameters, the difference is 0.3 ° C, which is still within the tolerance range of the temperature value, which is ± 1 ° C.

The lost data test results with an error of 0%, which indicates that the shipment uses the IoT Telemedicine technique is safe for sending medical data because the data sent is in accordance with the original.

The results of testing the delivery time show the average delivery time per sample data of 1 second, and it can be concluded that the delivery of IoT Telemedicine can send a small sample of data at one sampling per second. The development of monitoring samples with the amount of data sampling of more than 1 / second cannot be applied to IoT technology that is still developing at this time, such as monitoring ECG signals.

4. Conclusion

This research shows that the Internet of Things (IoT) technology can be utilized in the health sector, especially in patients' vital sign monitoring that is done in a Realtime manner (Telemedicine). This technology allows the process of monitoring care without being accompanied by medical personnel directly more effectively. The design of this tool is expected to be able to handle cases of shortages of medical personnel in an area and maximize the monitoring process of outpatient conditions.

References

[1] Jahan E, Barua T and Salma U 2014 An overview on heart rate monitoring and pulse oximeter system *Int. J. Latest Res. Sci. Technol* **3** 148-152

[2] Babiker S F, Abdel-Khair L E and Elbasheer S M 2011 Microcontroller based heart rate monitor using fingertip sensors *University of Khartoum Engineering Journal* **1**(2)

[3] Singh D G and Balram G 2018 IoT-Temperature and Heart Beat Detection over the Internet using Arduino *International Journal of Engineering Research in Computer Science and Engineering* **5**(4) 104–107

[4] Fitriyah H, Rachmadi A and Setyawan G E 2017 Automatic Measurement of Human Body Temperature on Thermal Image Using Knowledge-Based Criteria *Journal of Information Technology and Computer Science* **2**(2)

[5] Chakole S R, Giripunje V G, Dhak B and Sawarkar M IOT Based Healthcare System For Signal Monitoring Using Cloud Technology I*nternational Journal of Engineering Applied Sciences and Technology* **4**(2) 206–211

[6] Tahat A A 2009 Body temperature and electrocardiogram monitoring using an SMS-based telemedicine system *2009 4th International Symposium on Wireless Pervasive Computing* 1-5

[7] Uddin M S, Alam J B and Banu S 2017 Real time patient monitoring system based on Internet of Things *2017 4th International Conference on Advances in Electrical Engineering (ICAEE)* 516-521

[8] Naravani M M M 2017 A Smart Healthcare Monitoring System Based on IOT *International Journal of Engineering Applied Sciences and Technology* **2**(6) 39–43

[9] Rangasamy M, Balasubramaniam A Y Y A S A M Y, Krishnarajan D E V A R A J A N, Raviteja A, Kante N R and KUMAR N S 2011 Role of telemedicine in health care system: a review *Int J Recent Adv Pharma Res* **2** 1-10

[10] Rizal A, Suryani V, Jondri J and Hadiyoso S 2014 Development of Wireless Patient's Vital Sign Monitor Using Wireless LAN (IEEE. 802.11. b/g) Protocol *International Journal of Electrical and Computer Engineering* **4**(6) 893

[11] Pierleoni P, Pernini L, Belli A and Palma L 2014 An android-based heart monitoring system for the elderly and for patients with heart disease *International journal of telemedicine and application* 1–11

[12] Kakria P, Tripathi N K and Kitipawang P 2015 A real-time health monitoring system for remote cardiac patients using smartphone and wearable sensors *International journal of telemedicine and applications* 1-12

[13] Kaur A and Jasuja A 2017 Health monitoring based on IoT using Raspberry PI *2017*

International Conference on Computing, Communication and Automation (ICCCA) 1335-1340

[14] Kumar A, Balamurugan R, Deepak K C and Sathish K 2017 Heartbeat Sensing and Heart Attack Detection using Internet of Things: IoT *International Journal of Engineering Science and Computing* **7**(4) 6662–6666

IOP Conf. Series: Materials Science and Engineering **850** (2020) 012019 doi:10.1088/1757-899X/850/1/012019

Gateway system for portable finger scanners

T P Nugraha* and D Wahyudin

Departemen Pendidikan Teknik Elektro, Universitas Pendidikan Indonesia, Jl. Dr. Setiabudhi 207, Bandung 40154, Indonesia

*teguhprataman@student.upi.edu

Abstract. The attendance system serves as authentic proof of someone's presence, in this instance the system is meant to account for lecturer's presence. The current portable finger scanner attendance tool used does not yet have a system for processing data. The purpose of this study is to create a system that is able to receive, process and store data that has been sent by a portable finger scanner. The method carried out in this study was an experiment including designing hardware, software, and testing tools. The trial results show that the tool runs in accordance with the planned specifications. Data can be received through LoRa, stored in the database, and then converted into Excel format. Each mode is able to function properly, the planned distance of acceptance is achieved, and the final purpose of designing the tool is fulfilled.

1. Introduction

The presence system has a very important role as a authentic proof. This system is applied to various offices with different equipment [1,2]. Usually a presence system uses signatures as proof of attendance. However, in current era to store data of presence, using a portable finger scanner gives a benefit. The obtained finger print data needs further processing. Hence, data can be utilized to the maximum use. The use of databases is needed to store presence data that has been permanently processed for archiving purposes.

Various gateway technologies for presence systems using Finger Scanner have been made using different components and connectivity. Darmawane et al. creates a fingerprint-based attendance and reporting system and sms gateway [3]. Miftah create a web-based student presence system with PHP and MySQL [4]. While the author in Fakih et al. created a Fingerprint Authentication presence system for Automation of Lecture Presence using a desktop computer [5]. However, these studies still do not use devices and connectivity that are more sophisticated in terms of specifications, capabilities, and ease of manufacture. Each room has a Portable Finger Scanner to send attendance data to examiners. A gateway server will be placed in the room which functions as a receiver to store lecturer attendance data in the form of a database.

Therefore, the author tries to use Raspberry Pi and LoRa as the main components of this tool. The Raspberry Pi is as single board computer (SBC). The size of SBC of a credit card. The Raspberry Pi is equipped with all functions like a complete computer, using ARM system-on-a-chip (SoC) which is packaged and integrated on a printed circuit board (PCB). This Raspberry Pi is able to work like a computer in general with the ability to run a Linux operating system and its application [6]. LoRa (short for Long Range) is a wireless technology developed by Semtech, this technology can reduce power consumption using special modulation techniques. This technology can reduce power consumption and has a data transmission distance of up to 10 kilometers [7]. This tool will be made in a simple form so that it is easy to carry or move from one place to another. The author aims to create a gateway system design for portable finger scanners in the design of the lecturer attendance system. Discussion is only limited to the design of the gateway system that functions as a receiver and the results of testing.

2. Methods

The first step is to design the system by detailing the components of input, process, and output. The LoRa radio module used is the Semtech E32-TTL-100 module with the SX1278 chip as a connecting component between a server (receiver) and a portable finger scanner. It has a fixed mode format for transmitting data from multiple transmitters to a receiver. The radio frequency used is 433 MHz in accordance with government regulations. The screen used is a 3.5 inch TFT LCD to display data received from the transmitter. The data will be stored in the database and can be converted into csv format to make it easier for users to print attendance data. The hardware design is equipped with a 5V supply adapter with a maximum current of 2A. Figure 1 shows the block diagram of hardware design.

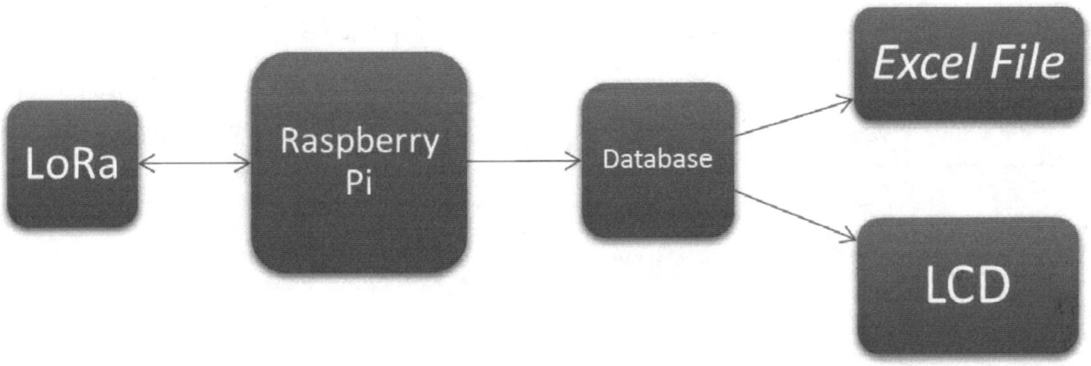

Figure 1. Hardware design diagram.

Program design is done after the hardware design is complete. So that programs are made in accordance with the components available. In the program design, there is a data reception section from LoRa. The data transmission feature via LoRa uses a fixed mode format where in one data reception there are six sub-data consisting of two initial data as the target address, the third data as the target channel, and the last three default data which means filled according to planning needs. So the author will make arrangements for this data format.

After the hardware is integrated with the program that has been created, then do a test tool. Trials include testing the suitability of the display on the hardware and programs that have been made. Then the next trial was to receive data sent by a portable finger scanner with a distance of more than 30 meters. This distance is chosen because the actual distance between the portable finger scanner server is 30 meters. If the trial does not match the purpose, it will be redesigned until the goal is reached.

3. Results and discussion

3.1. Hardware

The UART pin of the Raspberry Pi is used as a communication path with the LoRa E32-TTL 100 radio module. The pins M0 and M1 on the radio module are connected to the ground to set the radio module to the normal mode state. Under normal circumstances the data transmission activity mode can be done. The 3.5 "TFT LCD screen uses an HDMI connector (High - Definition Multimedia Interface) that is available on the Raspberry Pi as a data communication path. While for the supply used it is a 5V 2A adapter to meet the required power requirements of this series.

3.2. Program

This program is created in accordance with the flowchart planned and supported by each component module using the Python programming language 3. Format of data transmission using fix mode which is a data array containing six pieces of information, including the first and second is the target address or the default address of the radio module used, the third is the target channel which is the default address of the radio module channel used, the radio module on the portable finger scanner must have the same target address and target channel configuration as the radio module used at the gateway. While the fourth data is the tool number data, the fifth data is the operating mode code that is owned by the portable finger scanner and the latest data as the user's fingerprint ID.

3.3. Trials

The following are the results of the program design that is run in the hardware circuit.

3.3.1. Initial display. In Figure 2 is the initial display when we run the program on the Raspberry Pi, in the default condition the program created will be in matching mode. In the upper left corner there is a menu bar for selecting other modes in the program.

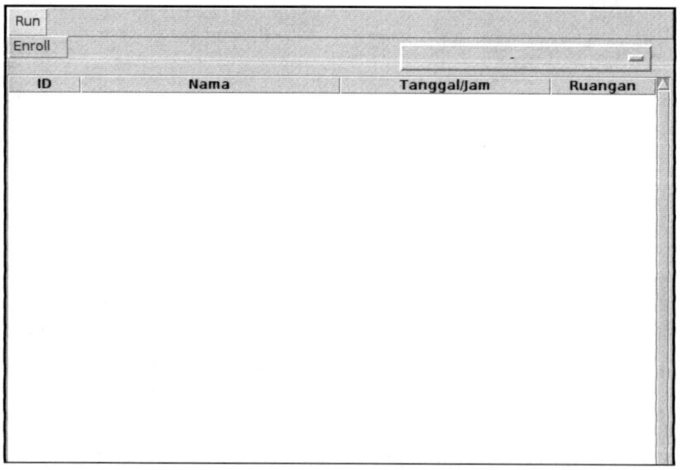

Figure 2. Initial display.

3.3.2. Enroll mode display. Figure 3 (A) is a display of evidence of data received from a portable finger scanner in the form of an ID number and class number in the shell. The display in Figure 4 (B) is the step of giving the user name to the ID number received, so that the ID number can be recognized by the program.

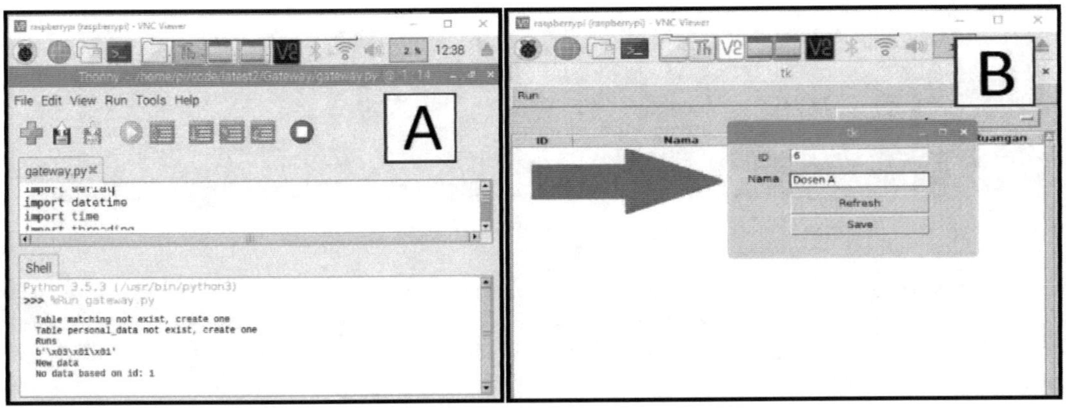

Figure 3. Data receive in shell .

3.3.3. Matching mode display. Matching mode is the name of the mode for the attendance system. In this mode there are 4 data displayed including, ID number, name, time of data received, and room number. The display in Figure 4 (A) is a display if the device receives an ID number data that has not been recognized. Registration of ID number data on enroll mode must be done first so that data can be identified. Figure 4 (B) is the display of ID number data from a portable finger scanner that was previously registered on the enroll mode, the ID number received has a name so that it can be identified who is the user of the ID number.

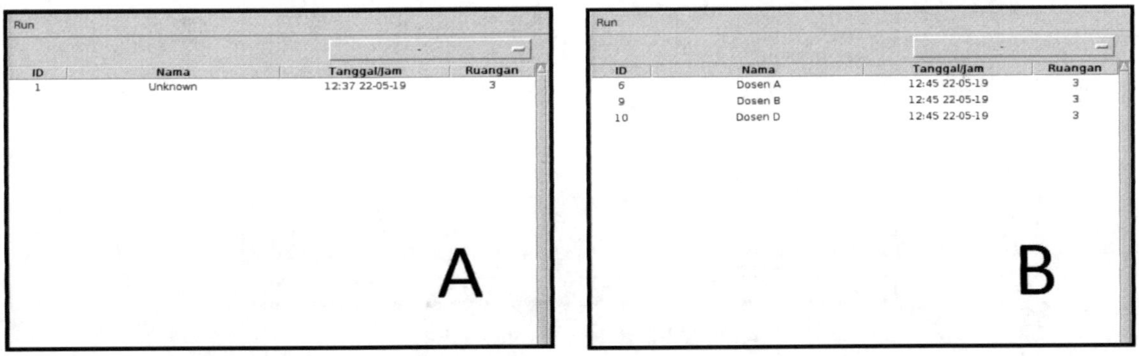

Figure 4. (A) Display of unregistered ID's **(B)** Display of registered ID's.

3.3.4. LoRa range. Distance of LoRa Measurements Measurements are carried out by placing the gateway server in the DPTE FPTK UPI room, while portable finger scanners are moved from one place to another that has been determined. Measuring distance testing between a gateway server and a portable finger scanner is drawn in a straight line.

3.3.5. Data conversion from database (.db) to excel (.csv). Data stored on the database has the format *.db,* so it needs to be converted to a format that can be opened by almost all computer devices. The format chosen is the format *comma-separated values (.csv).* Converting a .db file to a .csv format is done by entering the directory where the database is stored through the terminal, and typing special commands. Figure 5 is a terminal display image with a command to convert the *.db* file to a *.csv* file named *matching.csv.*

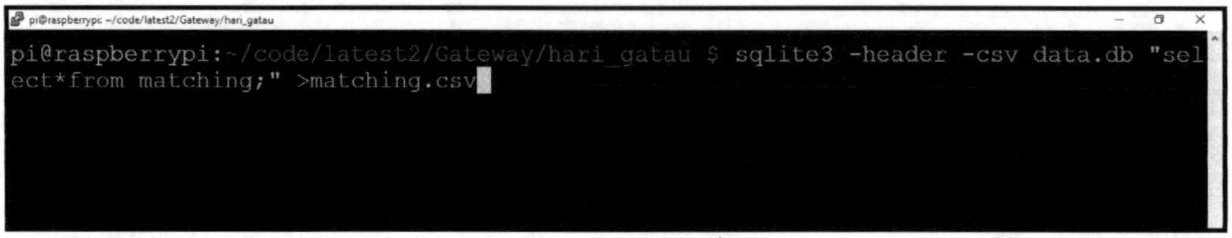

Figure 5. Terminal display.

3.3.6. Final data display. Figure 6 is a data display that has a .csv format, this data is the end result of a Gateway System tool for Portable Raspberry Pi and LoRa Based Finger Scanners for Lecturer Presence System on Design of Final Defense Session in DPTE FPTK UPI. The data that has been obtained can then be easily processed or printed, because the .csv file format can be opened in almost all computers and devices that are in the DPTE FPTK UPI.

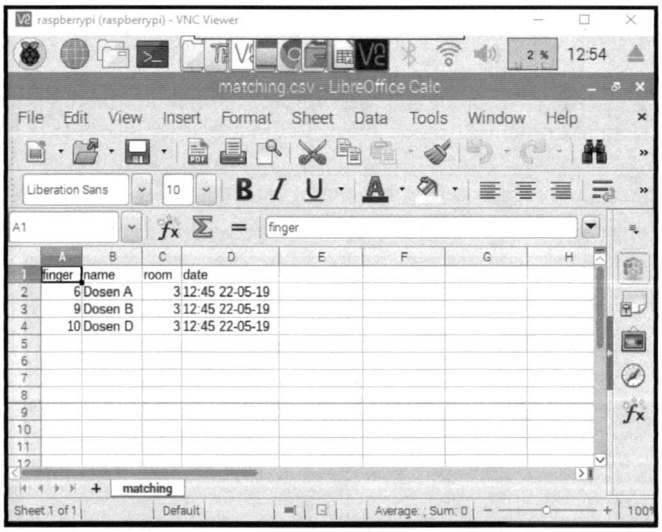

Figure 6. Final data display (file .csv).

4. Conclusion

Based on the results and discussion, it can be concluded that the design of the Gateway System for Raspberry Pi and LoRa-based Portable Finger Scanners was successfully made according to the plan. On the results of the trial shows that each mode can function properly, the planned distance of acceptance is achieved, and the final goal of designing a tool that is fulfilled.

References

[1] Veer N D and Momin B F 2016 An automated attendance system using video surveillance camera *2016 IEEE International Conference on Recent Trends in Electronics, Information & Communication Technology (RTEICT)* 1731-1735

[2] Oo S B, Oo N H M, Chainan S, Thongniam A and Chongdarakul W 2018 Cloud-based web application with NFC for employee attendance management system *2018 International Conference on Digital Arts, Media and Technology (ICDAMT)* 162-167

[3] Darmawan A, Yuliawati D, Marcella O and Firmandala R 2016 Sistem Absensi dan Pelaporan Berbasis Fingerprint dan SMS Gateway *Explore: Jurnal Sistem informasi dan telematika* **7**(2) 10
[4] Miftah A S 2014 Aplikasi SMS Gateway Presensi Siswa Berbasis WEB dengan PHP dan MYSQL (Tugas Akhir Teknik Informatika, Yogyakarta: UMY)
[5] Fakih A, Raharjana I K and Zaman B 2015 Pemanfaatan Teknologi Fingerprint Authentication untuk Otomatisasi Presensi Perkuliahan *Journal of Information Systems Engineering and Business Intelligence* **1**(2) 41-48
[6] Jain S, Vaibhav A and Goyal L 2014 Raspberry Pi based interactive home automation system through E-mail *2014 International Conference on Reliability Optimization and Information Technology (ICROIT)* 277-280
[7] Trinh L H, Bui V X, Ferrero F, Nguyen T Q K and Le M H 2017 Signal propagation of LoRa technology using for smart building applications *2017 IEEE Conference on Antenna Measurements & Applications (CAMA)* 381-384

IOP Conf. Series: Materials Science and Engineering **850** (2020) 012020 doi:10.1088/1757-899X/850/1/012020

A low-cost Holter monitor design equipped with external memory and Bluetooth connection

D Nofitasari*, I D G H Wisana, T Triwiyanto, E D Setioningsih, M R Mak'ruf and P C Nugraha

Department of Electromedical Engineering, Poltekkes Kemenkes, Jl. Pucang Jajar Timur No. 10, Surabaya, 60245.

*dewinofita51@gmail.com

Abstract. Monitoring the heart activity is critical for a patient with heart disease. Twenty-hour detecting of the ECG signal will help the clinician to diagnose the hearth. This device is usually called a Holter monitor. However, the price of this device is very expensive, therefore the development of a Holter is required. The purpose of this study is to develop a portable and low-cost Holter system to record the ECG signal in twenty-four hours. The mainboard is composed of a pre-amplifier, bandpass filter, notch filter, summing amplifier, Arduino microcontroller, SD Card Memory, and a Bluetooth transmitter. ECG signal is collected from the body based on the standard measurement of LEAD II. The ECG signal is sampled with a frequency sampling of 200 Hz. To record the raw data of the ECG signal, the SD Card Memory was used to save the data for further data analysis. Calibration was conducted by using an ECG phantom. The calibration was performed in order to make this design to be a standard ECG machine. The design was also equipped with a Bluetooth transmitter to transmit the data to the computer. The total price to develop this device is 460,000 IDR. After the evaluation, the BPM error of this Holter is ±0.0273% and the Holter able to operate in 24 hours. The developed ECG Holter is portable in dimension and low cost to be fabricated for mass production to help people with heart disease.

1. Introduction

Holter Monitor, or sometimes also referred to as an ambulatory electrocardiographic (ECG) device, is a portable device used to monitor various activities on the cardiovascular system continuously for 24 hours [1]. The principal work of this device is by attaching electrodes to the patient's chest while an instrument records the ECG [2]. This device usually works with the battery. This Holter can be used as an ECG recorder that can be bagged or carried everywhere with a belt or straps on the shoulder [3]. Holter monitor works by tapping the ECG's electrical signal on the heart and is reinforced so that the signal can be read and diagnosed by the medical doctor. Holter monitor function is to analyze and monitor a person's heart activity caused by abnormalities or abnormal changes in the heart [4]. However, the use of this device is only limited to the area in the hospital or medical institution, while the need for this Holter is also essential for monitoring ECG signals at home or in areas that far from the hospital. The function of this device is as early prevention of heart problems in a person, and it is critical to always monitor the heart health regularly by examining the heart signal recordings independently [5]. Therefore, this is very much needed a portable and practical Holter (easy to carry and easy to use) for a patient with heart disease. By adding a wireless monitoring system and utilizing a computer as a visualization of the ECG signal, the Holter can be used effectively to monitor ECG signals wherever they are without having to look at the ECG record in the treatment room [5]. Holter monitors are vital for checking the heart abnormalities. In cases of arrhythmia, the use of the Holter monitor is very emphasized because this device can record ECG signal activity for 24 hours and is portable [6].

Tai et al. [7] proposed an ECG recorder with an adaptive sampling algorithm and has been evaluated using the MIT / BIH database, but from this study, it can use a simpler method so that it can improve signal reconstruction better. Related research was also developed by Luciani et al. [8]; they made a portable ECG monitoring device with Bluetooth and the Holter that has the capability for telemedicine applications for remote monitoring. In this study, the data obtained can be stored on a flash memory card (FAT16) or sent via Bluetooth or USB to a local station or Access Point. The research can operate in two modes: Holter and on-line transmission. Jin and Miao [9] developed ECG Holter using the MSP430 microcontroller and USB flash. ECG signals are recorded and sent to the monitoring center via a USB flash disk using the SL811HS USB controller, but in this study monitoring still uses serial USB. Further research was developed by Carrasco et al. [10] with the title of design and development of Holter Prototype with Bluetooth Transmission, in this study using Bluetooth technology as wireless communication, using data storage and user interfaces with Lab View Software but the signal obtained still has noise. Lee et al. [11] develop Holter with noise artifact reduction using the Empirical Decomposition Mode and Statistical Approach. In this study, it has also been designed with a real-time method for detecting artifact and noise (MN) that often disrupts the rhythm of specific ECG signals. But from this research, there is still noise from the signal and PPG signal. Furthermore, the development of the Holter monitor was also has been developed by Putra et al. who modified the EKG into a Holter with a display on the computer, however, in those studies, the data could not be sent directly to the Computer. Furthermore, the Holter was also developed by Ananta Faxia, the Holter Monitor, which can be used as an Electrocardiograph (ECG) monitoring device that can be monitored the ECG signal via a computer screen using wireless system, but the weakness of those study is that the noise still exists and unstable when there is a movement of the subject [11,12].

Based on those problems, the objective of this study is to develop a Holter monitoring based on Arduino microcontroller equipped with SD card memory and Bluetooth connection [13]. The salient feature of this design is that the recorded ECG data can be sent wireless through the Bluetooth connection to the computer host. Furthermore, the Holter can be used as a standard electrocardiograph (ECG) monitoring device and portable ECG recorder. As a conventional ECG device, this proposed design can monitor the ECG signal wirelessly to display on the computer using Bluetooth communication [14]. The second option, the proposed model can be used to record the ECG signal for 24 hours, and the ECG data was saved to the SD card memory. In this study, a marker, in which to assign when a heart attack happens, was also equipped. Hence the system can measure the number of abnormalities [1]. This study implies that the design can be used personally at home for further diagnosed by a doctor and can be built at a low cost.

2. Materials and methods

2.1. Experimental setup

This study used ten normal subjects with the criteria the ages ranged between 22 and 27 years old and the weight is between 45 to 50 kg. The subjects were randomly sampled and the data collection is repeated 5 times.

2.1.1. Materials and tool. This study is used a disposable electrodes ECG (OneMed, Jayamas Medical Industri, Indonesia). The electrodes were attached to the right and left a hand on the human. The instrumentation amplifier was built based on TL084 OP-AMP [15]. The Arduino Nano microcontroller was used for ECG data acquisition and communication to the computer unit using Bluetooth module HC-05. SD Card Memory with a 16 Mbyte was used to save the ECG data in real-time. A digital storage Oscilloscope (Textronic, DPO2012, Taiwan) was used to test the analog circuit. An ECG phantom (Fluke, PS320, USA) was used to calibrate the analog circuit.

2.1.2. Experiment. In this study, after the design was completed then the frequency response of this device was tested using a function generator according to the specification of the ECG signal. In the calibration stage, the Holter ECG was tested using an ECG simulator (phantom) with all ranges (30, 60, 120, 180, and 240 BPM). Each setting, the output of the Holter was calculated to validate the result of this study. Then the Holter was tested on the human body [3].

3. The diagram blocks

In this research, the ECG recording was performed based on the placement of the three electrodes for LEAD II combination as shown in Figure 1. The ECG signal was then amplified using instrumentation amplifier, filtered using a bandpass filter, notch filter, non-inverting amplifier, and summing amplifier or adder for offset adjustment. A built-in analog to digital converter (A/D) in Arduino Nano was used to convert the analog ECG signal into digital [16]. The ECG recording was saved to the SD card memory for further data analysis. The ECG data was also can be delivered to the computer unit via the Bluetooth module.

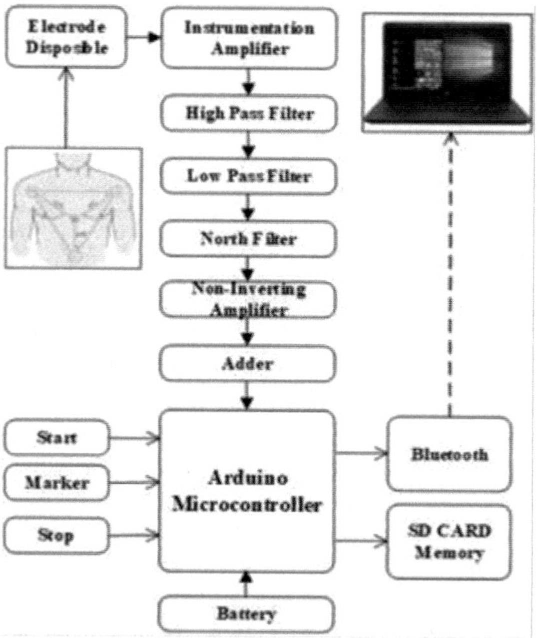

Figure 1. The diagram block of the Holter monitor.

3.1. The flowchart

The Arduino program was built based on the flowchart as shown in Figure 2. After the initialization of the Arduino, the program asks the Bluetooth whether the connection is active or not. If the Bluetooth is active, then the data was ready to send to the computer. The other option is the data can be directly saved to the SD card memory.

The computer program was built based on the flowchart as shown in Figure 3. The menu divided into 2 namely readings from Bluetooth serial and from SD card memory [17].

The data from Bluetooth connection active, the data is ready to send to the computer and then read ECG Signals from the Delphi program. While reading the data from SD Card Memory is initialization and then received data, Conversion data from SD card memory and final reading of the ECG signal.

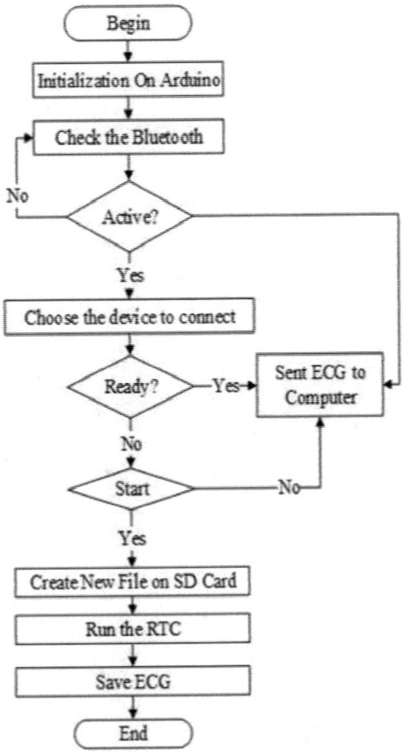

Figure 2. The Flowchart of the Arduino program.

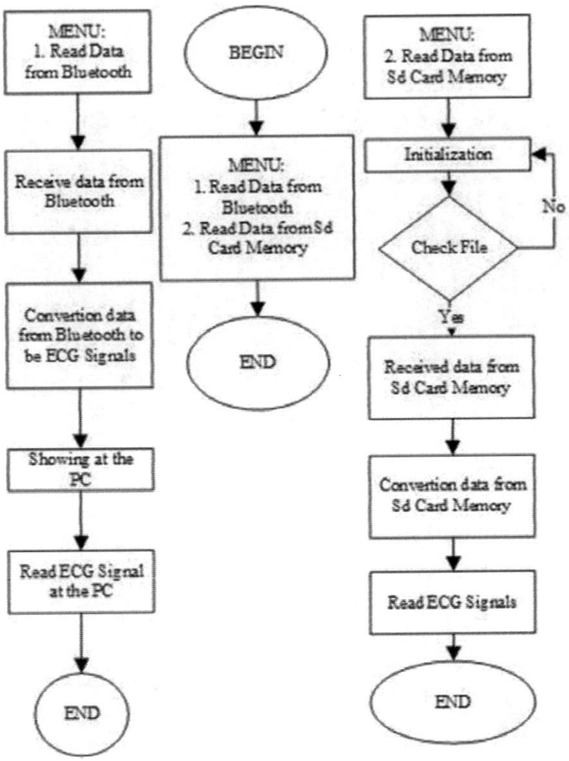

Figure 3. The flowchart of for the Delphi programming.

3.2. The analog circuit
The important part of this development is the analog circuit which describes in Figure 4 (instrumentation amplifier), Figure 5 (bandpass filter), Figure 6 (notch filter), and Figure 7 (summing amplifier). The circuit is used to process the ECG signal. Hence it will ready for digital processing using Arduino microcontroller.

3.2.1. Preamplifier. An Instrumentation amplifier circuit as shown in Figure 4 was the pre-amplifier for the ECG amplifier which takes an important part for analog processing. This circuit is composed of three OP-AMP and one additional OP-AMP for buffering the signal. The overall gain was adjusted by R1 (560 Ohms). The two leads of electrodes were connected to J16 and J17 as shown in Figure 4.

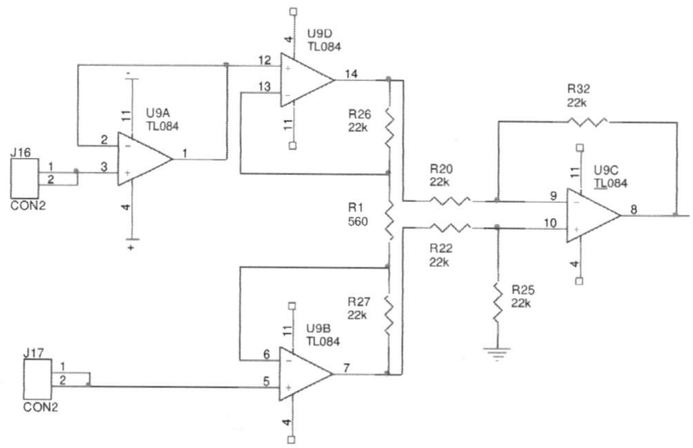

Figure 4. Instrumentation amplifier.

3.2.2. Bandpass filter. The bandpass filter is to limit the frequency of the signal; hence the processed ECG signal is only by characteristics of the ECG signal. The bandpass filter was designed for the cut off frequency of 0.05 to 100 Hz. The gain and the cut off value were calculated based on equation (1) and equation 2, respectively. In this study, the gain of the bandpass filter was 101 times.

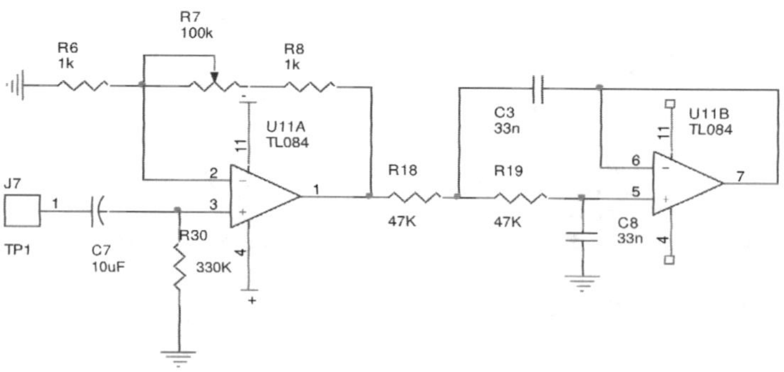

Figure 5. Bandpass filter.

$$\text{Gain} = 1+2(R_f/R6) \tag{1}$$

where R_f is the feedback gain, R6 is the divider resistance and the Gain is the amplification of the circuit

$$f_c=1/(2\pi RC) \tag{2}$$

where R is the resistance value, C is the capacitance value, and f_c indicated the cut off frequency.

3.2.3. Notch filter. The notch filter was designed to eliminate the line power interference from the environment in which the frequency of the bandstop filter is 50 Hz. This interference could dominate the main signal (ECG) due to the small amplitude of the ECG signal. The input of this circuit was connected to the output of the bandpass filter. The notch filter was designed as shown in Figure 6.

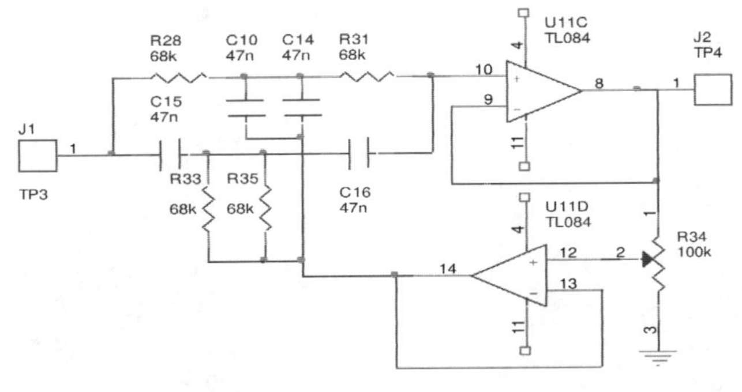

Figure 6. Notch filter 50 Hz.

3.2.4. Summing amplifier. The summing amplifier was composed of two operational amplifiers (Buffer and Inverting Amplifier). This circuit is to offset the level of the ECG signal so that the signal is fit to the digital part of Arduino Microcontroller which is the A/D converter. The ECG signal was added with a setpoint voltage from a variable resistor (R4) as shown in Figure 6. According to Figure 7, the buffer circuit is used to maintain the signal voltage.

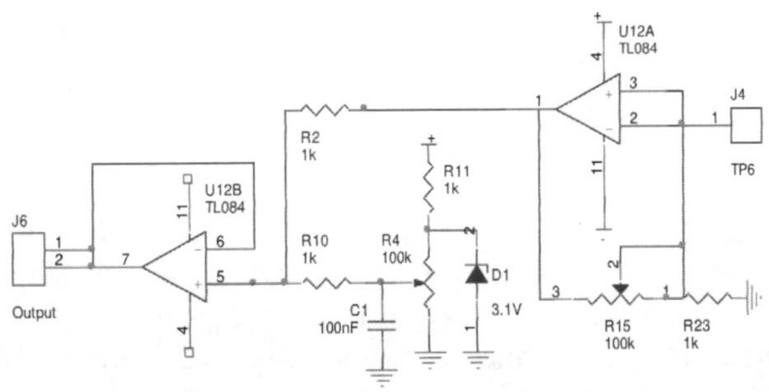

Figure 7. Summing amplifier.

3.2.5. Bluetooth module. The Bluetooth (BT) module used in this design was HC-05. This Bluetooth module was used to transmit the ECG signal from Holter monitor to the computer unit. For the communication between the Arduino board and the Bluetooth module, it needs two pins which are Tx (to transmit the ECG Data, and The Rx is to receive the information between the microcontroller and computer unit). The connection between the BT and Arduino module was shown in Figure 8.

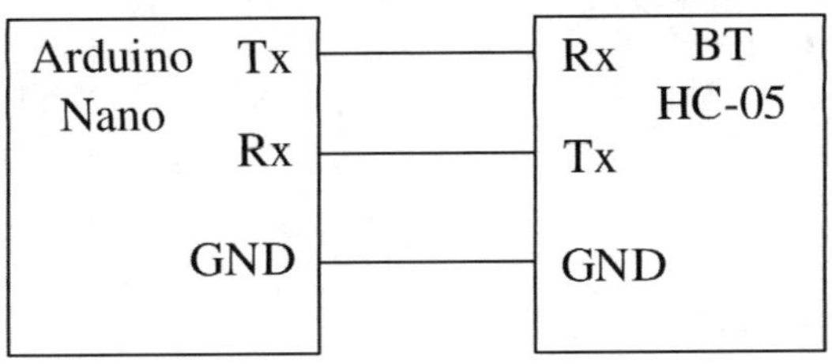

Figure 8. Bluetooth Connection.

4. Results

In this study, the Holter has been tested using an ECG phantom (Fluke, SP2002, USA) and ECG from the human body. The result shows that the recording is feasible to record the ECG signal from the human body [17].

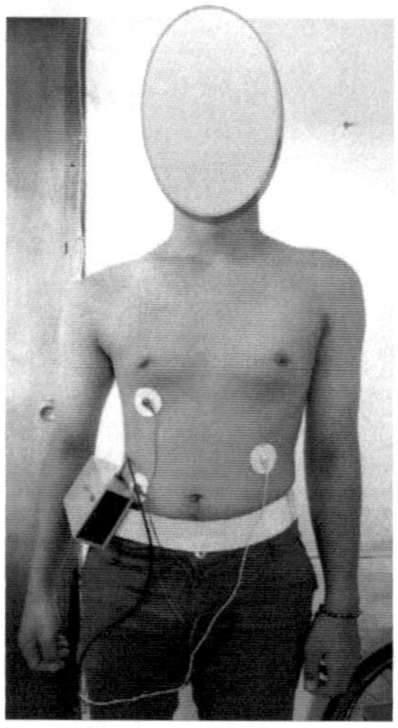

Figure 9. The Holter monitor electrode placement (front view).

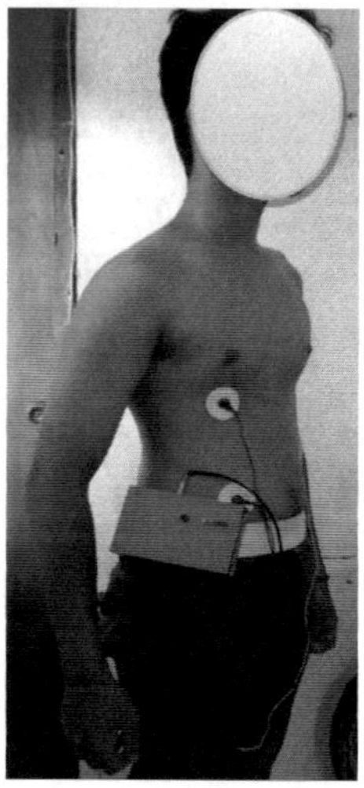

Figure 10. The Holter monitor electrode placement (side view).

Figure 9 and Figure 10 are examples of the placement of disposable electrodes that are affixed to RL, RA, LL. Holter Monitor can be attached to the respondent's waist area and can be taken anywhere by recording the respondent's heart rate results [5].

4.1. The Holter ECG design

The photograph of the analog and digital part of the Holter ECG was shown in Figure 11 and Figure 12, respectively. The analog part consisted of three of TL084 (OP-AMP) which each unit composed of four OP-AMP [6]. Some variable resistors were used to adjust the gain and offset of the ECG signal. The main part of the digital circuit is Arduino microcontroller, SD card memory, Bluetooth HC-05, and RTC DS1307.

Figure 11. The Holter ECG design

Figure 12. The Digital part of the Holter ECG

4.2. The listing program for Arduino Holter monitor

The listing program for Arduino was shown in the Listing Program 1. The program consisted of the program to send the data to the computer and to save the data into the SD card memory in which it can be chosen in the initial program. The ECG data was sent to the computer host using Bluetooth communication.

Listing program 1. Program to send the ECG data to a computer

```
void BT(){
  float sinyal2 = analogRead(A0);
  float tegangan = ((sinyal2/1023)*5);
  if (ref<=tegangan){
    ref=tegangan;
  }
  else{
    ref=ref;
    hold= (ref*0.7);
  Serial.print("a");
  Serial.print(sinyal2,0);
  Serial.print("b");
 if(digitalRead(9)==LOW){
    statebutton=false;
   }
  else{
     if(statebutton==false){
       Serial.print("c");
       Serial.print('1');
       Serial.print("d");  }
  statebutton=true;
  }
}
```

4.3. ECG Holter with the input from ECG simulator

Before the ECG Holter was tested on the human, the device was calibrated using artificial ECG generated from ECG Simulator. Figure 13 and Figure 14 were the examples of the recorded ECG signal from ECG simulator for 30 and 60 BPM, respectively.

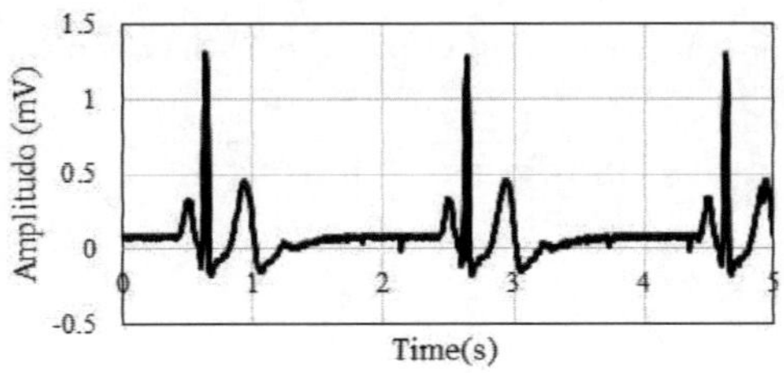

Figure 13. The recorded ECG signal from the ECG simulator (BPM 30).

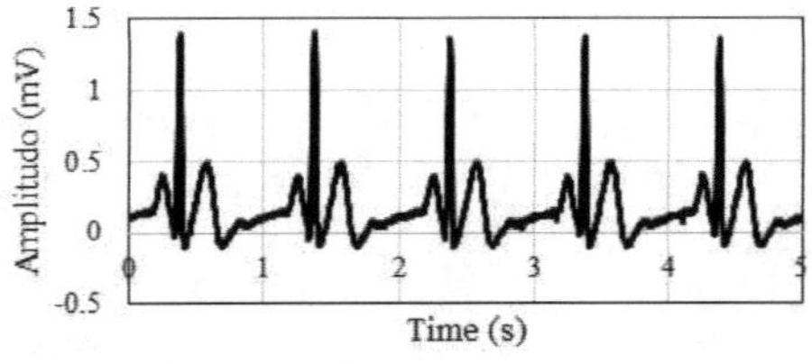

Figure 14. The recorded ECG signal from the ECG simulator (BPM 60).

4.4. The recorded ECG signal from the human body by LEAD II

In this study, the ECG Holter was also tested using the human body by attaching the three electrodes on the left hand, right hand and right leg as shown in Figure 15 to Figure 16.

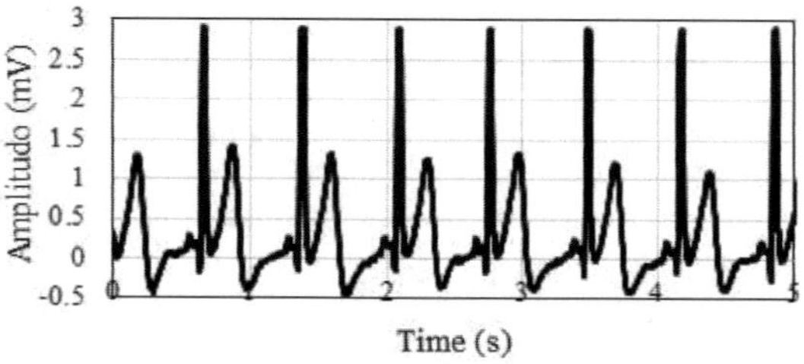

Figure 15. The recorded ECG signal from the human body (subject 1).

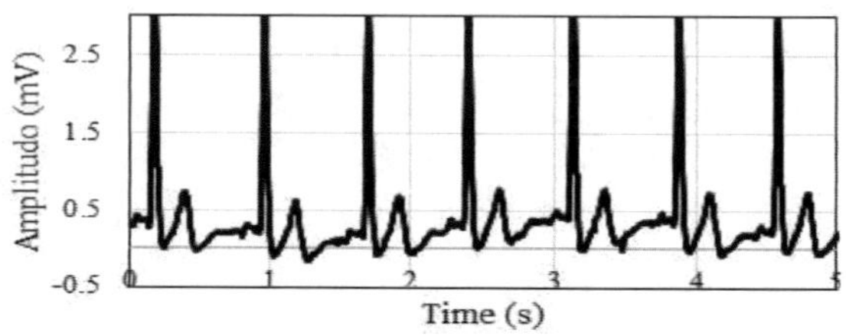

Figure 16. The recorded ECG signal from the human body (subject 2).

4.5. The error of BPM (beats per minutes)

The validation of the BPM value shown in the Delphi programming was compared with the Pulse Oximetry device. The error of measurement for BPM parameter between the design and calibrator, for different BPM setting, was showed in Table 1.

Table 1. The error of measurement for BPM parameter between the design and calibrator for various BPM.

Rate Setting (BPM)	Error (%)
30	0.00
60	0.00
120	0.00
180	0.01
240	0.00

Table 2. The error of measurement for the BPM parameter between the design and calibrator (Pulse Oximetry) for various subject.

Subject	Error (%)
P1	±0.0024
P2	±0.0000
P3	±0.0026
P4	±0.0044
P5	±0.0273
P6	±0.0084
P7	±0.0024

The measurement of BPM value in which the ECG signal is from the human body was also compared in this study, between the Holter design and Pulse Oximetry. The result was shown in Table 2.

4.6. Component prices list

Table 3 is the price to build the Holter monitor with a total price of IDR. 460,000. Mostly, all of the electronics components can be found in the local market. Further, at a low price, the public can build the Holter monitor by themselves in order to record the heart activities.

Table 3. The List of components price.

No.	Components	Vol.	Price (IDR)	The Totals (IDR)
1	Arduino Nano	1	30,000	30,000
2	SD Card Modul	1	20,000	20,000
3	SD Card Memory	1	30,000	30,000
4	Bluetooth Modul	1	60,000	60,000
5	RTC Modul	1	20,000	20,000
6	IC TL084	3	5,000	15,000
7	IC 7660	2	10,000	20,000
8	Resistor	100	250	25,000
9	Capasitor	10	1,000	10,000
10	Multitune	5	5,000	25,000
11	Jumper Cables	1	5,000	5,000
12	PCB	4	15,000	60,000
13	Elektrode	1	60,000	60,000
14	ECG Cables	1	20,000	20,000
15	Box	1	40,000	40,000
16	Belt	1	20,000	20,000
The Total				460,000

5. Discussion

The Holter design has been examined and test completely in this study. Based on the measurement of the Holter output, the resulted ECG signal when using the input from ECG simulator showed the right pattern of ECG signal which consisted of P, Q, R, S, and T waveform with the amplitude of 1 mV, for various BPM (30, 60, 120, 180 dan 240), and sensitivity of 1 mV .

By comparing the output of the Holter between the input using the ECG simulator and the human body, it was shown that there is a different pattern on the PQRST waveform. Each ECG recording for each subject showed a different amplitude. This is reasonable because each subject has different characteristics of the heart. The error of BPM between the design and the input from the ECG simulator showed a value of 0.01. The error of BPM (from the seven subjects) is ±0.0273%. This error value indicated that the Holter is feasible to be used as a medical device.

The main feature of this design is that the ECG data can be sent directly to the computer host by using Bluetooth communication. In this situation, one of the Bluetooth modules has to be configured as a master or slave. In the experiment, we have tested the communication as far as 3 m away between Holter and computer host. In communication, it was proved that there was no corruption in the ECG data. The performance of this work was also compared to other works. In order the Holter able to work in 24 hours then this design needs a battery with high capacity. Carrasco found that his Holter is able to operate in 16 hours without the intervention of the users [10,13]. Jin and Miao claimed that their Holter able to work stable more than 24 hours [9]. In the future work, some issues need to be implemented by using better technology, such as using a Bluetooth module with low power energy and increase the battery capacity but still with compact size.

6. Conclusion

This study has demonstrated the development of the Holter to monitoring the ECG signal from a subject in real-time. This project was built based on an Arduino microcontroller and some analog circuit and a Bluetooth transmitter-receiver to connect to the computer unit. The result has proofed that the accuracy is feasible to be used to monitor the ECG signal in real-time and the data recording can be read from the SD card memory. In the future, we may develop to fabricate and use it in the small clinic in the villages at a low cost.

IOP Conf. Series: Materials Science and Engineering **850** (2020) 012020 doi:10.1088/1757-899X/850/1/012020

References

[1] Tanaka S, Motoi K, Nogawa M and Yamakoshi K 2004 A New Portable Device for Ambulatory Monitoring of Human Posture and Walking Velocity Using Miniature Accelerometers and Gyroscope . 2283–6
[2] Gonzalez L, Walker K, Challa S and Bent B 2016 Monitoring a Skipped Heartbeat :
[3] Carozzi L, Carrara M, Moss T J, Cerutti S, Ferrario M, Lake D E and Moorman J R 2014 Heart Rate Dynamics Predict 2-Year Mortality Risk in Ambulatory Patients Undergoing Holter Monitoring UVa Holter database 89–90
[4] Dian Permana, Mada Sanjaya W.S. H A 2015 Desain dan Implementasi Perancangan Elektrokardiograf (EKG) berbasis Bluetooth **2** 407
[5] Nemati S, Ghassemi M M, Ambai V, Isakadze N, Levantsevych O, Shah A and Clifford G D 2016 Monitoring and Detecting Atrial Fibrillation using Wearable Technology 3394–7
[6] Sreekesh S, Zala A and G K C 2016 Customizable Holter Monitor using off-the-shelf components 2302–6
[7] S. C. Tai, C. W. Chang and C F C 1997 Designing Better Adaptive Sampling Algorithms for ECG Holter Systems *Pan Am. Heal. Care Exch. PAHCE 2011 - Conf. Work. Exhib.* **44**
[8] Lucani D, Cataldo G, Cruz J, Villegas G and Wong S 2006 A portable ECG monitoring device with Bluetooth and Holter capabilities for telemedicine applications *Annu. Int. Conf. IEEE Eng. Med. Biol. - Proc.* 5244–7
[9] Jin H and Miao B 2007 Design of Holter ECG system based on MSP430 and USB technology *2007 1st Int. Conf. Bioinforma. Biomed. Eng. ICBBE* 976–9
[10] Juarez-Carrasco A and Chong-Quero J E 2011 Design and development of a holter prototype with Bluetooth transmission *Pan Am. Heal. Care Exch. PAHCE 2011 - Conf. Work. Exhib. Coop. / Linkages An Indep. Forum Patient Care Technol. Support* 323–7
[11] Lee J, McManus D D, Merchant S and Chon K H 2012 Automatic motion and noise artifact detection in holter ECG data using empirical mode decomposition and statistical approaches *IEEE Trans. Biomed. Eng.* **59** 1499–506
[12] Jagelka M, Donoval M, Telek P, Horínek F, Weis M and Da M 2016 Wearable Healthcare Electronics for 24-7 Monitoring with Focus on User Comfort Institute of Electronics and Photonics , Slovak University of Technology in Bratislava , Ilkovi č ova 3 , 812 19 Bratislava , Slovakia Powertec Ltd ., Drotárska 19 / a , 811 04
[13] Mahdy L N and Tan Q 2018 Smart ECG Holter Monitoring System Using Smartphone *2018 IEEE Int. Conf. Internet Things Intell. Syst.* 80–4
[14] Veiga C, Paredes-galán E, Rivera D, Calvo F, Pazos P, Peña C, García E and Romo A I Defining requirements for an Atrial Fibrillation Screening system that implements tele-Holtering on a cloudified infrastructure .
[15] Costin H, Member S, Pasarica A, Alexa I, Ilie A C and Costin D 2017 Short-Term Heart Rate Variability using Wrist-Worn Pulse Wave Monitor Compared to a Holter ECG 635–8
[16] Parveen R Design and Feasibility Study of Health Related Devices using COTS Components *2018 IEEE Ind. Cyber-Physical Syst.* 529–33
[17] Marsili I A, Masè M, Pisetta V, Ricciardi E, Andrighetti A O, Ravelli F, Nollo G and Member I 2016 Optimized Algorithms for Atrial Fibrillation Detection by Wearable Tele-Holter Devices 16–9

IoT based wearable personal air monitoring for sinusitis

G N Amani and D Wahyudin*

Departemen Pendidikan Teknik Elektro, Universitas Pendidikan Indonesia, Jl. Dr. Setiabudhi 207, Bandung 40514, Indonesia

*deewahyu@upi.edu

Abstract. Sinusitis is inflammation of the cavities around the nose, caused by viruses or allergic reactions of bacteria, fungi, and ARI. Other influential factors are polluting the environment, cold air, and dry. There is no specific device that can help sinusitis sufferers in monitoring air condition. This research was developed to be easy to use for people with sinusitis to be able to display temperature and humidity in real-time using a BME280 sensor and providing action in the form of a humidifier when the humidity is below the set point. The trial results show that the data sent has been successfully received by the output and can turn on the humidifier. From the measurement data obtained, 2,571% error for temperature measurement and 5,898% for humidity measurement. The results of interviews with participants found that this device can help sufferers of sinusitis as a preventive measure if the air condition is not good.

1. Introduction

According to the Indonesian Ministry of Health, sinusitis ranks 25th out of 50 major ranking disease patterns or around 102,817 outpatients in hospitals. This proves that there are still many people who suffer from sinusitis to this day [1]. The prevalence of sinusitis in Indonesia is quite high. The results of research from the sub-section of the Department of ENT Rhinology FKUI-RSCM, of 496 outpatients found 50% of patients with crony sinusitis [2].

Sinusitis is inflammation of the cavities around the nose, caused by viruses or allergic reactions of bacteria, fungi, and ARI. Other influential factors are polluting the environment, cold and dry air, and smoking habit. Dry humidity and cold temperatures can irritate sinusitis and trigger nasal congestion. So that in patients with sinusitis, symptoms more easily appear [3]. The treatment for acute sinusitis is antibiotic therapy, whereas for those who are chronic antibiotic therapy is not enough [4]. Therefore, in order not to become chronic sinusitis, it is better to take preventive measures.

Research on personal water monitoring is carried out [5] to produce large-sized devices with the use of complex sensors. It is generally used to monitor the condition of the surrounding air by focusing on the condition of a particulate meter (PM), which affects the work of the brain and has not been specific to sufferers of certain diseases that are affected by air conditions. Esquiagola et al. [6] by using the same sensor displaying the results of air conditions in the room, not paying attention to the appearance of a wearable device for users and more emphasis as IoT Platform Monitoring Indoor Environments. By using the same sensor research Baca Gómez et al. also focuses on Air Monitoring as a whole with the use of several types of sensors [7]. Research Tastan using a DHT22 sensor produces a fairly large device with the addition of other sensors [8].

According to this issue, IoT Based Wearable Personal Air Monitoring for Sinusitis with Humidifier has been developed. Air condition monitoring equipment specifically for people with sinusitis equipped with a humidifier. The design of this device takes focus on two factors, i.e., cold and dry air, because the air temperature is too low, and the humidity of the dry air can increase sinusitis pain [3]. As a preventive measure, when the condition of the surrounding air is not in good condition, which will notify the user and activate the humidifier to moisturize the air.

2. Methods

This research was preceded by a literature study on sinusitis and its causal factors. Determine which sensors will be used and design what stages will be carried out during the study. The design starts with writing a program script, then designing the PCB layout, soldering, and hardware integration. In designing the program, there are two types of microcontrollers used. For the first program, the device is devoted to monitoring surrounding air conditions by publishing data using the MQTT protocol. The second program design as an output that gives action when the humidity is below the set point by subscribing data that has been previously published by the first program.

The next stage is the hardware design, which includes two types of hardware. The first hardware function is to send data temperature, humidity, air pressure, and altitude to the cloud that comes from the BME280 sensor readings [9]. ESP32WROOM microcontroller will display the measurement results on the OLED display and the FlexIoT platform Dashboard. The second hardware humidifier functions to receive humidity data from the cloud using the ESP32 microcontroller and provides water vapor spray output to the air to restore humidity above the set point. Figure 1 is a block diagram for the workings of the designed device.

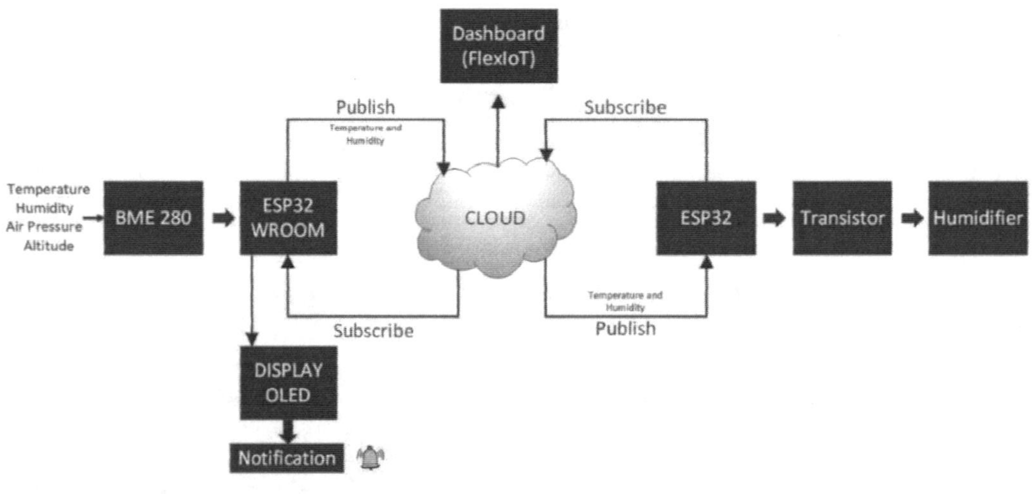

Figure 1. Design of the system.

Determination of the setpoint is taken from the book Goldstein [10] and research Wolkoff [3] where the good air condition approaches relative humidity, the more humidity, the higher the relative humidity. When the air temperature is low, the humidity will increase if the air temperature is high, the humidity will decrease. In general, when the relative humidity of a room rises above 70%, mites and dust will quickly develop [10]. While on research Wolkoff [3] states that dry humidity and cold temperatures can trigger nasal congestion and cause irritation to sinusitis. Ideal air humidity so that allergies do not recur is 45% -65%. So when the humidity is below 45%, the humidifier will turn on and restore the dry air to the ideal air condition.

After successful hardware and software design, testing is done by integrating hardware and software. The sensor measurement data is successfully sent to the cloud and received by the other hardware to turn on the humidifier.

How the user uses the developed system, interviews were conducted. The first participant was an ENT specialist Dr. C, and other participants were three people with sinusitis. Interviews were conducted by presenting the designed device and giving space for participants to express their opinions about this device, what should be added, and suggestions for further development of the device.

3. Result and discussion

3.1. System design

3.1.1. Hardware. The first hardware consists of ESP32WROOM as a microcontroller, BME280 as a sensor, Oled Display, MMBT2222A for automatic reset when the program is uploaded, and the supply is taken from a 3.7V LiPo 1s battery. The second hardware consists of a humidifier that will spray water vapor into the air if the humidity is below the set point. ESP32 as a microcontroller, transistor as a switch, and supply comes from the power bank. When the humidity is below 45%, the transistor will work as a switch that will turn on the humidifier until the humidity returns above 45%.

3.1.2. Software. The program is created using Arduino IDE software and data transmission format using JSON. The first program publishes data to the cloud using the MQTT protocol. The data sent is in the form of temperature, humidity, air pressure, and altitude data. The second program will subscribe to only one measurement result data, namely humidity from FlexIoT. Figure 2 is a display of data on the FlexIoT Platform with four measurement parameters.

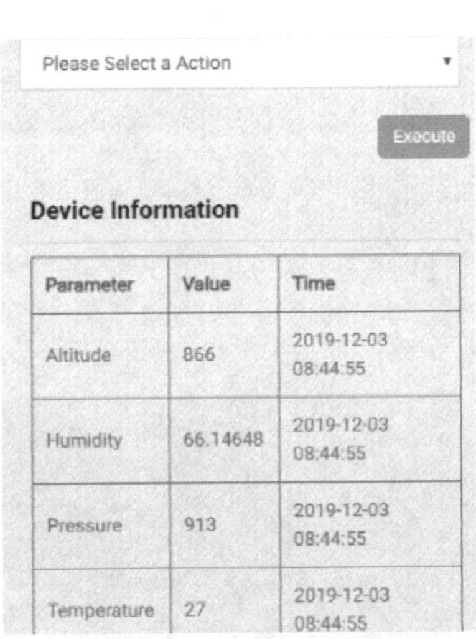

Figure 2. FlexIoT data.

When the hardware is turned on, ESP32 will initialize Wi-Fi. Once connected, the measurement data is displayed on an OLED display and published to the cloud. The second hardware subscribes to humidity measurement data coming from the cloud. A humidifier as an output will turn on if the humidity <45%.

3.2. Test

3.2.1. Hardware test. The results of the design of the device that has been made are tested by integrating the first hardware and the second hardware. Figure 3 displays current temperature (T), and humidity (H) data, and the user can see additional features in the form of a digital clock.

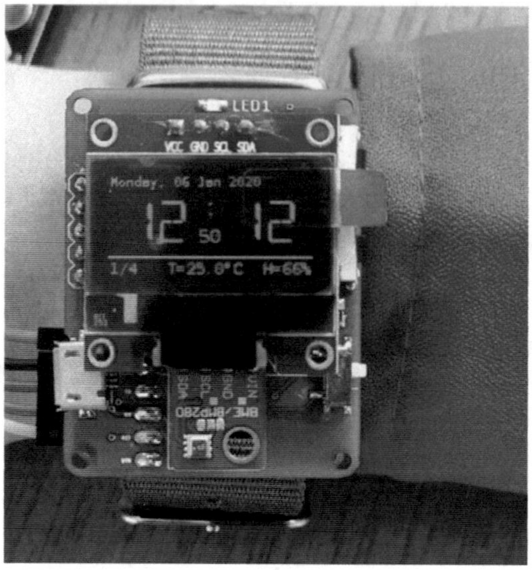

Figure 3. The designed hardware.

Figure 4 show the integration of the first and second hardware. Integration between the two hardware is successful when the humidity is below the set point, and the humidifier turns on. Data from FlexIoT has been published by the first hardware was successfully subscribed to the second hardware.

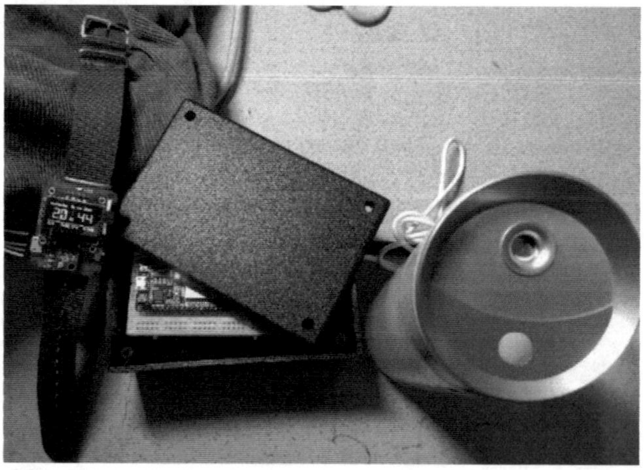

Figure 4. The hardware of personal air monitoring and humidifier

3.2.2. The comparison of hardware measurement.

- Measurements were made in several different rooms with a device that has been designed and make comparisons of measurements using a hygrometer. Temperature measurement error value [9]:

$$eT = \frac{TDesign - THygrometer}{THygrometer} \times 100\% \qquad (1)$$

eT	= Error of temperature measurement
TDevice	= the value of temperature
THygrometer	= the value of hygrometer temperature

Table 1. Comparative temperature of device and hygrometer.

No	Room	Measurement		Error
		Designed Temperature	Hygrometer	
1	A	25,3°C	24,8°C	2,016 %
2	B	26,2°C	25,7 °C	1,946 %
3	C	27°C	25,9 °C	4,247 %
4	D	26,1 °C	25,7 °C	1,556 %
5	E	26,7°C	25,9 °C	3,089 %
		Error Average		2,571 %

- Moisture measurement and comparison. Temperature measurement error value [9]:

$$eH = \frac{HDevice - HHygrometer}{HHygrometer} \times 100\% \quad (2)$$

eH = Error of humidity measurement
HDevice = the value of humidity
HHygrometer = the value of hygrometer humidity

Table 2. Comparative humidity of device and Hygrometer.

No	**Room**	**Measurement**		**Error**
		Designed Humidity	**Hygrometer**	
1	A	75%	80%	6,250%
2	B	70%	75%	6,667%
3	C	68%	72%	5,556%
4	D	69%	73%	6,849%
5	E	69%	72%	4,167%
		Error Average		5,898%

Through the measurement data in Table 1 and Table 2, the results show that the average error from the temperature measurement is 2.571%, and the average error from the humidity measurement is 5.898%.

3.2.3. Interviews with ENT specialists and sinusitis patients. To find out if this device can help sufferers of sinusitis conducted an interview. From the results of interviews conducted with ENT specialists, Dr. C said that things that must be shunned for people with sinusitis are cold, dust, and smoke. From these explanations, the cold, which is the measured part of this device, has been successfully displayed. For sinusitis sufferers, there were three speakers.

"This device is very useful when sinusitis is cured, but the air condition is not suitable for sufferers, so it is sure to interfere with the activity, while the suggestion is better to add dust conditions in the air, because according to personal interviewee A this dust greatly affects the condition of sinusitis sufferers" Participants A, Male, 20 years.

"This device can represent the condition of the surrounding air well" Participants B, Female 22 years.

"This device is very good to use sinusitis sufferers with a form of devices that are easy to use and equipped with a humidifier that can help maintain air condition, while the advice is given would be nice if added with a dust sensor" Participants C, Female 21 years old.

From the results of the interview above, it can be concluded that this device can help sinusitis sufferers to find a suitable air condition so that symptoms do not recur. The advice given tends to add dust sensors that really affect sinusitis sufferers.

4. Conclusion

Based on the results and discussion, it can be concluded that the design of IoT Based Wearable Personal Air Monitoring for Sinusitis was successfully carried out. The result of device test, the data from the sensor are sent to the cloud and successfully received by the second hardware. When the humidity is below the setpoint, the humidifier turns on. From the measurement, 2,571% error for temperature measurement and 5,898% for humidity measurement. The response of the interview results is known that this device can help sufferers of sinusitis as a preventive measure if the air condition is not good.

References

[1] Amelia N L, Zuleika P and Utama D S 2017 Prevalensi Rinosinusitis Kronik di RSUP Dr. Mohammad Hoesin Palembang *Maj. Kedokt. Sriwij.* **49** 75–82

[2] Kurniasih C and Ratnawati L M 2019 Distribusi penderita rinosinusitis kronis yang menjalani pembedahan di RSUP Sanglah Denpasar *Medicina (B. Aires).* **50** 133–7

[3] Wolkoff P 2018 Indoor air humidity, air quality, and health–An overview *Int. J. Hyg. Environ. Health* **221** 376–90

[4] Mustafa M, Patawari P, Iftikhar H M, Shimmi S C, Hussain S S and Sien M M 2015 Acute and chronic rhinosinusitis, pathophysiology and treatment *Int. J. Pharm. Sci. Invent.* **4** 30–6

[5] Zhuang Y, Lin F, Yoo E-H and Xu W 2015 Airsense: A portable context-sensing device for personal air quality monitoring *Proceedings of the 2015 Workshop on Pervasive Wireless Healthcare* pp 17–22

[6] Esquiagola J, Manini M, Aikawa A, Yoshioka L and Zuffo M 2018 SPIRI: Low Power IoT Solution for Monitoring Indoor Air Quality. *IoTBDS* pp 285–90

[7] Baca Gómez Y R, Estrada Esquivel H, Martínez Rebollar A and Villanueva Vásquez D 2019 A Novel Air Quality Monitoring Unit Using Cloudino and FIWARE Technologies *Math. Comput. Appl.* **24** 15

[8] Taştan M 2018 An IoT Based Air Quality Measurement and Warning System for Ambient Assisted Living *Avrupa Bilim ve Teknol. Derg.* 960–8

[9] Utama Y A K, Widianto Y, Sardjono T A and Kusuma H 2019 Perbandingan Kualitas Antar Sensor Kelembaban Udara Dengan Menggunakan Arduino Uno *Pros. SNST Fak. Tek.* **1**

[10] Goldstein MF 2017 *Understanding Sinusitis and Allergy: The Asthma Center Education and Research Fund Manual* Philadelphia: Professional Arts Building Suite 300

IOP Conf. Series: Materials Science and Engineering **850** (2020) 012022 doi:10.1088/1757-899X/850/1/012022

Mini tickets printing machine based on IOT with mobile information using a smartphone

S P Lestari* and S Elvyanti

Education Department of Electrical Engineering, University of Indonesia, Jl. Dr. Setiabudhi 209, Bandung 40154, Indonesia

*sripujilestari64@gmail.com

Abstract. Queue is an event that we often encountered in various places or in public services such as hospitals, banks, restaurants and many more. The restriction of the queuing process is usually due to a long queue thus the process is not enough effective and efficient to the public process. The rapid development of technology that is growing, need the integration between the engine and the mobile information system. So that, we can do a job quickly, and accurately. ESP8266 as a microcontroller and a thermal printer as a ticket printer, both of them can be made a mini machine ticket printer that is small and connected with the smart phone through the Internet connected to the database, so that the user can see directly through the queue number in the smartphone applications. In this paper conducted a mini-making ticket printing machine using ESP8266 based IOT by using a smartphone.

1. Preliminary

Along with the times, a lot of things that can be solved by technology. Information system has been developed which aims to provide convenience to users. Information system created in order to improve the performance of a company or individual. In the development of the necessary link between the machine with the mobile information system so that we can do a job more effectively and efficiently [1]. With the overall relationship between the system allows users to have easy.

Queuing process is an event that we often encounter in society or public services such as banks, hospitals etc. Queuing process is often time consuming and not effective, especially if there are many people taking queue. Queue system is still manual, it makes long waiting time until there is a long queue. In the development of making the application of queuing systems sometimes still find obstacles if the server suddenly died. IOT or the internet of things is already widely used in the development of the industrial world [2]. Advances technology make people want to be able to access or control of an object from far away without the need to come in. The IOT allows users to access the mobile information anytime and anywhere.

From several studies and projects that have been carried out previously, these studies mostly use a microcontroller that has a large power supply. Many studies and projects that have been made regarding the ticket printing machine, but no one has to use a microcontroller ESP8266 as controller. In this study a small ESP 8266 is used as a microcontroller and a data communications using Wi-Fi. Based on the background outlined above shows that the need for a ticket printing machine that has a feature such as small and easy to monitor through a smartphone. Thus, in this study made the ticket printing machine having a small size feature. This tool uses Firebase connected to the Internet as a store database, ESP8266 as microcontroller and Wi-Fi data communication.

2. Research methods

After conducting a review of research and the identification of the problem, the method used in this research is experiment. In order to make a ticket printer engine is able to be integrated with a smartphone, we need the other components. Figure 1 depict the design of a whole system. Power supply serves as the power supply to all components except smartphones. The power supply required is 8V DC to activate a thermal printer, then the power supply is also needed to give ESP8266 that can be lit without the need to connect with the laptop. Pushbutton as input to command the printer to issue a paper ticket.

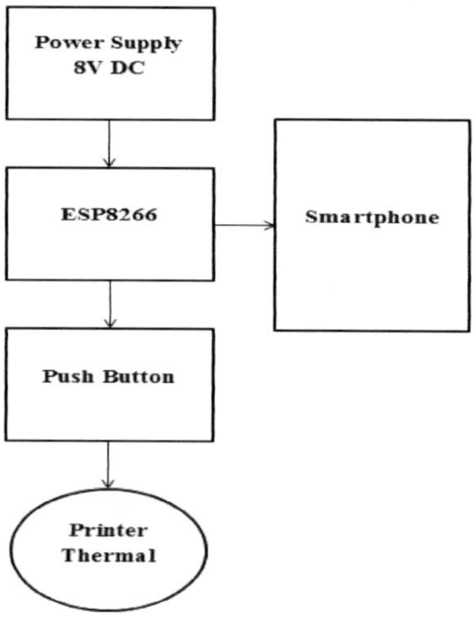

Figure 1. Block diagram.

2.1. *Installation schematic*

Having made the block diagram in Figure 1. then made the circuit design using Fritzing application in order to facilitate the author in assembling the circuit. Thermal printers work with 5-9 V DC but the greater the voltage applied, the result will be better. Because ESP8266 works at 3.3 V then VCC Power supply input at Vin pin on pin VCC ESP8266 not because Vin has a tolerance of up to 10V. Thermal printers have a 5 pin is GND, VCC, RX, TX, DTR yet in use only pin GND, VCC, RX.

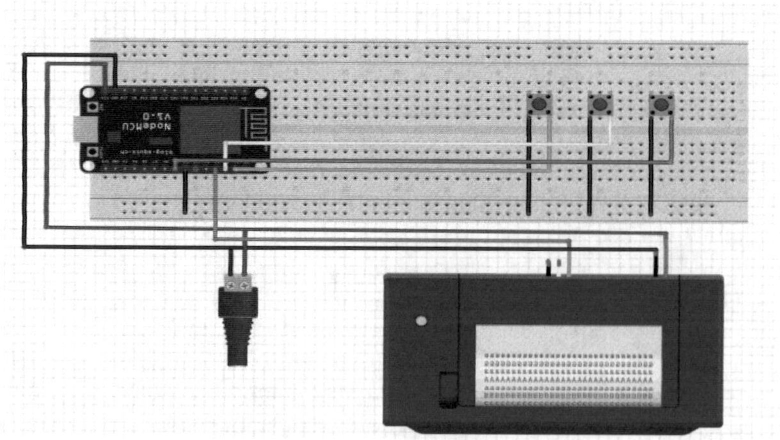

Figure 2. The circuit design.

Subsequent tools placement schemes can be seen in Table 1.

Table 1. Installation schematic.

ESP 8266	Thermal printer	Pushbutton	Research board	Power supply
Vin	VCC			+
Gnd	GND	-	-	-
digital 1		GND	-	
digital 2		GND	-	
digital 3	RX			
digital 5		GND	-	

ESP8266 as microcontroller received input power supply voltage of 8V DC, there are 3 pushbuttons. 1-foot pushbutton connected to the ground pin ESP8266 and the other leg is connected to pin Digital ESP8266, the pin D1, D2 and D5. RX pin on the thermal printer is connected to pin D3 at ESP8266. Then VCC and GND is connected to a power supply. VCC and GND on the power supply is connected to VCC and GND in thermal printer and ESP8266.

2.2. *Flowchart software and hardware*
There are 2 flowcharts or flow diagram in this study. Figure 3 (A) Shows hardware and Figure 3 (B) Represents software.

2.2.1. Hardware. A diagram showing the process from start to enter the Wi-Fi SSID and password on Arduino IDE program. Furthermore, if the Wi-Fi is already connected it will automatically connect to Firebase or Database. Once connected, the user can press the pushbutton to give the orders to the printer to print the ticket. After that Firebase will send data to the smartphone to be displayed in Realtime.

2.2.2. Software. B Diagram shows the process on Android in the Smartphone. This application must be connected to Wi-Fi in advance to retrieve the data on Firebase. After getting data on Firebase then the data will be displayed on the smartphone screen and will continue to change in accordance with the existing data on Firebase.

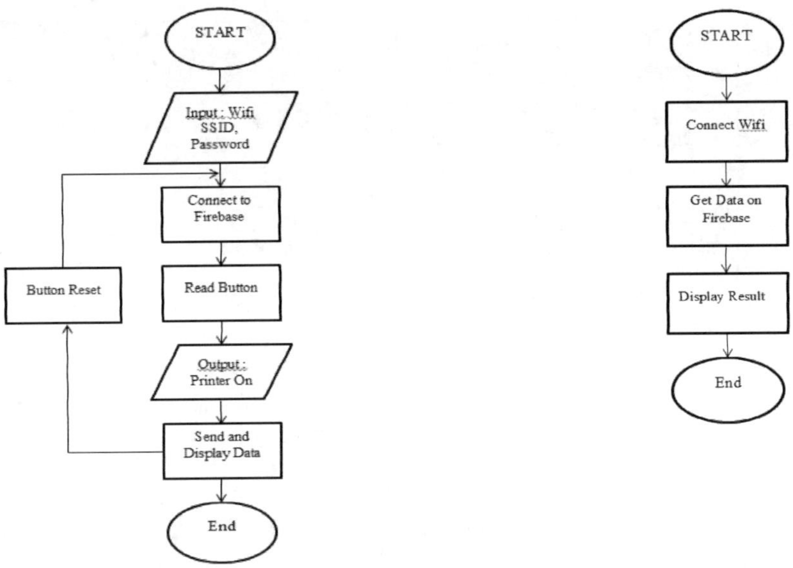

(A) (B)

Figure 3. Flow diagram.

2.3. Programming

The programming tool uses the Arduino IDE (Integrated Development Environment) is software developed by the Arduino is used to design various processes associated with the Arduino programming [3], Arduino IDE can be used as a programmer at ESP8266.

```
#include <ESP8266WiFi.h>
#include <FirebaseArduino.h>
#include <time.h>

#include <PString.h>
#include "Adafruit_Thermal.h"
#include <Wire.h>
```

Figure 4. Include program.

```
Firebase.set("antrian",hitung);
str.print(hitung_int);
printer.boldOn();
printer.justify('C');

printer.setSize('M');
printer.println(F("Nomor Antrian :"));
delay(1000);
printer.setSize('L');
printer.println(str);
delay(1000);
printer.setSize('S');
```

Figure 5. Setting posts.

```
if(!(digitalRead(buttonPin2))&& sekali2 == true){
    printer.wake();
    printer.setDefault();
    hitung--1;
    char buffer[40];
    PString str(buffer, sizeof(buffer));
    if(hitung<100)
    {
      str.print(0);
      if(hitung<10)
      {
        str.print(0);
      }
    }
    else if(hitung<10)
      {
        str.print(0);
      }
```

Figure 6. Input pushbutton.

Figure 4 is an input or input on the Arduino IDE program, input or the input of libraries required to run commands on a program that will be included. Figure 5 is a program for controlling the display the writing on the paper to be printed, such as adjust the font size and input the word you want to create on paper. Figure 6 is a program to execute commands pushbutton, where there are three pushbuttons which serves as the ups, downs and reset button.

```
import android.app.ProgressDialog;
import android.support.constraint.solver.widgets.Snapshot;
import android.support.v7.app.AppCompatActivity;
import android.os.Bundle;
import android.widget.TextView;
import android.widget.Toast;
import com.google.firebase.database.DataSnapshot;
import com.google.firebase.database.DatabaseError;
import com.google.firebase.database.DatabaseReference;
import com.google.firebase.database.FirebaseDatabase;
import com.google.firebase.database.ValueEventListener;
```

Figure 7. Input program android studio.

For the manufacture of android apps using Android Studio. Android Studio is an application to make the application on android. In Figure 7 there is some input in order to run the program on android studio. In addition, to make the machine can print a paper based on input from the pushbutton required a database to store and provide data to be displayed on the smartphone screen. Database using the platform of Firebase i.e. Realtime Database [4]. Once the tool is programmed using the Arduino IDE, then the tool associated with the smartphone with the aid of ESP8266 already connected with Wi-Fi connected with Firebase.

3. Results and discussion

Once all components are assembled, then performed a test by connecting the appliance to the power supply with a voltage of 8V. when the pushbutton is pressed then the ticket will be automatically logged out of a thermal printer. After that firebase will receive the data and sends it back to the application on the smartphone. Firebase using realtime of the data retrieval system. There is no delay on delivery of data, depending on the speed of the Wi-Fi connection. If during the use of the Wi-Fi connection is not stable then when the pushbutton is pressed, the printer will eject the paper and on the smartphone screen will not display data from firebase. Firebase also take time in real time on the Internet to be displayed on the print paper. In addition, to display the time, Firebase also take a date accurately based on the time and date that exist in the internet. Can be seen in Figure 8. Once the push button is pressed it will be out of paper, the number listed corresponds to the number that we press the push button.

Figure 8. Display thermal printer.

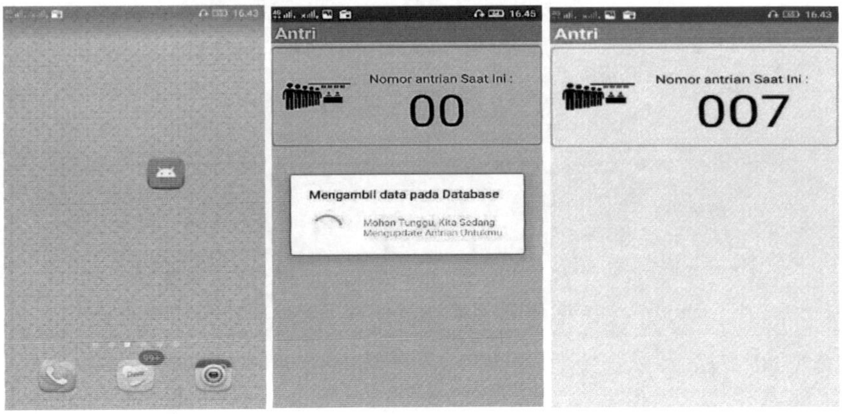

(A) (B) (C)

Figure 9. Display applications on smartphones.

Figure 9 (A) The initial view of the application in the home menu android smartphone before the application is pressed.

Figure 9 (B) Display applications after it opened. The initial view displays the data before the application is connected to the firebase. The application will connect to firebase using internet data connection. After receiving data from firebase application, the application will display the data that has been stored on the firebase.

Figure 9 (C) The appearance after application gets the data from firebase in accordance with the pushbutton which has been pressed. In view Figure 9 (C) a queue number will change according to the pushbutton is pressed, it may be increased or reduced in accordance with the last condition.

4. Conclusion

Based on research that has been done, ESP8266 can be used as a substitute for the Arduino Uno microcontroller which serves as the sender of the data using Wi-Fi networks. ESP8266 is programmed using the Arduino IDE uses Java Script programming language that is easily understood and used. Applications that have been created using android studio in this study can already be connected to the Internet and can be accessed by all users without having to log in first. Time and date are also already connected to the internet so accurately captured. Mini ticket printing machine is easy to put on all over the place and do not require a large power supply, requiring only 8V DC.

However, the results in this study, there are still shortcomings and can be redeveloped, for application on android just display the queue number for the user.

References

[1] Triwibowo D, Kridalukmana R and Martono K T 2015 Pembuatan Aplikasi Terintegrasi, Pendataan Barang di Gudang Berbasis Android *Jurnal Teknologi dan Sistem Komputer* **3**(2) 320-334

[2] Junaidi A 2015 Internet of things, sejarah, teknologi dan penerapannya *Jurnal Ilmiah Teknologi Infomasi Terapan* **1**(3)

[3] Louis L 2016 Working Principle of Arduino and Using IT as a Tool For Study and Reserch *Int. J. Control. Autom. Commun. Syst.* **1** 21-29

[4] Khawas C and Shah P 2018 Application of firebase in android app development-a study *International Journal of Computer Applications* **179**(46) 49-53

IOP Conf. Series: Materials Science and Engineering **850** (2020) 012023 doi:10.1088/1757-899X/850/1/012023

Designing a microcontroller based electric physics experiment board "Mini Serial Lab 01"

M Husni* and W Purnama

Electrical Engineering of Education Department, Indonesia University of Education, Bandung, Indonesia

*husnimuttaqin@student.upi.edu

Abstract. The importance of STEM (Science Technology Engineering and Mathematics) because of technology development are increase and STEM education's media are developed. Many company offer STEM education's media which integrated to data acquisition system and integrated to computer. But not every school or campus can afford that media because of price. To handle this problem in this paper, explain how to make such a STEM education's media which microcontroller based electric physics experiment board which integrated to data acquisition system. There are 5 experiments on this physics experiment board, the experiments are Series Load, Parallel Load, Series and Parallel Load, RC Natural Response, and Electric Induction. There are three main components which is: hardware, software, and firmware. After three components established and integrated to test the system we do some experiment and then compare the result from measurement and the result from equation. The conclusion was the result from the measurement was close from equation.

1. Introduction

The importance of STEM (Science, Technology, Engineering, and Mathematics) nowadays was very important. Electrical engineering, computer programming, biomedical engineering, and environment analyzer was a job which need STEM knowledge [1]. There are so many research which study physics experiment in this several years especially physics experiment that integrate to computing tools. There are so many company that offer physics experiment equipment that integrate to computing tools but the price is so expensive that not every school can afford that expensive tool.

Arduino is a tool that can read the input and control the output. Digital and analog sensor can drive motor, LED, and heater [2]. Arduino is a microcontroller that can use to educational purpose because can use for many application [3]. Focus in this paper is how designing a microcontroller based electric physics experiment board that has a data acquisition. topics of the experiment are: series load experiment, parallel load experiment, series parallel load, electric induction, and charge/discharge capacitor.

2. Method

2.1. Theory

2.1.1. Ohm's law. Ohm's law is made from mathematical analysis from galvanic circuit which is analogy form electric between heat. The Fourier transform from the heat in equation 1.

$$\frac{dQ}{dt} = -kA\frac{dT}{dl} \tag{1}$$

By following the Fourier equation for heat induction an assume as electric induction intensity and temperature gradient, electric current formula on equation 2.

$$\mathrm{I} = \frac{A}{\rho}\frac{dv}{dl} \tag{2}$$

Conductor has a prevalent surface, so the current equation is [4]:

$$\mathrm{I} = \frac{A}{\rho}\frac{V}{l} = \frac{V}{R} \tag{3}$$

2.1.2. Kirchoff's voltage law. Kirchoff's voltage law stated sum of voltage in close circuit on the network is zero [5]. Kirchoff's voltage law has a principle that is energy conversion which is energy transfer is equal than energy absorb [4].

2.1.3. Kirchoff's current law. Kirchoff's current law stated sum of current input is equal than current output [5]. Kirchoff's law is charge conversion. Sum of electron must same per second. If not same there is a charge explosion based of Coulomb's law, but the charge explosion is never happening [4].

2.1.4. Faraday's law. In 1851 Michael Faraday discover that magnetic field and electric field happened because of moving charge. Magnetic field can make an electric field and vice versa. That phenomenon is proved by Faraday because of experiment on figure1.

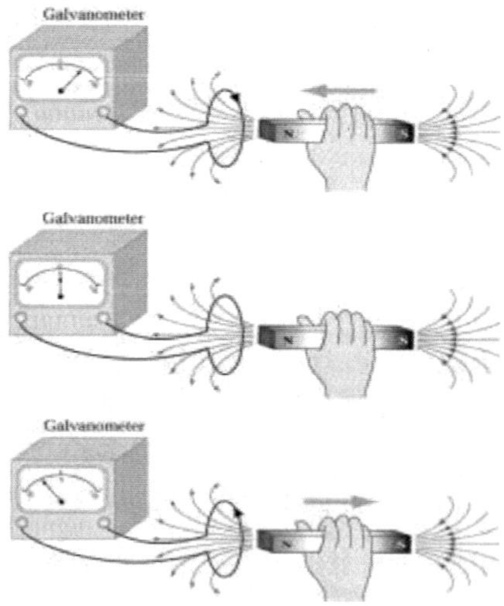

Figure 1. Illustration of Induction GGL experiment.

This experiment proved that galvanometer detect the direction of the field when the magnet close/far from the coil based on direction of electric current that produce by magnetic flux from the coil. The equation of voltage differences on equation 4 [6]:

$$\mathrm{e} = -N\frac{d\varphi}{dt} \tag{4}$$

2.1.5. RC first order transient. When the circuit has a difference from a condition to another condition there is a period when the voltage of every element changing from the value to another value by the time. The differential equation which happen in this circuit are complement and special solving. [5]. The example case is RC series circuit. Implementation of Kirchoff's Voltage Law on equation 5 [6]:

$$\text{Vs} = RC\frac{dv}{dt} + \text{v} \tag{5}$$

2.2. Designing system

Design on this system (Mini Serial Lab 01) there are several systems like collect data, process data, showing data, and save data. On this system there are three main components that is MSL 01 Board, MSL 01 Firmware, and MSL 01 Desktop on the figure 2.

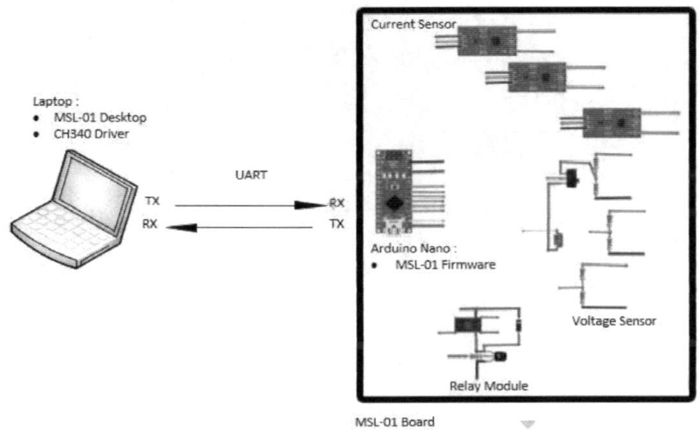

Figure 2. MSL 01 architecture.

2.2.1. MSL 01board. MSL 01 Board is a board that can be used to electric physics experiment. To measure the voltage and the current this board has voltage sensor and current sensor which is connected by arduino nano and the data transferred to PC (Personal computer) via serial data RS232. Design of the schematics on the figure.

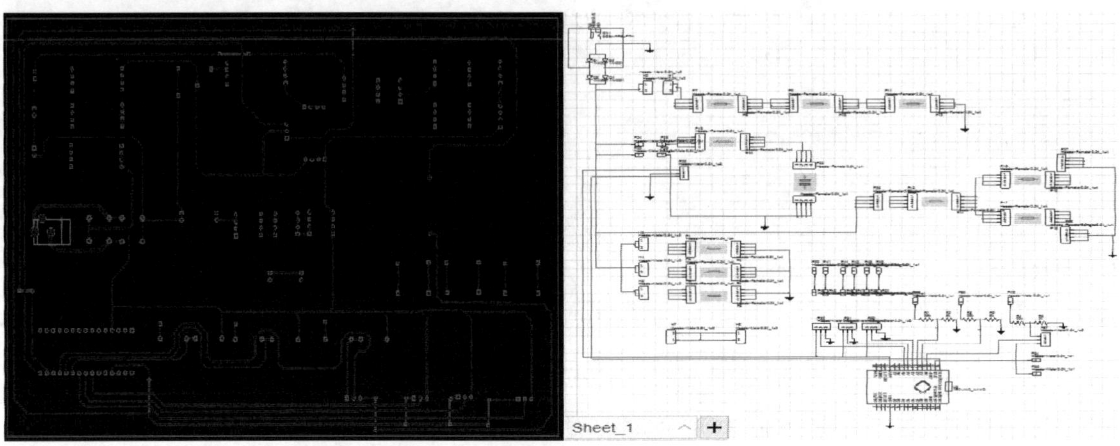

Figure 3. Schematics and PCB layout MSL 01 board.

2.2.2. MSL 01 firmware. MSL 01 Firmware is a program that run on arduino nano to convert ADC value to voltage and the current and transmitt to PC with packet data. Design of packet data on the figure 4.

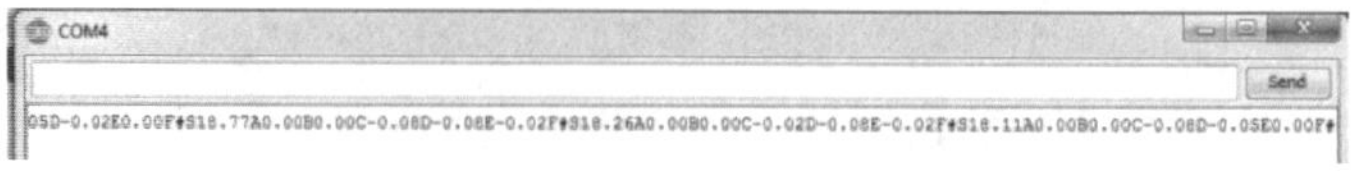

Figure 4. Transmitted packet data.

2.2.3. MSL 01 desktop. MSL 01 Desktop developed by Pascal programming language using Delphi7 IDE (Integrated Development Environment). MSL 01 Desktop has so many function which: displaying sensor measurement value, data parsing, show data in graph and table, data acquisition to excel. display MSL 01 Desktop in fig 5.

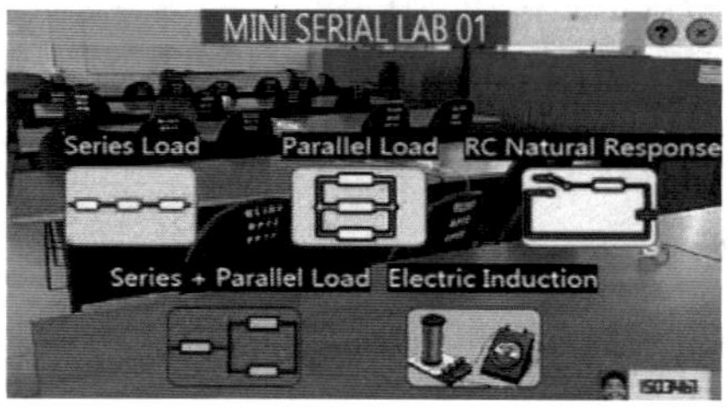

Figure 5. MSL 01 desktop menu.

2.3. Data collecting

Data collecting is from experimental. After an experiment the value compare by measurement value using proper measurement tools and the calculation value.

3. Results

3.1. Measurement result

3.1.1. Series load result. Voltage source is 8V, R1 = 100Ω, R2 = 100Ω, and R3 = 10Ω. The result on the figure 6 and the comparation between proper measurement tool and calculation on the table 1.

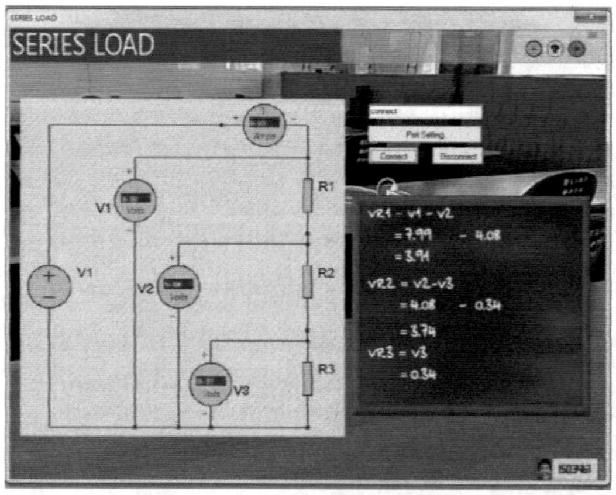

Figure 6. MSL 01 desktop on series load experiment.

Table 1. Comparison table.

Parameter	MSL-01	Proper measurement tool	Calculation
V1	8.02 V	7.57 V	8 V
V2	4.08 V	3.90 V	4.19 V
V3	0.37 V	0.37 V	0.38 V
A1	0.03 A	0.02 A	0.038 A

3.1.2. Parallel load result. Voltage source is 12V, R1= 56Ω, R2=56Ω, dan R3=56Ω. The result on the figure 7 and the comparation between proper measurement tool and calculation on the table 2.

Figure 7. Parallel load experiment.

Table 2. Comparison table.

Parameter	MSL 01	Proper measurement tool	calculation
I total	0.44A	0.34A	0.42A

3.1.3. Series and parallel load. Voltage source is 10V, R1=18Ω, R2=18Ω, dan R3=18Ω. The result on the figure 8 and the comparation between proper measurement tool and calculation on the table 3.

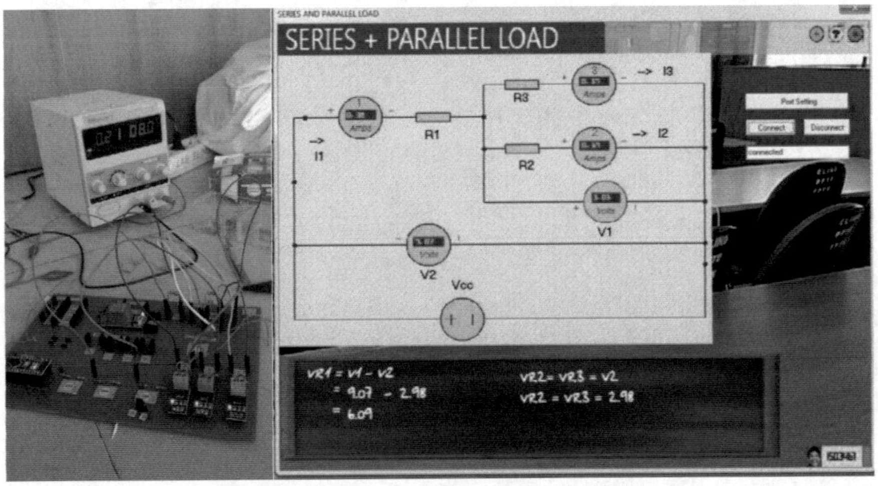

Figure 8. Series and parallel experiment.

Table 3. Comparison table.

Parameter	MSL 01	Proper measurement tool	Calculation
I total	0.38A	0.21A	0.37A

3.1.4. Charge and discharge capacitor. Resistor value is 1.3kΩ and capacitor value is 470μF. The result on the figure 9 and the comparation between proper measurement tool and calculation on the table 4.

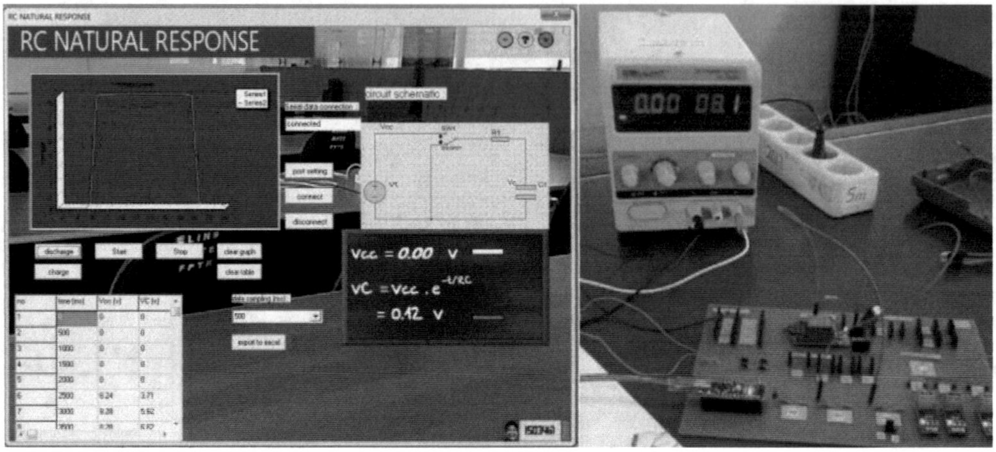

Figure 9. Charge and discharge capacitor experiment.

Table 4. Measurement result by time.

No	time (ms)	Vcc (v)	VC (v)
1	0	0	0
2	500	0	0
3	1000	0	0
4	1500	0	0
5	2000	0	0
6	2500	0	0
7	3000	0	0
8	3500	0	0
9	4000	0.68	0.68
10	4500	0.61	0.64
11	5000	8.19	3.84
12	5500	8.21	5.67
13	6000	8.24	6.62
14	6500	8.26	7.11
15	7000	8.26	7.38
16	7500	8.26	7.53
17	8000	8.28	7.6
18	8500	8.28	7.65
19	9000	8.28	7.67
20	9500	8.28	7.67
21	10000	8.16	7.5
22	10500	0	3.86
23	11000	0	1.98
24	11500	0	1
25	12000	0	0.51
26	12500	0	0.24
27	13000	0	0.12
28	13500	0	0.05
29	14000	0	0

3.1.5. Electric induction experiment. Electric induction use coil that has 601 turns and 10 neodinium magnets. The result on the figure 10 and table 5.

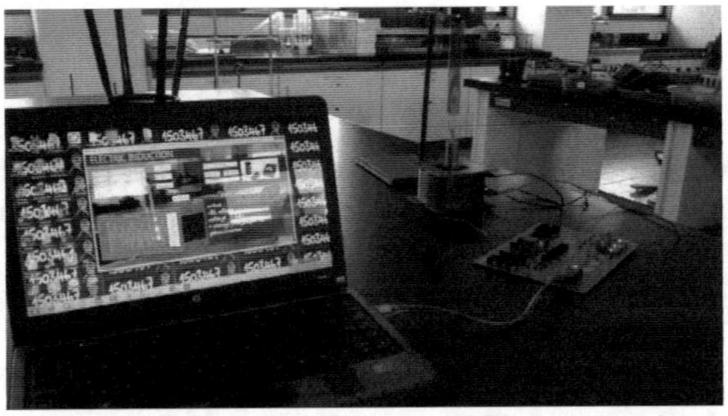

Figure 10. Electric induction experiment.

Table 5. Voltage difference and time measurement.

time (s)	voltage (V)
0	-0.06
0.5	-0.05
1	0.02
1.5	-0.08
2	-0.06
2.5	0.02
3	-0.08
.....	
28.5	-0.06

4. Conclusion

The conclusion of this research is the combination of the hardware, software, and firmware on this electric physics experiment board "Mini Serial Lab 01" works well and the measurement was close from the calculation and suitable for educational purposes.

References

[1] Garrigós A, Marroqui D, Blanes J M, Gutierrez R, Blanquer I and Canto M 2017 Designing Arduino electronic shields: Experiences from secondary and university courses *2017 IEEE Global Engineering Education Conference (EDUCON)* (pp. 934-937) IEEE

[2] Petry C A, Pacheco F S, Lohmann D, Correa G A and Moura P 2016 Project teaching beyond Physics: Integrating Arduino to the laboratory *2016 Technologies Applied to Electronics Teaching (TAEE)* (pp. 1-6) IEEE

[3] Galadima A A 2014 Arduino as a learning tool *2014 11th International Conference on Electronics, Computer and Computation (ICECCO)* (pp. 1-4) IEEE

[4] Sudirham S 2002 *Analisis Rangkaian Listrik* (Bandung: Penerbit ITB)

[5] Nahvi M and Edminister J 2011 *Schaum's Outline Teori dan Soal-soal Rangkaian Listrik* (Jakarta: Penerbit Erlangga)

[6] MIT 2019 *Chapter10 : Faraday's Law of Induction* [online]. Available FTP:web.mit.edu/viz/EM/visualization/coursenote/modules/guide10.pdf

Prosthetic hand with 2-dimensional motion based EOG signal control

M R Pratomo*, B G Irianto, T Triwiyanto, B Utomo, E D Setioningsih and D Titisari

Department of Electromedical Engineering, Poltekkes Kemenkes, Surabaya, Indonesia

*rizky.pratomo97@gmail.com

Abstract. A prosthetic hand is an artificial device that resembles a human hand which can help the human with a physical disability. Previously, the development of a prosthetic hand is designed in various method, from passive to bionic. Electrooculography (EOG) is a technique for measuring potential differences between the front (positive pole formed by the cornea) and the back (negative pole formed by the retina) of the eyeball which can be used to detect eye movements. The purpose of this study is to design a prosthetic hand with two degrees (2D) of freedom using EOG based control. This system consists of electrodes, EOG amplifier, Bluetooth transmitter-receiver, servo motors, and hand prosthetics. In this study, the system will recognize the eye movements, namely front, right, left, up, and down. The system will recognize the motion based on a threshold value. In the hardware implementation, the system was composed of five electrode sensors which installed around the eye, instrumentation amplifier, high pass filter, low pass filter, noninverting amplifier, summing amplifier, a notch filter circuit, and Arduino UNO microcontroller. In EOG data acquisition, this study involved ten healthy subjects. After the evaluation with five trial for each motion, the error for each eye movement is 0%, 0%, 36%, 4% and 16% for right, left, top, bottom, and front, respectively. This study provided an alternative method to control a prosthetics hand with good performance.

1. Introduction

Electrooculography is a technique for measuring the resting potential of the retina. The resulting signal is called an electrooculogram. An electrooculography is a device that measures the voltage between two electrodes placed on the face of the subject so that it can detect eye movements [1]. Electrooculography (EOG) is a method for measuring potential differences between the front (positive pole formed by the cornea) and the back (negative pole formed by the retina) of the eyeball and thus can be used to detect eye movements and blink, when the eyes are fixed in the future, a stable base line potential is measured by electrodes placed around the eyes [2]. Electrodes must be placed as close as possible to the eye to maximize measured potential. Potential changes are detected when the pole approaches or moves away from the electrode when moving the eyes. The sign of change depends on the direction of movement. EOG measurements can be influenced by artifacts arising from muscle potential and small electromagnetic disturbances due to wires or disturbances in the surrounding power lines [3].

Some studies related to EOG signals, Chacko, et al. conducted a study on "Microcontroller Based EOG Guided Wheelchair" on the results of the EOG signal successfully applied and shown for microcontroller-based wheelchair control [4]. Similar research is also conducted by Naga Rajesh, et al. has carried out research on "EOG Controlled Motorized Wheelchair for Disabled Persons" in the results of the study that EOG motorized wheelchair control for persons with disabilities was successfully implemented, but in this study the wireless EOG acquisition system has not been implemented [5]. In studies related to the implementation of control systems and wireless EOG acquisition, Sanjaya, et al. has conducted a research on "Design and Experiment of Electrooculogram (EOG) System and Its Application to Control Mobile Robot" on the results of the research navigation control system for robot cars using a microcontroller with serial data communication via Bluetooth HC-05 successfully applied [6].

Based on the identification of the above problems, EOG signal detection has been successfully made and implemented for the control system using wireless serial data communication, the authors will develop the use of EOG 2 channel signals (horizontal and vertical), for left, right, up and down to control the movement of the open cap of the robot claws and the movement of the pronation robot arm using the Arduino microcontroller, sending and receiving serial data via Bluetooth [7]. Therefore, the author wants to make the module with the title of the thesis "2 Dimensional Motion Prosthetic Hands with EOG Signal Control" [8].

2. Materials and methods

2.1. Experimental setup
In this study, researchers looked at results in 5 respondents with normal eye conditions over the age of 18 years. Respondents were taken randomly and data collection for each respondent was repeated 5 times.

2.1.1. Materials and tool. This study uses disposable ECG electrodes. The placement of the electrodes as shown in Figure 1. Using the Arduino UNO Board as a microcontroller for EOG signal data processing, uses Bluetooth HC-05 as the sending and receiving of data and using a MG995 servo motor as a prosthetic hand drive.

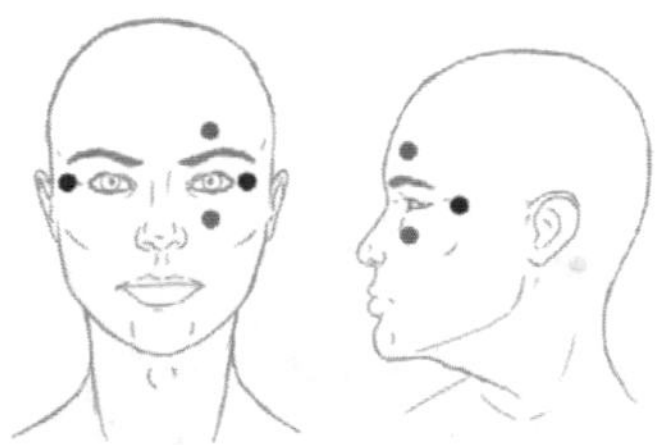

Figure 1. Electrode placement.

In figure 1, is the point of mounting electrodes around the face, black marks are placement of vertical channel electrodes, red marks are placement of horizontal channel electrodes, and yellow marks are placement of reference electrodes or ground [9,10].

2.1.2. Experiment. In this study, the modules that have been made will be applied to 5 respondents and will be tested for success rates and error rates in controlling prosthetic hand movements 5 times in each respondent. Tests carried out consist of upward eye movement signals to control pronation motion, downward eye movement signals to control supination motion, right eye movement signals to control close motion, left eye movement signals to control open motion and forward eye movement signals.

3. Diagram block and flowchart
In this study, the researcher made 2 modules consisting of a transmitter module and a receiver module [1]. The block diagram and flowchart from the transmitter module and receiver module are shown in figure 2 and figure 3.

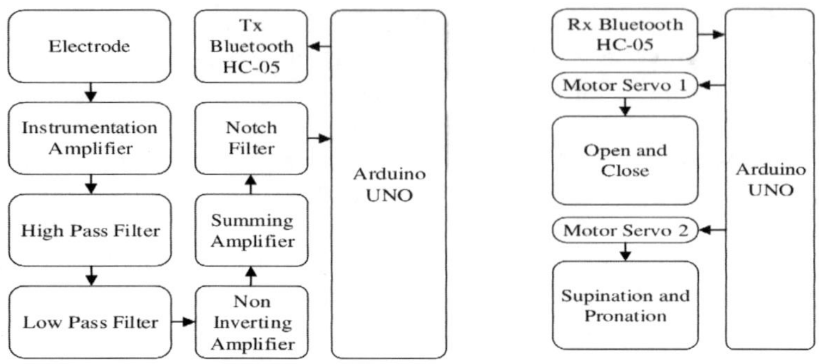

(a) (b)

Figure 2. Diagram block, (a) Transmitter module, (b) Receiver module.

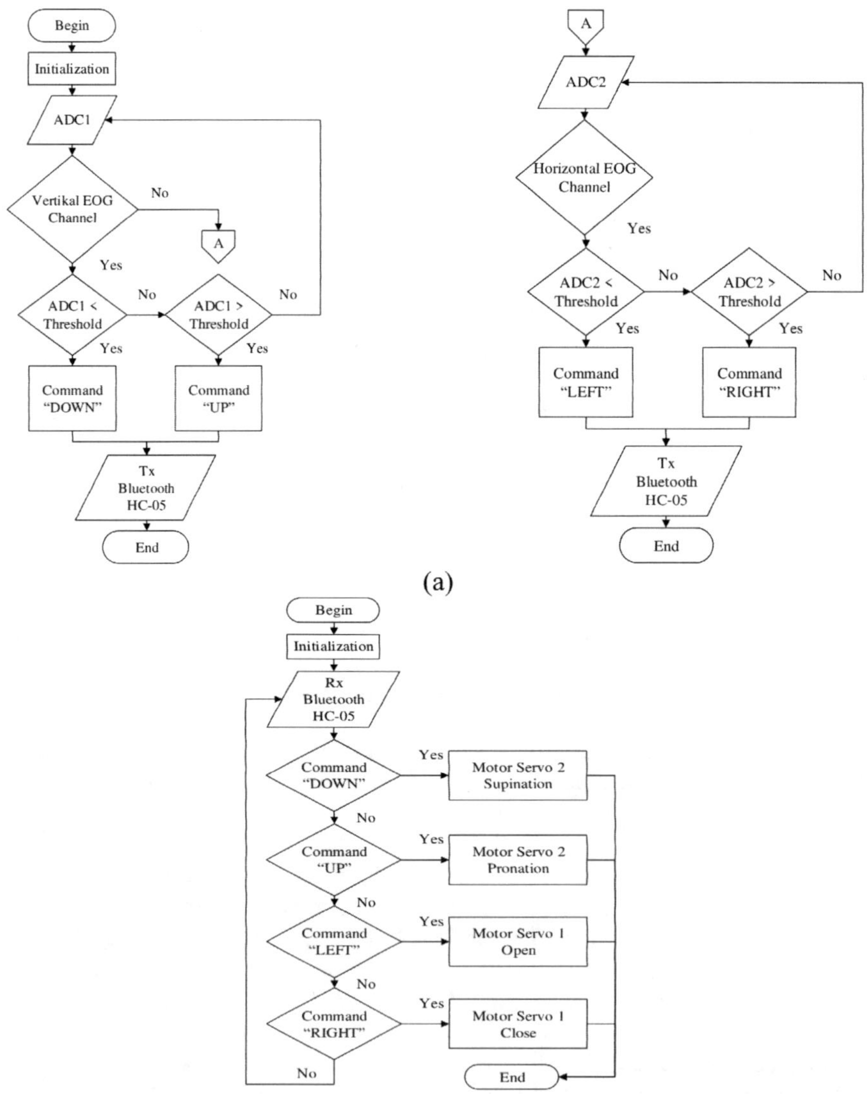

(b)

Figure 3. Flowchart, (a) Transmitter module, (b) Receiver module.

In figure 2 and figure 3 can be explained from the diagram block and flowchart the transmitter module is input through five electrodes that will be installed around the eyes to measure potential differences from the front eye view, right eye view, left eye view, eye view up and down eyes. Input from the five electrodes will enter the instrumentation amplifier circuit, so that the voltage obtained will be more maximal and can know the range of changes in eye movement. A filter circuit is needed for processing the desired signal using a 0.1 Hz High Pass Filter and a 30 Hz Low Pass Filter. The output of the vertical channel EOG circuit will enter the ADC1 pin and the horizontal channel instrumentation output will enter the ADC2 pin on the Arduino Uno for processing. The results of decisions of the right, left, up and down eye movements will be processed using the Arduino software design. In the processed output, a command will be sent via the Tx Bluetooth HC-05 module through the control of the Arduino UNO Board.

3.1. Analog circuit

3.1.1. Instrumentation amplifier. The instrumentation amplifier circuit based on figure 4 serves as an initial gain from the Horizontal Channel Electrode and Vertical Channel inputs to get the shape of the EOG signal. The shape of the EOG signal requires amplitude amplification [11-13].

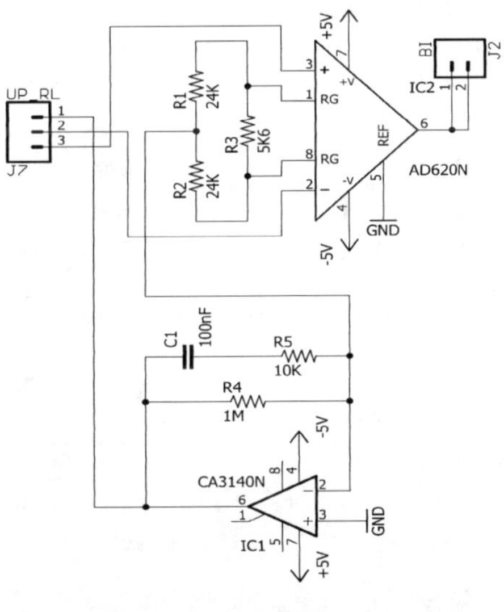

Figure 4. Instrumentation amplifier.

The R3 resistance is a resistor gain (Rg) which is used to regulate its gain. By setting the resistor value on R3, gain can be adjusted to adjust the output voltage of the EOG signal. In this EOG signal an initial gain of 9.8 times will be set. To calculate the magnitude of reinforcement can be determined in the following equation, Eq. (1):

$$G = \frac{49.4K}{R_3} + 1 \tag{1}$$

4. Results and analysis

4.1. The design of prosthetic hand with 2-dimention motion based EOG signal control

The photograph of the analogue and digital part of the Prosthetic Hand with 2-Dimention Motion Based EOG Signal Control was shown in figure 5 and figure 6, respectively. The analogue part consisted of AD620 and three of CA3240 (OP-AMP). The digital part consisted of the Uno Arduino microcontroller which is the main board of Prosthetic Hand with 2-Dimention Motion Based EOG Signal Control device,

IOP Conf. Series: Materials Science and Engineering **850** (2020) 012024 doi:10.1088/1757-899X/850/1/012024

Bluetooth module which used to communicate the data between the microcontroller and The Design Hand Prosthetic Robotic.

Figure 5. The analogue design module.

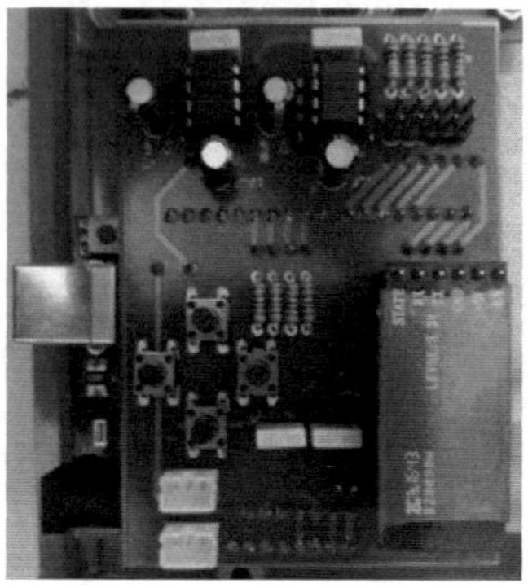

Figure 6. The digital design module.

4.2. The result and measured of respondent

In this study, the authors measured the Horizontal and Vertical EOG Channel signals to the respondents and tested the respondent's eye movements if there was an order to move the 2-dimensional prosthetic hand simulation [14,15].

Figure 7. Placement of Disposable EOG Electrodes.

Figure 8. Placement of Disposable EOG Electrodes.

Table 1. Test results on subject 1.

Eye Movement	Command	Test result					The Error
		1	2	3	4	5	
Right	Supination	✓	✓	✓	✓	✓	0%
Left	Pronation	✓	✓	✓	✓	✓	0%
Up	Open	×	✓	✓	✓	×	40%
Down	Close	✓	✓	✓	✓	✓	0%
Front	No Movement	✓	✓	✓	×	✓	20%

Table 2. Average results of testing errors 5 respondent.

Eye Movement	Command	Test Result (%)					Average Error
		S1	S2	S3	S4	S5	
Right	Supination	0	0	0	0	0	0%
Left	Pronation	0	0	0	0	0	0%
Up	Open	40	60	40	20	20	36%
Down	Close	0	20	0	0	0	4%
Front	No Movement	20	40	20	0	0	16%

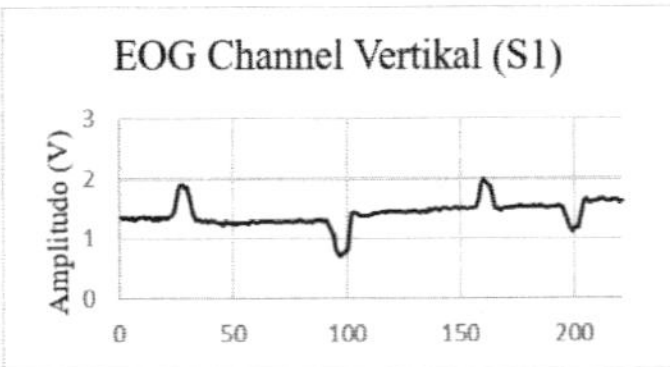

Figure 9. Vertical EOG channel (S1).

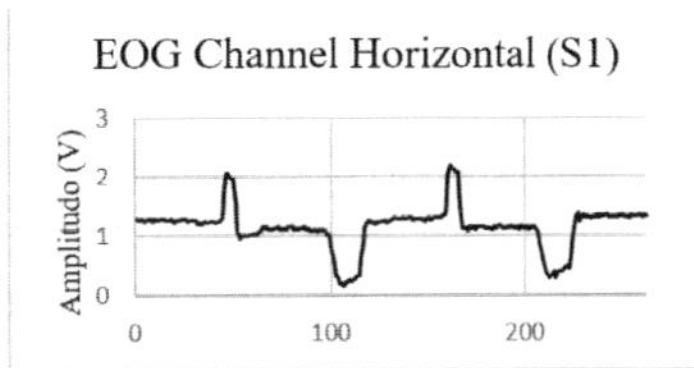

Figure 10. Horizontal EOG channel.

5. Conclusion

This study shows that the prosthetic wireless Bluetooth hand control technology can facilitate users with hand disability with eye movements from the EOG signal. Electrooculography (EOG) is a technique for measuring potential differences between the front (positive pole formed by the cornea) and the back (negative pole formed by the retina) of the eyeball which can be used to detect eye movements. The EOG signal is tapped by attaching 5 disposable electrodes which are placed around the eyes. The results of the study show that the EOG signal has a cut off frequency of 0.1-30 Hz.

Based on the results of measurements and testing of 5 respondents, the average error in each movement is Supination 0%, Pronation 0%, Open 36%, Close 4% and no move 16% by testing 5 times.

References

[1] Siddiqui U and Shaikh A N 2013 An Overview of " Electrooculography " **2** 4328–30

[2] Fang F and Shinozaki T 2018 Electrooculography-based continuous eye-writing recognition system for efficient assistive communication systems *PLoS One* **13** 1–20

[3] Banerjee A, Datta S, Pal M, Konar A, Tibarewala D N and Janarthanan R 2013 Classifying Electrooculogram to Detect Directional Eye Movements *Procedia Technol.* **10** 67–75

[4] Chacko J K, Oommen D, Mathew K K, Sunny N and Babu N 2013 Microcontroller Based EOG Guided Wheelchair **7** 746–9

[5] A. Naga Rajesh, S. Chandralingam, T. Anjaneyulu K S 2014 EOG Controlled Motorized Wheelchair for Disabled **8**

[6] Sanjaya W S M, Anggraeni D, Multajam R, Subkhi M N and Muttaqien I 2017 Design and Experiment of Electrooculogram (EOG) System and Its Application to Control Mobile Robot Design and Experiment of Electrooculogram (EOG) System and Its Application to Control Mobile Robot

[7] Cotton D P J, Cranny A, Chappell P M, White N M and Beeby S P 2007 Control strategies for a multiple degree of freedom prosthetic hand *Meas. Control* **40** 24–7

[8] Dave A, Muthu P, Karthikraj V and Latha S 2018 Structural Integration and Control of Peerless Human-like Prosthetic Hand *J. Phys. Conf. Ser.* **1000**

[9] He S and Li Y 2017 A single-channel EOG-based speller *IEEE Trans. Neural Syst. Rehabil. Eng.* **25** 1978–87

[10] Barea R, Boquete L, Mazo M and Lopez E 2002 Wheelchair Guidance Strategies Using EOG *Entomol. Exp. Appl.* **103** 239–48

[11] Arthi S V and Norman S R 2015 Analysis of Electrooculography signals for the Interface and Control of Appliances *Int. J. Multidiscip. Curr. Res.* **3** 87–90

[12] Brahmaiah V P, Sai Y P and Prasad M G 2017 Data Acquisition System of Electrooculogram *2017 IEEE 7th International Advance Computing Conference (IACC)* (pp. 716-721) IEEE

[13] Golparvar A J and Yapici M K 2018 Electrooculography by wearable graphene textiles *IEEE Sensors Journal* **18**(21) 8971-8978

[14] Xu W, Chen N, Han X and Sun J 2018 Research on wheelchair robot control system based on EOG *AIP Conf. Proc.* **1955**

[15] Clement R G E, Bugler K E and Oliver C W 2011 Bionic prosthetic hands: A review of present technology and future aspirations *Surgeon* **9** 336–40

IOP Conf. Series: Materials Science and Engineering **850** (2020) 012025 doi:10.1088/1757-899X/850/1/012025

Design of baby growth monitor system in the Posyandu for nutrition status analysis

B Utomo[1,*], H Gumiwang[1], L Soetjiatie[1], T Triwiyanto[1] and D S Oswarida[2]

[1] Department of Electromedical Engineering, Politeknik Kesehatan Kemenkes Surabaya, Surabaya, Indonesia
[2] Nutrition Installation Hospital Dr Soetomo Surabaya, East Java, Indonesia

*bedjoutomo123@gmail.com

Abstract. Generally, the reporting and recording a nutrition status of infants and baby, which conducted at the Posyandu, is still using the manual system. The problem in the manual system is that the local Health Office cannot monitor the status of infants at the Posyandu. Therefore to support government programs, especially in maternal and child health, an application system is needed that helps to monitor the growth and development of infants and baby. The purpose of this study is to design a system to monitor the infant and baby growth at the Posyandu for Health Office of Surabaya, which further it can be accessed online. This research method was Retrospective and perspective designed, where the development of the growth of infants and baby can be monitored and analysed for its impact. The population and samples were taken from infants and baby in the local Posyandu. The growth monitoring of infants and baby used indicators of nutritional status with the index of weight for ages (BB/U), weight for length (BB/TB), and length for ages (TB/U). In this research, the designs of baby growth monitoring system can be applied well at many Posyandu. In this study, the MySQL and PHP database system data can be integrated into other database systems to support existing reporting systems. Furthermore, this research application can be applied to monitor baby growth in the city and this country.

1. Introduction

Information systems are a combination of components that work together to collect, process, store and provide information to support decision making, coordination, control, problem analysis and virtualization in an organization [1]. The information system is a network of interconnected, gathered procedures to carry out an activity or complete a certain goal. Therefore, the information system is a collection of interconnected components with its task to process data into information to be used for certain purposes. So it can be concluded that the information system is a collection of interconnected components and processing data into information so that the information can be used for specific purposes,one of them is the monitor of the growth and development of infants in Posyandu. This monitor system uses a web-based information system that can inventory all data on the growth and development process of babies. Expectations of this system can be used later to assist Posyandu activities and assist community service activities, especially baby growth and development monitor. Furthermore, the data can be integrated into the database system of nutrition services at the Puskesmas, especially monitor the incidence of malnutrition, stunting in infants from 0 months to 12 months.

In Indonesia, the average coverage of children under-5 years measurement is 49.4% with an indicator big as 65%. Meanwhile, KMS (*Kartu Menuju Sehat* – Growth Chart) is only used by 30.5% of the total population of children under 5-years [2]. KMS coverage and monitoring is an indicator to determine the growth and development of children under 5-years. Impaired growth during pregnancy can lead to low birth weight. Research in Nepal shows that babies with low birth weight have a higher risk of stunting [3].

In the course of developing the nutritional conditions of children, according to the Indonesian Ministry of Health, nutrition conditions in Indonesia have improved in the last five years. This can be seen from the results of the 2018 Basic Health Research (Riset Kesehatan Dasar - Riskesdas), which showed that the prevalence of malnutrition, stunting, and excess nutrition had declined compared to 2013. Malnutrition dropped from 19.6% to 17.7%, and stunting decreased from 37.2% to 30.8%. However, according to the health policy observer and Chair of the Indonesian Health Economics Association (IHEA), Prof. Hasbullah Thabrany, regardless the fact that nutrition problems have decreased in number, the basic concern is the risk of non-communicable diseases, malnutrition and stunting that are correlated with the level of Intelligence Quotient (IQ) [4]. The research results showed that small born length (OR=4,091; CI=1,162-14,397), the absence of breastfeeding (OR=4,643;CI=1,328-16,233), low family income (OR=3,250; CI=1,150-9,187), mother's low education (OR=3,378; CI=1,246-9,157), and low knowledge on nutrient (OR=3,877; CI=1,410-10,658) are among the factors related to stunting among the children under-5 years [5]. On the other hand, overweight and obesity are other issues needed to be addressed [6]. Based on Riskesda data by the Ministry of Health of the Republic of Indonesia, the prevalence of overweight and obesity in children aged 5–12 years reached 18.8% in 2013 [2]. Overweight and obesity not only affect physical health, but also mental condition.

Several studies on the condition of children under-5 years who are suffered from malnutrition, stunting and overweight and obesity, there are several advantages in implementing computer-based, website and phone applications. However, it is still not widely used in supporting health programs. Scholars from Bangalor, India, Sohas Hola, Mahima M Katti, developed the design of Android-Based Mobile Application Development [7]. An open-source application is recommended for nutritional status measurement. However, the web-based database processing system to store high-capacity data is still needed to monitor continuous development monitoring. Therefore, a study on database development is still important, especially in monitoring infant's growth in Posyandu to support the attempt of developing the health of mothers and their children in Indonesia.

Malnutrition and stunting show the interrelated relationship. Stunting in children is the impact of nutrient deficiency during the first thousand days of their lives. Stunting is a state of height index according to age below minus two standard deviations based on WHO standards. Stunting is the long-term manifestation influenced by a low-quality diet, recurring infectious disease, and environment [2]. Based on the background, this article is aimed at designing research to monitor baby growth by implementing a database-system. The use of database-system will allow wider data service in the future: e-cloud [8]. In general, this article is focused on designing a system to integrate with Android-based baby growth monitoring system, especially to identify the children's nutritional status based on body weight for age (BB/U), bodyweight for body height (BB/TB), and body height for age (TB/U).

2. Methods and designs

2.1. Development stages

Survey and system design development were conducted in baby measurement post in Posyandu before the initialization of research. The initial research includes: a) learning input and report system, b) analysing problems with the support of related literature and article journal, c) developing the application system and preparing the infrastructure such as computer, software, and hardware, d) composing the concept framework, e) testing the application, and f) issuing reports. The stages are the following:

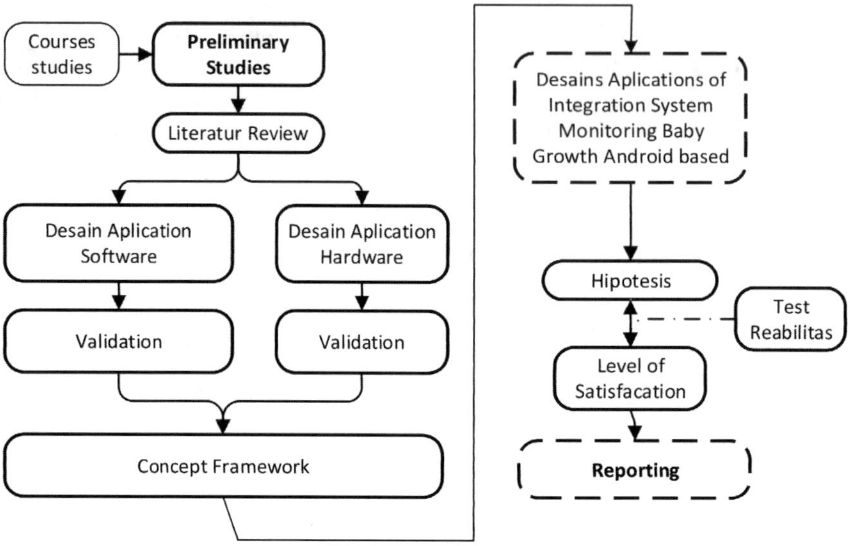

Figure 1. Stages of research.

2.2. Working concept framework

Working concept framework is composed of several methods to support the application system of baby growth monitoring, including: a) user as administrator and operator officer, b) data input that include data of the baby, anthropometry, and the parents, c) media, or the development of database-based application system of baby growth monitoring using MySQL, PHP, and Java Program [9], d) database [10], e) service output, individual, and institution. Later, the data can be developed for multiuser as explained in the following chart [11]:

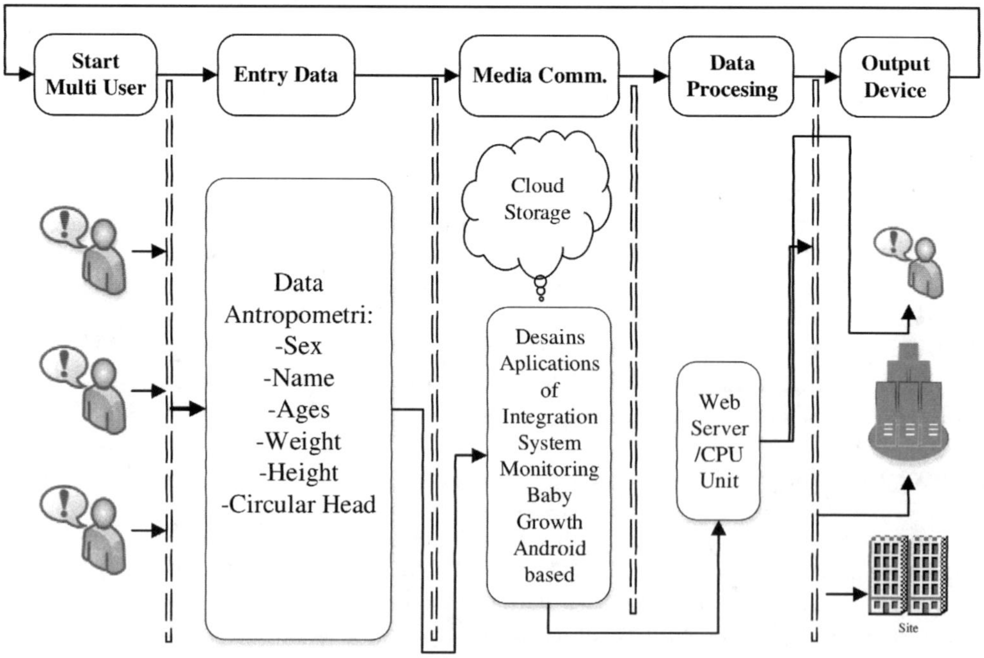

Figure 2. Block diagram system monitoring application design baby growth.

3. Results and discussion

3.1. System development and application

The following figures show the designed interface of the monitoring application system. It includes a menu for username and password and input working stages.

Figure 3. Interface design.

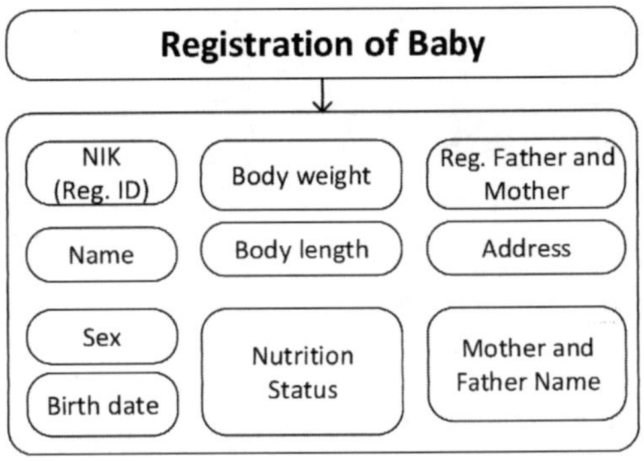

Figure 4. Working stages of data input system.

3.2. Design of infant monitoring menu

The development of the infant monitoring menu is the main activity of the Posyandu, which is the monitoring of infant height and weight measurement. The data is inputted by the entry form of baby examination to be used as the monitoring of infant growth. The following is the design of the infant monitoring menu.

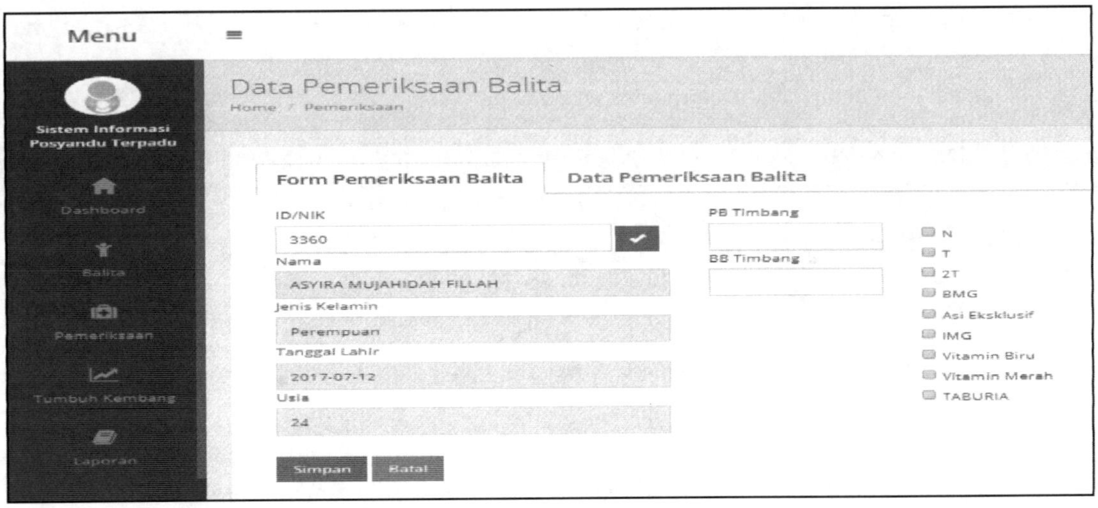

Figure 5. Menu interface toddler's inspection data.

Next, the output of the report design involved baby weight (kg), and age (month). It is shown in the following figure.

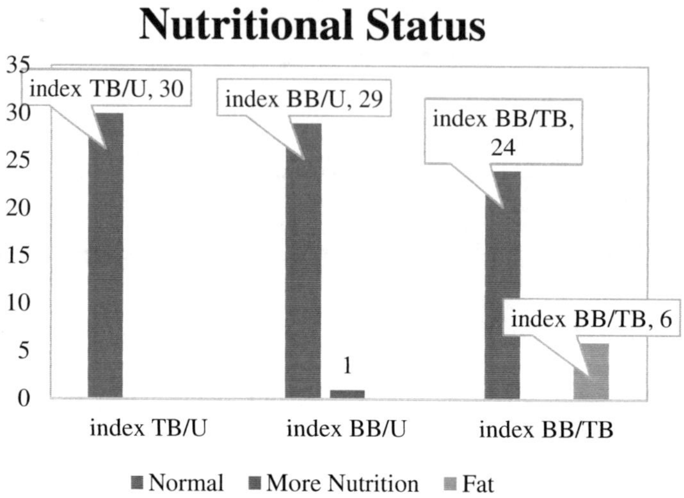

Figure 6. Distribution of children under-5 year for status nutritional in Srikandi Posyandu.

The above figure is the result of the database application system to be downloaded and processed to make a report on the monitoring of baby growth according to nutritional status.

3.3. ZScore analysis

The Z-Score of the recorded data is then analysed to determine the baby's nutritional status [12]. The following is the example of the Z-Score calculation [13] for 15-months old male baby and 15-months old female baby with the BW/age measurement of 10 kilograms:

$$ZScore = \frac{10-10.3}{10.3-9,2} = \quad -0,27 \gg \quad Good\ nutritional\ status \tag{1}$$

$$ZScore \quad = \frac{Subject\ Individual\ Score - Reference\ Median\ Score}{Reference\ Standard\ Intersection\ Value} \tag{2}$$

Table 1. Standard value of body weight for age (BW/age) in male child < 60 month.

Age (month)	Body Weight (Kg)						
	-3SD	-2SD	-1SD	Median	1SD	2SD	3SD
15	7.4	8.3	9.2	10.3	11.5	12.8	14.9

Nutritional status of 15-month old female baby with bodyweight of 10 kilograms:

$$z\ Score = \frac{10-9.6}{10.9-9.6} = \ 0.30 \gg Good\ nutritional\ status \tag{3}$$

Table 2. Standard value of body weight for age (BW/age) in female child < 60 month.

Age (month)	Body weight (Kg)						
	-3SD	-2SD	-1SD	Median	1SD	2SD	3SD
15	6.7	7,6	8.5	9.6	10.9	12.4	14.1

The Z-Score analysis is also conducted using software applications. Furthermore, the results of the examination of toddlers' data according to Zscore calculation based on the BB / U index, BB / TB can be seen its distribution in the chart below.

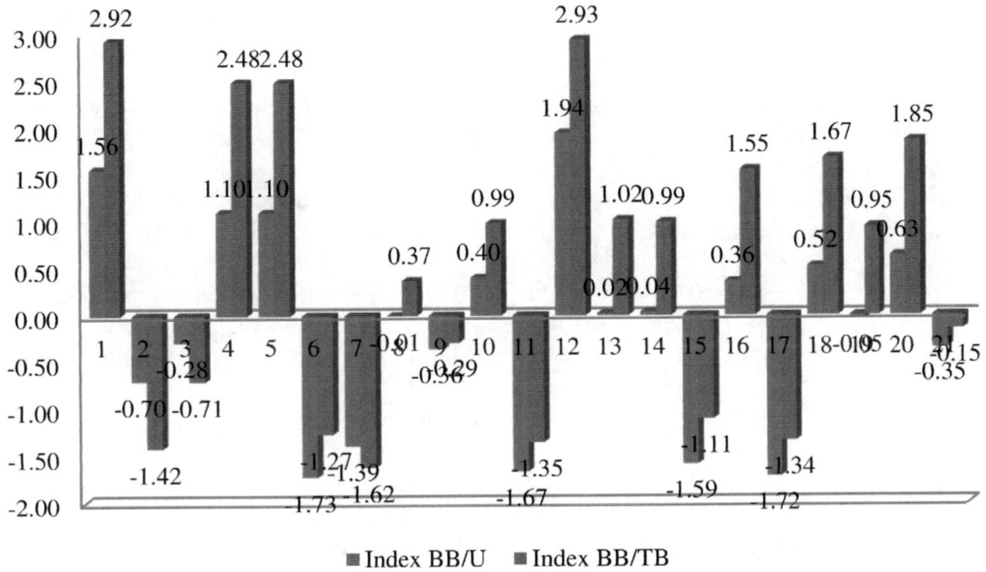

Figure 7. Analysis of body weight for ages (BB/U) and body weight for body length (BB/TB).

The following program code is used to determine the baby's nutritional status:

```
$this->load->model('m_tbref_bb_u');
$w = array(
'umur' => $usia,
'jnskel' => $jnskel);
$data_ref = $this->m_tbref_bb_u->read($w);
$data_x = $data_ref->row();
$m = $data_x->median;
$sd1 = $data_x->sd1;
$_1sd = $data_x->_1sd;
$sd2 = $data_x->sd2;
$_2sd = $data_x->_2sd;
$sd3 = $data_x->sd3;
$_3sd = $data_x->_3sd;
```

```
if($bb>$m){
$z = ($bb - $m) / ($sd1 - $m);
} else {
$z = ($bb - $m) / ($m - $_1sd);}
$status = '' ;
if($z < -3){
$status='Gizi Buruk';
} else if(($z >= -3) and ($z < -2)){
$status='Gizi Kurang';
} else if(($z >= -2) and ($z <= 2)){
$status='Gizi Baik';
} else if($z > 2){
$status='Gizi Lebih';}
```

4. Conclusion

The development of the database-based application using the programming language of PHP and Javascript and application program of XAMPP and MySQL can be used to monitor the baby growth in Posyandu. It can help the government's program in developing the health of mothers and their children. The data-based application system allows faster and more effective measurement of children's nutritional status. The application can later be developed by adding access rights for all members of Posyandu and processing the data of baby nutritional status to produce information for the public on the website and Android devices.

References

[1] Laudon K C and Laudon J P 1999 *Management information systems* (US: Prentice Hall PTR)
[2] Badan Penelitian dan Pengembangan Kesehatan 2013 *Riset kesehatan dasar (Riskesdas) 2013* (Jakarta: Kemenkes RI)
[3] Paudel R, Pradhan B, Wagle R R, Pahari D P and Onta S R 2012 Risk factors for stunting among children: a community based case control study in Nepal *Kathmandu University Medical Journal* **10**(3) 18-24
[4] Renyoet B S and Nai H M E Estimasi potensi kerugian ekonomi akibat wasting pada balita di indonesia *Jurnal Gizi Indonesia (The Indonesian Journal of Nutrition)* **7**(2) 127-132
[5] Kemenkes RI 2011 *Program Kesehatan dan Gizi Berbasis Masyarakat* (Jakarta: Kemenkes RI)
[6] Tochie J N, Mbonda A, Fonkwo V and Aletum V 2019 Childhood Overweight and Obesity in Sub-Saharan Africa: Current Definition, Prevalence and Risk Factors *Acta Scientific Paediatrics* **2** 37-40
[7] Holla S and Katti M M 2012 Android based mobile application development and its security *International Journal of Computer Trends and Technology* **3**(3) 486-490
[8] Tripathi V 2017 Adaptive Computation Offloading in Mobile Cloud Computing *CLOSER* 524-529
[9] Mühleisen H, Bertram A and Kallen M J 2018 Database-inspired optimizations for statistical analysis *Journal of Statistical Software* **87**(4) 1-20
[10] Akherfi K, Gerndt M and Harroud H 2018 Mobile cloud computing for computation offloading: Issues and challenges *Applied computing and informatics* **14**(1) 1-16
[11] Ma H, Yu S, Gabbouj M and Mueller P 2018 Guest Editorial Special Issue on Multimedia Big Data in Internet of Things *IEEE Internet of Things Journal* **5**(5) 3405-3407
[12] Depkes RI 2010 *Kepmenkes RI No. 1995 Tahun 2010 Tentang Standar Antropometri Penilaian Status Gizi Anak* (Jakarta: Depkes)
[13] World Health Organization 2017 *The Z-score or standard deviation classification system* (Geneva, Switzerland: World Health Organization) [Online] Retrieved from https://www.who.int/nutgrowthdb/about/introduction/en/index4.html

Analysis of estimated busses arrival time on public transportation using real-time monitoring

R N Yuniar, P A Kencan, E Ruth and A H S Budi*

Department of Electrical Engineering Education, Universitas Pendidikan Indonesia, Jl. Dr. Setiabudhi 207, Bandung 40154, Indonesia

*agusheri@aol.com

Abstract. The purpose of this study is to create a system that can inform the bus position and provide information about the arrival time of the bus. We propose real time monitoring of the position and estimated time of arrival of Damri buses using the Haversine formula. GPS devices that are applied to each bus are monitored by utilizing passengers' cellular applications. The results show information about the position of the bus and estimated time of arrival of the bus at each stop. This research can help estimate the arrival time, especially in providing information about the certainty of bus arrival to prospective passengers who are waiting at each stop according to the bus destination.

1. Introduction

To realize infrastructure development in urban areas, it is necessary to create a reliable transportation system for the entire community. Damri Bus is one of the public transportation vehicles in Indonesia, mainly used in the city of Bandung. With this, Damri is expected to be able to meet the needs of community movements and overcome congestion by combining several urban transportation routes into Damri bus transportation routes. However, the level of Damri bus transportation services is still considered not optimal at this time. In some cases there are complaints with public transport performance, including frequent overload, inadequate vehicle quality, and slow travel speeds [1].

It was found that based on the results of score analysis [1], the value of user perceptions of time travel performance stated that it was sometimes late Damri bus AC at 88.10 which shows a negative value of 11.90%, with the value of this variable getting a score of 1 or unsatisfactory performance. Whereas for Damri non-AC bus transportation there is no difference with AC Damri bus transportation, with a perception value of a delay performance of 67.24%, which shows a negative value of 32.76%, with this variable gain score 1 or unsatisfactory performance.

This shows that the travel time for Damri AC and Non AC buses is no different [2]. The second type of bus from Damri buses has very poor or unsatisfactory performance. The poor performance of travel time is greatly influenced by road routes that have high activity and are often jammed, also caused by the behaviour of vehicle crews who often drop passengers in any place. In addition to traffic jams in the middle of the road, this is also because Damri buses do not have their own lanes like Jakarta Busway which already has its own lanes.

To overcome the problem of unexpected travel times from Damri buses and to help Damri passengers know the arrival time of Damri buses is easy; a reliable method of estimating travel time is needed [3]. We propose using REMOT (Real-Time Monitoring) to determine the location of Damri buses. By utilizing a GPS device that is applied to each bus and then the data will be processed in such a way that it can be monitored by utilizing a cellular application by the passenger. With this system, it is expected to help provide information about bus arrivals to prospective passengers waiting at each stop to the destination bus.

2. Methods

The research location is in the Bandung city area and uses Damri buses as the object of study where the route studied is Ledeng-Leuwipanjang. The process of research methods carried out can be seen in the following diagram.

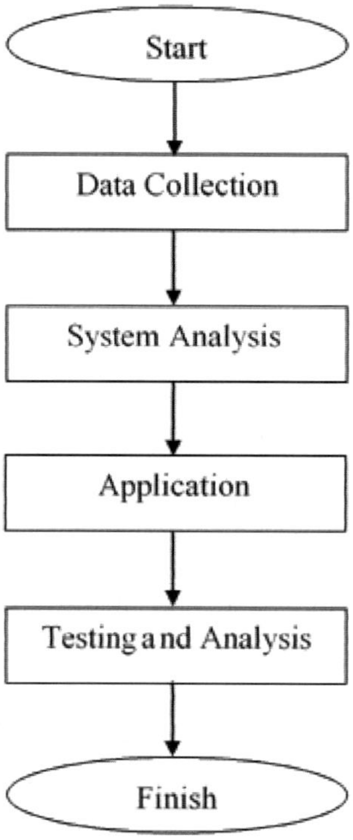

Figure 1. Flowchart of research method.

The source of research data is obtained through observation and interview. The observation process is done by making direct observations to observe the conditions in several Damri stops. Observations were made, namely at the bus stop that passed the Damri Ledeng-Leuwi Panjang bus route. While a direct survey by conducting an interview process at the Damri stop on the Ledeng-Leuwi Panjang route. Interviews were conducted with officers and prospective passengers. Interviews conducted to bus stop staff are about conditions at the bus stop and problems faced by bus stop staff. Interviews conducted with prospective passengers are about their waiting time and satisfaction in using Damri buses.

Real-time monitoring of Damri buses uses GPS technology which consists of several stages. The stage starts from GPS satellites that send data to the GPS module, then the system will receive data and parse data to obtain latitude, longitude and speed data. Data will be displayed on a special application for vehicle tracking.

3. Results and discussion

In this research, our tracking process uses GPS Tracker TK110 and to see the bus position in real-time we use the android and web applications that can be used free of charge by any GPS device, namely IDTrack and http://server-gps.com. The tracking process can only be done in real-time, information obtained from the GPS server application is the coordinates of the vehicle's position, the speed of the vehicle's speed, and the address of the position of the vehicle's position located.

Figure 2. GPS tracked TK110.

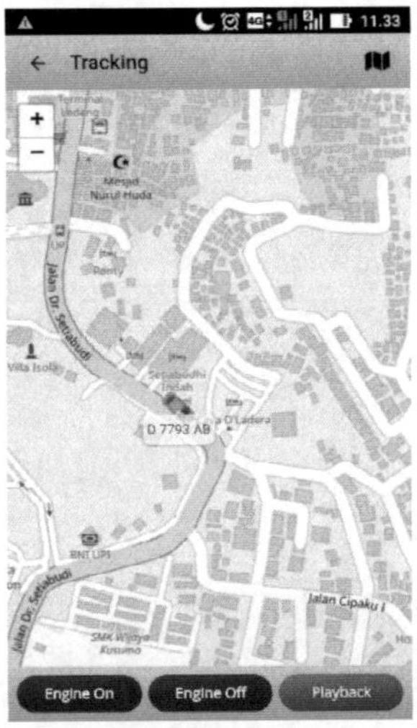

Figure 3. IDTrack application for tracking process.

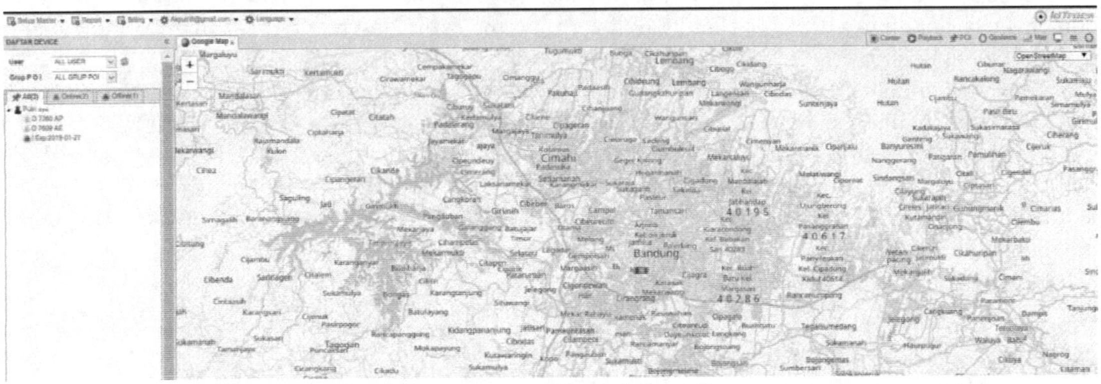

Figure 4. Tracking process with web http://server-gps.com.

Data collection was carried out by physical measurements, namely taking latitude and longitude coordinates from the stop location on the Damri Ledeng-Leuwi Panjang bus route, namely the distance between stops, estimated travel time data, and average bus speed obtained from direct survey results in the field. Latitude and longitude, shown in Table 1.

Table 1. Latitude and Longitude Bus Stop.

Bus Stop	Latitude	Longitude
Ledeng	-6.85961	107.59515
Setiabudhi	-6.874403	107.59613
Sukajadi	-6.881009	107.5964
RSHS	-6.898963	107.59752
Braga	-6.914474	107.60906
Otto Iskandar Dinata	-6.933925	107.60328
Leuwi Panjang	-6.85961	107.59515

After seven tracking processes are carried out on Damri buses, the average Damri bus travel time is obtained from one stop to another. This process can be seen in Table 2.

Table 2. Average of destination time.

Departure Stop	Destination Stop	Time Passed of Bus							Average of Destination Time (Minutes)
		1	2	3	4	5	6	7	
Ledeng	Setiabudhi	10	12	15	13	11	10	11	11,71
Setiabudhi	Sukajadi	3	3	2	2	2	5	1	2,57
Sukajadi	RSHS	7	9	11	8	18	13	19	12,14
RSHS	Braga	9	13	10	9	12	15	12	11,43
Braga	Otto Iskandar Dinata	14	11	12	7	19	19	19	14,43
Otto Iskandar Dinata	Leuwi Panjang	17	10	14	14	13	14	13	13,57

Table 2 shows that the average travel time for Damri buses has been obtained from one bus stop to another. Data can be used as a prediction to determine the estimated bus arrival time for each event, to find out the bus arrival time.

Before getting an estimated bus arrival time at each stop, we need to know the distance from one stop to another using the Haversine method. The distance calculation process using the Haversine method is shown in Table 3. Haversine is a method for determining the distance between two points by considering that the earth is not a flat plane but a plane that has a degree of curvature. The Haversine method calculates the distance between two points based on a long straight line between two points at latitude and longitude [4,5].

Haversine method is formulated as follows.

$$\Delta lat = (lat_2 - lat_1)$$
$$\Delta long = (long_2 - long_1)$$
$$a = sin^2\left(\frac{\Delta lat}{2}\right) + cos(lat_1) \cdot \cos(lat_2) \cdot sin^2\left(\frac{\Delta long}{2}\right)$$
$$c = 2 \cdot atan \cdot 2(\sqrt{a}, \sqrt{1-a})$$
$$d = c \cdot R$$

With the following caption:
ΔLat = The amount of change in latitude
$\Delta Long$ = The amount of change in longitude
D =Distance (km)
R = Earth radius = 6371 km
1 degree = 0.0174532925 radians
c = Calculation axes intersect

Table 3. Calculation of distances based on the Haversine method.

Departure stop	Latitude 1	Longitude 1	Destination stop	*d* (°)	*d* (km)
Ledeng	-6.85961	107.59515	Setiabudhi	94,39065419	1,6474276
Setiabudhi	-6.874403	107.59613	Sukajadi	42,11166363	0,7349872
Sukajadi	-6.881009	107.5964	RSHS	114,5355623	1,9990226
RSHS	-6.898963	107.59752	Braga	115,4467029	2,014925
Braga	-6.914474	107.60906	Otto Iskandar Dinata	127,3932698	2,2234319
Otto Iskandar Dinata	-6.933925	107.60328	Leuwi Panjang	87,36354528	1,5247815
		Total of Distance			**10.144576**

After getting the distance from one stop to another, we can calculate to find the estimated bus arrival time based on the distance known as the haversine method and the average speed obtained from the average speed of Damri buses every day. Estimated bus arrival times based on the haversine method are shown in Table 4.

Table 4. The travel time of bus based on the Haversine method.

Departure Stop	Destination Stop	Bus Travel Time (Minutes)
Ledeng	Setiabudhi	9,6
Setiabudhi	Sukajadi	4.2
Sukajadi	RSHS	12
RSHS	Braga	12
Braga	Otto Iskandar Dinata	13,2
Otto Iskandar Dinata	Leuwi Panjang	9

Table 5 shows two bus trip time results obtained based on the tracking process and based on distance calculation using the Haversine method. From Table 5, it can be seen when there are differences in Damri bus trips at each stop. This happens because in the process of tracking data obtained in the process of tracking direct trips through the route that is indeed passed by Damri buses. While the results of calculating the bus travel time using the Haversine method depends on the results of distance calculations, while drawing a straight line from one stop to another to get the distance. But on the route, the Damri bus route is winding, a straight line cannot be drawn to find out the actual distance.

Table 5. The travel time of the bus based on the tracking process and based on distance calculations using the Haversine method.

Departure Stop	Destination Stop	Bus Travel Time (Minutes)		Time Difference (Minutes)
		Tracking Process	Haversine Method	
Ledeng	Setiabudhi	11,71	9,6	2,11
Setiabudhi	Sukajadi	2,57	4.2	1,63
Sukajadi	RSHS	12,14	12	0,14
RSHS	Braga	11,43	12	0,57
Braga	Otto Iskandar Dinata	14,43	13,2	1,23
Otto Iskandar Dinata	Leuwi Panjang	13,57	9	4,57

From the two research processes conducted above, we have obtained data on Damri bus travel times from one bus stop to another. Based on the Damri bus travel time, we can make it as a prediction to determine the estimated arrival time of the bus. Thus, we can find out the position of the bus directly by using the IDTrack application, while knowing the estimated arrival time of the bus can be adjusted to the process of tracking bus trips through the IDTrack application that can be seen in real-time.

4. Conclusion

Average travel time in the research table can be used as a prediction of bus arrival time at each stop. The expected results of research on analyzing the estimated arrival time of Damri buses in real-time is to be able to provide information to the public about the whereabouts of the bus through the application and find out travel time from one stop to another and to find out the estimated time of arrival of the bus. The results of this study are expected to be a reference for further research in order to display predictions of bus arrival times in an application.

References

[1] Herdiana S and Gustamola W 2013 Evaluasi Kinerja Pelayanan Angkutan Bus Damri Kota Bandung Berdasarkan Persepsi Pengguna dan Pengelola *Reka Loka* **1**(1)

[2] Intari D E, Fathonah W and Ferdiansyah M A 2018 Studi Kinerja Pelayanan Bus Damri Non Ac Kota Serang (Studi Kasus Serang-Cikotok) *Jurnal Fondasi* **7**(1)

[3] Coifman B (002 Estimating travel times and vehicle trajectories on freeways using dual loop detectors *Transportation Research Part A: Policy and Practice* **36**(4) 351-364

[4] Winarno E, Hadikurniawati W and Rosso R N 2017 Location based service for presence system using haversine method *2017 International Conference on Innovative and Creative Information Technology (ICITech)* 1-4

[5] Alam C N, Manaf K, Atmadja A R and Aurum D K 2016 Implementation of haversine formula for counting event visitor in the radius based on Android application *2016 4th International Conference on Cyber and IT Service Management* 1-6

IOP Conf. Series: Materials Science and Engineering **850** (2020) 012027 doi:10.1088/1757-899X/850/1/012027

Implementation of cuckoo search algorithm for support vector machine parameters optimization in pre collision warning

A Puspaningrum*, A Suheryadi and A Sumarudin

Department of Informatics, Politeknik Negeri Indramayu, Jl. Lohbener Lama No. 08 Lohbener, Indramayu 45252, Indonesia.

*alifia.puspaningrum@polindra.ac.id

Abstract. Nowadays, pre-collision warning is one of the substantial aspects of the transportation sector. One of the steps to detect the collisions is by classifying and predicting the collisions. There are many supervised machine learning algorithms used, such as Least Square – Support Vector Machine (LS-SVM). Radial Based Function (RBF) is one of the LS-SVM kernels, which is a well-known method to support reliable performance. However, C and Gamma of its parameters are chosen randomly. This makes the performance of the classifier less optimal. To overcome that problem, this paper proposed a cuckoo search algorithm for optimizing two parameters to get optimal accuracy. The proposed approach is applied to 8437 transportation records and evaluated by using Accuracy. In addition, the performance of the proposed method is compared to other well-known meta-heuristic optimization algorithms, namely: Bat Algorithm (BA-SVM) and Firefly Algorithm (FA-SVM). Experimental results show that the Cuckoo Search Algorithm (CSA-SVM) yields the best performance for each of the 10-folds cross-validation by reaching 84.817% for accuracy, compared to the Bat Algorithm and Firefly Algorithm.

1. Introduction

Nowadays, pre-collision warning is one of the substantial aspects of the transportation sector. It provides many advantages to reduce and mitigate collisions by using various kinds of approaches, such as visual and audio recommendations [1,2]. Since the 1990s, many automotive companies have proposed many algorithms to assist the driver and ensure safety in driving [3].

One of the steps to detect the collisions is by classifying and predicting the collisions. There are several machine learning methods have been proposed to predict collisions, such as Least Square Support Vector Machine (LS-SVM), Random Forests (RF), and Artificial Neural Networks (ANN) as supervised machine learning algorithms [4-6]. Among those methods, LS-SVM provides faster computation compared to other machine learning methods. LS-SVM has several well-known kernels, namely: linear, polynomial, radial basis function (RBF), etc. Usually, linear and polynomial kernels consume less time and get less accuracy compared to RBF. In addition, RBF is one of the LS-SVM kernels which support reliable performance. There are two parameters of LS-SVM, which are chosen randomly, such as C and Gamma. So, the performance of the classifier becomes less optimal.

Wherefore, many researchers have conducted various nature-inspired metaheuristic methods to optimize the value of LS-SVM parameters. Finding the optimal solution is the aim of estimating the value of LS-SVM parameters. Some of the metaheuristic methods include Particle Swarm Optimization, Genetic Algorithm, and Bat Algorithm [7-9]. On the other hand, in 2009, the Cuckoo Search Algorithm (CSA) was developed by Yang and Deb to overcome optimization problems. CSA was inspired by Cuckoo Bird's life behavior. Levy Flight of CSA combines local and global explorative random walk. It, therefore, helps the system not to be trapped in local optima while searching the best solution [10]. Moreover, CSA has been implemented in many sectors, for instance, water distribution networks, software effort estimation, wireless sensor networks, contrast enhancement of medical images [11-14].

Thus, to optimize the value of C and Gamma parameters, this paper proposed a cuckoo search algorithm to get the optimal performance. The performance is evaluated by Accuracy then. After implementing the method to 8437 transportation records and finding the optimal performance, the parameters can be implemented widely in the classification process in order to mitigate the collisions.

The structure of research paper is organized as follows: Section 1 contains the introduction, Section 2 contains a literature review, Section 3 explains the proposed method which has been done to improve the accuracy, result of the experiment is analyzed in section 4, and Section 5 presents the conclusion and future work of this research.

2. Related works

LS-SVM is one of robust intelligent technique which has faster computation time compared to other supervised techniques [15]. However, a random value of the C and Gamma parameter in LS-SVM can influence the classification performance. So, some researches have been conducted to optimize the parameters. Huang C and Wang C proposed parameter optimization by using the Genetic Algorithm in 2006 [8]. The experiment was evaluated in 11 real-world datasets of the UCI repository. The experiment shows that the best performance of accuracy achieved in fewer features of classification.

Therefore, it has widely implemented by using Particle Swarm Optimization (PSO) Algorithm in 2008. This paper has evaluated the same dataset as the research [8] used. But the result shows that the proposed method cannot yield better performance compared to Genetic Algorithm performance. These two types of research were also tested by Tharwat A in 2018 [9]. Tharwat A used the Bat Algorithm to optimize the parameters of LS-SVM in nine UCI Machine Learning dataset [9]. The results demonstrated that Bat Algorithm can achieve optimal performance and has a lower classification error compared to the Genetic Algorithm and PSO.

In 2009, CSA was developed by Yang and Deb as one of the modern metaheuristic methods to answer optimization problems [12]. This method also has been implemented widely in many kinds of research, for example, water distribution networks, software effort estimation, wireless sensor networks, contrast enhancement of medical images [11-14].

From the literature above, most of those researches attempt to achieve the best optimization performance. To the best of knowledge, the proposed method of this paper is the only research that attempts to optimize the parameter values of LS-SVM by using a real transportation dataset. The dataset implemented consists of six attributes, as Huang C and Wang C recommended to use fewer features in the previous research [8].

3. Method

In optimizing LS-SVM parameters, CSA is implemented to find the optimal performance. CSA-SVM was implemented in a PC with the following features: Intel(R) Core (TM) i7-8550U CPU @ 1.80GHz, 16G RAM, a Windows 10 Operating System, and MATLAB 2014a.

Table 1. Dataset information.

Dataset Information		
0 (Not Danger)	4907	8437
1 (Danger)	3530	

As shown in Table 1, the label of the classification consists of two labels, danger and no danger. The experimental data used in this research consists of six attributes, as seen in Table 2.

Table 2. Parameters of CSA.

Input	:	Transportation Dataset (id, speed, distance, period, warning, pair, label)
Output	:	Accuracy, Parameter C and Gamma

K-fold cross-validation has also been used in all experiments. The value of k was set to 10. The dataset then divided into training and testing data. The input and output of the dataset are described in Table 2.

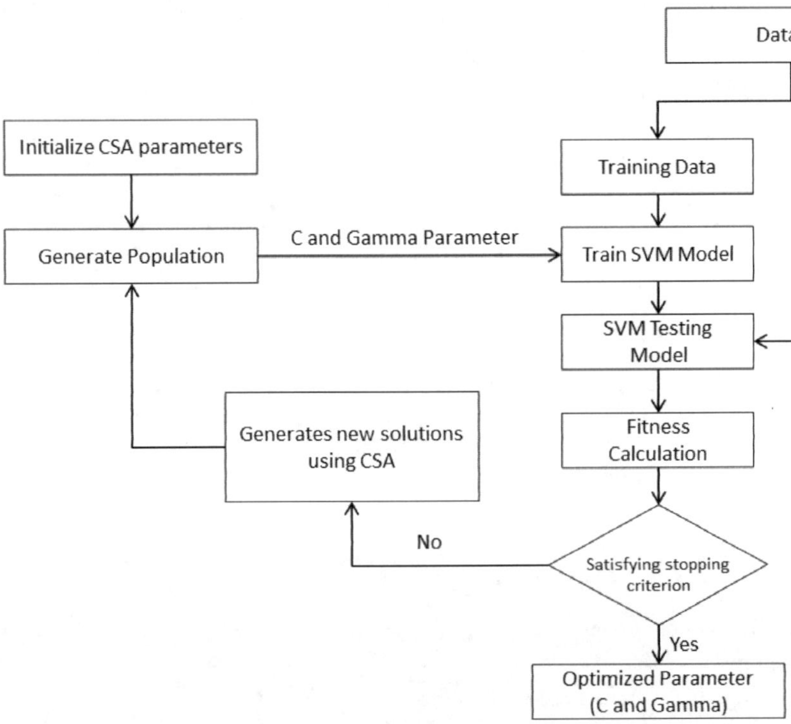

Figure 1. CSA-SVM flowchart.

Actually, the proposed method generally consists of two main processes, namely optimization, and classification process, as shown in Figure 1. The description of implemented steps as described in this following steps:

- Preparing the dataset
- Splitting dataset into training and testing data
- The training data is used to construct the classification model. The value of C and Gamma parameters are taken from CSA optimization include CSA parameters and Generate Population.
- By using the classification model constructed, the testing data is classified using SVM then
- The accuracy value is being the fitness value
- If the stopping criterion is satisfied, the optimized parameters are achieved
- If it is not, CSA will generate new solutions and generate population then
- Step 3-5 will be repeated until the stopping criterion is achieved

Table 3. Parameters of CSA-SVM.

Parameters	Values
Number of Cuckoos	5
Number of Eggs	15
Iteration	2
Position	Minimal = 50 Maximal = 125
Pa	0.25

Table 3 shows the parameters used in CSA-SVM. Then, the performance is evaluated by using Accuracy. It's also used to calculate the fitness value of CSA-SVM. The equation of the accuracy is shown in Equation 1

$$Accuracy = \frac{TP + TN}{TP + TN + FP + FN} \quad (1)$$

According to Equation 1, some cases with True Positive (TP) define the danger class, which correctly classified as dangerous. But, when the danger class is classified as not dangerous, it will be classified as False Negative (FN). In addition, when 'no danger' class is correctly classified as 'no danger,' the record will be classified as True Negative (TN). However, when 'no danger' is classified as 'danger,' it will be classified as False Positive (FP).

4. Experimental result

In this section, the proposed method has been implemented for LS-SVM parameters optimization. In addition, the Cuckoo Search Algorithm (CSA-SVM) is compared with other recent metaheuristic methods such as Bat Algorithm (BA-SVM) and Firefly Algorithm (FA-SVM). We considered 8437 records to be tested in five experiments. Each experiment tests the performance of each method.

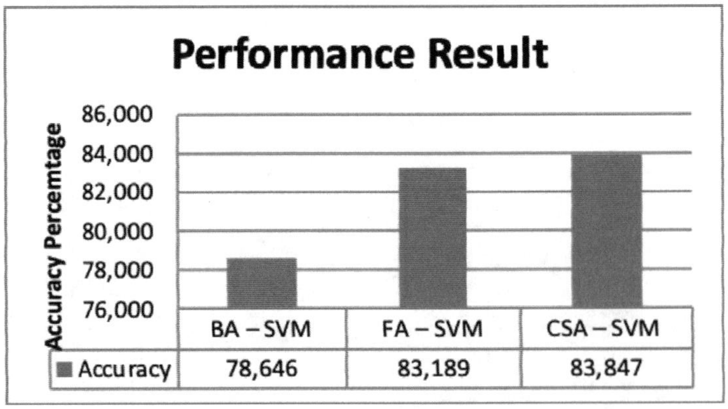

Figure 2. Average performance result.

According to Figure 2, the average value of CSA-SVM yields 83.847% as the best performance compared to BA – SVM and FA – SVM. The average performance is achieved based on five scenarios that have been done. The detail performance for each experiment is described in Table 4.

Based on experimental results show in Table 4, those three methods give accuracy that is not much different for each experiment. CSA-SVM becomes the best method except in the first experiment. FA-SVM gives accuracy that is not much different. On the contrary, BA-SVM gives the worst performance compared to other benchmarking methods. The aim of the proposed method is to know whether CSA-

SVM can optimize the value of each parameter. So, the value of CSA-SVM accuracy in Table 4 has been analyzed deeper, as shown in Table 5.

Table 4. Accuracy of experimental results.

Experiment	BA – SVM	FA – SVM	CSA – SVM
1	77.657	83.217	82.659
2	78.726	83.205	83.655
3	76.034	83.229	83.561
4	81.368	83.146	84.544
5	79.447	83.146	84.817
Average	78.646	83.189	83.847

Table 5. Parameters performance of CSA-SVM.

Experiment	Gamma	C	CSA – SVM
1	11.354	68.248	82.659
2	55.492	120.809	83.655
3	34.464	97.491	83.561
4	38.889	93.936	84.544
5	62.254	108.307	84.817

C and Gamma are two parameters optimized in this proposed method. When the optimal value of these two parameters is found, the classification performance of LS-SVM is going to be achieved, and the accuracy will be increased. However, not all performance of CSA-SVM experiments shows the best result. According to Table 5, the higher value of C can reduce the misclassified sample. On the other hand, the higher value of Gamma tends to increase the misclassified data, as shown in the value of accuracy of the performance. In LS-SVM, parameter C influence the margin width of the classification boundary for each class. Thus, the classifier will help to get correct classification data easier. However, the larger value of gamma can maximize the kernel value. So, incorrect data will be difficult to differentiate more. However, discussing these two parameters, under-fitting, is an issue for this topic.

Table 6. Time execution.

BA – SVM	**FA – SVM**	**CSA – SVM**
11.148	65.878	61.457

Besides comparing the accuracy value of each method, this paper also observed the time execution, which is described in Table 6. It shows that BA-SVM reached the fastest time execution compared to, and FA-SVM gets the slowest time execution. Indeed, the performance of time execution for each method is determined by how the method literally works. Moreover, Table 7 shows the best optimized CSA-SVM parameters in the traffic records dataset.

Table 7. Optimized parameters of CSA.

Coefficient	**Value**
C	62.25
Gamma	108.31

To sum up, the parameters optimized in this research are C and Gamma. The value of these parameters is balancing parameters to change the random number in calculating the accuracy. So, when the optimal parameters are found, the optimal performance of classification is going to be achieved. CSA-SVM shows the best performance for accuracy, but the time execution is in contrast. However, there is no guarantee that this method also gives the same performance when implemented in another sector.

IOP Conf. Series: Materials Science and Engineering **850** (2020) 012027 doi:10.1088/1757-899X/850/1/012027

There're also several factors which influence the performance, such as dimension, iteration, number of data, and the quality of the data itself.

5. Conclusion and further work

A Cuckoo Search Algorithm is proposed for optimizing LS-SVM coefficients. The experimental result has done by using 8437 transportation records to verify the proposed method's reliability. Cuckoo Search Algorithm yields the best performance for each of the 10-folds compared to the other methods by reaching 84.817% for accuracy, compared to the Bat Algorithm and Firefly Algorithm. Regardless of the success of the proposed method, the exploitation of parameter quality still can be explored more. So that, hybrid optimization method can be one of option to get better quality for future research.

Acknowledgments

This research was fully supported by Pusat Penelitian dan Pengabdian Masyarakat, Politeknik Negeri Indramayu, as Penelitian Terapan Dana Lokal Polindra 2019.

References

[1] Reagan I J, Cicchino J B, Kerfoot L B and Weast R A 2018 Crash avoidance and driver assistance technologies – Are they used ? *Transp. Res. Part F Psychol. Behav.* **52** 176–90

[2] Eichelberger A H and Mccartt A T 2015 Toyota drivers' experiences with Dynamic Radar Cruise Control, Pre-Collision System, and Lane-Keeping Assist

[3] Wang J, Wu J, Zheng X, Ni D and Li K 2016 Driving safety field theory modeling and its application in pre-collision warning system *Transp. Res. Part C* **72** 306–24

[4] Dogru N and Subasi A 2018 Traffic Accident Detection Using Random Forest Classifier *2018 15Th Learn. Technol. Conf.* 40–5

[5] Gomes S L, Rebouças E de S, Neto E C, Papa J P, Albuquerque V H C d., Rebouças Filho P P and Tavares J M R S 2017 Embedded real-time speed limit sign recognition using image processing and machine learning techniques *Neural Comput. Appl.* **28** 573–84

[6] Park H S, Kim D J, Kang C M, Kee S C and Chung C C 2018 Object detection in adaptive cruise control using multi-class support vector machine *IEEE Conf. Intell. Transp. Syst. Proceedings, ITSC* **2018-March** 1–6

[7] Lin S, Ying K, Chen S and Lee Z 2008 Particle swarm optimization for parameter determination and feature selection of support vector machines **35** 1817–24

[8] Huang C and Wang C 2006 A GA-based feature selection and parameters optimization for support vector machines **31** 231–40

[9] Tharwat A, Hassanien A E and Elnaghi B El 2016 A BA-based algorithm for parameter optimization of Support Vector Machine A BA-based algorithm for parameter optimization of support vector machine

[10] Yang X, Cb C and Deb S 2009 Cuckoo Search via L ´ evy Flights

[11] Naidu M N, Boindala P S, Vasan A and Varma M R R *Optimization of Water Distribution Networks Using Cuckoo Search Algorithm* (Springer Singapore)

[12] Parwita I M M, Sarno R and Puspaningrum A 2017 Optimization of COCOMO II Coefficients using Cuckoo Optimization Algorithm to Improve The Accuracy of Effort Estimation pp 99–104

[13] Razzaque M A and Hussein M 2015 A Novel Cuckoo Search Based Clustering Algorithm for Wireless Sensor Networks

[14] Daniel E and Anitha J 2016 Optimum wavelet based masking for the contrast enhancement of medical images using enhanced cuckoo search algorithm *Comput. Biol. Med.* 1–7

[15] Siuly Siuly and Yan Li 2012 Improving the Separability of Motor Imagery EEG Signals Using a Cross Correlation-Based Least Square Support Vector Machine for Brain–Computer Interface *IEEE Trans. Neural Syst. Rehabil. Eng.* **20** 526–38

Investigation of characteristics of bamboo fiber for composite structures

M Martijanti[1,2,*], A L Juwono[1] and S Sutarno[3]

[1] Department of Physics, Faculty of Mathematics and Natural Science, Universitas Indonesia (UI), Jl. Gedung F, Fakultas MIPA, Kampus UI Depok, Pd. Cina, Beji, Kota Depok, Jawa Barat 16424
[2] Department of Mechanical Engineering, Faculty of Manufacturing Technology, Universitas Jenderal Achmad Yani (UNJANI), Jl. Jenderal Gatot Subroto PO. Box 807 (PT PINDAD) Bandung 40285
[3] Department of Metallurgical Engineering, Faculty of Manufacturing Technology, Universitas Jenderal Achmad Yani (UNJANI), Jl. Gatot Subroto PO Box 807 (PT PINDAD) Bandung 40285

*martijanti@lecture.unjani.ac.id

Abstract. Bamboo fibers have been used for reinforcements in structural composites because of their environmental friendly properties. The purpose of this study was to investigate the characteristics of three types of bamboo, namely *Tali* bamboo, *Gombong* bamboo and *Haur Hejo* bamboo to meet the requirements for a reinforcement in structural composites. The three types of bamboo fibers were alkaline treated by varying the concentrations of NaOH (4% and 6%), and immersion times (1, 2, and 3 hours). The alkaline pre-treatment was conducted at the temperature of 120°C and the pressure of 0.5 MPa using a steam explosion method. The treated bamboo fibers were characterized by FTIR and measured their tensile strengths. The tensile test results showed that the tensile strengths of Tali, Gombong, and Haur Hejo bamboss were 710 MPa, 418 MPa, and 457 MPa respectively. Whereas, the FTIR analysis indicated that there was still a peak of lignin with small intensity marked by aromatic C = C (alkyne) at the wavelength of 1500-2600 cm^{-1} for all types of bamboo. There was also a peak of cellulose with an increase in the intensity marked by the aromatic group C = H (alkene) at the wavelength of 1900-2400.

1. Introduction

Sources of producing polymer fibers can be divided into 4 groups, namely 1) Animal fibers; 2) Synthetic polymer fibers made from petroleum fractions; 3) Ceramics/silica/glass fibers, and 4) Plant fibers. Natural polymer fibers refer to fibers obtained from extraction from plants such as hemp, straw, wood, sugar cane and bamboo, kenaf, hemp, oil palm empty bunches, sisal, coir, water hyacinth, banana fiber [1]. In addition to renewable sources, natural fiber is also biodegradable, abundant with fast and inexpensive production cycles and have the potential to replace synthetic polymer fibers as well as solutions to environmental impacts [2-8].

Bamboo is a source of natural fiber with abundant availability with limited application to meet needs such as houses, walls, ornaments, traditional equipment and even consumed in the form of bamboo shoots. The economic value and function of bamboo can be increased through the diversification of natural polymer fiber products into bio-composites [3,5] for various lightweight structure applications [9], sound vibration dampers [10-12], biofilters [13], and thermal [14]. From various research and development of this diversification potential it is very prospective because of the superior mechanical properties of bamboo, namely low density i.e. 0.8-1.5 gr/cm^3, tensile strength of 500-575 MPa, high elastic modulus of 27-40 GPa [11,15].

Bamboo fiber has a chemical composition of 55% fiber consisting of 34.5% cellulose and 2.5% hemicellulose, 26% lignin and the remaining volatile content and is hydrophobic. The bond of lignin with cellulose (lignocellulose) and hemicellulose and other extractive compounds forms a structure blocking the adhesion of cellulose polymer fibers to the matrix in producing composites. Separation of lignin-cellulose (lignocellulose) and hemicellulose and other extractive compounds was carried out by dissolving bamboo fiber NaOH solution with steam explosion method [16,17], and kenaf fiber into NaOH solution [18]. The results of fiber extraction can increase adhesion [19], also the bond of adhesion between fibers and matrices [20]. Tensile strength and composite modulus of fibers extracted by steam explosion increased by 15-30% with polypropylene matrix [16,17]. In other words, critical to quality (CTQ) natural fiber composites are the tensile strength of a single polymer fiber and adhesion to the matrix, either in the form of a thermoplastic matrix and/or thermosetting.

The extraction process of lignocellulose with NaOH solution is known as an alkalization process or extraction process wherein the lignocellulose chain breaks into cellulose and lignin [16-20] as shown in Figure 1. Based on figure 1, OH- (results of dissociation of NaOH into Na^+ and OH^-) reacts with the H group on lignin, then forms H_2O. Cluster O forms free radicals and reacts with C to form epoxy rings (C-O-C). The reaction produces two separate benzene rings, where each ring has a reactive O group. This reactive group O reacts with Na^+ and dissolves in alkaline solutions so that lignin is removed when rinsed [16-21].

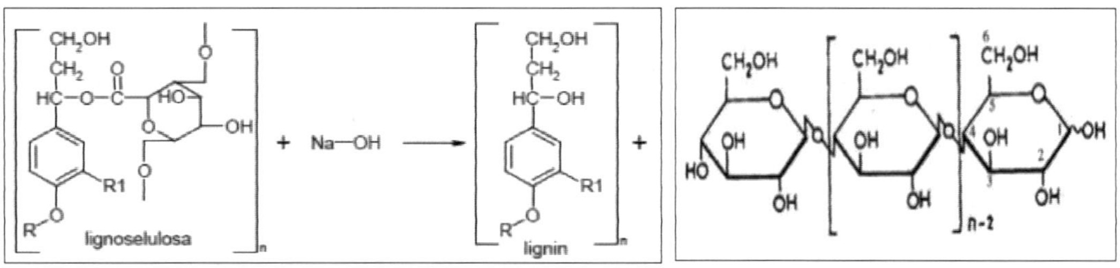

Figure 1. Lignocellulose reaction in NaOH solution [3,21].

Based on figure 1, it can be seen that lignin has a functional group OH, C-O-C, C-C, C = aromatic, CH aliphatic, CH2, while cellulose OH and C = O. The effectiveness of the lignin extraction process from lignocellulose with NaOH solution is influenced by the concentration of NaOH solution, pressure and extraction time, and diameter of bamboo fiber. The effect of the extraction process with NaOH solution can be detected by FTIR spectrometry [22].

This FTIR spectrometry characterizes organic compounds based on their constituent functional groups, where each functional group of an organic molecule has energy and rotation levels with the value of wave numbers absorbed by the bonds of a different functional group. A typical infrared scan is generated in the mid infrared region of the light spectrum. The mid-infrared region is 400-4000 cm-1 wave number, which is equal to the wavelength 2.5 to 25 microns. The correlation between wave numbers indicated by peaks with functional groups of natural fiber compounds are shown in table 1.

Table 1. Correlation of wave numbers to functional groups on natural fibers [23].

Sources of fiber	Functional groups.	-OH	C-O-C	C-C	C-H	C-H aliphatic	C-H2	C=C	C=C ring aromatic
Bamboo [24]	Weave number-WOT Weave number-WT	1590	1161	1027	896	2900	1420	1200-1300	1238.97
Ketandan sawit [10]	Weave number-WOT Weave number-WT	1593.46 1592.74 3322.84	1161.32	1027.49 1028.63	895.57	2918.62	1420.32	1200-1300	1238.97
[12]	Weave number-WOT Weave number-WT	1627	1250				1319	1735	

Note: WOT: without extraction; WT: with extraction.

To calculate the tensile strength of a single polymer cellulose fiber, which determines the tensile strength of the resulting composite, it is calculated using a tex unit [12],

$$1\ tex = \frac{1000\ x\ W}{0.05} \tag{1}$$

Where W is the weight of the test fiber sample in milligrams (W). Furthermore, the calculation of tensile strength of fiber is expressed as gram/tex, where the value of a is in milligrams,

$$(\text{gram/tex}) = \frac{a}{W\ (in\ unit\ tex)} \tag{2}$$

The gram/tex value is converted to gram/denier, where 1 denier = 1/9 tex, the calculation of tensile strength (σ) is stated [12],

$$\sigma = 130\ x \frac{gram}{dinier}\ (MPa) \tag{3}$$

The effectiveness of the extraction process of lignin and hemicellulose from lignocellulose bamboo fiber polymer with NaOH solution can be measured by FTIR spectrometry where the functional groups are identified [22-25], meanwhile the morphology of fiber can be seen by SEM results [12,26,27].

2. Experiment procedure

2.1. Bamboo fiber preparation

The stages of the extraction process and fiber characterization are shown in figure 2. Fiber is extracted from 3 types of bamboo, namely bamboo Tali, *Haur Hejo* bamboo and *Gombong* bamboo. The three types of bamboo samples are made of strips size 300-500 x100 x 50 mm^3 (l x w x t). The fiber extraction process was carried out in an autoclave at a temperature of 120°C with a pressure of 0.5 MPa with a variation of 4 and 6% NaOH solution, and extraction time of 1, 2, and 3 hours. Furthermore, the bamboo strip sample was neutralized by using HCl and rinsed the water at a temperature of 90-95°C, and dried it in an oven at a temperature of 105°C for 4 hours.

Characterization of single fibers includes tensile strength, lignin and cellulose levels, morphology and identification of functional groups of wavenumber-based fiber compounds using FTIR spectrometry. Data compilation, analysis and evaluation related to the influence of extraction process parameters, lignin and cellulose levels and morphology are related to tensile strength. The functional groups are related to the levels of lignin and cellulose which have an impact on tensile strength. Based on the concept of measurement, analysis and evaluation, improvement can be determined key process parameters that produce maximum tensile strength and answered whether bamboo fibers have good prospects to replace synthetic fibers. Flowchart of bamboo extraction process and characterization it can be seen in figure 2.

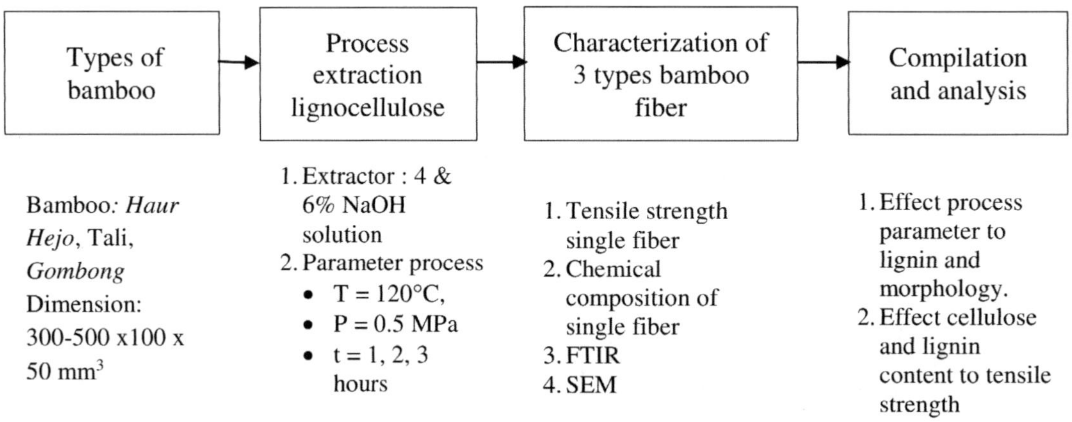

Figure 2. Flow chart of bamboo extraction process and its characterization.

2.2. *Testing the pull of single bamboo fibers*

The single tensile strength of bamboo polymer fibers is a determinant of the composite strength of the three types of bamboo fibers. Preparation of tensile test samples consisted of 150-200 mm long fiber bundles and 5 mm band width, and fiber weight 400-600 mg/bundle. The two ends of the fiber bundle are given adhesive paper so that the distance between the two adhesive papers is 50 mm. The tensile test sample as shown in figure 3 is mounted on a Tenso Lab-168 E tensile testing machine, with a cross head speed of 1.3 mm/minute to break up within 20 + 3 seconds according to ASTM D3379 or Indonesian National Standard (SNI) 08-1112-1989.

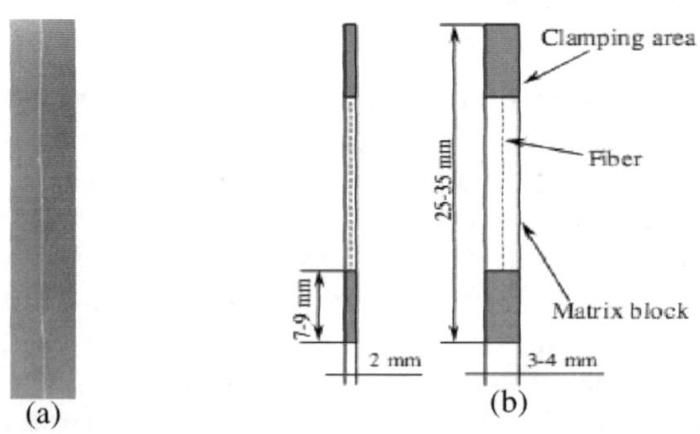

Figure 3. Illustration of preparation tensile strength testing: (a) single fiber; (b) mounting single fiber [28].

2.3. Testing for hemicellulose, cellulose and lignin
Tests for lignin levels using the SNI 0492: 2008 procedure, while for cellulose and hemicellulose levels using the SNI 0444: 2009 procedure.

2.4. FTIR testing
FTIR spectrometer bonds of bamboo fibers of *Gombong (Gigantochloa pseudoarundinacea), Haur Hejo (Bambusa tuldoides),* and *Tali (Gigantochloa apus)* by using the standard of ASTM E1252. FTIR's specimen is a pellet consisting of a mixture of cellulose and KBr. FTIR testing was performed by using Shimadzu IRAffinity-1. A typical infrared scan is generated in the mid infrared region 400-4000 cm-1 wave number, which is equal to the wavelength 2.5 to 25 microns of the light spectrum.

2.5. SEM testing
To identify the effect of fiber extraction process on morphology SEM testing was carried out using the JEOL JSM-6360, 15 kV with magnifying of 1,500 times.

Compilation of process parameters and test results data such as tensile strength, levels of lignin and cellulose, identification of functional groups and morphology then analysed and evaluated which answers whether bamboo fibers can replace synthetic fibers.

3. Results and discussion

3.1. Effect NaOH concentration and extraction time to tensile strength of single fiber
Data compilation of tensile testing results from *Tali, Haur Hejo* and *Gombong* bamboss are presented in table 2.

Table 2. Effect of fiber extraction process parameters on tensile strength of single fiber.

Type of Bamboo	Process Parameters Autoclave (T=120°C, and P= 0,5 MPa)		Test result of Tensile strength of single fiber (MPa)					Average Tensile strength (MPa)
	Concentration of NaOH (%)	Extraction time (hour(s))	1	2	3	4	5	
Haur Hejo	4	1	284	256	336	344	413	327
		2	298	408	405	367	279	351
		3	397	557	560	413	357	457
Tali	4	1	529	560	474	549	552	533
		2	512	577	558	667	357	534
		3	502	495	595	562	519	535
Gombong	4	1	441	343	493	470	341	418
		2	120	211	208	104	255	180
		3	172	57	79	58	62	86
Haur Hejo	6	1	105	124	169	73	211	136
		2	160	181	238	260	279	224
		3	183	213	246	254	304	240
Tali	6	1	614	507	541	499	621	556
		2	840	700	624	743	642	710
		3	758	592	640	598	652	648
Gombong	6	1	222	216	201	81	175	179
		2	120	211	208	104	255	180
		3	43	79	81	76	54	67

The results of the single fiber tensile strength test of the three types of bamboo as shown in Table 2 are then processed in the form of histograms to facilitate the analysis of the effect of extraction time and NaOH concentration as shown in Figure 4. Based on Figure 4, it can be seen that the single fiber tensile strength average 543 MPa with a range of 357-667 MPa at 4% NaOH concentration and a 1-3 hour extraction process time, and the tensile strength of a single fiber bamboo *Tali* averaging 638 MPa with a range of 499-840 MPa at 4% NaOH concentration with a range extraction process time 1-3 hours. Meanwhile for bamboo *Haur Hejo* and *Gombong* at the time of extraction process 1-3 hours, the increase in NaOH concentration causes a decrease in the tensile strength of a single fiber adding concentration. The results of other researchers indicate that the tensile strength of a single bamboo fiber is 420 MPa, 383 - 862 MPa [4,28].

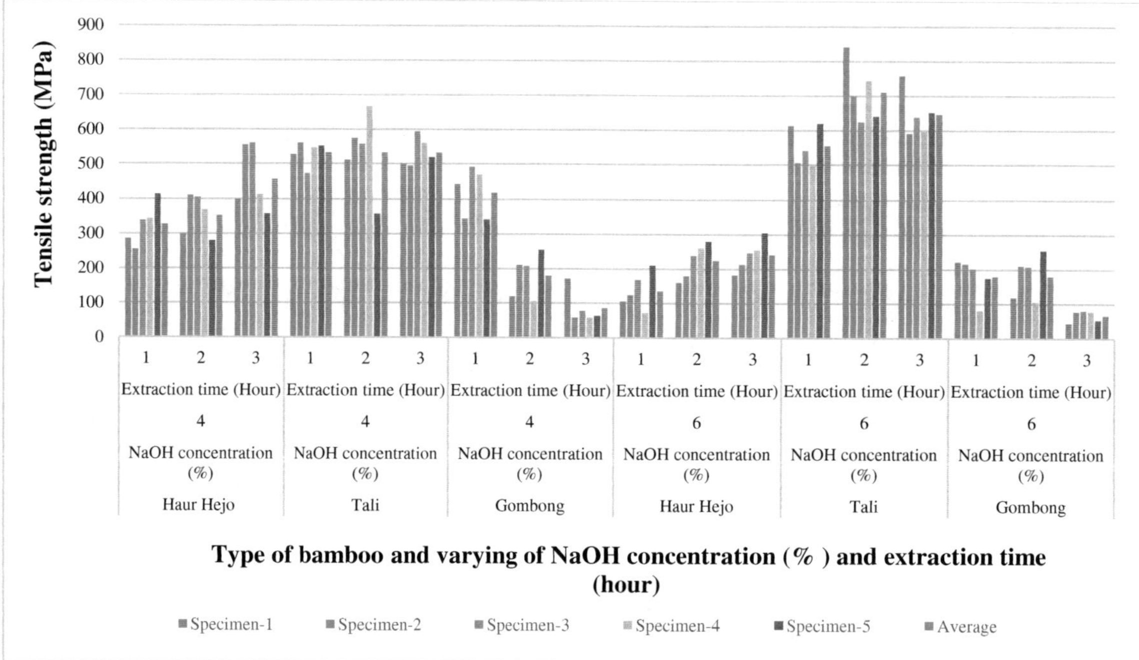

Figure 4. Effect of varying NaOH concentration (%) and extraction time (hour) to tensile strength of single fiber.

3.2. Content of hemicellulose, cellulose, and lignin

Data compilation results of cellulose and lignin testing on *Tali* and *Haur Hejo* bamboo for before and after the extraction process in NaOH solution are presented in table 3.

Table 3. Chemical composition of bamboo fiber.

Extraction process (in NaOH solution)	Chemical composition of bamboo fiber (%)		
	Hemicellulose	α-cellulose	Lignin
1. Bamboo *Tali* without extraction	17.29	46.91	21.30
2. Bamboo *Haur Hejo* without extraction	19.15	52.3	20.01
3. Bamboo *Tali* with extraction	16.21	55.02	16.93
4. Bamboo *Haur Hejo* with extraction	16.38	55.68	18.28

Based on table 3, it was seen that bamboo *Tali* had a reduction in lignin levels of 20.51% and an increase in α-cellulose levels of 17.29%, while *Haur Hejo* bamboo had a reduction in lignin levels of 8.65% and an increase in α-cellulose levels of 6.46%. The hemicellulose for both types of bamboo has a not significant decrease.

3.3. Effect cellulose content and SEM examination to tensile strength of single fiber
Table 1, which is clarified by figure 4 and table 2, it can be understood that the tensile strength of a single bamboo polymer fiber is inversely bound to the level of lignin. In other words, the smaller the level of lignin in fiber will increase its tensile strength. Based on the SEM examination as shown in figure 5, it can be seen that the morphology of the fiber surface with the extraction process is rougher. Fiber surface roughness is predicted to occur in lignin extraction, which will increase cellulose levels and adhesion to fiber-matrix composites.

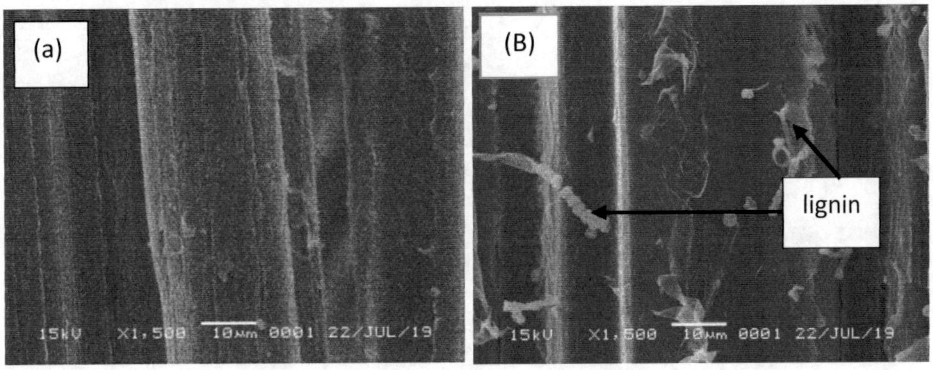

Figure 5. Morphology fiber (a) without process extraction; (b) with process extraction.

3.4. FTIR test
The FTIR test results as shown in figure 6, showed that the wave number spectrum of *Tali*, *Haur Hejo* bamboo and *Gombong* bamboo between 3000 - 3500 cm^{-1} showed a 0-H bond to change due to the extraction process of lignocellulose in a solution of NaOH to cellulose and lignin. The hydrosil group reacts with NaOH which causes an increase in OH^- concentration. The wave number spectrum area between 2000-2800 shows changes in the stretching peak between bamboo without extraction processes and with the extraction process, which is indicated by the increase in cellulose levels in bamboo as a result of the extraction process. The FTIR spectrum shows a large peak range because of the vibration of the O-H stretch at 3700–3200 cm^{-1}, the peak of the C-H stretches around 2900 cm^{-1} and the short peak of 1647.21 cm^{-1} representing the stretch of OH and C-H from the cellulose group. The FTIR results are very in accordance with Li et al. [19] and Sedan et al. [24].

Lignin is indicated in a range of wave numbers between 1500-1600 cm^{-1} with an aromatic C = C group. The peak point in this area turns out that the lignin extraction process is only reduced by 20.51% for *Tali* bamboo. Changes in the peak point in the source between 1000-1100 cm^{-1} and a decrease in the stem point, the results of the extraction process in the area of 896 cm^{-1} have an increase in the peak point in the bamboo extraction process, this shows that the extracellular process of stretching C-O and C-O-C bonds which means there is a stretch of the acetyl hemicellulose group and the formation of cellulose glycosides bonds.

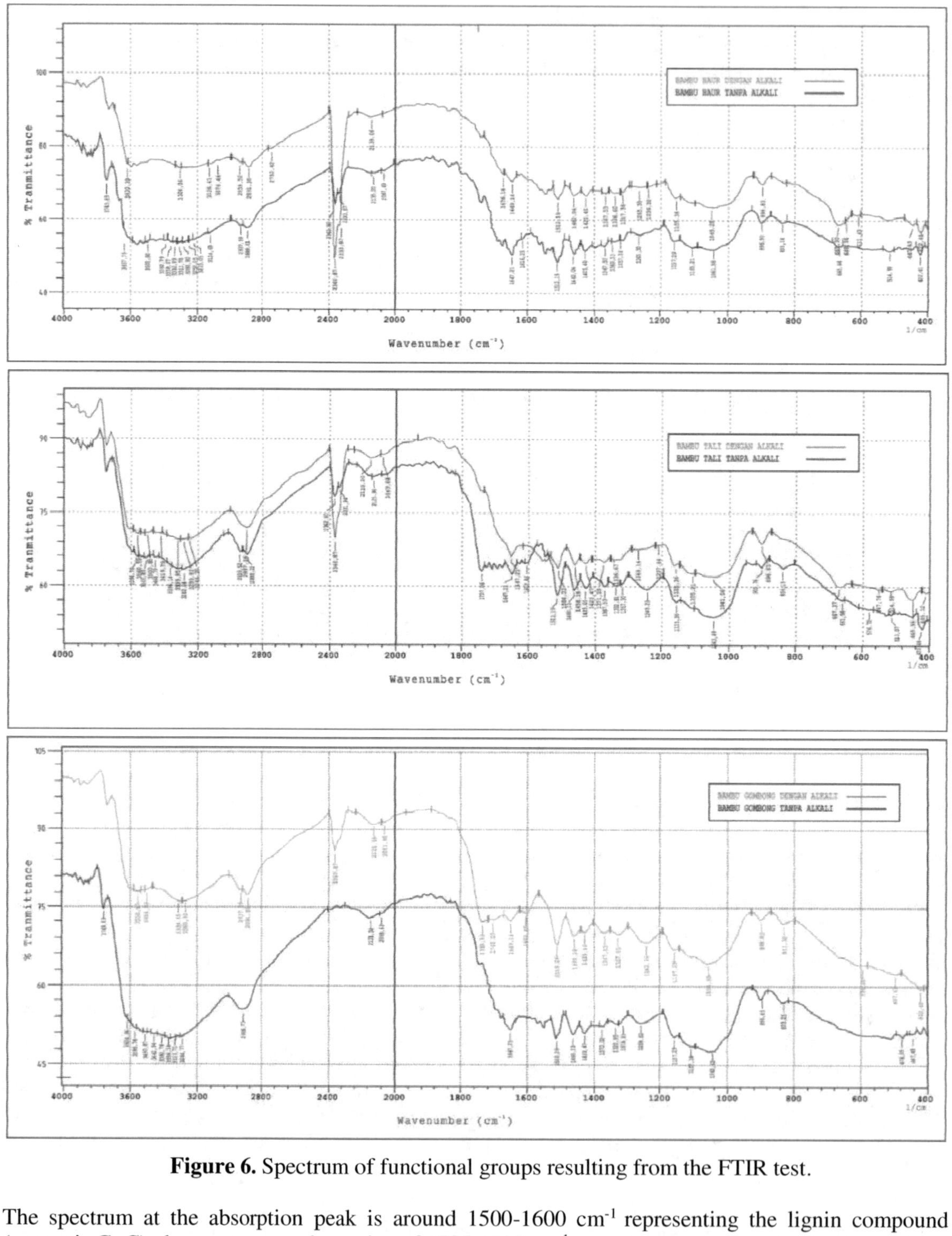

Figure 6. Spectrum of functional groups resulting from the FTIR test.

The spectrum at the absorption peak is around 1500-1600 cm^{-1} representing the lignin compound (aromatic C=C), the spectrum at absorption of 1735-1739 cm^{-1} represents the carbonyl (C=O) stretching the acetyl hemicellulose group. The peak C-O stretch of about 1000-1100 cm^{-1} from the cellulose group and the C-O-C peak around 896 cm^{-1} stretching cellulose β-glycosidic bonds similar to Cai et al. [22].

4. Conclusion
The results of the characteristics of 3 types of bamboo showed that *Tali*, *Haur Hejo* bamboo had the potential to become natural fibers which replaced synthetic fibers. Diversification of bamboo natural fibers into composite products for various applications is directly able to increase bamboo's added value and at the same time provide solutions to environmental problems as a result of the application of synthetic fibers. The parameter of lignin extraction process from lignocellulose to produce cellulose with NaOH solution was influenced by the concentration of NaOH solution, extraction time and temperature and initial fiber diameter. The effectiveness of the lignin extraction process can be shown by the results of FTIR and fiber morphology using SEM.

Acknowledgments
The authors gratefully acknowledge the financial support of the research has been supported by DRPM UI Grant No.: NKB-0145/UN2.R3.1/HKP.05.00/2019.

References
[1] Ramesh M, Palanikumar K and Reddy K H 2017 Plant Fibre Based Bio-Composites: Sustainable and Renewable Green Materials *Renewable and Sustainable Energy Reviews* **79** 558-584
[2] Venkatesha B K, Saravanan R and Bavan D S 2017 Review on Mechanical properties and Fatigue life of E-Glass/Bamboo Fiber Reinforced Polymer Composites *ICAMS International Journal of Engineering Sciences & Management* 52-57
[3] Gaceva G B, Avella M, Malinconico M, Buzarovska A, Grozdanov A, Gentile G and Errico M E 2007 Natural Fiber Eco-Composites *Polymer Composites* **28**(1) 98-107
[4] Trujillo E, Moesen M, Osorio L, Van Vuure A W, Ivens J and Verpoest I 2014 Bamboo fibres for reinforcement in composite materials: Strength Weibull analysis *Composites Part A: Applied Science and Manufacturing* **61** 115-125
[5] Joshi S V, Drzal L T, Mohanty A K and Arora S 2004 Are natural fibers composites environmentally superior to glass fiber composites? *Composites Part A: Applied Science and Manufacturing* **35**(3) 371-376
[6] Cheung H Y, Ho M, Lau K, Cardona F and Hui D 2009 Natural fibre-reinforced composites for bioengineering and environmental engineering applications *Composites Part B: Engineering* **40**(7) 655-663
[7] Kinoshita H, Kaizu K, Fukuda M, Tokunaga H, Koga K and Ikeda K 2009 Development of green composite consists of woodchips, bamboo fibers and biodegradable adhesive *Composites Part B: Engineering* **40**(7) 607-612
[8] Saheb D N and Jog J P 1999 Natural fiber polymer composites: A review *Advances in Polymer Technology: Journal of the Polymer Processing Institute* **18**(4) 351-363
[9] Gibson R F 2010 Review of recent research on mechanics of multifunctional composite materials and structures *Composite Structures* **92**(12) 2793-2810
[10] Pradana M A, Ardhyananta H and Farid M 2017 Pemisahan Selulosa dari Lignin Serat Tandan Kosong Kelapa Sawit dengan Proses Alkalisasi untuk Penguat Bahan Komposit Penyerap Suara *Jurnal Teknik ITS* **6**(2) 413-416
[11] Khalil H P S A, Bhat I U H, Jawaid M, Zaidon A, Hermawan D and Hadi Y S 2012 Bamboo fibre reinforced biocomposites: A review *Materials and Design* **42** 353-368
[12] Aji P, Farid M and Nurdiansyah H 2017 Analisis Morfologi dan Spektroskopi Infra Merah Serat Bambu Betung (Dendrocalamus Asper) Hasil Proses Alkalisasi Sebagai Penguat Komposit Absorbsi Suara *Jurnal Teknik ITS* **6**(2) 441-444
[13] Reddy N and Yang Y 2015 Natural Cellulose Fibers from Corn Innovative Biofibers from Renewable Resources *Innovative Biofibers from Renewable Resources* 5-8
[14] Chattopadhyay S K, Khandal R K, Uppaluri R and Ghoshal A K 2011 Bamboo fiber reinforced polypropylene composites and their mechanical, thermal, and morphological pTali rties *Journal of Applied Polymer Science* **119**(3) 1619-1626

[15] Hongbo Li and Shen S 2011 The mechanical pTali rties of bamboo and vascular bundles *Journal of Materials Science Research* **26**(21) 2749-2756
[16] Nguyen T P, Fujii T, Chuong B and Okubo K 2011 Study on how to effectively extract bamboo fibers from raw bamboo and wastewater treatment *Journal of Materials Science Research* **1**(1) 144-155
[17] Subash S, Stanly J R B and Edwin R D J 2017 A Review on Extraction of Bamboo Fibres and Its PTali rties *International Journal of Advanced Chemical Science and Applications (IJACSA)* **5**(2) 22-27
[18] Hashim M Y, Amin A M, Marwah O M F, Othman M H, Yunus M R M and Huat N C 2017 The effect of alkali treatment under various conditions on physical pTali rties of kenaf fiber *Journal of Physics: Conference Series* **914**(1) 012030 1-15
[19] Li M, Zhou S and Guo X 2017 Effects of alkali-treated bamboo fibers on the morphology and mechanical pTali rties of oil well cement *Construction and Building Materials* **150** 619-625
[20] Zakikhani P, Zahari R, Sultan M T H and Majid D L 2014 Extraction and preparation of bamboo fibre-reinforced composites *Materials and Design* **63** 820-828
[21] Bakri M K B, Jayamani E, Hamdan S, Rahman M R, Soon K H and Kakar A 2016 Fundamental Study on The Effect of Alkaline Treatment on Natural Fibers Structures and Behaviors *ARPN Journal of Engineering and Applied Sciences* **11**(14) 8759-8763
[22] Cai Q, Fan Z, Chen J, Guo W, Ma F, Sun S, Hu L and Zhou Q 2018 Dissolving process of bamboo powder analyzed by FT-IR spectroscopy *Journal of Molecular Structure* **1171** 639-643
[23] ASTM_E1252 *Standard Practice for General Techniques for Obtaining Infrared Spectra for Qualitative Analysis* (Philadelphia, USA: American Society for Testing and Materials)
[24] Sedan D, Pagnoux C, Chotard T, Smith A, Lejolly D, Gloaguen V and Krausz P 2007 Effect of calcium rich and alkaline solutions on the chemical behaviour of hemp fibres *J. Mater. Sci.* **42**(22) 9336-9342
[25] Osorio L, Trujillo E, Van Vuure A W and Verpoest I 2011 Morphological aspects and mechanical pTali rties of single bamboo fibres and flexural characterisation of bamboo/epoxy composites *Journal of Reinforced Plastics Composites* **30**(5) 396-408
[26] Yudhanto F, Jamasri and Rochardjo H S B 2018 Application of taguchi method for selection parameter bleaching treatments against mechanical and physical pTali rties of agave cantala fiber *IOP Conf. Series: Materials Science and Engineering* **352**(1) 012002 1-7
[27] Kim H, Okubo K, Fujii T and Takemura K 2013 Influence of fiber extraction and surface modification on mechanical pTali rties of green composites with bamboo fiber *Journal of Adhesion Science and Technology* **27**(12) 1348-1358
[28] Loiwatu M and Manuhuwa E 2018 Chemical Component and Anatomical Feature of Three Bamboo Species from Seram Maluku *Agritech* **28**(2) 76-83

A non-empirical lattice-relaxation estimation of TM^{3+} doped α-Al_2O_3

M Novita[1,*], D Nuvitalia[2], N Cholifah[3] and K Ogasawara[4]

[1] Faculty of Engineering and Informatics, Universitas PGRI Semarang, Jl. Sidodadi-Timur No.24, Semarang, Central Java 50232, Indonesia
[2] Faculty of Mathematics, Natural Science and Information Technology Education, Universitas PGRI Semarang, Jl. Sidodadi-Timur No.24, Semarang, Central Java 50232, Indonesia
[3] Lembaga Penelitian dan Pengabdian Kepada Masyarakat, Universitas PGRI Semarang, Jl. Sidodadi-Timur No.24, Semarang, Central Java 50232, Indonesia
[4] School of Science and Technology, Kwansei Gakuin University, 2-1 Gakuen Sanda, Hyogo 669-1337 Japan

*novita@upgris.ac.id

Abstract. We investigated the effect of lattice relaxation in the trivalent transition metal ions doped α-Al_2O_3 crystal (α-Al_2O_3: TM^{3+}). The transition metal ions are Sc^{3+}, Ti^{3+}, V^{3+}, Cr^{3+}, Mn^{3+}, Fe^{3+}, and Co^{3+}. The lattice relaxation effect was considered by comparing two different methods such as Shannon's crystal radii methods and Cambridge Serial Total Energy Package (CASTEP) methods. In the Shannon's crystal radii methods, the lattice relaxation effect was estimated as the ratio of the crystal radii of the substituted ion and the replaced ion. On the other hand, in the CASTEP method, the lattice relaxation effect was estimated by geometry optimization. The detail results show that in the case of Shannon's crystal radii method, the lattice relaxation ratio was ca. 111.0 - 100.5%. Whereas in the case of CASTEP method, the lattice relaxation ratio was ca. 107.8 - 102.1%. However, the lattice relaxation ratio of α-Al_2O_3: Sc^{3+}, Ti^{3+}, V^{3+}, Cr^{3+}, Mn^{3+}, Fe^{3+}, and Co^{3+} is unique depends on the ion type.

1. Introduction

Transition metal (TM) ions doped compounds are one of important materials in the luminescence technologies such as light emitting diodes (LEDs). LEDs are made of semiconductors, but the conductivity of the semiconductors in LEDs has been specially enhanced to allow for the unique properties of LEDs. The conductivity of semiconductors can be increased by adding small, controlled amounts of "impurities". This process is known as doping. However, when an impurity ion (in this case TM ion) substitutes for a metal (M) ion, the transition metal-ligand (TM-L) bond lengths change due to the nature of lattice relaxation. It can be shorter or longer than the original metal-ligand (M-L) bond lengths. For example, in the case of ruby or α-Al_2O_3: Cr^{3+}. The trivalent chromium (Cr^{3+}) is one of TM ion which has 3 electrons in the 3d orbital ($3d^3$). When a Cr^{3+} ion is substituted into α-Al_2O_3 crystal, the Cr-O bond lengths are longer than the Al-O bond length. Previously, we have studied the bond length of α-Al_2O_3: $3d^3$ ions such as V^{2+}, Cr^{3+}, Mn^{4+} using a combination method of Shannon's crystal radii and Extended X-ray absorption fine structure (EXAFS) methods [1]. The results show that the TM-L bond lengths increase when the atomic number increases. In the other study, we have also shown that in order to theoretically reproduce the optical properties that agree to the experimental values, creating model close to the real material cannot be neglected [1-4]. Since the consideration of lattice relaxation effect is especially important to mimic the real material, in this work we want to perform a study on the lattice relaxation estimation of α -Al_2O_3: TM^{3+} ions such as Sc^{3+}, Ti^{3+}, V^{3+}, Cr^{3+}, Mn^{3+}, Fe^{3+}, and Co^{3+}. Several approaches are compared. The lattice relaxation effects were estimated using a non-empirical approach i.e., Shannon's crystal radii and the geometry optimization using Cambridge Serial Total Energy Package (CASTEP) methods.

2. Methods

We constructed three types of model namely no-relax model, Shannon's model and CASTEP model. The no-relax models are obtained from the experimental crystallographic information file (CIF) of pure α-Al2O3 crystal [5]. The Shannon's models are obtained by modifying the no-relax model according to the Shannon's crystal radii. On the other hand, the CASTEP models are obtained from the geometry optimization process using CASTEP code.

2.1. Shannon's crystal radii method

Shannon's crystal radii method is one of the simplest approach to estimate the lattice relaxation effect. We have reported some studies employing this method for oxide and fluoride materials [1-4]. Table 1 shows the lattice relaxation ratio of α-Al_2O_3: TM^{3+} estimated using Shannon's crystal radii method. The values of crystal radii were obtained from Refs. 9 and 10 It is defined by the ratio between the crystal radii of the system after substituting the dopant (TM^{3+}-O^{2-}) and the crystal radii of system before substituting the dopant ion (Al^{3+}-O^{2-}). Therefore, the estimated bond length with considering lattice relaxation effect can be easily calculated by multiplying the lattice relaxation ratio with the original bond length as shown in the equation (1). *S* indicates the Shannon's crystal radii value, while *d* indicates the bond length.

$$d(\mathbf{TM-O})_{\text{relaxed}} = \frac{S(\mathbf{TM}^{3+}) + S(\mathbf{O}^{2-})}{S(\mathbf{Al}^{3+}) + S(\mathbf{O}^{2-})} \times d(\mathbf{Al-O})_{\text{unrelaxed}} \quad (1)$$

Table 1. Lattice relaxation ratio of α-Al_2O_3: TM^{3+} estimated using Shannon's crystal radii method [6,7].

Atom	Atomic number	Ionic charge	Coordination number (CN)	Crystal radii	Relaxation ratio (%)
Al	18	+3	6	0.675	-
O	8	-2	4	1.240	-
Sc	21	+3	6	0.885	110.97
Ti	22	+3	6	0.810	107.05
V	23	+3	6	0.780	105.48
Cr	24	+3	6	0.755	104.18
Mn	25	+3	6	0.785	105.74
Fe	26	+3	6	0.785	105.74
Co	27	+3	6	0.685	100.52

2.2. Geometry optimizations

The first-principles band-structure calculations were carried out using CASTEP code [8-10] in Material Studio 6.0. The Vanderbilt [11] ultrasoft pseudopotentials were used. Here we consider the Generalized Gradient Approximation (GGA) exchange-correlation potential proposed by Perdew, Burke, and Ernzerhof (PBE) [12]. It has been known for its general applicability and gives rather accurate results for a wide range of systems. In the first step, the structural optimizations on the pure α-Al_2O_3 crystal were carried out. We have reported some studies employing this method for oxide and fluoride materials [13-15]. The procedure employed here is similar with those used in [16]. Next, one TM^{3+} such as Sc^{3+},

Ti^{3+}, V^{3+}, Cr^{3+}, Mn^{3+}, Fe^{3+}, or Co^{3+} ion was substituted in the Al^{3+} site and then followed by the geometry optimization step.

3. Results and discussion

Figure 1 shows the estimated bond length of α-Al_2O_3: TM^{3+} based on (a) no-relax model, (b) Shannon's model and (c) CASTEP model. In the case of pure α-Al_2O_3 crystal, there are two different bond lengths called d1 and d2 for the shorter and longer bond length respectively. Here we calculated the d1, d2 and the average of d1-d2. Figure 1a shows that the TM-O bond lengths are the same for all kind of ions. The d1 was 1.855 Å, d2 was 1.971 Å and the average of d1-d2 was 1.913 Å. Figure 1b shows that the TM-O bond lengths are decreased in the order of Sc^{3+}, Ti^{3+}, V^{3+}, Cr^{3+}, and then increased for Mn^{3+}, Fe^{3+} and decreased again for Co^{3+}. On the other hand, Figure 1c shows that the TM-O bond lengths depend on ion type. However, if we see carefully, the average of TM-O bond length estimated from Shannon's model was found to be from ca. 2.12 – 1.92 Å for Sc^{3+} to Co^{3+}. On the other hand, the average of TM-O bond length estimated from CASTEP model was found to be from ca. 2.06 – 1.96 Å for Sc^{3+} to Co^{3+}. Therefore, the ranges of the average TM-O bond length were ca. 0.2 and 0.1 Å for Shannon's and CASTEP models respectively.

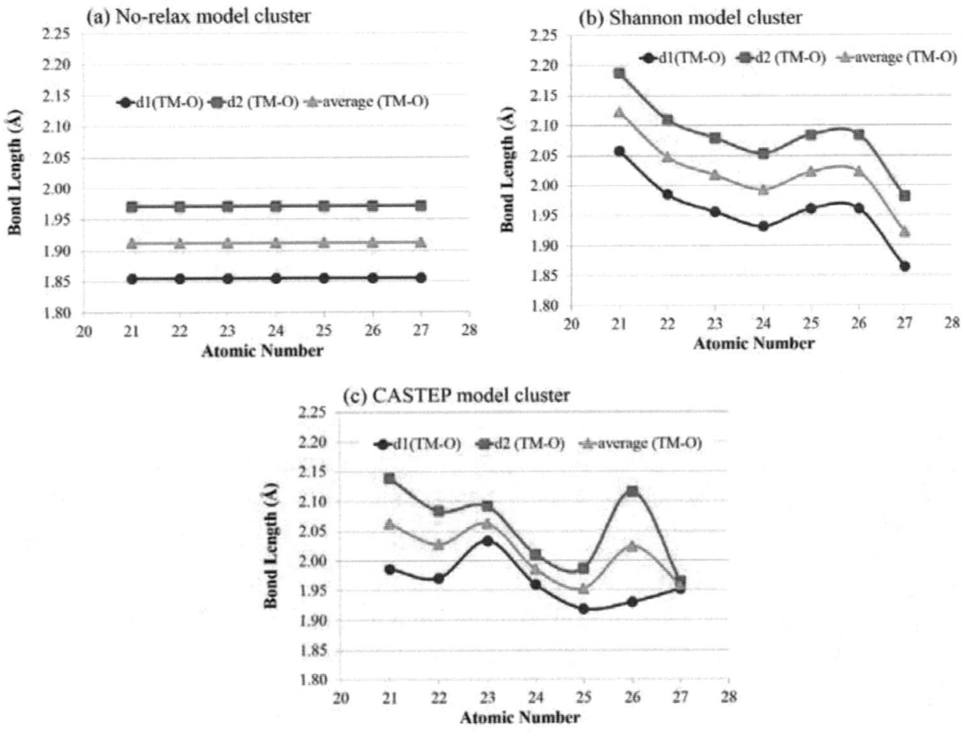

Figure 1. The TM-O bond length of α-Al_2O_3: TM^{3+} estimated using no-relax, Shannon, CASTEP models. The shorter TM-O bond length (d1) are indicated by circles (●). The longer TM-O bond length (d2) are indicated by squares (■). The averages of d1-d2 are indicated by triangle (▲).

Figure 2 shows the lattice relaxation ratio of α-Al_2O_3: TM^{3+} estimated using Shannon's and CASTEP models. Although it depends on the ion type, they were relatively decreased. In the case of Sc^{3+}, Ti^{3+}, and Mn^{3+}, the lattice relaxation ratios estimated from Shannon's model were overestimated. On the other hand, in the case of V^{3+} and Co^{3+}, the lattice relaxation ratios estimated from Shannon's model were underestimated. Whereas in the case of Cr^{3+} and Fe^{3+}, lattice relaxation ratios estimated from Shannon's model were almost the same by those estimated from CASTEP model.

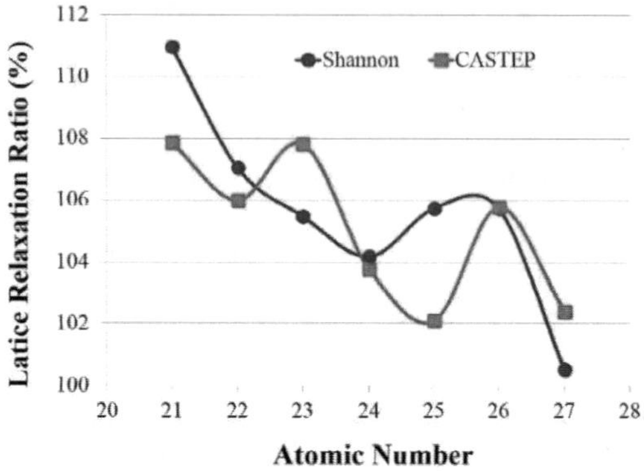

Figure 2. Lattice relaxation ratio of α-Al_2O_3: TM^{3+} estimated using Shannon's crystal radii and CASTEP method are indicated by circles (●) and squares (■) respectively.

In both cases, α-Al_2O_3: Sc^{3+} has the highest lattice relaxation ratio compared to the other TM^{3+} ions. They are 111.0 and 107.8% estimated from Shannon's and CASTEP model respectively. On the other hand, the lowest lattice relaxation ratios are different. the lowest lattice relaxation obtained from Shannon's model is α-Al_2O_3: Co^{3+} 100.5%. Whereas the lowest lattice relaxation obtained from CASTEP model is α-Al_2O_3: Mn^{3+} ca. 102.1%.

4. Conclusion

In this work, the lattice relaxation effect was investigated thoroughly by comparing two different methods using Shannon's crystal r dii method and first-principles band structure calculations CASTEP. The results show that the lattice relaxation estimated from Shannon's crystal radii method and those estimated from CASTEP method strongly depends on the ion type. This results should be investigated further such as compare the calculated lattice parameter with the observed Extended X-Ray Absorption Fine Structure (EXAFS) data.

Acknowledgments

The authors gratefully acknowledge the support from Direktorat Riset dan Pengabdian Masyarakat, Direktorat Jenderal Penguatan Riset dan Pengembangan, Kementrian Riset, Teknologi, dan Pendidikan Tinggi, Indonesia.

References

[1] Novita M and Ogasawara K 2012 Comparative Study of Absorption Spectra of V2+, Cr3+, and Mn4+ in α-Al2O3 Based on First-Principles Configuration–Interaction Calculations *Journal of the Physical Society of Japan* **81**(10) 104709

[2] Novita M and Ogasawara K 2012 Comparative study of multiplet structures of Mn4+ in K2SiF6, K2GeF6, and K2TiF6 based on first-principles configuration–interaction calculations *Japanese Journal of Applied Physics* **51**(2R) 022604

[3] Novita M, Honma T, Hong B, Ohishi A and Ogasawara K 2016 Study of multiplet structures of Mn4+ activated in fluoride crystals *Journal of Luminescence* **169** 594-600

[4] Novita M, Yoshida H and Ogasawara K 2013 Investigation of Ion Dependence of Electronic Structure for 3d3 Ions in Mg2TiO4 Based on First-Principles Calculations *ECS Transactions* **50**(41) 9-17

[5] Sawada H 1994 Residual electron density study of α-aluminum oxide through refinement of experimental atomic scattering factors *Materials research bulletin* **29**(2) 127-133

[6] Shannon R T and Prewitt C T 1969 Effective ionic radii in oxides and fluorides. *Acta Crystallographica Section B: Structural Crystallography and Crystal Chemistry* **25**(5) 925-946

[7] Shannon R T and Prewitt C T 1969 Effective ionic radii in oxides and fluorides *Acta Crystallographica Section B: Structural Crystallography and Crystal Chemistry* **25**(5) 925-946

[8] Payne M C, Teter M P, Allan D C, Arias T A and Joannopoulos A J 1992 Iterative minimization techniques for ab initio total-energy calculations: molecular dynamics and conjugate gradients *Reviews of modern physics* **64**(4) 1045

[9] Segall M D, Lindan P J, Probert M A, Pickard C J, Hasnip P J, Clark S J and Payne M C 2002 First-principles simulation: ideas, illustrations and the CASTEP code *Journal of Physics: Condensed Matter* **14**(11) 2717

[10] Clark S J, Segall M D, Pickard C J, Hasnip P J, Probert M I, Refson K and Payne M C 2005 First principles methods using CASTEP *Zeitschrift für Kristallographie-Crystalline Materials* **220**(5/6) 567-570

[11] Vanderbilt D 1990 Soft self-consistent pseudopotentials in a generalized eigenvalue formalism *Physical review B* **41**(11) 7892

[12] Perdew J P, Burke K and Ernzerhof M 1996 Generalized gradient approximation made simple *Physical review letters* **77**(18) 3865

[13] Novita M and Ogasawara K 2014 Study on Multiplet Energies of V2+, Cr3+, and Mn4+ in MgO Host Crystal Based on First-Principles Calculations with Consideration of Lattice Relaxation *Journal of the Physical Society of Japan* **83**(12) 124707

[14] Novita M, Nagoshi H, Sudo A and Ogasawara K 2018 Ab-initio study on the absorption spectrum of color change sapphire based on first-principles calculations with considering lattice relaxation-effect *IOP Conference Series: Materials Science and Engineering* **299** 1 012060

[15] Novita M, Wibowo S, Nada N Q and Ogasawara K 2019 Comparative Study on R-line and U-band Energies of Ruby Estimated from One-Electron and Many-Electron First-Principles Approaches *Journal of Physics: Conference Series* **1179** 1 012104

[16] Novita M, Nagoshi H, Sudo A and Ogasawara K 2018 Ab-initio study on the absorption spectrum of color change sapphire based on first-principles calculations with considering lattice relaxation-effect *IOP Conference Series: Materials Science and Engineering* **299** 1 012060

IOP Conf. Series: Materials Science and Engineering **850** (2020) 012030 doi:10.1088/1757-899X/850/1/012030

Optimal Power Flow the Sulselrabar 150 KV system before and after the penetration of wind power plants considering power loss and generation costs

A M Ilyas[1,2,*], A Suyuti[1], I C Gunadin[1] and A Siswanto[1]

[1] Department of Electrical Engineering, Faculty of Engineering, University of Hasanuddin, Jl. Perintis Kemerdekaan Km. 10 Makassar, 90245, Indonesia.
[2] Electrical Engineering Study Program, Faculty of Engineering, Khairun University, Kelurahan Gambesi, 97719, Ternate, North Maluku, Indonesia.

*aamilyas@gmail.com

Abstract. Optimal Power Flow (OPF) is to optimize power flow in interconnection systems to decrease production costs and increase system reliability, quality, and stability. This research was conducted on power generation 150 KV systems in South Sulawesi, Southeast Sulawesi, and West Sulawesi, abbreviated as Sulselrabar. The number of buses is 44, 15 generator buses before penetration with the Sidrap Wind Power Plant (PLTB), and 16 bus generators after penetration with PLTB. The method used in this study is Improved Particle Swarm Optimization (IPSO) and Lagrange. The results showed that IPSO showed better results compared to Lagrange and the results of existing systems.

1. Introduction

Optimal power flow aims to determine the best way to operate the power system [1-15]. Today more and more optimization methods and algorithms are developed for use in optimal power flows, including; Lambda Iteration [16], Particle Swarm Optimization (PSO) [17-19], Biogeography-based Optimization (BBO), Artificial Bee Colony algorithm (ABC) [20], Krill Herd Algorithm (KHA) [21], Cultural Algorithm (CA) [22], Charged System Search (CSS), Bat Algorithm (BA), Novel Bat Algorithm (NBA) [23], Genetic Algorithm (GA) [24].

Previous research used the Novel Bat Algorithm (NBA) method to discuss some Sulserabar systems, namely 29 buses, 5 interconnect generators for PLTB penetration thermal plants. In this study, the IPSO and Lagrange methods were used to solve optimal power flow in all Sulselrabar systems, which is 44 buses, 16 generators. Optimal power flow includes various optimization variables such as power balance constraints, generator capacity limits, voltage limits, and power losses. The mathematical equation in solving optimal power flow is as follows:
The Economic Dispatch Formula is:

$$F\left(P_g\right) = \sum_{n=1}^{n}(a_i P_{gi}^2 + b_i P_{gi} + c_i) \quad (1)$$

The Economic Dispatch limits are as follows.
Power balance limit:

IOP Conf. Series: Materials Science and Engineering **850** (2020) 012030 doi:10.1088/1757-899X/850/1/012030

$$\sum_{n=1}^{n} P_i = P_d + P_L \tag{2}$$

Optimal Power Flow Formula:

$$P_{Ga} - P_{Da} = V_a \sum_{n=1}^{NB} V_b \left(G_{ab}\cos(\theta_a - \theta_b) + B_{ab}\sin(\theta_a - \theta_b)\right) \tag{3}$$

$$Q_{Ga} - Q_{Da} = V_a \sum_{n=1}^{NB} V_b \left(G_{ab}\sin(\theta_a - \theta_b) + B_{ab}\cos(\theta_a - \theta_b)\right) \tag{4}$$

Limitation of generating capacity:

$$P_{Gi,min} \leq P_{Gi} \leq P_{Gi,max} \tag{5}$$

$$Q_{Gi,min} \leq Q_{Gi} \leq Q_{Gi,max} \tag{6}$$

Voltage limit:

$$V_{m,min} \leq V_m \leq V_{m,max} \quad \theta \tag{7}$$

Information:

$F(P_g)$	=	Total generation costs
a_i, b_i, c_i	=	The coefficient of fuel costs
P_d	=	Beban dayaPower load
P_L	=	Losses
P_{Gi}	=	Active power generated on the bus i
Q_{Gi}	=	Reactive power generated on the bus i
G_{ab}, B_{ab}	=	Admittance of line a, b
V_a, V_b	=	Voltage of bus a,b
θ_a, θ_b	=	Voltage bus phase angle a, b

In general, power losses as in the following formula.

$$P_L = I^2 R \tag{8}$$

Where:

P_L = power losses
I = current flowing on line (Ampere)
R = resistance on line (Ω)

2. Materials and methods

2.1. Sulselrabar system

The Sulselrabar 150 KV system consists of 44 buses and 14 conventional generator buses and 1 new renewable energy bus, namely the Wind Power Plant (PLTB) located in Sidrap Regency. The bus generators in question are: Bakaru as a slack bus and generator buses are Pinrang, Suppa, Sidrap (PLTB), Barru, Tello, Borongloe, Tellolama, Sungguminasa, Tallasa, Punagaya, Sinjai, Sengkang, Makale, Palopo, Poso, the remaining 28 the bus is a load bus. The results of this optimization are compared between before and after penetration with the Wind Power Plant PLTB considering voltage, emissions, generation costs, and power losses on the electricity network.

The data used in the simulation is the peak day load data at 14.00 WITA, December 4, 2018. The measuring parameters used in this study are bus data, generation data, and load data. The generator data and single line Sulselrabar system can be seen in Figure 1.

2.2. Optimization methods

The optimization methods used are Improved Particle Swarm Optimization (IPSO), and Lagrange methods.

2.3. IPSO algorithm

$$V_i^{k+1} = \omega V_i^k + c_1 rand_1(Pbest_i^k - X_i^k) + c_2 rand_2(Gbest^k - X_i^k) \tag{9}$$

Where:

V_i^k	=	velocity individual i in the iteration k
w	=	weight parameter
c_1, c_2	=	acceleration coefficient
$rand_1$, $rand_2$	=	random number between 0 and 1
X_i^k	=	individual position i in the iteration k
$Pbest_i^k$	=	individual i to iteration k
$Gbest^k$	=	*Gbest* group until iteration k

For more details, see Flowchart's research in Figure 2.

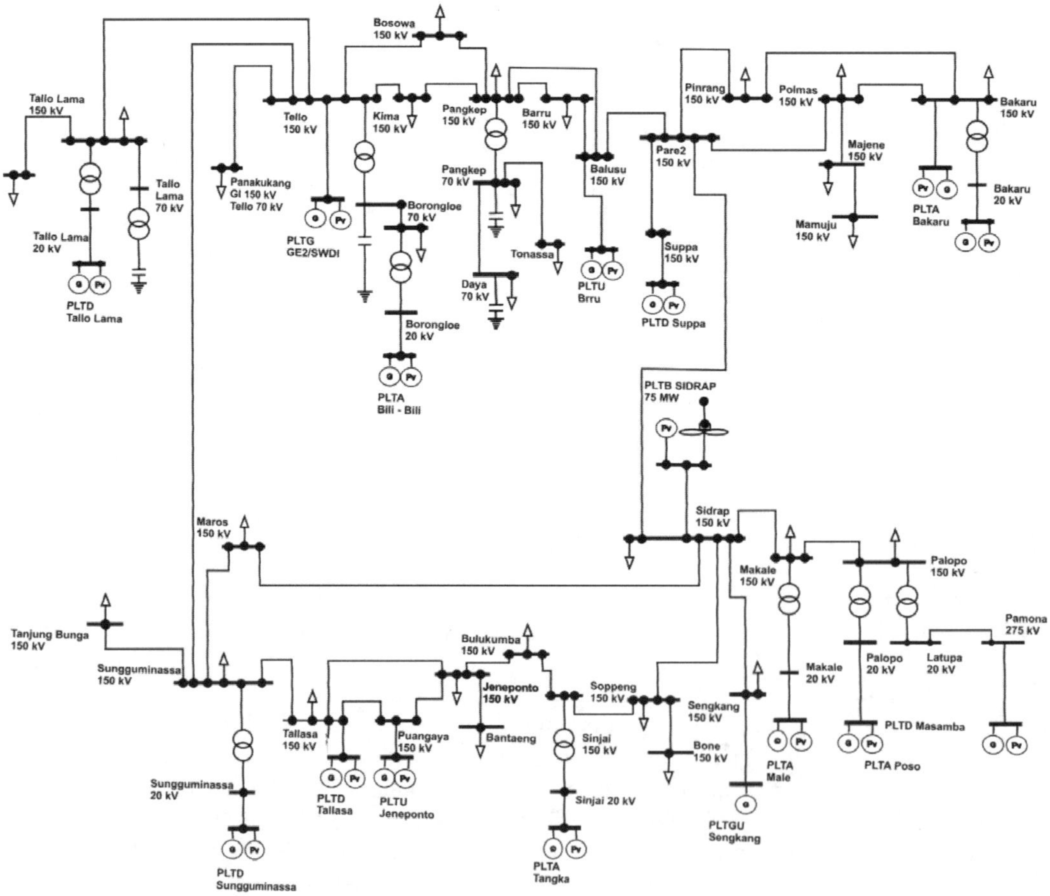

Figure 1. Sulselrabar single line system.

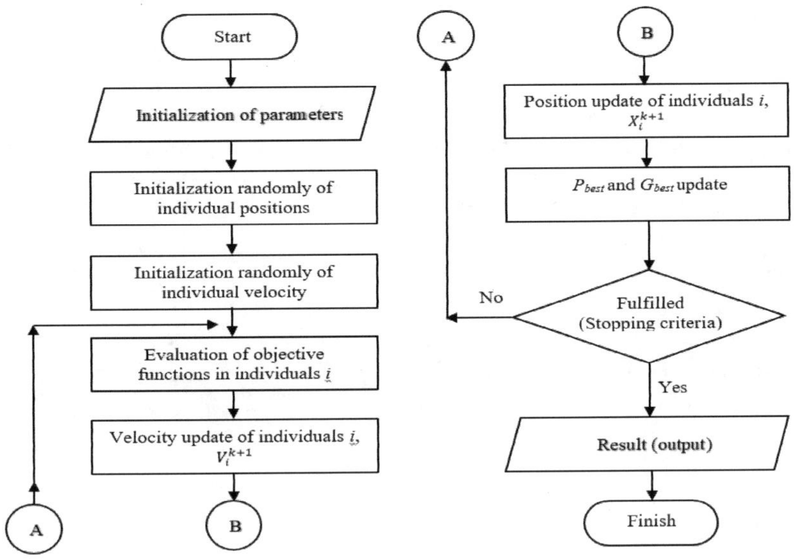

Figure 2. Research flowchart.

3. Results and discussion

3.1. Optimal power flow results, system 44 Sulselrabar bus before integration with PLTB wind power plants

The results obtained from the IPSO method, compared with the Lagrange method and the results of the existing system before penetration with the PLTB data are seen in Table 1 and Figure 3.

Table 1. Optimal power flow results of Sulselrabar system before PLTB integration.

Bus Number	*Existing System (MW)*	Lagrange (MW)	IPSO (MW)
1	62.10	50.05	78.69
5	14.30	32.51	14.32
7	31.10	35.77	29.25
9	60.30	74.61	67.46
16	21.00	66.76	23.55
19	15.20	16.07	18.30
24	12.60	22.82	11.40
27	20.00	24.38	21.59
29	79.00	36.49	73.78
31	193.10	228.50	197.00
34	14.00	11.21	6.83
37	265.20	218.47	251.41
38	8.20	7.05	7.69
39	4.00	6.59	3.75
41	190.00	157.75	184.08
Total Load (MW)	990.10	989.02	989.10
Losses (MW)	150.30	119.95	80.13
Load (MW)	839.80	869.07	908.97
Losses (%)	15%	12%	8%
Total Generation Cost (IDR/hr.)	995,709,690	962,519,367	945,924,206

IOP Conf. Series: Materials Science and Engineering **850** (2020) 012030 doi:10.1088/1757-899X/850/1/012030

The simulation results using the IPSO method show cheaper results compared to the results using the Lagrange method and the results of the existing system. The generation cost using the IPSO method is Rp. 945,924,206 per hour and the cost of the Lagrange method is Rp. 962,519,367 per hour, while the results of the existing system are Rp. 995,709,690 per hour. Generation costs that can be reduced using the IPSO method are Rp.49,785,485 per hour or around 5% when compared to the results of existing systems. Using the IPSO method, the power losses obtained are 80.13 MW or 8%, while those from the existing system are 150.30 MW or 15% and the Lagrange method is 119.95 MW or 12%.

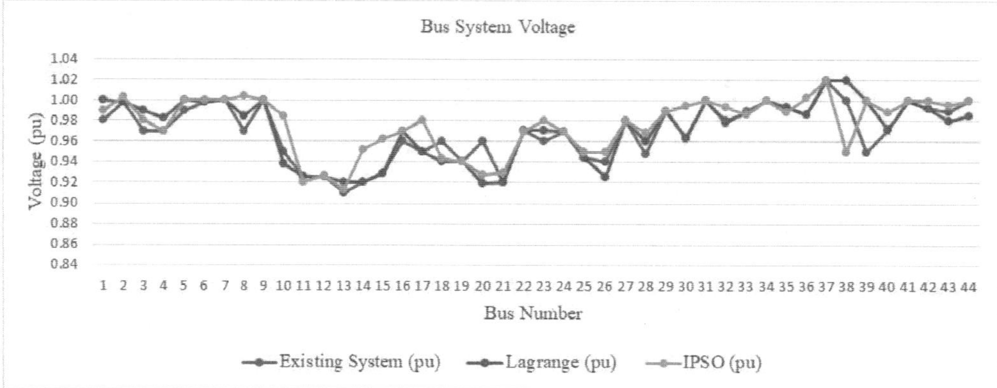

Figure 3. Comparison of bus voltage systems without integration of PLTB.

After observing the optimal results of power flow, voltage conditions are also considered. The stability of the voltage using IPSO shows quite good results because it is following PLN's regulatory limit, which is 0.9 to 1.04.

3.2. Optimal results of power flow system 44 Sulselrabar buses after integration with PLTB wind power plants

The integration of PLTB in the Sulselrabar system has a good impact on the system, especially concerning generation costs. The results obtained using the IPSO method are compared with the Lagrange method and the results of the existing system. The bus generator on this system increases, from 15 bus generators to 16 generator buses. The load bus in serial number 8 is converted into a bus generator by integrating Sidrap PLTB on the system. For more details about optimal power flow in systems integrated with PLTB can be seen in Table 2, and Figure 4.

The simulation results of the 44 bus 16 generator system, after PLTB integration there was a decrease in the cost of generation and power loss on the network. The generation cost using the IPSO method is Rp. 927,005,722 per hour, while the yield from Lagrange is Rp. 933,643,786 per hour and the proceeds from the existing system are Rp. 975,795,496 per hour. The generation cost is reduced in the IPSO method by Rp. 48,789,774 per hour or about 5% when compared to the results of the existing system. By using the large IPSO method power losses of as much as 49.28 MW or 5%, while the Lagrange method is 109.08 MW or 10%, and the results of the existing system are 149.07 MW or 14%.

Table 2. Optimal power flow results of Sulselrabar system after PLTB integration.

Bus Number	Existing System (MW)	Lagrange (MW)	IPSO + Wind (MW)
1	62.07	55.00	57.53
5	14.30	32.51	32.53
7	31.10	35.77	56.79
8	75.00	70.00	36.81
9	60.40	74.61	94.53
16	22.00	66.77	65.45
19	15.20	32.07	30.24
24	12.56	22.82	21.83
27	20.00	24.38	16.05
29	79.00	76.49	4.56
31	196.10	135.58	192.55
34	5.00	11.22	4.38
37	265.17	239.54	256.75
38	8.20	27.05	5.85
39	4.00	6.59	5.47
41	195.00	148.75	183.58
Total Load (MW)	1,065.10	1,059.13	1,064.88
Losses (MW)	149.07	109.08	49.28
Load (MW)	916.03	950.05	1,015.60
Losses (%)	14%	10%	5%
Total Generation Cost (IDR/hr.)	975,795,496	933,643,786	927,005,722

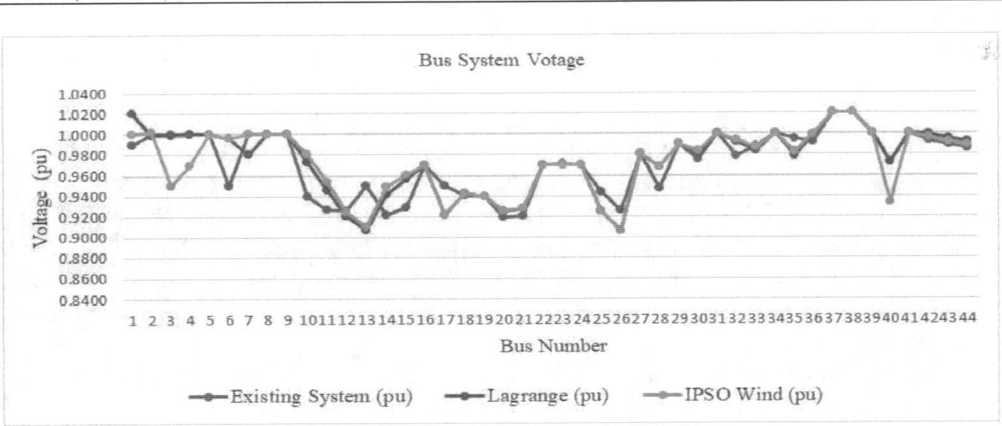

Figure 4. Comparison of Bus System Integration Voltage with PLTB.

The voltage profile based on the optimal results of power flow using the IPSO method shows better results than the results of the existing system. The stability of the voltage using IPSO shows good results because it is in the range that is under PLN's regulatory limit, which is 0.9 to 1.04.

4. Conclusion

The results showed that IPSO showed better results compared to Lagrange and the results of the existing system. Before integration with PLTB, IPSO was able to reduce costs by IDR 49,785,485 per hour or about 5% compared to the results of the existing system results. The loss of power using IPSO is 80.13 MW or 8%, using Lagrange is 119.95 MW or 12% and the results of the existing system are 150.30 MW or 15%.

After integration with PLTB, IPSO can reduce costs by Rp. 48,789,774 per hour or 5% of the results of the existing system, Lagrange reduced costs by IDR 42,151,710 per hour or 4% of the results of the existing system. While the IPSO simulation power loss is 49.28 MW or 6% and Lagrange is 109.08 MW or 13%.

The difference or decrease in power loss between before and after integration with PLTB uses the IPSO method of 30.85 MW, the Lagrange method of 10.89 MW, and the results of the existing system 1.23 MW.

References

[1] Berahim H 2011 *Teknik Tenaga Listrik Dasar* (Yogyakarta: Graha Ilmu)

[2] Riffonneau Y, Bacha S, Barruel F and Ploix S 2011 Optimal power flow management for grid connected PV systems with batteries *IEEE Transactions on sustainable energy* **2**(3) 309-320

[3] Cekdin C and Barlian T 2013 *Transmisi Daya Listrik* (Yogyakarta: Penerbit Andi)

[4] Wood A J, Wollenberg B F and Sheblé G B 2013 *Power generation, operation, and control* (Canada: John Wiley & Sons)

[5] Stevenson W D 1984 *Analisis System Tenaga Listrik* (New York: McGraw-Hills)

[6] Gunadin I C, Muslimin Z and Sudrajat E 2014 Studi Keandalan Ketersediaan Daya Perencanaan *Seminar Nasional Riset Inovatif* **2**

[7] Gunadin I C, SAID S and Irsan M 2016 Determination of Stability Index of Electrical Power System Using Rei-Dimo Methods *Journal of Theoretical & Applied Information Technology* **90**(1) 161-167

[8] Gunadin I C, Soeprijanto A and Penangsang O 2010 Real power generation scheduling to improve steady state stability limit in the Java-Bali 500 kV interconnection power system *World Acad. Sci. Eng. Technol* **72** 1-5

[9] Gunadin I C, Muslimin Z and Siswanto A 2017 Transient stability improvement using allocation power generation methods based on moment inertia *2017 International Conference on Electrical Engineering and Informatics (ICELTICs)* 147-152

[10] Siswanto A, Gunadin I C, Said S M and Suyuti A 2018 Stability improvement of wind turbine penetrated using power system stabilizer (PSS) on South Sulawesi transmission system *AIP Conference Proceedings* **1941**(1) 020036

[11] Asmi R Y, Siswanto A and Mahmudi I 2018 Stability issues in presence variable distributed generation into radial distribution network *MATEC Web of Conferences* **218** 01005

[12] Suyuti Y A, Kitta I and Akil Y S 2017 The impact of the operation planning of power plants for environmental emissions in south Sulawesi *ARPN J. Eng. Appl. Sci* **12**(11) 3440-3444

[13] Darusman B M, Suyuti A and Gunadin I C 2018 Small Signal Stability Analysis of Wind Turbine Penetration in Sulselrabar Interconnection *System Journal of Physics: Conference Series* **1090**(1) 012034

[14] Siswanto A, Suyuti A, Gunadin I C and Said S M 2019 Steady state stability limit assessment when wind turbine penetrated to the systems using REI approach *przegląd elektrotechniczny* **95** 53-57

[15] Karim A N, Said S M and Gunadin I C 2018 Impact of Penetration Wind Turbines on Transient Stability in Sulbagsel Electrical Interconnection System *Journal of Physics: Conference Series* **979**(1) 012028

[16] Kristianto D and Suyono H 2014 Operasi Ekonomis Pembangkit Tenaga Listrik Dengan Metode Iterasi Lambda Menggunakan Komputasi Paralel *Jurnal Mahasiswa TEUB* **2**(6)

[17] Kundur P, Balu N J and Lauby M G 1994 *Power system stability and control* Vol 7 (New York: McGraw-Hill)

[18] Humena S 2016 *Optimasi Ekonomis System Interkoneksi 150 kV Sulselbar Menggunakan MIPSO* (Makassar: Universitas Hasanudin)

[19] Ilyas A M and Rahman M N 2012 Economic dispatch thermal generator using modified improved particle swarm optimization *Telkomnika* **10**(3) 459

[20] Pratama D A 2016 *Economic and Emission Dispatch pada System Transmisi Jawa Bali 500KV Berdasarkan Ruptil 2015-2024 Menggunakan Modified Artificial Bee Colony Algorithm* (Doctoral dissertation, Surabaya: Institut Teknologi Sepuluh Nopember)

[21] Chaphekar S N, Karad P R and Dharme A A 2016 Optimal power flow for power management in Microgrid *2016 IEEE 1st International Conference on Power Electronics, Intelligent Control and Energy Systems (ICPEICES)* 1-5

[22] Thamrin F 2012 *Study Inference Fuzzy Tsukamoto Untuk Penentuan Faktor Beban* (Thesis, Semarang: Universitas Diponegoro)

[23] Fitri S N, Akil Y S and Gunadin I C 2018 Economic Dispatch using Novel Bat Algorithm Constrained by Voltage Stability *2018 2nd East Indonesia Conference on Computer and Information Technology (EIConCIT)* 163-167

[24] Hanna B and El-Shahat A 2017 Optimal power flow for micro-grids *2017 IEEE Global Humanitarian Technology Conference (GHTC)* 1-3

IOP Conf. Series: Materials Science and Engineering **850** (2020) 012031 doi:10.1088/1757-899X/850/1/012031

An affordable mobile robot soccer (ReROSO): A student's project of robotics trainer for sophomore

M Renaldy[*] and D Wahyudin

Departemen Pendidikan Teknik Elektro, Universitas Pendidikan Indonesia, Indonesia

*muhammad.renaldy@student.upi.edu

Abstract. Nowadays, the education system in Indonesia as an emerging country is on the advanced stage. There are many private high schools include the robotics system in their extra curriculum. However, robotics is a complex system. Therefore, it is difficult to do so for the nonengineering school's students. Hence, developing a robotics kit to assist students on how to learn robotic system will give some benefits. This paper presents the final project of a student of Electrical Engineering Education, that is an affordable robot trainer for learning robotics system in HS. It has features that learners have the flexibility to modify, and easy to program as well as to operate. The testing of ReROSO gave the evidence that event though it was constructed using a low-cost of robotics part. The result also proves that the design of ReROSO is adequate to teach robotics for the sophomore.

1. Introduction

In recent decades, researchers argued that robotics is very important because it has a role to solve some problems that existed in the manufacturing as well as in daily life [1]. In education, robotics can play an additional role to stimulate learners' sensitivity when facing life problems [2]. It can also to be used on developing cognitive and social skills [3]. Therefore, it is necessary to include the robotics learning curriculum into various level of education. Hence, the robotics system could be taught not only in university education but also in high school (HS).

In the process of learning, how to understand the programming algorithm always makes a constraint. This occurred due to the lack of instructional media can be used by teachers [4]. The development of learning media using the robot has been running rapidly, as evidenced by some of the research that uses robots as learning media, among others: robot balance [3], soccer robot [5], humanoid robot [6], dance robot [7], etc. But most of the robot learning media is limited because it uses only certain sensor modules only and must also use additional devices.

Learning media in this study using an SBC (Single Board Computer) Raspberry Pi model 3 (Raspi3). Raspi3 in its programming can be done without the use of enhancements making it easier and not necessarily driver installation to the computer [8]. Raspi3 has an affordable price so in making the robot does not spend a lot of money [9]. SBC in its programming can use several types of programming languages like C, C ++ and Python. Learners can choose the programming language according to the student's ability. In this SBC there is GPIO (General Pin Input / Output) and a USB port that can help learners create creative sensors as desired.

IOP Conf. Series: Materials Science and Engineering **850** (2020) 012031 doi:10.1088/1757-899X/850/1/012031

The purpose of this paper is to present the design and development of affordable robot soccer (ReROSO) platform intended as a robotic trainer for the sophomore. ReROSO is a final project work of a student of Electrical Engineering education. The ReROSO have features, i.e., easy to modify, easy to program, and easy to operate. Therefore, learners are expected to have a good understanding of the robot programming algorithm and system work. The examination of ReROSO gave the evidence that it can be used in robotics extra curriculum in HS.

2. Design of ReROSO

2.1. Platform requirement and analyses

The process to build the ReROSO started by analysing the robotics features. ReROSO is intended to be used as a robotics learning kit. Hence, it should provide the features as follows.

2.1.1. Easy to modify. To assemble robot soccer, the student needs to understand some basic requirement, i.e., electronic circuit and microcontroller, and also mechanical knowledge. Hence, because of this issue, ReROSO should provide learners with the flexibility of learning robotics. As such, ReROSO designed by a modular approach that could be easy to modify. Each part of ReROSO could be attached and detached easily, and it also supports the future expansion not only as robot soccer but also for other robot types such as robot of the crawler, fire-fighting, and obstacle avoidance.

2.1.2. Easy to program. Robotics programming similar to the other system, that sometimes getting harder for the sophomore. Learners should understand basic programming algorithm and mastery in one of the programming languages, such as C or C++. ReROSO designed to help learners construct and debug robotics program without serious obstacle. It is the advantage of ReROSO. How the robot can be programmed described as follow. Programming is using predefined library embedded into Raspi3. Learners just need to call such a library in the programming IDE installed in a PC/laptop. The robot could be programmed directly using Raspi3, desktop computer or even a smartphone. For direct programming, the student just adds the monitor screen, keyboard and mouse connected to the Raspi3. Programming ReROSO can also be done using a desktop computer. In this ability, ReROSO becomes a hotspot. So, desktop computers used to program the robot can connect to ReROSO through the Raspi3 Wi-Fi. Furthermore, it will provide the learners with an easy way to program the robot, because the computer connects to the SSID of the robot at all time. After a laptop connected to the robot, the student uses the Remote Desktop Connection. Hence all windows in Raspi3 can be seen in a seamless operation. Besides, the programming can be done through an Android smartphone using the RemoteToGo-RDP / VNC application for Android. Initially the same as using a computer, the smartphone is connected to a robot hotspot then connect via an android application.

2.1.3. Easy to operate. As an educational tool, ReROSO designed to have a feature of easy to operate. Furthermore, ReROSO could be played by everyone without educational background limitation. With such feature, ReROSO could also be used not only for an extra-curricular program of sophomore but also for demonstration and exhibition of robotic for elementary, and also for short training of how a robot could assist human in daily life for university student and adults.

2.2. Construction of ReROSO

ReROSO is designed to have a lightweight concept as shown in Figure 1. The body platform of ReROSO using acrylic to achieve the concept of lightweight but strong enough, hence it could support the durability robot trainer kit. The main processor of the kit using Raspi3. For the movement, ReROSO used four wheels drive of DC motors. The motor is controlled through the motor driver connected to the Raspi3 GPIO.

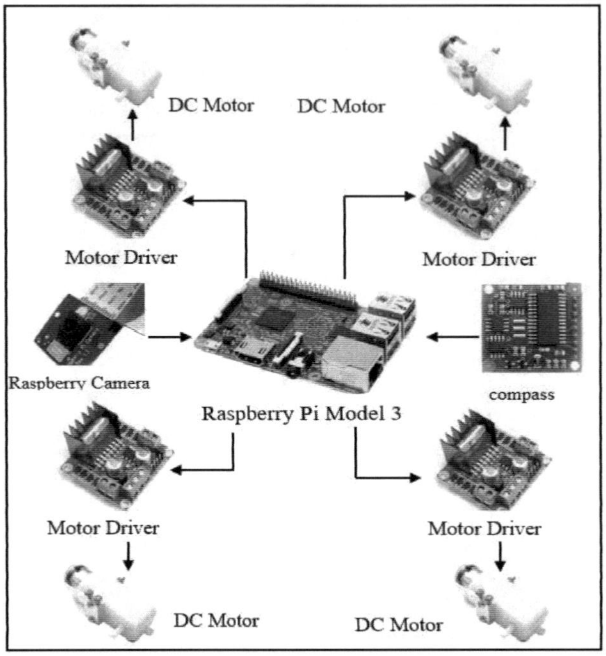

Figure 1. ReROSO system design.

To sensing the environment, there are two robot sensors. The first sensor is to determine the position using a compass sensor (CommSEN). The use of CommSEN is to exactly measure robot position. CommSEN connected to Raspi3 through I2C (Integrated to Integrated Circuit) communication. The second sensor is Raspberry camera (RaspCAM) used to recognize the soccer ball with orange colour. To do so, ReROSO detects the soccer ball by image recognition through the camera. The camera can be directly connected to the camera port of Raspi3. However, the power required by Raspi3 is 5V. Because of the used battery is 7.4 V. Furthermore the supply voltage must be lowered than the battery voltage. So, a module of Ultimate Battery Eliminator Circuit (UBEC) is used to achieve this purpose. The specific function of ReROSO components describes as follows.

2.2.1. Robot movement. The movement of ReROSO using four-wheel drives of Omni wheel with X configuration as shown in Figure 2. Actually, using four Omni wheels will be more difficult to control the robot. However, it will allow the robot to accelerate with the same speed in all directions [10]. Furthermore, Omni wheel construction has some advantages to the system, i.e., low strain, high acceleration, and less risk of motor burn [10]. In addition, the current of DC motors of robot wheel control by UBEC to provide various speed of robot movements. The drive system uses a dc motor and L298 motor driver, then the interface between the motor driver and raspberry pi directly uses GPIO from raspberry pi.

Figure 2. X configuration of ReROSO wheel drives.

2.2.2. Robot controller. To control ReROSO movement, Raspi3 has been used. It utilized Raspbian Jessie as a common Raspi3 operating system for a robotics project. However, some libraries should be included in the OS installation, such as OpenCV for image processing, C++ module for Raspi3 programming, and Python language for communication between ReROSO and Raspi3. To make clear how ReROSO is programmed, its programming algorithm could be explained as follow. The first step is to activate CommSEN. The specific function of CommSEN is to determine the position of the goal woodwork and crossbar. The next step is activating the camera to scan the orange ball position by image recognition. If the robot could not find the orange ball position, the robot wheels will move and rotate the body of the robot to find the exact position — the way to find the target ball combined with the movement of servo motor where the camera attached. When the ball position detected, the wheels will move to align the body of the robot to the ball and the goal crossbar position. Then, the robot will dribble the ball to the target.

2.2.3. Robot eye. ReROSO equipped with a robot eye by image processing using RaspCAM Version 1.3. However, due to the viewing angle received on the camera is so small, the robot could not reach the surrounding area in wide view. Hence, to tackle the problem, the RaspCAM was modified to have a wide lens. By doing so, the range of the robot's vision extended to wider angle. Figure 3a shows the modified RaspCAM into wide angle camera using smartphone camera accessories, and Figure 3b and 3c show a considerable difference of images produced by the original and modified RaspCAM.

a b c

Figure 3. (a) Modification of RaspCAM into wide angle, (b) the image captured by the original camera, and (c) the image captured by the modified camera.

IOP Conf. Series: Materials Science and Engineering **850** (2020) 012031 doi:10.1088/1757-899X/850/1/012031

3. Results and discussion

3.1. Economic analyses

Robot Soccer platform is not a new technology use in education. Most of such platforms were used for competition involved university students in the globe. There are many commercial robot platforms ready to use. However, to have such robots platform is not easy because the majority of the platform are pricey. To gauge the economic benefit of building its own robot soccer training kit, the following are economic analyses of ReROSO.

From Table 1, it can be calculated that to build a usable mobile soccer robot only spent IDR1,694,800 (USD111,19). It is worth to spend this amount of money because of the mobile robotics kits in the market is more expensive. For instance, MICROSOT-100 could be bought at prices above USD1000. Hence, an HS establishing an extra-curricular of robotics can build more low-cost mobile robot soccer kit like ReROSO.

Table 1. Pricelist of ReROSO parts.

No.	Components	Qty	Function	Price (IDR)
1	Raspi3	1	The main processor and learning server	550,000
2	Raspi Camera	1	Image recognition for measuring the ball and the goal crossbar pattern	200,000
3	Compass sensor	1	Sensing the position of goal crossbar	45,000
4	DC motor and gearbox	4	Robot movement	150,000
5	DC motor Driver Module	4	Driving DC motors for robot movement	35,000s
6	Servo motor	1	Camera rotator	17,800
7	Lippo battery 3,500 mAh	1	Power Supply	675,000
8	UBEC Module	4	Control the DC motors supply	220,000

3.2. Performance of ReROSO

3.2.1. Image recognition test. The image processing system is used to filter images with 3 variables, that are hue, and saturation value (HSV). Initially, the image was taken using modified RaspCAM and will be processed by the OpenCV library. The image that consisted of red, green, and blue (RGB) elements will convert into HSV format. Figure 4 shows the difference between the original image (RGB) and the image with the HSV format.

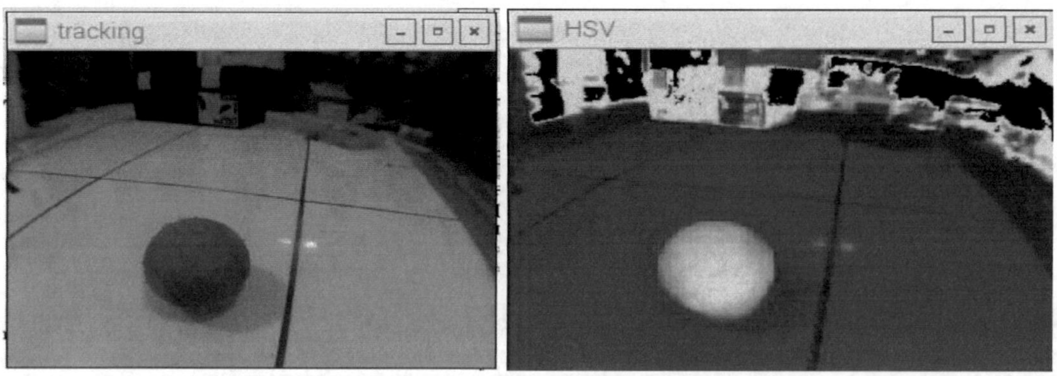

Figure 4. The comparison of the original source image and HSV format.

After the image converted to HSV format, then image is filtered separately, between hue, and saturation value (HSV). For each variable, the tolerance range is given. Hence, the robot can adjust to the difference in light intensity.

The hue value shows a comparison with the actual colour. Figure 5a you can see the hue colour filter shown as 0-16. Figure 5b shows the saturation value in the range 189-255. This value shows the strength or purity of colour. Figure 5c shows the value of values in the range 70-129. This value shows the brightness of the colour, the higher the value, the brighter the desired colour.

From the results of the filter image hue, saturation, and separate values, the value is in logic and so that the combined value between hue, saturation and value is obtained as shown in Figure 5d. This value is the final value that can be used as a benchmark for finding robot positions ball.

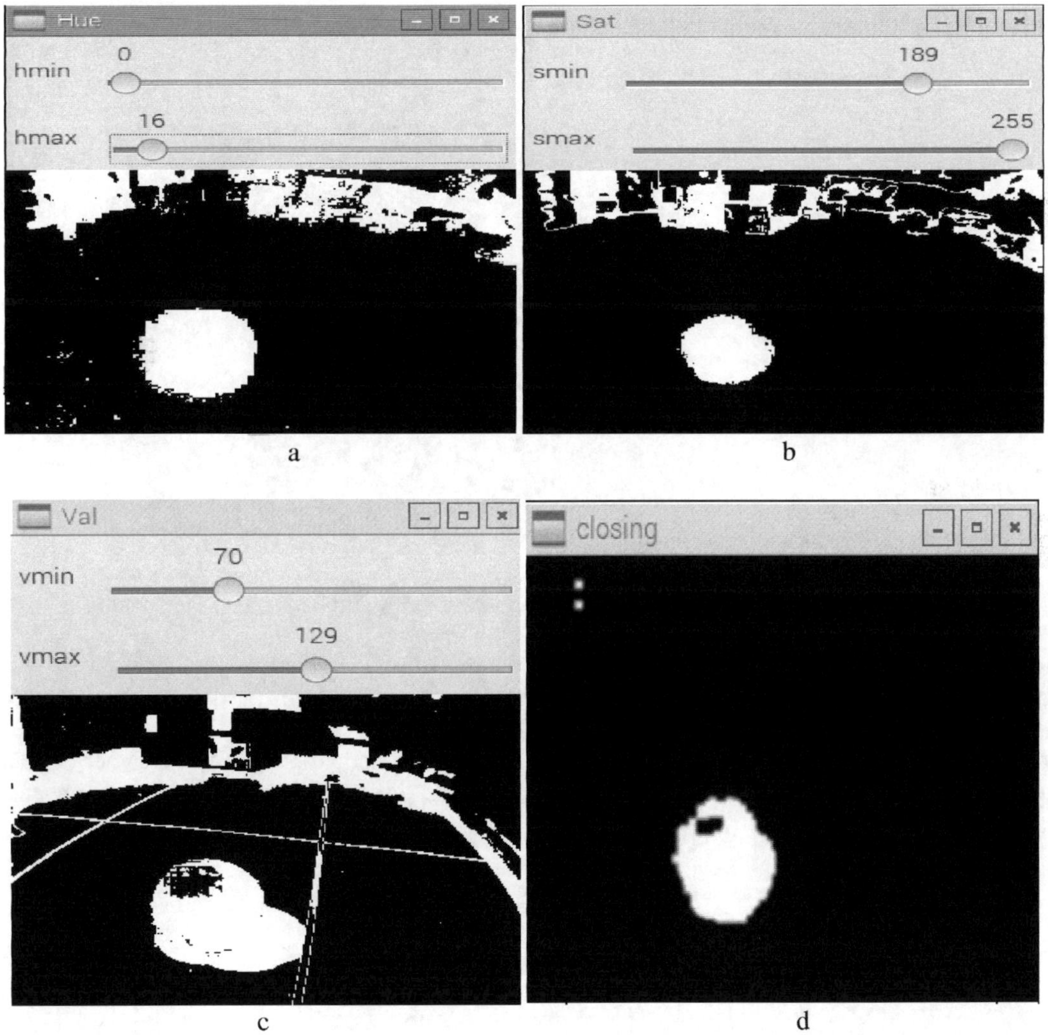

Figure 5. (a) Hue filter, (b) saturation filter, (c) value filter and (d) result of combination hue, saturation and value.

However, there is a problem that should be concerned. Because, the light surrounding the robot influences the image processing. Hence, to have accurate detection of the ball, ReROSO's eye should be calibrated several times.

3.2.2. PID control and movement. The previous section explained how the robot could recognise the ball and the goal woodwork and crossbar. Based on the results of filtering the HSV format, the position coordinates of the object are taken. So, the position and coordinate of the ball against the robot are known.

However, the x and y coordinate on this robot is the point where the value is 140 degree. So this value is set as a set point. By using the Proportional, Integral, Derivatives algorithm (PID), the output can be generated in the form of pulse width modulation (PWM). In consequence, the robot can adjust its position to the ball. Figure 6 shows the coordinate position of the spherical object. The motor control system uses PWM so that the motor speed can be controlled. Because the robot's movement system uses an Omni wheel, the robot can move in all directions without turning the robot head. However, because the Omni wheel used is made of nylon, the wheel will be slippery if run on the sleek surface, therefore it should be run in a rather rough like on a carpet.

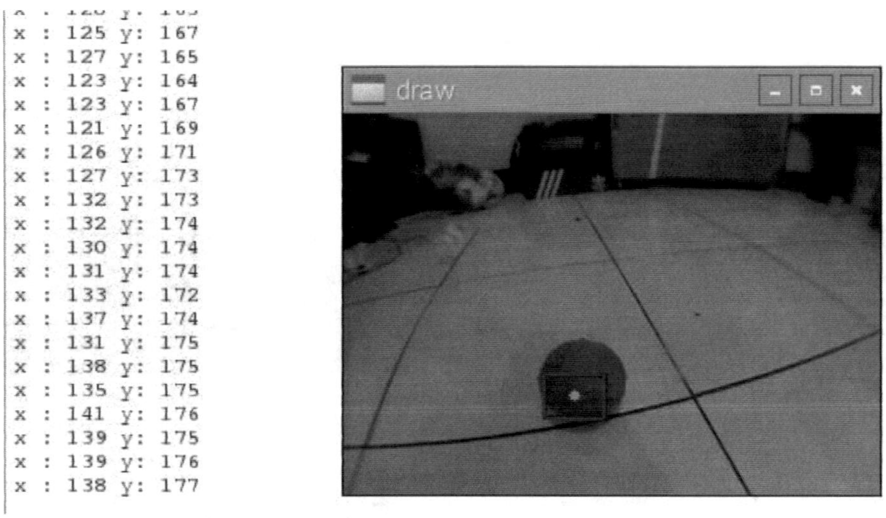

Figure 6. Spherical coordinate position.

Besides, the PID system in the robot is also used for controlling compass sensors. The set point value is obtained when initializing the robot. This value is processed through the PID algorithm, and the output value is produced to position the robot with the goal.

In the ReROSO, two servo motors are installed as a camera actuator. Hence, the camera can move without rotating the robot. Furthermore, it can facilitate the robot in the process of scanning the ball position.

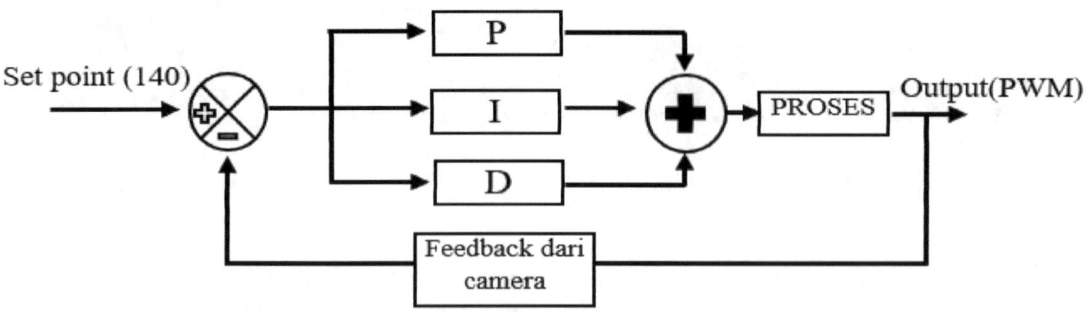

Figure 7. Block diagram of PID system.

3.2.3. Programming algorithm. The programming algorithm is as follows. First, the compass is activated to determine the position of the opponent's goal; then image processing is activated to scan the ball. If the ball has not been found the robot will continue to rotate. If the ball is found, then the robot straightens with the ball position then the robot straightens its position with the goal. Finally, the robot kick the ball into the opponent's goal.

3.3. Advantages of ReROSO as an educational robotics platform

As an educational robotics platform, ReROSO designed to facilitate the learners to learn practical robotics easily. A student could study the following advantage features of ReROSO, that are: kinematics of robot movement, Omni wheel control, PWM, image processing, compass sensor programming and data analyses, PID, and Robotics integrative function. This ability can be seen in Figure 8.

Figure 8. Direct Programming of ReROSO.

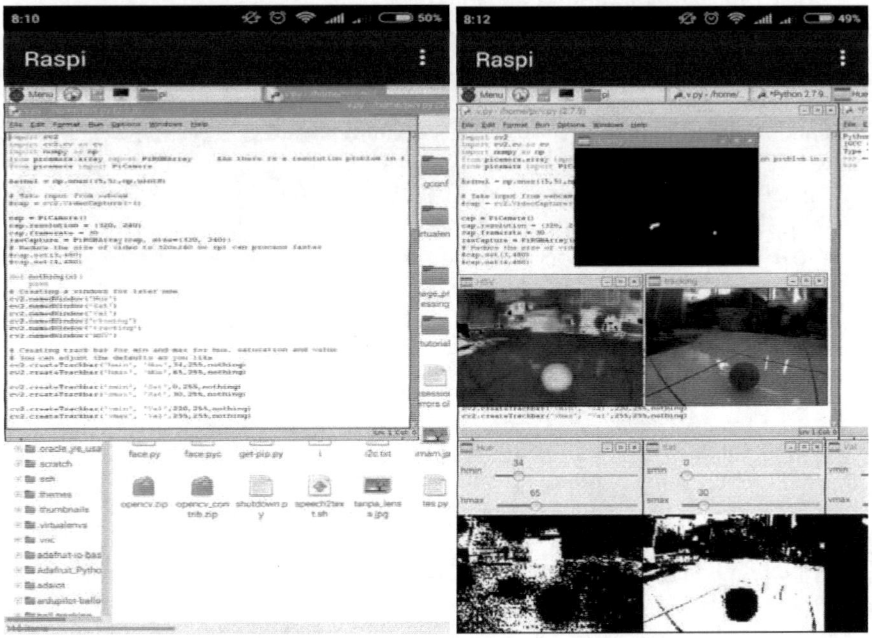

Figure 9. ReROSO programming using a smartphone.

IOP Conf. Series: Materials Science and Engineering **850** (2020) 012031 doi:10.1088/1757-899X/850/1/012031

4. Conclusion and future work

The development of ReROSO has implemented image processing algorithms where the principle is to filter ordinary images with hue, saturation and value variables. From the results of the filter, coordinates can be obtained and then can be set into values as the input of PID. Hence, ReROSO can move to the target as expected. ReROSO test results show that as robot soccer, it can move and dribble the ball to target successfully. As a learning platform, ReROSO can be used for student how to study robotics. A student can easily program the robot directly on Raspberry pi, and through Wi-Fi of PC/Laptop or smartphone. Future development of ReROSO is intended for students with less experience of programming. This is a practical way how-to program robotics platform easily. To achieve this purpose, ReROSO will provide another approach of programming by developing a plugin for graphical or node programming such as MIT-Scratch and NodeRed. Beside, researcher will also held an examination of ReROSO implementation in some high schools, hence it can measure the usability and learning effectiveness.

References

[1] Baez H, Perez K, Rojas E, Rodriguez S, Lopez J, Quintero C and Calderón J M 2013 Application of an educational strategy based on a soccer robotic platform *2013 16th International Conference on Advanced Robotics (ICAR)* 1-6

[2] Martins F N, Gomes I S and Santos C R F 2015 Junior soccer simulation: providing all primary and secondary students access to educational robotics *2015 12th Latin American Robotics Symposium and 2015 3rd Brazilian Symposium on Robotics (LARS-SBR)* 61-66

[3] Binugroho E H, Pratama D, Syahputra A Z R and Pramadihanto D 2015 Control for balancing line follower robot using discrete cascaded PID algorithm on ADROIT V1 education robot *2015 International Electronics Symposium (IES)* 245-250

[4] Fraanje R, Koreneef T, Le Mair A and de Jong S 2016 Python in robotics and mechatronics education *2016 11th France-Japan & 9th Europe-Asia Congress on Mechatronics (MECATRONICS)/17th International Conference on Research and Education in Mechatronics (REM)* 014-019

[5] Mansoor S B, Razzaq S, Rehman A and Shah A A 2011 Robot soccer framework for learning. *2011 IEEE 14th International Multitopic Conference* 50-55

[6] Kirstein F and Risager R V 2016 Social robots in educational institutions they came to stay: Introducing, evaluating, and securing social robots in daily education *2016 11th ACM/IEEE International Conference on Human-Robot Interaction (HRI)* 453-454

[7] Eguchi A and Almeida L 2013 A proposal for RoboCupJunior in Africa: Promoting educational experience with robotics *2013 Africon* 1-5

[8] Szabó R and Gontean A 2016 Industrial robotic automation with Raspberry PI using image processing *2016 International Conference on Applied Electronics (AE)* 265-268

[9] Krauss R 2016 Combining Raspberry Pi and Arduino to form a low-cost, real-time autonomous vehicle platform *2016 American Control Conference (ACC)* 6628-6633

[10] Lynch N 2005 *Conceptual Design of a Robot for the RoboCup competition* (Bethlehem, Pennsylvania Final report)

Automatic metal sorting conveyor machine based on Programmable Logic Controller

M Eriyadi[1,*], S B Mulia[2] and I R Purnomo[1]

[1] Electrical Engineering Department, Politeknik Enjinering Indorama, Jatiluhur 41152, Purwakarta, Indonesia
[2] Automation Engineering Department, Politeknik Manufaktur Negeri Bandung, Dago 40153, Bandung, Indonesia

*mindit.eriyadi@pei.ac.id

Abstract. The clothing and food industry had to require for each product which is produced by them is free of dangerous metal. To handle the problem, the industry did anyways to decreasing metal contents that might be involved in the product. One of the methods that the industry able to do is metal sorter automatically system that used to do when the product is on the conveyor. This method is counted so efficiently if we know that the conveyor is used for moving an item from a place to another place by using an induction motor as a mover. This research purposes of presenting an example of a Programmable Logic Controller (PLC) uses as an electronic device that able to control the conveyor metal sorter system automatically. We used a photoelectric sensor to detect an item on the conveyor. Meanwhile, inductive proximity is used to detect metal content behind the item. This research also is written on how this PLC is more efficient if we compared it with using the conventional relay.

1. Introduction

Demand for high-quality objects, better efficiency, and automatic machines continues to grow as globalization develops [1]. To meet these needs, we need a tool that can control various types of work tools that are universal [2].

PLC is a system controller that is widely used in industrial automation [3]. PLC stands for Programmable Logic Controller, which is a device that implements logic algorithms in its application in the industry [4]. Controllers that have been widely applied in the industrial world are used instead of complex conventional relays to be simpler by adding software to them [5]. PLC is widely used in the industrial world because of the effectiveness and efficiency of the installation, financing, and modification of the circuit.

Automation is a technology that refers to the use of hands-on less in the production process [6]. In the process, the industry is increasingly innovative in production automation. One of them is the use of high-tech automation machines in sorting objects with great attention to accuracy and improving efficiency [7]. Even for the food industry, sorting food from a mixture of pollutants and metal elements that are harmful to digestion is the main focus of industry standards so that the innovation of detecting metal elements in foodstuffs was carried out in the past few years [8].

A conveyor is a medium between goods that move based on the rotation of an induction motor. The conveyor rate can be controlled using a PLC [9]. The control of the induction motor through deceleration of the motor rotating speed is done using VFD. This VFD can reduce motor speed and improve the efficiency of electric power consumption [10].

2. Material and methods

The electrical material required for making this machine is 11 kinds. The electrical material needed to make this machine is shown in table 1.

Table 1. Electrical Material.

Component	Specification	Amount
PLC [11]	Omron Type CPM1A–30CDR–A–V1 Input18 Output 12	1 unit
Inductive Proximity	range 8 mm NPN 3 cable with NO type	1 unit
Photoelectric Sensor	Emitter = E3S-2LE4 Receiver = E3S-2DE4 *Range* 2 meter NPN	3 unit
Limit Switch	220 VAC/24 VDC	2 unit
Motor Induction 220 VAC	3 Phase, 1710 Rpm, 4 Pole, ½ HP. 50 HZ.	1 unit
MCB	4A	2 unit
VFD 1 HP	Danfoss input 24 VDC, Voltage Source 220 VAC	1 unit
Relay [12]	24 VDC for forward and reverse	2 unit
PWS	24 VDC, 2 A	1 unit
Cable	NYAF 0.75 mm^2	sufficiently
Terminal and cable duct	standard	sufficiently
Panel Box	40 x 60 x 50 cm^2	1 unit

Figure 1 shows the conveyor frame design that the author made. They are using L 4 x 4 mm2 with the welding method.

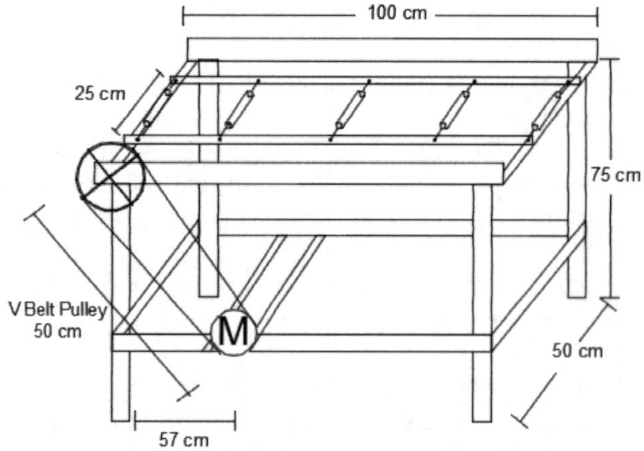

Figure 1. Conveyor frame.

Components installed on the conveyor consist of input and output. Inputs are in the form of 3 photoelectric sensors, one inductive proximity sensor, and two pieces of limit switches, and 1 unit 24 VDC motor as a wiper drive. As a roller drive to be able to rotate, an induction motor is placed under the conveyor with an additional pulley and transmission in the form of a v-belt. Figure 2 is the layout of the components installed on the conveyor frame.

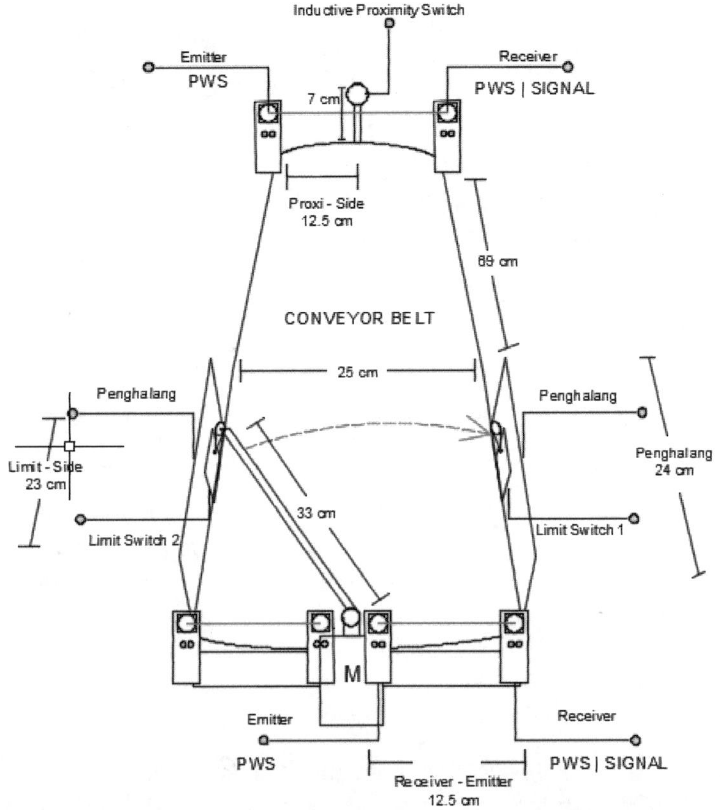

Figure 2. The layout of the components installed on the conveyor frame.

The automation method uses the CPM1A PLC OMRON CPU-30 using the ladder program language with CX-Programmer software. The work system begins with a photoelectric sensor1, and inductive proximity is placed at the beginning of the conveyor. Photoelectric 1 detects the presence of objects that cut laser light. The output of photoelectric 1 is in the form of a conveyor drive induction motor with 8.5 Hz and 14.5 Hz rotational speed. The induction motor is set by its rotation speed by the Danfoss variable frequency drive (VFD) with a capacity of 1 HP. If the object placed above the conveyor contains metal elements, then the 24 V DC motor actuator in the form of a wiper will move to the left. The rotation of the wiper will touch the limit switch one, which is mounted on the conveyor barrier. The touch of the limit switch by the wiper causes the cessation of the DC motor to the left. Figure 3 will explain the statement. The object will automatically enter the right side of the conveyor. Moving objects on the right side of the conveyor will be detected by photoelectric 2. Detection of objects by photoelectric two causes rotation of DC motor to the right (back to its original state). The wiper motor will rotate until it pushes the limit switch two, which is on the conveyor barrier. When pressed, the wiper motor will stop. The limit switch is in NC condition so that it can cut the current on the motor. Photoelectric 2 counts up to 5 objects that are permitted. On the count of 5 objects, the conveyor will stop automatically.

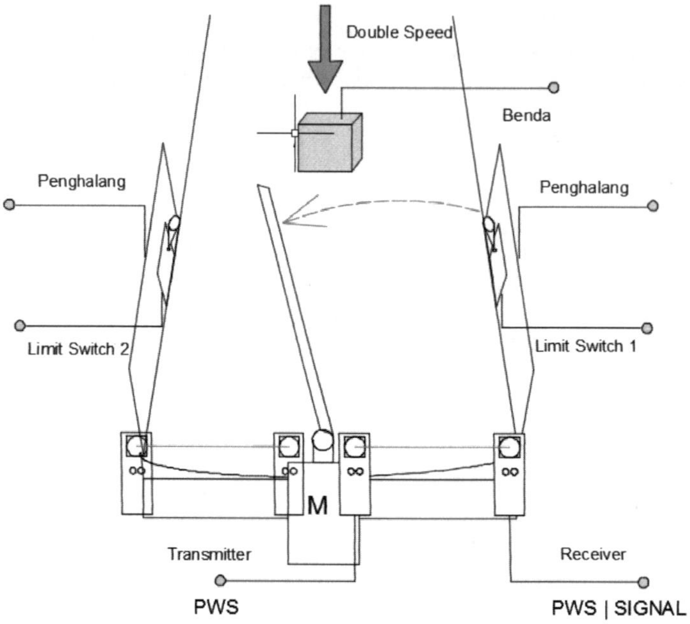

Figure 3. Conveyor conditions on metal objects.

Another condition is when the metal on the conveyor is not detected, there will be no change in the wiper motor. Rotating the wiper motor is caused by the detection of metal by an inductive proximity sensor. Photoelectric 3 on the conveyor on the left will detect objects passing through it. Photoelectric 3 will count many objects that pass it up to 5 x. after the object is detected five times, the system will die. In this system, the condition of the system is when the user presses the emergency button, objects calculated on photoelectric 2 and 3, and forcibly turns off by removing the plug from the source. Counting will reset again when the user presses the emergency button off.

Table 2. PLC Input output addressing.

I/O Type	Component	Function	PLC Address
Input	Emergency OFF Button	OFF System	0.00
	Photoelectric 1	Moving motor conveyor	1.01
	Inductive Proximity	Metal detector	0.11
	Limit Switch 1	DC motor forward stopper	0.03
	Photoelectric 2	Detect objects with reverse DC motor output	1.05
	Limit Switch 2	DC motor reverse stopper	0.06
Output	Induction Motor ½ HP / VFD	Moving the conveyor	10.00
	Motor DC Wiper	Moving the wiper to the right (forward)	10.01
		Moving the wiper left (reverse)	10.02
	Motor induksi ½ HP / VFD	Double-speed induction motor in the state of the object above the metal-conveyor	10.04

Figure 4 is wiring carried out on the PLC. Use the 0.75 mm2 NYAF type cable attached to 0.75 mm2 ferules.

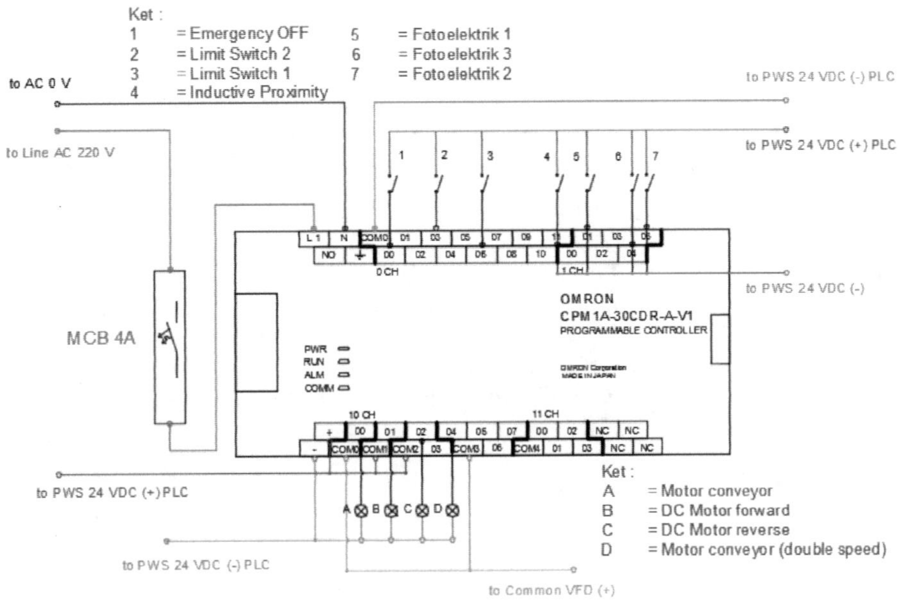

Figure 4. I/O wiring on PLC.

Information:

- L1 enters MCB 220 VAC and Neutral PLC to Neutral source.
- COM up to (-) PWS 24 VDC.
- The input channel depends on the type of input. Three-wire sensor, foot + and - to PWS 24 VDC and signal to PLC channel. Switches, feet + to PWS 24 VDC and feet - to PLC channel.
- The input channel is adjusted in table 1.
- Channel Output is adjusted in table 1.
- COM 1, COM 2 down to PWS (+) 24 VDC.
- COM 0, COM 3 below to (+) VFD.
- MCB to Source Phase 220 VAC.

The program used is a ladder diagram with available software is CX-Programmer and as an offline simulation using CX Designer. As a process viewer, I use a laptop for simulation and CIF02 cable for data communication between PLCs and programs.

3. Results and discussion

After testing the usefulness of the tool on several electrical parameters, the authors obtain data as presented in tables 3, where the√ sign means to function properly.

Table 3. Functional test result.

I/O	Component	Number of experiment				
		1	2	*3*	*4*	*5*
Input	Emergency OFF	√	√	√	√	√
	Fotoelektrik 1	√	√	√	√	√
	Inductive Proxi	√	√	√	√	√
	Limit Switch 1	√	√	√	√	√
	Limit Switch 2	√	√	√	√	√
	Fotoelektrik 2	√	√	√	√	√
	Fotoelektrik 3	√	√	√	√	√
Output	Motor *Conveyor*	√	√	√	√	√
	Forward DC	√	√	√	√	√
	Reverse DC	√	√	√	√	√
	Double Speed	√	√	√	√	√
	Indicator Lamp	√	√	√	√	√

Table 4 is the measurement data based on electrical system parameters.

Table 4. Measurement result.

No	Component	Parameter	Average Measurement Results 10 Times
1	Emergency Button	Voltage (v)	23.5
		Current (mA)	11.4
2	Limit Switch	Voltage (v)	23.5
		Current (mA)	4.9
3	Photoelectric	Voltage (v)	23.5
		Current (mA)	104.5
4	Inductive Proximity	Voltage (v)	23.5
		Current (mA)	2
5	Motor DC FWD	Voltage (v)	23.8
		Current (mA)	1.16
6	Motor DC RVS	Voltage (v)	23.9
		Current (mA)	1.15

Automatic metal sorting conveyor machine based on programmable logic Controller is more efficient by consuming only 20 W maximum for DC and 60 VA maximum for AC compared to conventional relays, which consume 9 Watt Coil power on DC with switching capacity 120 W for DC and 1100 VA for AC.

4. Conclusion

PLC performance is controlling the inputs that enter through the input module, processed according to the program being planted and issuing output to the output module. The use of PLC is much more efficient and does not require a lot of space in the panel. The author has made several changes to the program to adjust I / O as planned work principles. But in these changes, the author did not add to costs, nor did he suffer losses. This is inversely proportional to the use of conventional relays. In addition to the need for additional cables at the time of testing, the use of conventional relays is also at risk of errors during installation. Unlike the PLC, which can be simulated before the installation is done.

References

[1] Ali M S and Ali M S R 2017 Automatic multi machine operation with product sorting and packaging by their colour and dimension with speed control of motors *2017 International Conference on Advances in Electrical Technology for Green Energy (ICAETGT)* 88-92
[2] Sahroni I, Dewi D M and Kunia E 2016 Perancangan dan Pembuatan Mesin Pengisian Air Minum Untuk Resto (Café) Dengan Menggunakan PLC Sebagai Kontrolnya *ENERGY* **6**(1) 23-27
[3] Basile F, Chiacchio P and Gerbasio D 2012 On the implementation of industrial automation systems based on PLC *IEEE Transactions on Automation Science and Engineering* **10**(4) 990-1003
[4] Otto A and Hellmann K 2009 IEC 61131: A general overview and emerging trends *IEEE Industrial Electronics Magazine* **3**(4) 27-31
[5] Sanver U, Yavuz E, Eyupoglu C and Uzun T 2018 Design and implementation of a programmable logic controller using PIC18F4580 *2018 IEEE Conference of Russian Young Researchers in Electrical and Electronic Engineering (EIConRus)* 231-235
[6] Smeu G A 2013 Automatic conveyor belt driving and sorting using SIEMENS step 7-200 programmable logic controller *2013 8th International Symposium On Advanced Topics In Electrical Engineering (ATEE)* 1-4
[7] Peilin L, Zhen Y, Wenlong Z and Hong L 2017 An automatic sorting system for sorting metal cylindrical workpiece based on machine vision and PLC technology *2017 2nd International Conference on Robotics and Automation Engineering (ICRAE)* 446-450
[8] Liu B and Zhou W 2011 The research of metal detectors using in food industry *Proceedings of 2011 International Conference on Electronics and Optoelectronics* **4** V4-43
[9] Howimanporn S, Thanok S, Chookaew S and Sootkaneung W 2016 Speed control technique for conveyor using PSO based PID with programmable logic controller *2016 IEEE/SICE International Symposium on System Integration (SII)* 670-675
[10] Agamloh E B 2016 Power and efficiency measurement of motor-variable-frequency drive systems *IEEE Transactions on Industry Applications* **53**(1) 766-773
[11] Omron Electronics n.d *PLC CPM1A Data Sheet* (Japan: Omron Electronics)
[12] OMRON Corporation 2016 *Miniatur Power Relay Data Sheet* (Japan: OMRON Corporation)

IOP Conf. Series: Materials Science and Engineering **850** (2020) 012033 doi:10.1088/1757-899X/850/1/012033

Design and realization of (7,4) Hamming code channel encoder trainer using Arduino Mega 2560

N Karostiani and A B Pantjawati*

Departemen Pendidikan Teknik Elektro, Universitas Pendidikan Indonesia, Jl. Dr. Setiabudhi 207, Bandung 40154, Indonesia

*arjunib@upi.edu

Abstract. Channel encoder is a part of transmitter block in a digital communication system which is capable of protecting information data from errors by adding parity bits from bits of information. The channel encoder receives bits of information from the source encoder and produces a new code called the codeword. Hamming code is one of the coding mechanisms used to protect information data from errors, with the ability to correct an error. This study explains the design of (7.4) Hamming encoder code channel trainer by using Arduino Mega 2560. The purpose of this research is producing a product that synchronizes a theory with the realization of how the channel encoder blocks work. Important parameters in the encoder channel are information bits, matrix generator, and parity bits. The form of the matrix generator affects the parity bits and the generated codeword. Therefore, the trainer is made and designed that can be used in the form of different matrix generator and qualify of (7.4) *Hamming code channel encoder*. This trainer is made in two stages: the simulation stage using Proteus 8 Professional and the hardware manufacturing stage. The results showed that the trainers worked in accordance with the theory (7.4) Hamming code channel encoder.

1. Introduction

The digital communication system is a system of information exchange from a transmitter (Tx) to a receiver (Rx) through transmission channels using certain procedures, and information data transmitted in the form of digital code as a discrete signal consists of bits 0 and 1. Error is possible to occur in data during transmission from the transmitter to the receiver due to noise factors or environmental disturbances along the transmission channel [1]. Errors will damage the shape of the data and cause differences between the data sent by the transmitter and the data received by the receiver. Digital communication systems must have the ability to control information data sent by the transmitter or received by the receiver to resolve errors. Coding data before transmission is a method used by the transmitter to protect data from errors. The part of the transmitter that has this function is the channel encoder [2].

Channel encoder is a part that forms error control coding as a data protector and for detecting and correcting errors in data received by receivers. The channel encoder converts information data into a codeword, by adding parity bits to the information bits to ensure that the transmitted bits do not change in value after arriving at the receiver [3].

Hamming code is one of the coding mechanisms used by channel encoders, which are easy to use and have the ability to correct the one-bit error and include the linear type of block code. Linear block code is a code that has a structure (n, k) where n represents the codeword bit length $(\vec{v})$, k information bit length $(\vec{u})$, $(n-k)$ the length of the parity bit (γ) and 2^k is the number of possible codeword combinations formed. The systemic code of the generated codeword is shown in Figure 1.

$n-k$	k
Bit parity	Bit information

Figure 1. Codeword systematics.

The codeword is generated by multiplying the number of bits of information $(\vec{u})$ with a dimension of $1\times k$ with a pixel dimension (G) generator whose lines form a linear code (n, k), where the matrix generator and bits of information are 0 and 1 bits. The results of the codeword are indicated by equations $\vec{v} = \vec{u} \cdot G$ The parameters of the coding are listed in Table 1 [4].

Table 1. Dimensions of basic coding parameters.

No.	Parameters	Dimension
1.	Information bits $(\vec{u})$	$\vec{u} = (u_0, u_1, \ldots.., u_{k-1})$
2.	Matrix Generator (G)	$G = \left[P_{k\times(n-k)} \vdots I_k \right] = \begin{bmatrix} p_{0,0} & p_{1,0} & \cdots & p_{0,n-k-1} & 1 & 0 & 0 & \ldots & 0 \\ p_{1,0} & p_{1,0} & \cdots & p_{1,n-k-1} & 0 & 1 & 0 & \ldots & 0 \\ \vdots & & & & & & & & \\ p_{k-1,0} & p_{k-1,1} & \cdots & p_{k-1,n-k-1} & 0 & 0 & 0 & \ldots & 1 \end{bmatrix} = \begin{bmatrix} g_0 \\ g_0 \\ \vdots \\ g_{k-1} \end{bmatrix}$
3.	Parity bits (γ)	$\gamma_j = \sum_{i=0}^{k-1} P_{i,j}$
4.	Codeword $(\vec{v})$	$\vec{v} = (v_0, v_1, \ldots., v_{n-k-1}, v_{n-k}, \ldots., v_{n-1}) = (\gamma_0, \gamma_1, \ldots., \gamma_{n-k-1}, u_0, u_1, \ldots., u_{k-1})$

In this study, the author made a design (7,4) Hamming code channel encoder trainer using Arduino Mega 2560. The trainer designed to do codeword formation on the channel encoder using different matrix generators. The study was conducted by comparing the results of experiments using two forms of matrix generators. Usman, et al. Conducted a similar study entitled Design of (7,4) Hamming Encoder and Decoder using VHDL (Very High integrated circuit hardware Description Language) [5].

2. Research methods

The study was conducted using an experimental methodology. The (7,4) Hamming code channel encoder trainer is made using Arduino Mega 2560 through three stages, that is the stage of design, simulation and fabrication. The flow chart of the research procedure is shown in Figure 2.

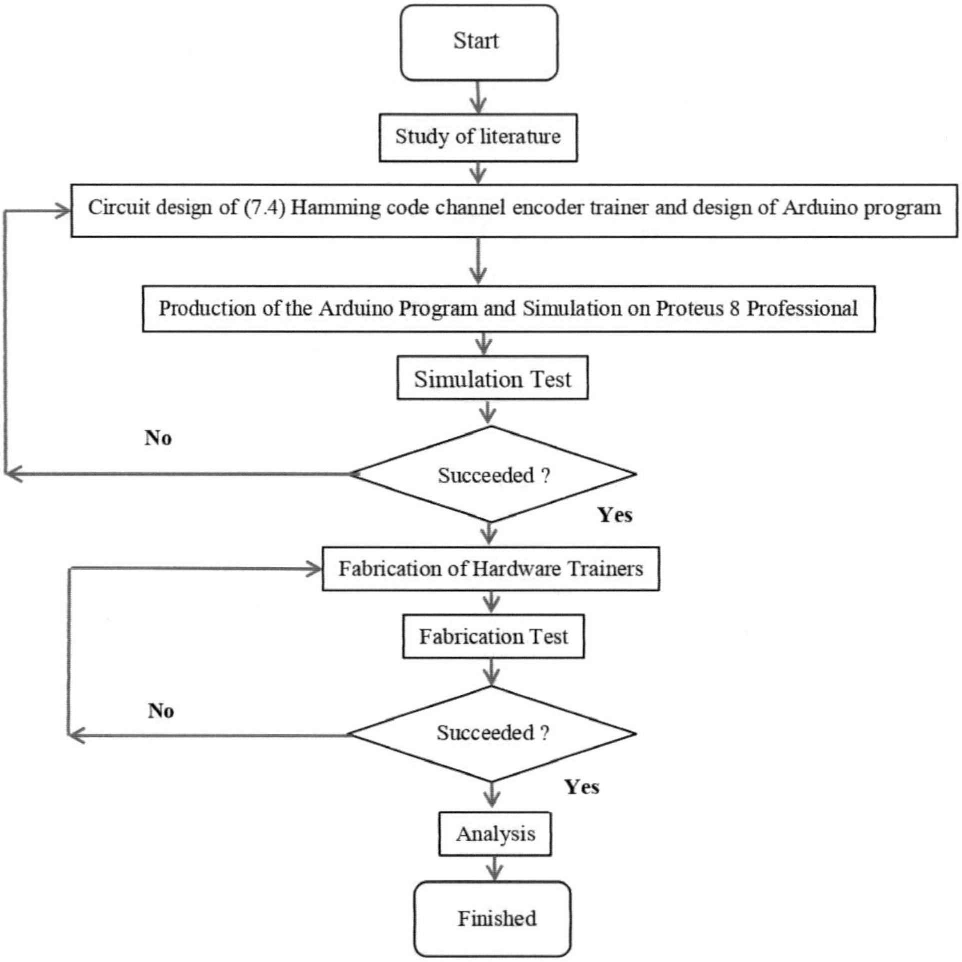

Figure 2. Flow chart of research procedures.

2.1. Trainer design

The trainer design stage is the first stage of trainer production which consists of determining specifications, designing the work system of the trainer and designing a trainer circuit scheme based on the reference design. The channel encoder trainer that is designed has the specifications of having 4 information bits (k), 3 parity bits $(n-k)$, 7 bits of the codeword (n) and 16 possible combinations of codewords (2^k). The design of the trainer system is shown in the following Figure 3.

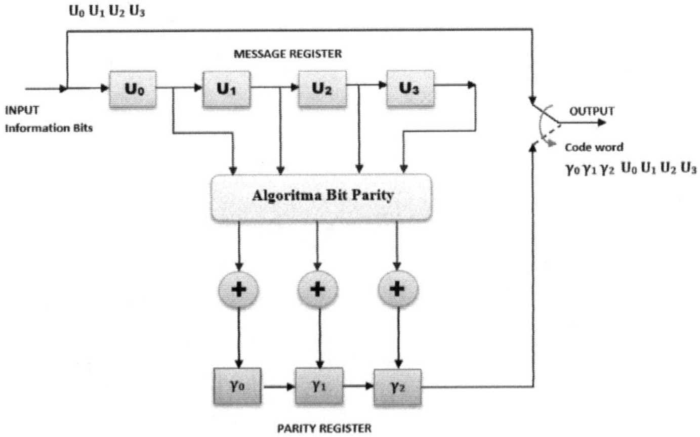

Figure 3. Design of the trainer system.

The trainer can be designed using a different matrix generator. In this study, two-sample matrix generators were used to show that the trainer is work properly. The type of matrix generator used affects the parity bit algorithm. The parity bit algorithm based on two matrix generator samples is as follows:

2.1.1. Sample 1. The parameters used in sample 1 are listed in Table 2.

Table 2. Parameters for sample 1.

Information Bits	Matrix Generator	Parity Bit Algorithm
$\vec{u} = u_0\ u_1\ u_2\ u_3$	$G = \begin{bmatrix} 101 & 1000 \\ 011 & 0100 \\ 110 & 0010 \\ 111 & 0001 \end{bmatrix}$	$\gamma_0\ \gamma_1\ \gamma_2 = \left(u_0\ u_1\ u_2\ u_3\right)\begin{bmatrix} 1\,0\,1 \\ 0\,1\,1 \\ 1\,1\,0 \\ 1\,1\,1 \end{bmatrix}$ $then,\ \gamma_0 = u_0 + u_2 + u_3$ $\gamma_1 = u_1 + u_2 + u_3$ $\gamma_2 = u_0 + u_1 + u_3$

2.1.2. Sample 2. The parameters used in sample 2 are listed in Table 3.

Table 3. Parameters for sample 2.

Information Bits	Matrix Generator	Parity Bit Algorithm
$\vec{u} = u_0\ u_1\ u_2\ u_3$	$G = \begin{bmatrix} 110 & 1000 \\ 101 & 0100 \\ 011 & 0010 \\ 111 & 0001 \end{bmatrix}$	$\gamma_0\ \gamma_1\ \gamma_2 = \left(u_0\ u_1\ u_2\ u_3\right)\begin{bmatrix} 1\,1\,0 \\ 1\,0\,1 \\ 0\,1\,1 \\ 1\,1\,1 \end{bmatrix}$ $then,\ \gamma_0 = u_0 + u_1 + u_3$ $\gamma_1 = u_0 + u_2 + u_3$ $\gamma_2 = u_1 + u_2 + u_3$

2.2. Trainer simulation

At this stage, the circuit scheme was made in the Proteus 8 Professional software and the Arduino program was built using the Arduino EDA software that was tailored to the trainer system that had been designed. Modification of circuits and programs is done as a form of optimization if the simulation results do not match the specifications that have been designed.

2.3. Trainer fabrication

The fabrication stage consists of PCB printing stage by making layouts using Easy EDA with consideration of the use of Arduino Mega 2560, installation of components according to design, case installation and testing.

3. Results and discussion

3.1. Simulation results

The scheme of the trainer circuit used for simulation is shown in Figure 4. From the figure, it can be seen that the circuit for simulation trainers uses rotary switch components to adjust the adder input according to the type of matrix generator parameters used.

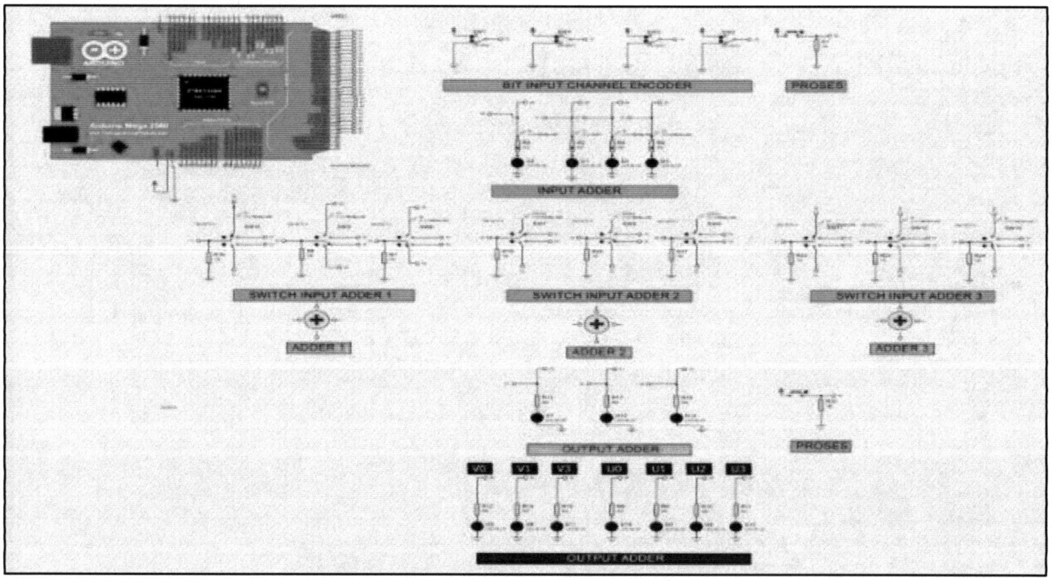

Figure 4. Schematic for simulation.

The results of the simulation experiments based on a two-sample matrix generator are listed in Table 4 and Table 5. Base on Table 4 and 5 it can be seen that the codeword generated in sample 1 and sample 2 is in accordance with the theory.

3.2. Fabrication results

The fabricated trainer device consists of PCBs which have been installed according to the circuit shown in Figure 6, jumper and adapter cables. The layout of the PCB is shown in Figure 5 and the physical shape of the trainer is shown in Figure 7. The fabricated trainer utilizes the header and jumper cable to adjust the input according to the parameters of the matrix generator used. The results of fabrication trials based on two-sample matrix generators that have been designed are listed in Table 6 and 7. Based on Table 6 and 7, it is known that all codewords generated by fabricated trainers use matrix generators sample 1 and sample 2 is in accordance with the theory.

Table 4. Simulation result of sample 1.

No.	Information Bits ($\vec{u}$)	Parity Bits (γ)		Codeword ($\vec{v}$)	
		Theory	Simulation	Theory	Simulation
1.	0000	000	000	0000000	0000000
2.	0001	111	111	1110001	1110001
3.	0010	110	110	1100010	1100010
4.	0011	001	001	0010011	0010011
5.	0100	011	011	0110100	0110100
6.	0101	100	100	1000101	1000101
7.	0110	101	101	1010110	1010110
8.	0111	010	010	0100111	0100111
9.	1000	101	101	1011000	1011000
10.	1001	010	010	0101001	0101001
11.	1010	011	011	0111010	0111010
12	1011	100	100	1001011	1001011
13	1100	110	110	1101100	1101100
14	1101	001	001	0011101	0011101
15	1110	000	000	0001110	0001110
16	1111	111	111	1111111	1111111

Table 5. Simulation result of sample 2.

No.	Information Bits ($\vec{u}$)	Parity Bits (γ)		Codeword ($\vec{v}$)	
		Theory	Simulation	Theory	Simulation
1.	0000	000	000	0000000	0000000
2.	0001	111	111	1110001	1110001
3.	0010	011	011	0110010	0110010
4.	0011	100	100	1000011	1000011
5.	0100	101	101	1010100	1010100
6.	0101	010	010	0100101	0100101
7.	0110	110	110	1100110	1100110
8.	0111	001	001	0010111	0010111
9.	1000	110	110	1101000	1101000
10.	1001	001	001	0011001	0011001
11.	1010	101	101	1011010	1011010
12	1011	010	010	0101011	0101011
13	1100	011	011	0111100	0111100
14	1101	100	100	1001101	1001101
15	1110	000	000	0001110	0001110
16	1111	111	111	1111111	1111111

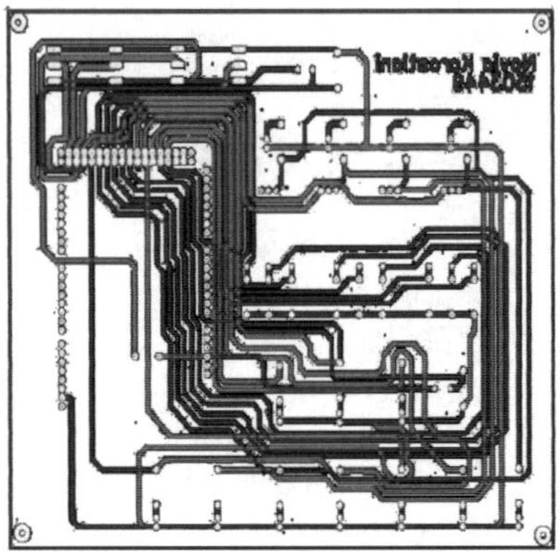

Figure 5. PCB layout.

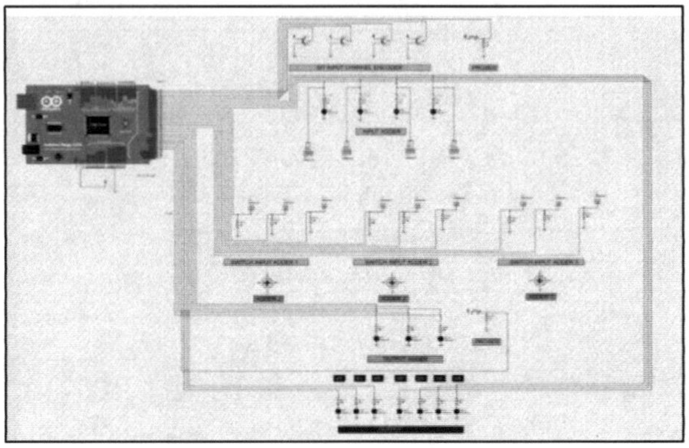

Figure 6. Schematic for fabrication.

Figure 7. Physical of trainer.

Table 6. Fabrication result of sample 1.

No.	Information Bit ($\vec{u}$)	Parity Bits (γ)		Codeword ($\vec{v}$)	
		Theory	Fabrication	Theory	Fabrication
1.	0000	000	000	0000000	0000000
2.	0001	111	111	1110001	1110001
3.	0010	110	110	1100010	1100010
4.	0011	001	001	0010011	0010011
5.	0100	011	011	0110100	0110100
6.	0101	100	100	1000101	1000101
7.	0110	101	101	1010110	1010110
8.	0111	010	010	0100111	0100111
9.	1000	101	101	1011000	1011000
10.	1001	010	010	0101001	0101001
11.	1010	011	011	0111010	0111010
12	1011	100	100	1001011	1001011
13	1100	110	110	1101100	1101100
14	1101	001	001	0011101	0011101
15	1110	000	000	0001110	0001110
16	1111	111	111	1111111	1111111

IOP Conf. Series: Materials Science and Engineering **850** (2020) 012033 doi:10.1088/1757-899X/850/1/012033

Table 7. Fabrication result of sample 2.

No.	Information Bits ($\vec{u}$)	Parity Bits (γ)		Codeword ($\vec{v}$)	
		Theory	Fabrication	Theory	Fabrication
1.	0000	000	000	0000000	0000000
2.	0001	111	111	1110001	1110001
3.	0010	011	011	0110010	0110010
4.	0011	100	100	1000011	1000011
5.	0100	101	101	1010100	1010100
6.	0101	010	010	0100101	0100101
7.	0110	110	110	1100110	1100110
8.	0111	001	001	0010111	0010111
9.	1000	110	110	1101000	1101000
10.	1001	001	001	0011001	0011001
11.	1010	101	101	1011010	1011010
12	1011	010	010	0101011	0101011
13	1100	011	011	0111100	0111100
14	1101	100	100	1001101	1001101
15	1110	000	000	0001110	0001110
16	1111	111	111	1111111	1111111

4. Conclusion

It has been designed (7,4) Hamming code channel encoder trainer using Arduino Mega 2560. Simulation and fabrication test results produce 7-bit codewords, 3 bits parity with shapes that match the theory. Then (7,4) Hamming code channel encoder trainer using Arduino Mega 2560 can be used to generate a codeword using different matrix generator.

References

[1] Choudhury D R and Podder K 2015 Design of Hamming Code Encoding and Decoding Circuit Using Transmission Gate Logic *International Research Journal of Engineering and Technology* **2**(7) 1165-1169

[2] Lin S and Costello D J 2010 *Error Control Coding* 2 edition (Noida, India: Pearson Education)

[3] Rhee M Y 1989 *Error-Correcting Coding Theory* First Edition (New York: McGraw-Hill)

[4] Nayak V S P, Madhulika C and Pravali U 2017 Design of low power hamming code encoding, decoding and correcting circuits using reversible logic *2017 2nd IEEE International Conference on Recent Trends in Electronics, Information & Communication Technology (RTEICT)* 778-781

[5] Sani U S and Shahono I H 2015 Design of (7, 4) Hamming Encoder and Decoder Using VHDL *1st International Engineering Conference 2015* 103–106

IOP Conf. Series: Materials Science and Engineering **850** (2020) 012034 doi:10.1088/1757-899X/850/1/012034

Automatic farmer pest repellent with Arduino ATmega2560 based on sound displacement technique

S R Ningsih, A H S Budi*, A T Nugraha and T Winata

Department of Electrical Engineering Education, Universitas Pendidikan Indonesia, 40154, Bandung, West Java, Indonesia.

*agusheri@aol.com

Abstract. Increasing pest population is the main cause of reduced crop yields and causes farmers to suffer heavy losses. The purpose of this study is to realize a tool that can help repel bird and mouse pests automatically. The prototype of this tool is controlled using the Arduino ATmega 2560 which will regulate the performance of sensors and servo motors as a mechanical system that pulls bells producing sound to repel birds and controls the buzzer to repel rice field pests. This tool works if the ultrasonic sensor or PIR sensor detects the presence of pests. The power source used is in the form of an accumulator with an output voltage of 12 volts, a servo motor as a motor driver for bells, and a buzzer to repel bird and mouse pests. The testing process carried out on this prototype shows that overall this tool can perform its functions properly.

1. Introduction

Indonesia has long been known as one of the agrarian countries with an area of agricultural land, especially rice farming, reaching millions of hectares. According to the Central Statistics Agency (BPS), Indonesia's paddy land area in 2015 was 8,087,393 hectares. In 2018 the area of rice harvested land was 10,990,007 hectares with a productivity of 51.84 Ku/ha, resulting in a total of 56,974,642 tons of rice. This should make Indonesia the largest rice producing country in the world [1,2].

But despite the abundance of rice crop yields, currently there are still some obstacles that can be the cause of losses in the profession of farming. One of the causes of these losses is the presence of rice pests that are difficult to control. In Indonesia, the losses suffered by rice farmers are 200,000-300,000 tons per year. Various ways have been carried out by farmers to control rice pests. However, the methods carried out by farmers apparently have not been effective enough in the process of pest control [3].

Starting from the background of these problems we have developed a tool in the form of a prototype of rice pest repellent. Automatic Farmer Pest Repellent is a rice pest repellent in the form of birds and rats that works automatically based on Arduino technology and sensors using servo motors that are connected with bells as the main tool to repel rice pests and use the buzzer feature. Buzzer is an additional feature to optimize the performance of tools in repelling rice pests.

2. Literature review

2.1. Pests and its repellent techniques

Pests are organisms that can damage plants. One of the rice pests that cause crop losses is birds and rats. Farmers in Indonesia in general still use traditional methods to repel these pests, for example by installing scarecrows, or by stretching ropes suspended by plastic or other objects that can emit sounds to repel pests such as birds and the like. The system still works manually, the rope can move if moved by the farmer. This method is deemed ineffective because the farmer must always be in the location of the rice field to move the repellent rope directly. At this time began to be developed eviction techniques by using sounds with frequencies that can interfere with hearing pests. Rat have a greater hearing threshold than humans, which is between 1KHz-70 KHz. Birds have almost the same hearing threshold as humans, but their most sensitive hearing limit is at the 1KHz-4Kz frequency [4].

2.2. Ultrasonic sensor
An ultrasonic sensor is a sensor that works by utilizing reflections of sound waves. In simple terms, this sensor will work by issuing ultrasonic waves until the wave meets an object. The object is caused by the sensor. This sensor has a working frequency of 40 kHz, a maximum range of 4m, and has an effective angle of 15°. **Figure 1** is an image of an ultrasonic sensor [5].

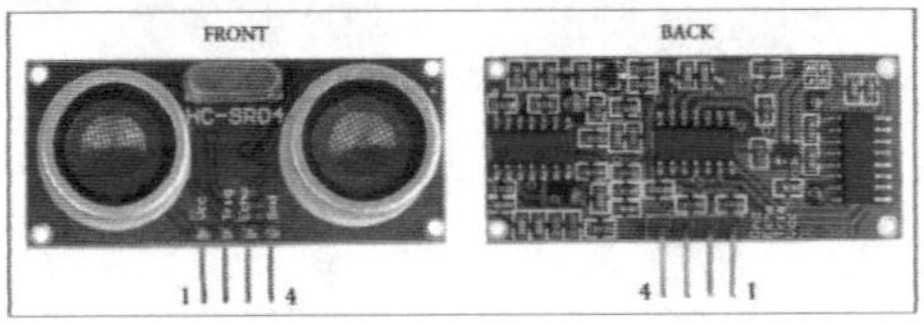

Figure 1. Ultrasonic sensor.

2.3. PIR sensor
PIR sensor (Passive Infra-Red) is a sensor used to detect infrared light from an object. This sensor is passive, meaning that the sensor does not emit infrared light but only receives infrared radiation from the outside. PIR sensors have a maximum range of 7m with an effective angle <110°. Images from the PIR sensor can be seen in **figure 2** [6].

Figure 2. PIR sensor.

2.4. Arduino ATmega2560
Arduino ATmega2560 is a microcontroller based on the ATmega2560 datasheet. This microcontroller has more digital I/O Pins and analog input pins than Arduino UNO, namely 54 I/O pins, and 16 analog inputs. Arduino ATmega 2560 can be seen in Figure 3. The programming language used by Arduino is C language [7].

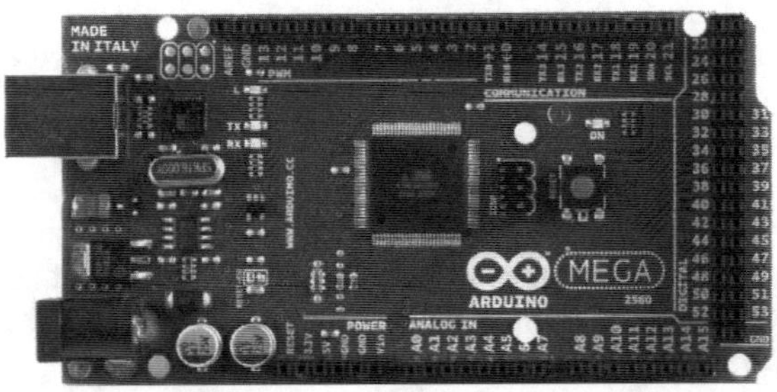

Figure 3. Arduino ATmega 2560.

IOP Conf. Series: Materials Science and Engineering **850** (2020) 012034 doi:10.1088/1757-899X/850/1/012034

2.5. Buzzer
Buzzer devices that can emit sound waves. In general, the buzzer is used as an indicator of something, for example for security sensor indicators and so on. The buzzer can work well in producing frequencies in the range of 1 - 5 kHz to 100 kHz for Ultrasound applications. The buzzer operational voltage is usually in the range of 3Volt-12 Volt [8].

3. Method
We implemented Design and Development approach with the following stages of research:

3.1. Study of literature
Literature studies are conducted by looking for references that are relevant to the tools to be developed. The literature used is in the form of journals, theories, and articles that support the creation of the automatic farmer pest repellent.

3.2. Make of tool design
At this stage, the design of the tool will be developed. Automatic Farmer Pest Repellent is a rice pest repellent device in the form of birds and rat that work automatically based on Arduino technology and sensors using a servo motor which is connected with a bell thread as the main tool for rice pest repellent and uses the buzzer feature. The bell is an additional feature to optimize the performance of the device to repel pests. For the energy source, this tool uses an accumulator with 12Volt output.

3.3. Prepare tools and materials
At this stage, the procurement of tools and materials will be used following the design that has been made before.

3.4. Tool assembly
In the process of designing the outline, two design processes are carried out. The first design process is software design in the form of Arduino programming. After the program is complete then hardware design is adjusted to the design that has been planned.

3.5. Trials
At the last stage, a trial phase was carried out on products that had been developed previously to find out whether the product could work well or not. Tests carried out include testing the ultrasonic sensor and the PIR sensor is giving a signal to drive the servo motor so that the pest repellent bells sound and turn on the buzzer to repel the pest. If the results of the trial still show that there is an error in the performance of the tool then the process returns to the design process to then be tested again until the tool can work properly.

4. Results and discussion

4.1. Prototype design
The tool developed is a prototype designed to repel mouse pests and bird pests on rice plants. In this prototype, two sensors were used to detect pests, namely ultrasonic sensors, and PIR sensors. The prototype dimension that will be made has a width of 40cm and a length of 50cm with a scale of 1: 100. This means that the prototype created can reach an area of 4mx5m for the actual area.

If an ultrasonic sensor has a maximum range of 4m and an effective angle of 15°, the width of the sensor range is 1.04m. Whereas the PIR sensor has a maximum range of 7m and an effective angle of <110° if the desired range is 4m the width of the sensor range is 11.4m. The range width for the two sensors is obtained by the following equation:

- Wide ultrasonic sensor range

$$\tan\theta = \frac{y}{x}$$

$$\tan 82.5^o = \frac{4m}{x}$$

$$x = \frac{4m}{\tan 82.5^o} = 0.52m$$

The broad picture of the ultrasonic sensor can be seen in **Figure 4.**

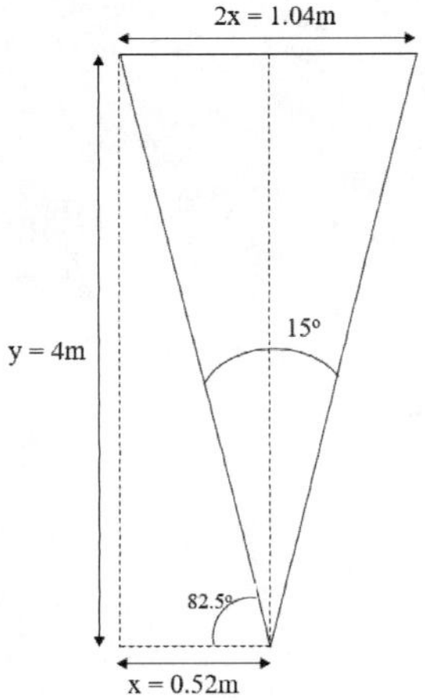

Figure 4. Range of ultrasonic sensor.

- Wide PIR sensor range

$$\tan\theta = \frac{y}{x}$$

$$\tan 35^o = \frac{4m}{x}$$

$$x = \frac{4m}{\tan 35^o} = 11.4m$$

The broad picture of the PIR sensor can be seen in **Figure 5.**

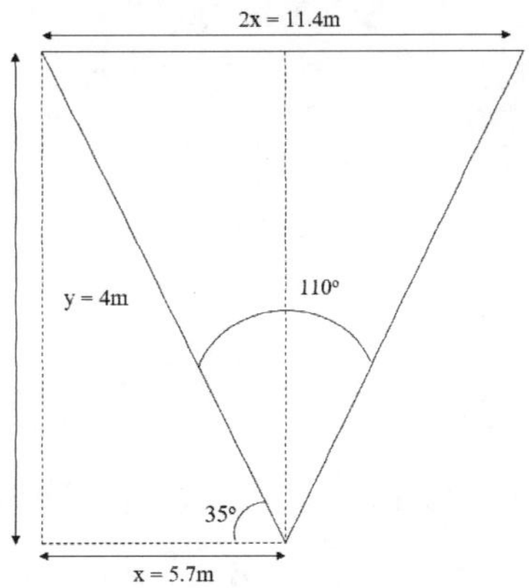

Figure 5. Range of PIR sensor.

The sensor specifications as described, then for an area of 4m x 5m, 10 ultrasonic sensors and 3 PIR sensors are needed to reach the entire area. The range of the entire sensor can be seen in Figure 6 and Figure 7.

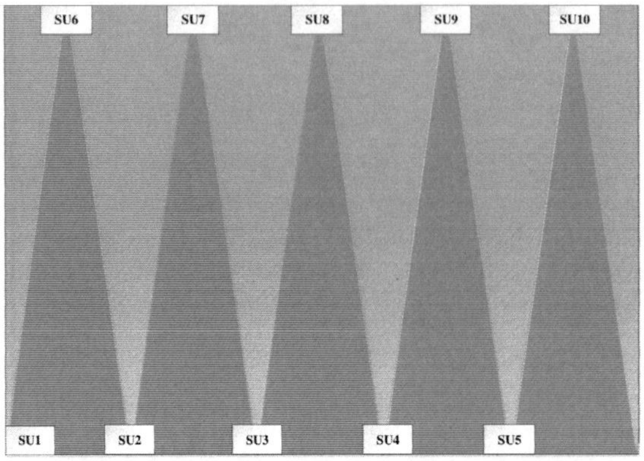

Figure 6. Placement and scope of ultrasonic sensors.

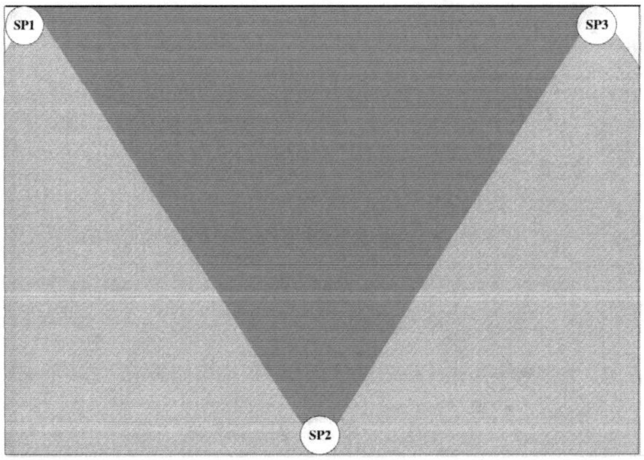

Figure 7. Placement and scope of PIR sensors.

Then the design of the prototype developed can be seen in Figure 8.

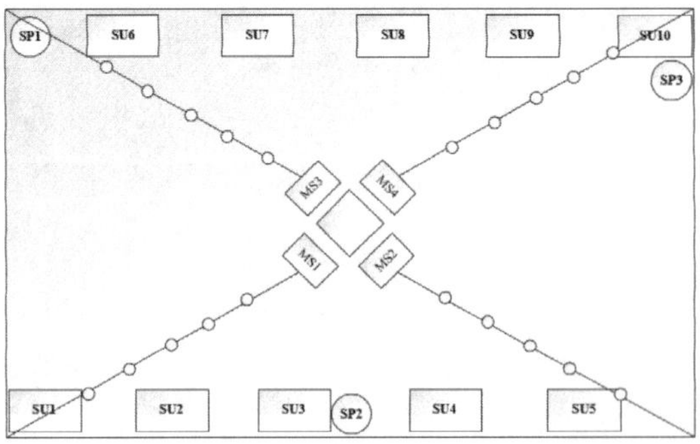

Figure 8. Top-view prototype design.

Information:
MS : Servo motor
SU : Ultrasonic sensor
SP : PIR sensor

In this prototype, Arduino ATmega2560 is used as a microcontroller in controlling the system. The sensors used to change are approved in this prototype, namely the ultrasonic sensor used for protection of bird pests, and the PIR sensor for the protection of rat pests that use servo motors as the driving motor of nylon thread, and activates the buzzer. Threaded bells installation to repel birds, and installation of buzzers to repel rats.

The number of sensors installed is 10 HC-SR04 ultrasonic sensors and 3 HC-SR501 PIR sensors. In determining the number of sensors it must be adjusted to the maximum coverage and effective angle of each sensor so that the entire area of the prototype can be properly covered. The power source of this tool uses an Accumulator with 12V output.

4.2. Flowchart of prototype

The system flow chart of this prototype can be seen in Figure 9 and Figure 10.

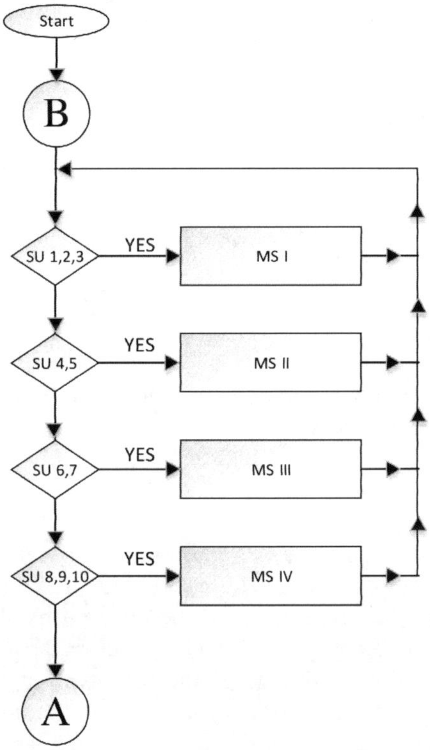

Figure 9. Ultrasonic sensor flow chart.

The diagram in Figure 9 shows how the system works in arranging ultrasonic sensors on the prototype. Previously it was mentioned that in this prototype we used 10 HC-SR04 ultrasonic sensors. From this system we can see that:

- Ultrasonic sensors 1, 2 and 3 are connected to the first servo motor,
- Ultrasonic sensors 4 and 5 are connected with a second servo motor,
- Ultrasonic sensors 6 and 7 are connected to the third servo motor,
- Ultrasonic sensors 8, 9, and 10 connected to the fourth servo motor.

Each servo motor has been connected with a bell so that when an ultrasonic sensor reads the existence of a pest it will move the servo motor connected to the sensor, and the bell on the thread will produce a sound that can repel the bird pest. The system will continue to loop until there are no sensors that read the existence of pests.

IOP Conf. Series: Materials Science and Engineering **850** (2020) 012034 doi:10.1088/1757-899X/850/1/012034

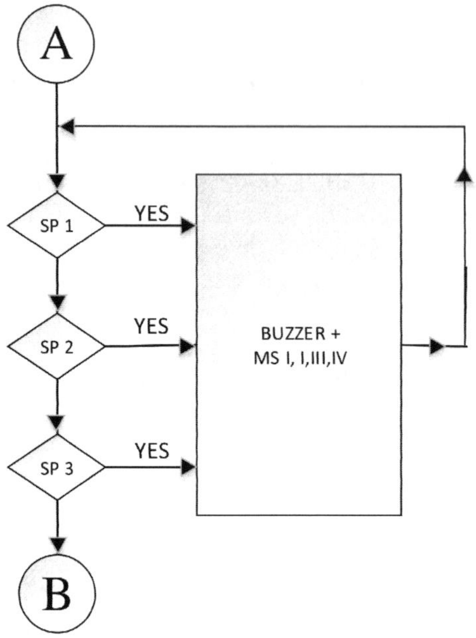

Figure 10. Flow chart sensor PIR.

The flow chart in Figure 10 shows how the system works in managing the HC-SR501 PIR sensor on the prototype. The number of PIR sensors used in this prototype is 3 pieces. From this system, we can see that each PIR sensor is connected to the buzzer and all servo motors. When there is a sensor that reads the presence of pests, the system will turn on the buzzer to repel the rat pests, and the servo motor will move and move the bellows thread to avoid the rat pests so as not to go further into the area. The system will continue to loop until there are no sensors that read the presence of pests.

4.3. Design result

The prototype that has been developed can be seen in Figure 11. The point of placement of the ultrasonic sensor and placement of the PIR sensor has been adjusted to the maximum range and the effective angle of each sensor so that the entire prototype area can be covered by the sensor. The bells we use for prototypes are 26mm in size, but if this prototype will be developed back into a larger scale, the bell size can be adjusted back as needed. The small size of the prototype made us decide the location of the accumulator and Arduino installation on the prototype stored at the bottom of the tool. If this tool will be developed into a ready-made tool, the installation of accumulators and other main tools can be installed in the middle of the prototype area.

Figure 11. Prototype.

4.4. Prototype testing
Testing on the prototype includes testing the sensor function in moving the motor, each of which is done 10 times for each sensor. The test results for ultrasonic sensors can be seen in Table 1, and the test results for PIR sensors can be seen in Table 1.

Table 1. Ultrasonic sensor testing in performing servo motor functions.

Ultrasonic Sensor	Testing (Functioning/Not Functioning)									
	1	2	3	4	5	6	7	8	9	10
SU1	√	√	√	√	√	√	√	√	√	√
SU2	√	√	√	√	√	√	√	√	√	√
SU3	√	√	√	√	√	√	√	√	√	√
SU4	√	√	√	√	√	√	√	√	√	√
SU5	√	√	√	√	√	√	√	√	√	√
SU6	√	√	√	√	√	√	√	√	√	√
SU7	√	√	√	√	√	√	√	√	√	√
SU8	√	√	√	√	√	√	√	√	√	√
SU9	√	√	√	√	√	√	√	√	√	√
SU10	√	√	√	√	√	√	√	√	√	√

Information, SU : Ultrasonic sensor

From Table 1, it can be seen that based on the results of the tests that have been carried out as many as 10 times all ultrasonic sensors installed on the prototype can function properly and can run servo motors that are connected with bell.

Table 2. PIR sensor testing in performing servo motor and buzzer functions.

PIR	Testing (Functioning/Not Functioning)																			
	1		2		3		4		5		6		7		8		9		10	
	m	b	m	b	m	b	m	b	m	b	m	b	m	b	m	b	m	b	m	b
SP1	√	√	√	√	√	√	√	√	√	√	√	√	√	√	√	√	√	√	√	√
SP2	√	√	√	√	√	√	√	√	√	√	√	√	√	√	√	√	√	√	√	√
SP3	√	√	√	√	√	√	√	√	√	√	√	√	√	√	√	√	√	√	√	√

Information:
SP : PIR sensor
m : Servo motor
b : Buzzer

From Table 2, it can be seen that based on the results of tests carried out all sensors can function properly and can perform servo motor functions and buzzer functions on the prototype.

5. Conclusion
The tool developed in this study is a prototype of pest repellents, especially pests of birds and rats that can work automatically. Expulsion techniques are carried out by hearing a pest. It is expected that later this prototype can be developed into a tool on a large scale so that it can be installed and used directly by farmers and pests that can work automatically. Repel techniques are carried out by generating sounds that can interfere with hearing a pest. It is expected that later this prototype can be developed again into a tool on a larger scale so that it can be installed and used directly by farmers.

References

[1] Central Bureau of Statistics. Rice Field Area by Province (ha) 2003-2005. [Online]. Retrieved: https://www.bps.go.id/dynamictable/ 2015/09/10/ 895/luas-lahan-sawah-menurut-provinsi-ha-2003-2015.html

[2] Central Bureau of Statistics. Harvested Area, Production and Productivity of Rice by Province 2018, [Online]. Retrieved dari : https://www.bps.go.id/dynamictable/ 2019/04/15/1608/luas-panen-produksi-dan-produktivitas-padi-menurut-provinsi-2018.html

[3] Alfarisi M S, Bintang C A and Ayatillah S 2018 Exsys (drone security with audio and expert system) village for exploiting birds and identifying rights and rice diseases keeping food security and enhancing food development in Indonesia. *J. Appl. Agric. Sci. Technol.* **2** 35–50

[4] Beason R C. 2004. What can birds hear? *Proc.21st Vertebr. Pest Conf.* 92-96

[5] Retnowati Y and Budi A H S 2018 Smart Cane Using ESPectro with GPS Tracking System *IOP Conference Series: Materials Science and Engineering* vol 384 (IOP Publishing) p 12054

[6] Pratama M F, Hasibuan M S and Arya D 2019 Smart System with Rice Pest Controlling IOT *J. Online Jar. Pengaj. Seni Bina* 37–44

[7] Tuluk E, Buyung I and Soejono A W 2017 Implementasi Alat Pengusir Hama Burung di Area Persawahan dengan Menggunakan Gelombang Ultrasonik Berbasis Mikrokontroler Atmega168 *Respati 7.*

[8] Riyuska A and Wildian W 2016 Rancang Bangun Sistem Identifikasi Data Pasien pada Rekam Medis Elektronik Menggunakan Teknologi RFID *J. Fis. Unand* 5 59–64.

IOP Conf. Series: Materials Science and Engineering **850** (2020) 012035 doi:10.1088/1757-899X/850/1/012035

Small Vessel Tracking Based on Multi Gateway LORA for Indramayu Traditional Fishery

A Sumarudin[1,*], W P Putra[1], A Rifai[1] and A E Putra[2]

[1]Department of Informatics, Politeknik Negeri Indramayu, Jl. Lohbener Lama No. 08 Lohbener Indramayu 45252.
[2]Department Computer Science, Universitas Gadja Mada, Sleman, Yogyakarta 55281.

*shumaru@polindra.ac.id

Abstract. In this study, we use Lora-based data communication for communication between boats and database management systems for vessel. The maximum distance from the Lora is 10 km, and in order to be able to monitor fishermen along the fisherman's coast in fishing, it is necessary to have a multi-gateway Lora to increase fishermen's tracking coverage. The system is built using multi gateway Lora installed on the coast. Devices built with a Lora client are installed on a boat with GPS input, the technique of sending data on the device to the gateway uses a time base with a 1-minute period. To minimize data on the gateway that intersects the range of other gateways, a time flag is made to determine the data at a certain time to determine the trajectory of the boat. The result this research is system-monitoring vessel, which is successfully built and can provide fishing boats along the coast with long-range coverage. Based on experiment, multi gateway can be implemented by creating an identical ID gateway, the Lora client will broadcast to the nearest gateway. The maximum distance obtained is 1.35Km NLOS and to cover 20 Km of beach length is needed 15 gateways.

1. Introduction

One-day fishing is fisherman (small fishermen who got out in the afternoon and go home early in the morning) in search of fish ranges from radius of 6 miles (9.65 km) from the shore. In this issue, data for boat tracking needs with data on the needs of the fishing boat by increasing the coverage of the tracking area of the boat using multiple gateways. Gateway is installed on the beach with a distance between gateways with the maximum distance the gateway distance. Several previous studies conducted tracking using a 2.4 GHz radio with a maximum distance of 8 KM [1] use WLAN outdoor radio.

Using the Long-Range Radio is a high-power and high-power LPWAN. The length of Indramayu beach is 114 KM [2] from Sukra beach to Krangkeng. Some AIS designs are built using advanced land observing satellite (ALOS) in vessel detection [3] and use synthetic radar aperture (SAR) [4]. The technology is too expensive for the implementation of fishing boats. For the implementation of fishermen's communication successfully made with a maximum distance of 3.73 Km [5] from coast and for in NLOS the maximum distance is 2 Km [6] but this is only for the implementation of 1 gateway. AIS itself is used to increase the level of security in shipping [7] for vessel fishermen.

In this study using Lora-based data communication for communication between boats and database vessel management systems. The long-range distance is possible with Lora, and in order to be able to monitor fisherman's coast in fishing, it is necessary to have a multi-gateway Lora to increase fishermen's tracking coverage.

2. System design

The design of our research propose is multigateway with the placement of gateways along the shore. Setting this can have an impact on the results. The topology design can be showed as in Figure 1.

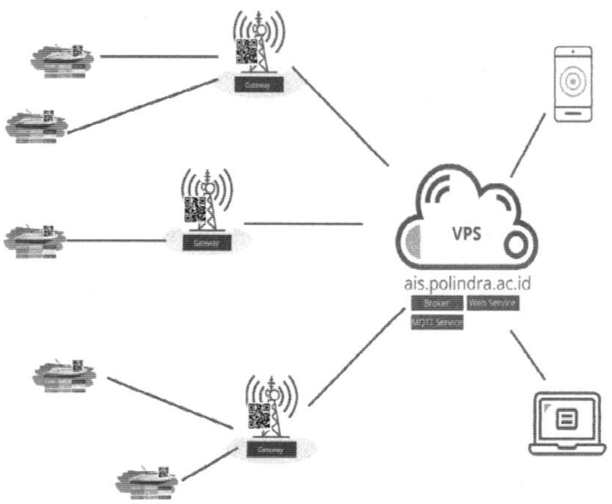

Figure 1. Topology multigateway lora system.

In the topology above, each lora node will send to the nearest gateway using a data format that allows the lora node to be able to be reliable even though it switches the gateway to connect. Every Lora Node has a device ID for the identity of the boat that is implemented. For Lora Node configuration, using tracking data consists of an accelerator sensor to detect speed, GPS for the position of Lora Node and RTC as time to synchronize time with the gateway. Lora Node can be seen in Figure 3.

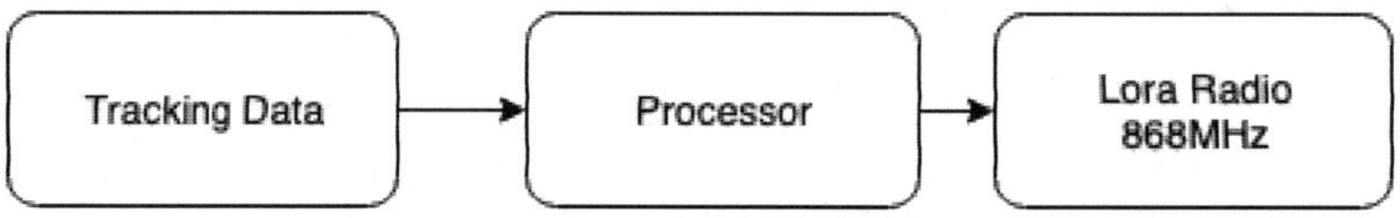

Figure 2. Lora node block.

From Lora this node sends the appropriate format from the tracking data in 73 Bytes format consisting of 6 Byte Device Id, 2 Byte Status, Latitude data 11 bytes, longitude 11 bytes, 5 Bytes RSSI (receive signal strength indication), 7 bytes Roll, 7 pitch, 5 Bytes Volt, 6 Bytes, 7 Bytes Speed, 8 Bytes times. This data is sent from the GPS sensor, accelerometer and ADC.

	Device ID	Status Online	Lat	Long	RSSI	Roll	Pitch	Volt	Compass	Speed	Time
Data Format	6 byte	2 byte	11 byte	11 byte	5 byte	7 byte	7 byte	5 byte	6 byte	7 byte	8 byte

Figure 3. Lora data format.

Data from Lora Node is sent to the gateway using wireless lora. Gateway can be seen in figure 4. Gateway consists of three main components, namely lora wirelss module, microcontroler and Linux. Microcontroller with Linux communicating SPI (Serial Peripheral Interface).

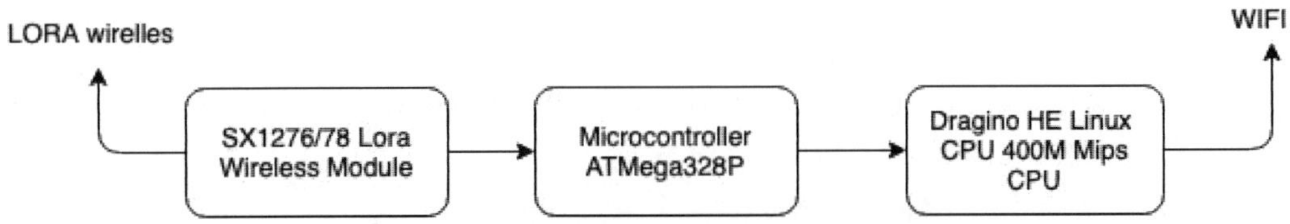

Figure 4. Gateway lora.

Each gateway is used with each gateway id. In research using the format <name> _gw_xxxx can be seen in figure 5. This format allows for 4 digit gateways 0-9999 gateways.

Table 1. Data format on gateway.

Host broker	ais.polindra.ac.id
Port broker	xxxx
Client id	{id_gateway}
Topic	channel/vessel/{id_device}
Payload	{id_device},{online_status},{lat},{lon},{rssi},{roll},{pitch},{battery _volt},{compass},{speed},{time}

The payload data is obtained from Lora Node which is sent broadcast to the gateway. For connections from multi gateways, use the 4 way MQTT protocol and publish according to the topic created by the MQTT server. Tpoic is made in this research channel / vessel / {id_device}.

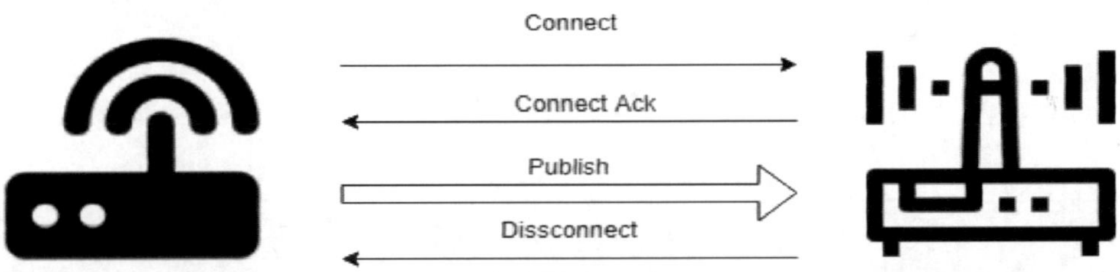

Figure 5. Data Connection gateway to Server based on MQTT.

3. Results and discussion

Based on our design, it is implemented using the Lora Node according to figure 6. Lora radio uses the L80-m39 Arduino Lora shield with NMEA data format with a maximum accuracy of 2.5 meters [8] with CEP -130dBm.

Figure 6. Lora node.

Whereas Gateway uses the Dargino LG01 Gateway Outdoor (figure 7) with Lora configuration on Arduino Yun, we connect the MQTT server using Linux to enable MQTT Protocol. By configuring the gateway id settings can be seen in Figure 8. To configure multiple gateways by providing identification of each gateway and setting on the MQQT Server side.

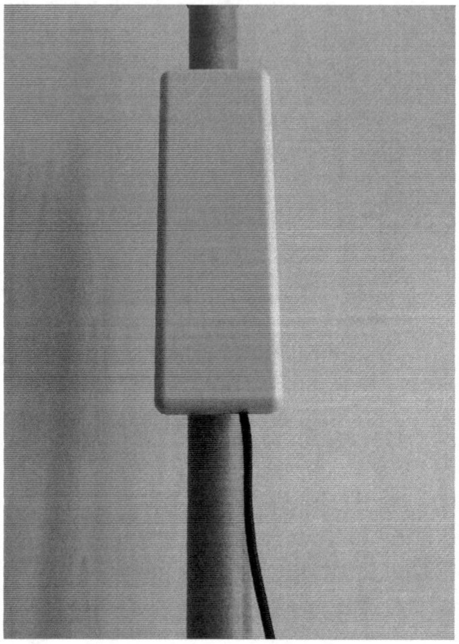

Figure 7. Gateway lora.

```
void sendDataMQTT(String id, String m){
  Process p;
  p.begin("mosquitto_pub");
  p.addParameter("-d");
  p.addParameter("-h");
  p.addParameter(_BROKER_HOST);
  p.addParameter("-p");
  p.addParameter(_BROKER_PORT);
  p.addParameter("-i");
  p.addParameter(_CLIENT_ID);
  p.addParameter("-t");
  p.addParameter("channel/vessel/"+id);
  p.addParameter("-m");
  p.addParameter(m);
  p.run();

  while (p.available() > 0) {
    char c = p.read();
    Console.print(c);
  }
  // Ensure the last bit of data is sent.
  Console.flush();
}
```

Figure 8. Gateway configuration.

The format of the Mosquitto MQTT Client service for sending data using the Command Line is as follows.

```
Mosquitto_pub –d
–h ais.polindra.ac.id
–p xxxx
–I id_gateway
–t channel/vessel/yyyyyy
–m D12345,1,-6.40801,108.27531,-75,-172.24,21.29,3.61,231.74,0.00,50
```

From the Lora Server side, check the boat id, if a boat id is found the data is parsed and stored.

```
function broadCastDevice(data){
  var array = data.split(',');
  Device.find().where('no_seri').equals(array[0]).exec(function(err, devices){
    if(devices.length !== 0){
      Perahu.find().where('id_device').equals(devices[0]._id).exec(function(err, perahus){
        ..............................................................................
          }
          var data = {
            .......................................................................
            last_information: devices[0].last_information
          }

          client.publish('channel/perahu/' + array[0], JSON.stringify(data));
        }
```

Connection data from the Lora client to the Lora Gateway is not based on the id gateway, with binary data streams this will allow the Lora client to communicate with any gateway closest to it to communicate. From the lora side the server will identify the gateway based on the id. When there is data sent from the two gateways this does not happen because the Lora clilent side can provide the data time information in real-time. Delay communication is not a problem for the reliability of the data sent.
Data from the gateway is successfully sent with 4-way (figure 9), the gateway id sends connect then the broker server Lora sends the ACK Connection. After the ACK connection has been received, the payload sender will be done using the publish format in table 1.

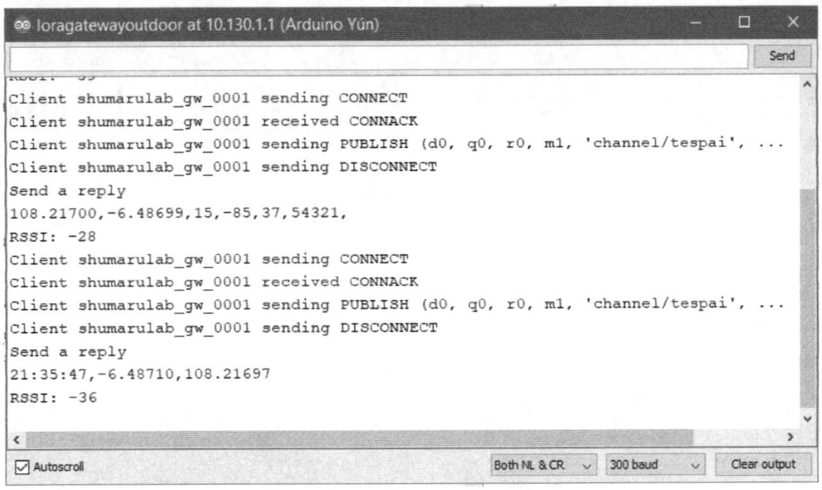

Figure 9. Data gateway receive.

Server-side data as an example can be seen in table 2. The data is carried out by json parsing and stored in the database as vessel tracking data.

Table 2. Data receive on lora server.

DEVICE ID	STAT ONLINE	LAT	LON	RSSI	ROLL	PITCH	VOLT	COMP ASS	SPEED (Km/s)	TIME
D12345	1	-6.41	108.28	-94	-179.53	48.74	3.48	236	0.31	4:51:34
D12345	1	-6.41	108.28	-93	-177.73	43.74	3.43	234	0.22	4:51:42
D12345	1	-6.41	108.27	-91	177.53	23.02	3.46	227	0.72	5:1:17

For testing carried out near the campus, with NLOS. Gateway is installed 10 meters seen in figure 11. From the results of this test, the maximum distance of 1.3 KM can be seen in table 3.

Figure 10. Range test.

Table 3. Testing range result.

No	Max Range [KM]	Derajat Radius [°]
1	1,38	78

Based on the results of the above tests if it is to cover 20 km of beach length, then it is possible with 15 gateways.

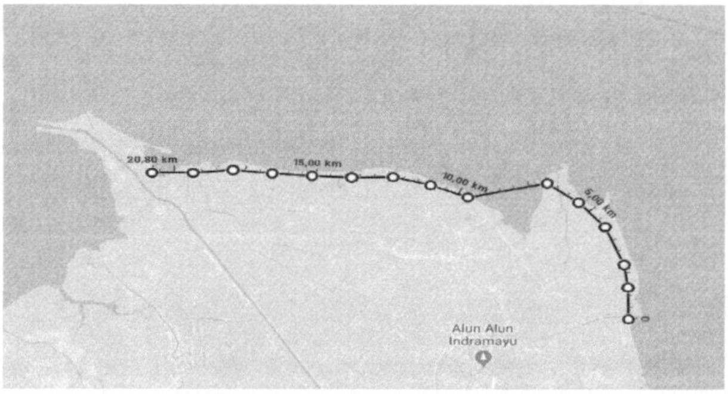

Figure 11. Multi gateway.

4. Conclusion

The system is built using multi gateway Lora Installed on the coast. Devices built with a Lora client are installed on a boat with GPS input, the technique of sending data on the device to the gateway uses a time base with a 1-minute period. To minimize data on the gateway that intersects the range of other gateways, a time flag is made to determine the data at a certain time to determine the trajectory of the boat. Berdasarkan percobaan diatas, multiGateway can be implemented by creating an identical ID gateway, the Lora client will broadcast broadcast to the nearest gateway. The maximum distance obtained is 1.35Km NLOS and to cover 20 Km of beach length is needed 15 gateways.

Acknowledgment

This research was Grandt using the Ministry of Research and Technology and Higher Education (MRTHE) PKPT Research Grant Scheme 2019.

References

[1] Zainuddin Z, Wardi and Nantan Y 2018 Applying maritime wireless communication to support vessel monitoring *Proc. - 2017 4th Int. Conf. Inf. Technol. Comput. Electr. Eng. ICITACEE 2017* **2018** 158–61

[2] Kusnida D and Astjario P 2008 Dinamika garis pantai kabupaten indramayu, jawa barat, berdasarkan penafsiran citra satelit **XVIII** 55–63

[3] Watagawa M, Kobayashi E and Wakabayashi N 2012 Monitoring of vessel traffic using AIS data and ALOS satellite image *Progr. B. - Ocean. 2012 MTS/IEEE Yeosu Living Ocean Coast - Divers. Resour. Sustain. Act.* 8–11

[4] Achiri L, Guida R and Iervolino P 2018 SAR and AIS Fusion for Maritime Surveillance *IEEE 4th Int. Forum Res. Technol. Soc. Ind. RTSI 2018 - Proc.* 1–4

[5] Adiono T, Dawani F, Adinugraha E, Rifai A, Arijal M, Fuada S, Purwanda I G and Ahmad Samhany H 2019 Development of Long-range Communication System for Fishermen: An Initial Study *Proceeding - 2018 Int. Conf. ICT Rural Dev. Rural Dev. through ICT Concept, Des. Implic. IC-ICTRuDEv 2018* 12–7

[6] Fuada S, Anindya S F, Dawani F, Rifai A, Adinugraha E, Purwanda I G and Adiono T 2019 Prototype of Long-Range Radio Communication for e-Nelayan Devices using LoRaWAN *J. Infotel* **10**

[7] Proud R, Browning P and Kocak D M 2017 Small Vessel Tracking using AIS for Enhanced Maritime Domain Awareness

[8] Quectel 2014 L80 Hardware Design GPS Module Series

IOP Conf. Series: Materials Science and Engineering **850** (2020) 012036 doi:10.1088/1757-899X/850/1/012036

Design and application of double layer bowtie antenna and single layer bowtie antenna as UHF sensors for partial discharge detection

J P Uwiringiyimana* and U Khayam

School of Electrical Engineering and Informatics, Institut Teknologi Bandung, Bandung, Indonesia

*upeter1@yahoo.fr

Abstract. The reliability of the power system cannot be achieved without good insulation quality of the high voltage apparatus. The Ultra-high frequency (UHF) method has been applied and proven to be accurate in detecting partial discharge (PD) signals emitted from the PD source in the form of electromagnetic waves. This paper discusses the design and application of double layer bowtie antenna and single layer bowtie antenna as UHF sensors for partial discharge detection in air insulation. The PD measurement results obtained by using the designed double layer bowtie antenna were compared with PD measurement results obtained by using the single layer bowtie antenna. The PD characteristics of induced EM waves detected by double layer bowtie antenna and single layer bowtie have been analysed. Based on the simulation results, the double layer bowtie antenna has a minimum return loss values of -31.54 dB and -33.43 dB, with corresponding resonance frequency values of 1.7 GHz and 2.7 GHz, with a total bandwidth of 602 MHz, and corresponding VSWR values of 1.04 and 1.05 while the measurement results by using Vector Network Analyser (VNA) shows that the designed double layer bowtie antenna has a minimum return loss of -20.2 dB , at a resonance frequency of 1.7 GHz, a bandwidth of 576 MHz and corresponding VSWR of 1.2. The single layer bowtie antenna has a bandwidth of 240 MHz, a minimum return loss of -24 dB, at a resonance frequency of 1.28 GHz, and a VSWR of 1.13 based on the simulation results while the single layer bowtie antenna has a bandwidth of 298 MHz, a minimum return loss of -23.32 dB, at a resonance frequency of 1.3 GHz, and a VSWR of 1.14 based on the measurement results by using VNA. The bowtie antennas were designed using the 3D-HFSS software with FR4-epoxy substrate type and later on were printed on the Printed Circuit Board (PCB). The PD measurement data were performed by using the applied voltage level of 7 kV. According to the antennas simulation results, antennas measurement results by VNA and PD measurement results, the double layer bowtie antenna has proven to be more sensitive than the single layer bowtie antenna.

1. Introduction

High voltage transmission has been chosen in an electric power system due to its capability to transfer a large amount of electrical power. Several key types of equipment such as the generator, transformer, GIS, cable and insulator play an important role in determining the reliability of the power system. The high voltage equipment in general contains gas, liquid or solid insulation. The failure of high voltage equipment may significantly contribute to the failure of the power system. Defects and a particular condition such as voids and protrusion in the insulation may lead to an extremely high electric field which can further result in the insulation failure. Particularly at interfaces between gas and solid and liquid insulations, the excessive electric field may cause the breakdown of the gas while the liquid and solid insulations are still able to withstand the electric field. This phenomenon is referred to as partial discharge (PD). Partial discharges in high voltage equipment for a long time may degrade the quality of the insulation and at the same time, the appearance of discharges can be used as a diagnostic signal for assessing the condition of the insulation in the high voltage equipment [1]. Diagnosis of PD in the early stage is needed to prevent the breakdown of the high voltage equipment. Partial discharge has been an important subject investigated by many researchers and discussion in many international conferences as well as international journals or transactions. In order to maintain the reliability and availability of the power system, it is needed to design the online measurement system by using the ultra-high frequency (UHF) method. UHF sensors are applied to detect the partial discharge phenomenon which releases energy in the form of electromagnetic (EM) wave during the discharge [1-2].

Ultra-High Frequency antenna is one of the UHF methods used to determine the quality of the insulation system of power apparatus. The main advantage of UHF PD detection system by using antennas is the capability to detect PD signal without prior shutdown of the power apparatus [2-11]. Therefore, it is advisable to perform the PD measurement system based on EM waves induced by PD. To do so, the suitable UHF sensor which is able to detect PD induced electromagnetic (EM) waves in ultra-wideband needs to be designed, manufactured and implemented in order to detect the released energy in the form of electromagnetic waves emitted from the PD source. In addition, the characteristics of PD induced EM wave measured by the sensors need to be investigated [2]. This paper discusses the design and application of double layer bowtie antenna and single layer bowtie antenna as UHF sensors for partial discharge detection. The experiment was conducted in air insulation by using the needle-plate electrode model as a PD source. The bowtie antennas were designed, simulated, fabricated and applied for partial discharge detection in air insulation. The 3D-HFSS software with FR4_Epoxy substrate type was used to design the bowtie antennas and later on the bowtie antennas were printed on the Printed Circuit Board (PCB).

Various types of bowtie antennas have been developed in the previous researches for PD induced EM wave measurement [4-9]. The bowtie antenna has been chosen for PD detection in this research because it has broadband, good characteristic parameters, small shape, simple design, high radiation efficiency and it is cheap [5-6]. The basic characteristics of PD induced EM waves such as PDIV and PD waveform, PD patterns are reported in this paper. The PD measurement data were performed by using the applied voltage level of 7kV. The double layer bowtie antenna has shown a better performance antenna characteristic parameters and higher sensitivity in detecting partial discharge compared to the single layer bowtie antenna.

2. Design and simulation of double layer bowtie antenna

2.1. The geometry of double layer bowtie antenna

The double layer bowtie antenna was designed by using 3-D HFSS software and later on was printed on double layer PCB with FR4-epoxy substrate of the relative permittivity of 4.4 and substrate thickness of 1.6. The designed double layer bowtie antenna has a wing radius of 36mmand a flare angle of 120°. It was designed of a rectangular substrate whose dimension is 110 mm x 92 mm x 1.6 mm. The stub length (antenna prolongation) used is 46 mm. The double layer bowtie antenna was designed by modifying the edge of the antenna in order to improve the current density distribution on the antenna surface. The modification of the edge of the antenna was performed by creating a radius of curvature in the wing of the antenna and later on by cutting out the part whose low current density at the tip of the antenna. Figure 1 shows the initial design of a bowtie antenna with wing radius *R1*, the radius of curvature *R2* and stub length *h*. The figure 2 shows the current distribution for the new design of double layer bowtie antenna with edge modification. Figure 3 shows the variation of radius of curvature against return loss of the designed double layer bowtie antenna. By changing the radius of curvature with sizes 21.5 mm, 22 mm, 22.5 mm and 23 mm, the optimum design was achieved by using the radius of curvature of 22.5 mm.

Based on the simulation results, the optimal design of double layer bowtie antenna has a minimum return loss values of -31.54 dB and -33.43 dB, with corresponding resonance frequency values of 1.7 GHz and 2.7 GHz, with a total bandwidth of 602 MHz, and corresponding VSWR values of 1.04 and 1.05.

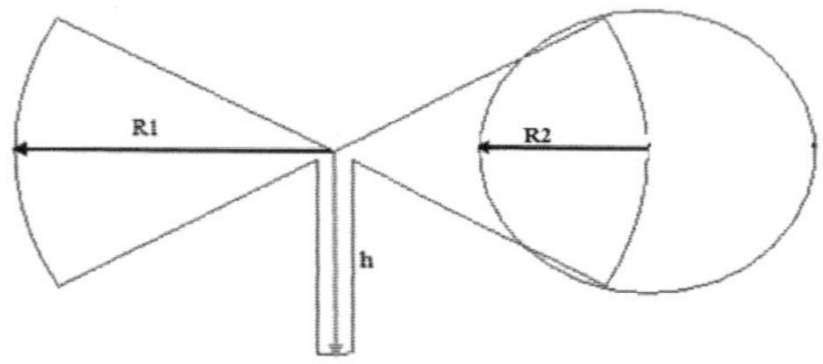

Figure 1. The initial design of double layer bowtie antenna in the previous research.

Figure 2. The new design of double layer bowtie antenna with edge modification.

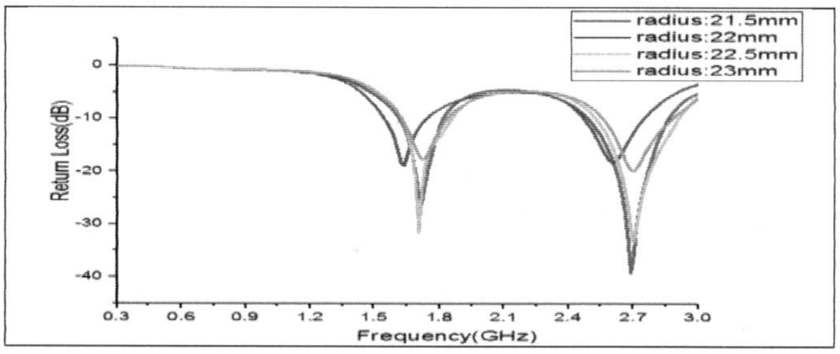

Figure 3. The radius of curvature against return loss.

2.2. Fabrication of double layer bowtie antenna

Figure 4 shows the fabricated double layer bowtie antenna. The double layer bowtie antenna was fabricated on double layer PCB with FR4-epoxy substrate.

Figure 4. Fabricated double layer bowtie antenna.

2.3. Testing double layer bowtie antenna by using Vector Network Analyzer (VNA)

After design and fabrication of double layer bowtie antenna, the antenna was tested by using the vector network analyzer (VNA) in order to measure some basic antenna parameters such as Return Loss (RL), Vector Standing Wave Ratio (VSWR), bandwidth, and the resonance frequency. Figure 5 shows the measurement setup of the designed double layer bowtie antenna by using the VNA. Figure 6 and Table 1 show the comparison between the simulation results of the designed double layer bowtie antenna and measurement results by using the VNA.

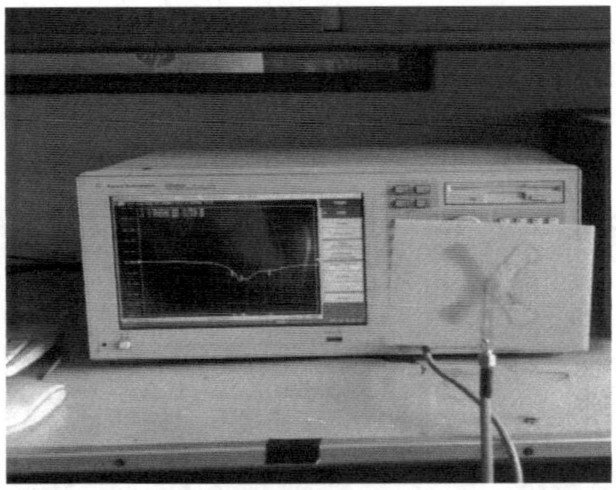

Figure 5. Measurement setup of double layer bowtie antenna by using VNA.

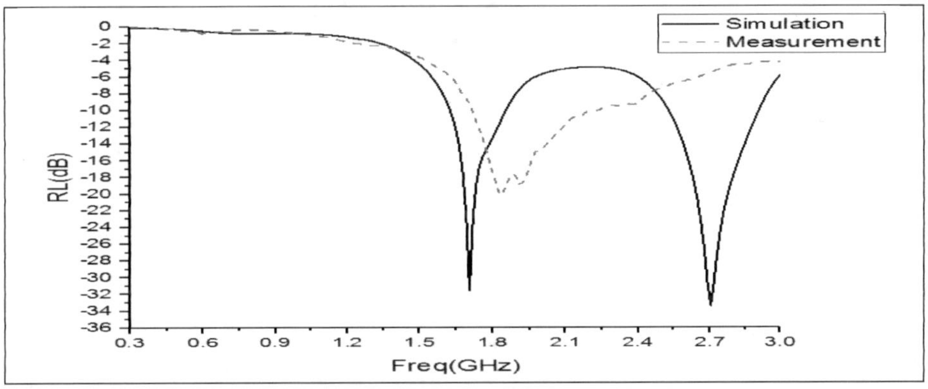

Figure 6. Comparison between simulation and measurement results for double layer bowtie antenna.

Table 1. Comparison between simulation results and measurement results for double layer bowtie antenna.

Results	RL1 (dB)	RL2 (dB)	BW1 (MHz)	BW2 (MHz)	FR1 (GHz)	FR2 (GHz)	VSWR1	VSWR2
Simulation	-31.54	-33.43	222	380	1.7	2.7	1.054	1.043
Measurement	-20.2		576		1.7		1.21	

3. Design and simulation of single layer bowtie antenna

3.1. The geometry of single layer bowtie antenna

The single layer bowtie antenna is designed on only one side of the substrate while the double layer bowtie antenna is designed on both sides of the substrate. In this research, the single layer bowtie antenna has been designed as a comparison sensor to the double layer bowtie antenna in order to see which bowtie antenna is more sensitive than the other in detecting partial discharge. The single layer bowtie antenna was also designed and simulated in a 3-D HFSS software with FR4-epoxy substrate whose relative permittivity is 4.4 and whose thickness is 1.6 mm. Later on the single layer bowtie antenna was fabricated on the single layer circuit board (PCB). The single layer bowtie antenna was designed by optimizing the antenna parameters such as wing radius, flare angle, and gap distance between the two patches of the antenna. The optimal designed was obtained by using a wing radius of 36 mm, a flare angle of 60°, a gap distance of 2 mm with a circular substrate whose diameter 84 mm. Figure 7 shows the geometry of the designed single layer bowtie antenna. Figure 8 shows the variation of wing radius while figure 9 shows the variation of flare angle in the design optimization for the single layer bowtie antenna.

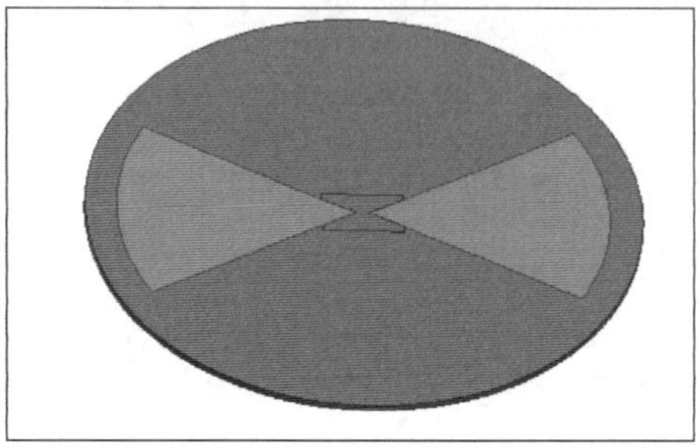

Figure 7. Single layer bowtie antenna.

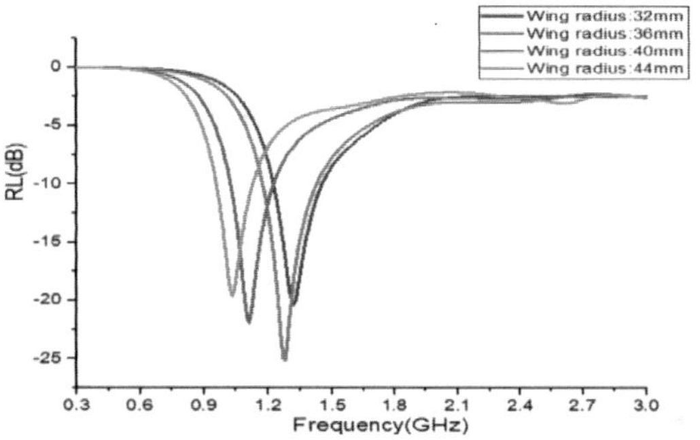

Figure 8. Variation of wing radius against return loss.

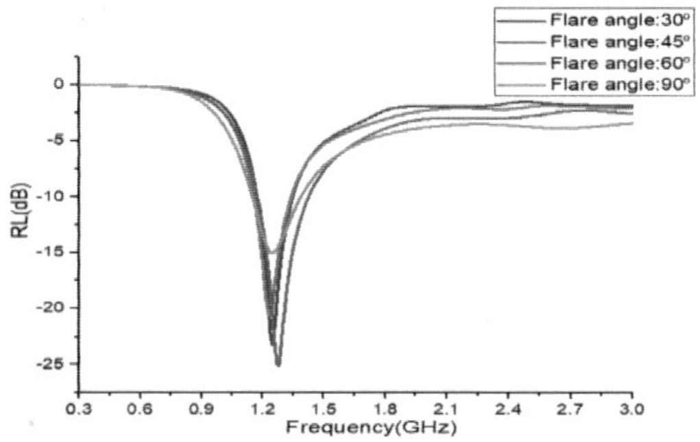

Figure 9. Variation of flare angle against return loss.

3.2. Fabrication of single layer bowtie antenna

After the simulation of the designed single layer bowtie antenna, the next step was to fabricate the bowtie antenna on single-layer PCB with FR4-epoxy substrate. Figure 10 shows the fabricated single layer bowtie antenna.

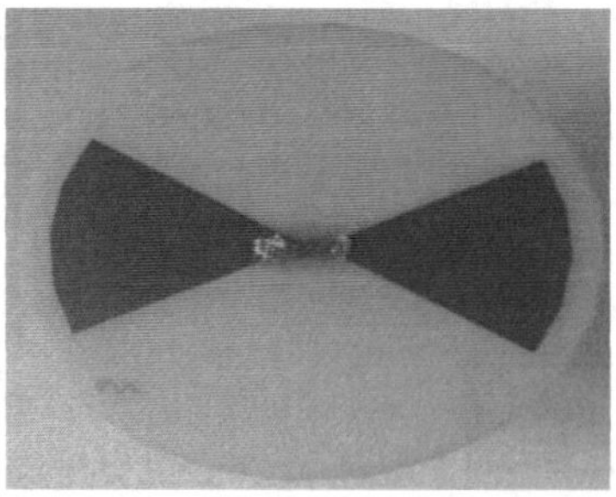

Figure 10. Fabricated single layer bowtie antenna.

3.3. Testing single layer bowtie antenna by using the Vector Network Analyzer (VNA)

After design and fabrication of single layer bowtie antenna, the antenna was test by using the Vector Network Analyser (VNA) in order to measure some basic antenna parameters such as Return Loss (RL), Vector Standing Wave Ratio (VSWR), bandwidth, and the resonance frequency. Figure 11 shows the measurement setup of the designed single layer bowtie antenna by using the VNA.

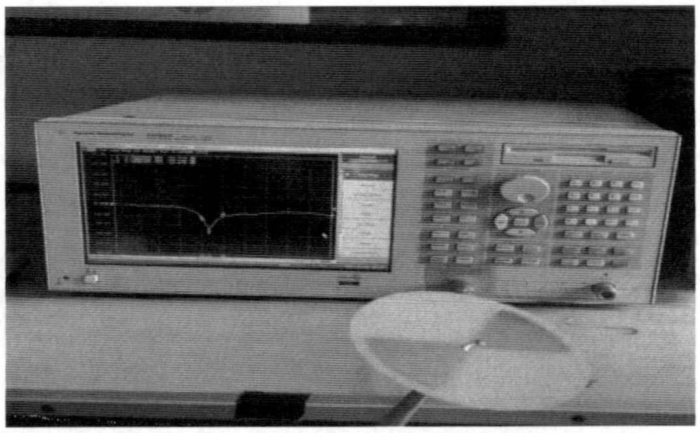

Figure 11. Measurement setup of single layer bowtie antenna by using VNA.

Figure 12 and Table 2 show the comparison between the simulation results of the designed single layer bowtie antenna and measurement results by using the VNA. As can be seen, there is a little difference between the simulation and the measurement results.

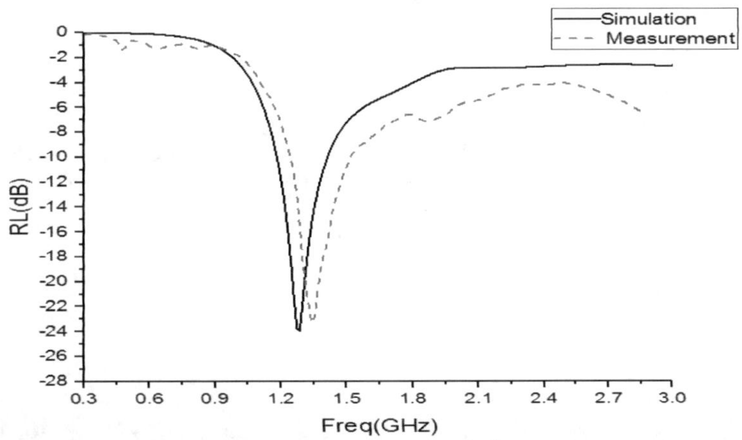

Figure 12. Comparison between simulation and measurement results.

Table 2. Comparison between Simulation results and measurement results for single layer bowtie antenna.

Results	Return Loss (dB)	Bandwidth (MHz)	Resonace Frequency (GHz)	VSWR
Simulation	-25.15	240	1.28	1.13
Measurement	-23.32	298	1.30	1.4

4. Application of designed double layer bowtie antenna and single layer bowtie antenna as UHF sensors for partial discharge detection in air insulation

After testing the designed double layer bowtie antennas by using the VNA, the bowtie antennas were implemented in detecting the PD in air insulation. The needle-plate electrode model was used as a PD source to generate the electromagnetic waves induced from the PD source. After that, the designed double layer bowtie and single layer bowtie antenna were placed near the PD source to capture the induced EM waves. The experiment for PD measurement in air insulation was conducted at the applied voltage level of 7 kV. Figure 13 shows the PD measurement circuit used to experiment.

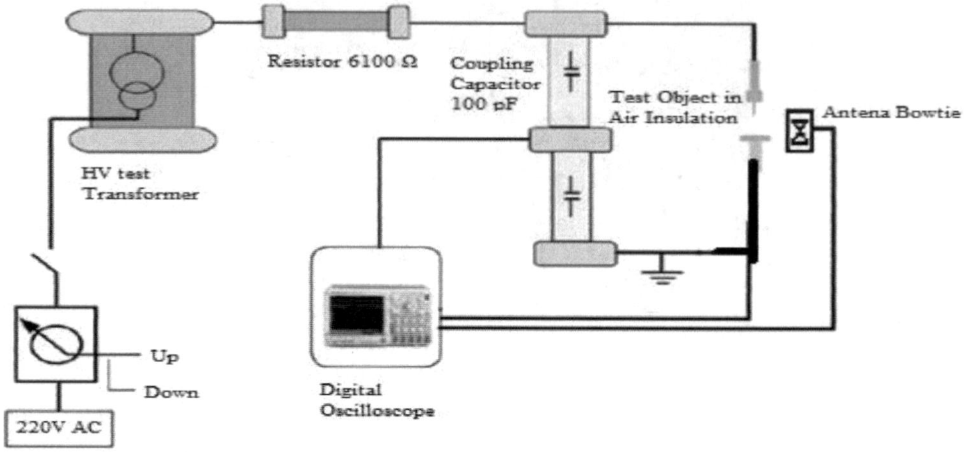

Figure 13. PD measurement circuit.

4.1. Background noise

The measurement of background noise is conducted in order to obtain accurate data during the PD measurement experimental by distinguishing the noise signal from the PD signal [3]. Figure 14 shows the background noise ON detected by the designed double layer bowtie antenna, while Figure 15 shows the background noise detected by the single layer bowtie antenna. The peak-to-peak voltage (Vpp) for BGN ON detected by double layer bowtie antenna is 20mV while the Vpp magnitude for BGN ON detected by the single layer bowtie antenna is 16.8 mV.

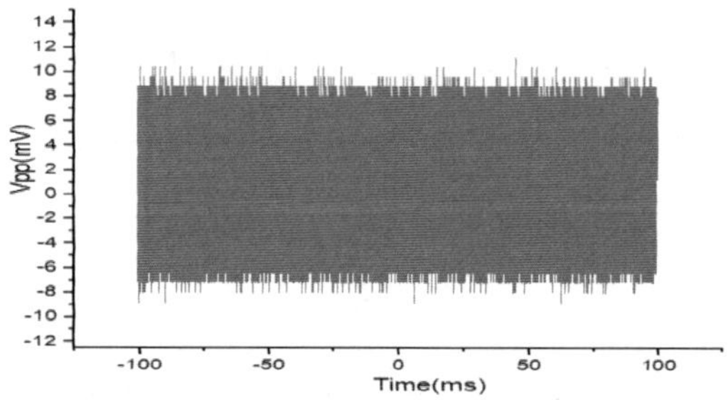

Figure 14. BGN ON detected by double layer bowtie antenna.

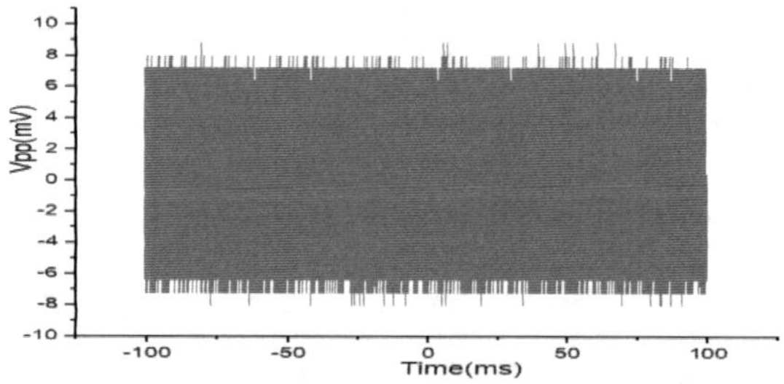

Figure 15. BGN ON detected by single layer bowtie antenna.

4.2. Partial discharge inception voltage

The partial discharge inception voltage (PDIV) is another parameter for PD that is defined as the voltage at which the first PD will appear. The Figure 16 and 17 show the negative PDIV in a negative half cycle while the Figure 18 and 19 show the positive PDIV in a positive half cycle for both double layer bowtie antenna and single layer bowtie antenna. The applied voltage for negative PDIV detected by both double layer bowtie antenna and single layer bowtie antenna are 3.21 kV and 3.42 kV respectively while the applied voltage for the positive PDIV detected by both double layer bowtie antenna and single layer bowtie antenna are 4.24 kV and 4.47 kV respectively. The Vpp magnitude for negative PDIV detected by double layer bowtie antenna and single layer bowtie antenna are 26.13 mV and 14.02 mV respectively while the Vpp magnitude for positive PDIV detected by double layer bowtie antenna and single layer bowtie antenna are 44.9 mV and 19.33 mV respectively.

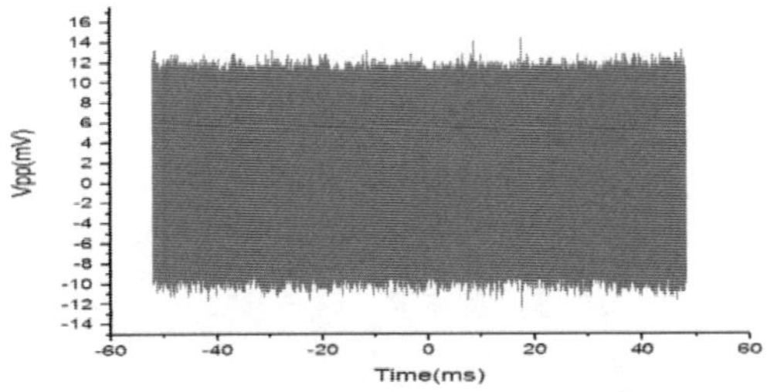

Figure 16. Negative PDIV detected by double layer bowtie antenna.

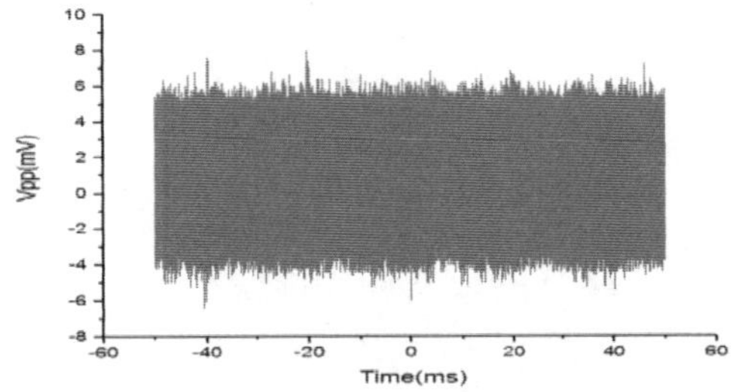

Figure 17. Negative PDIV detected by single layer bowtie antenna.

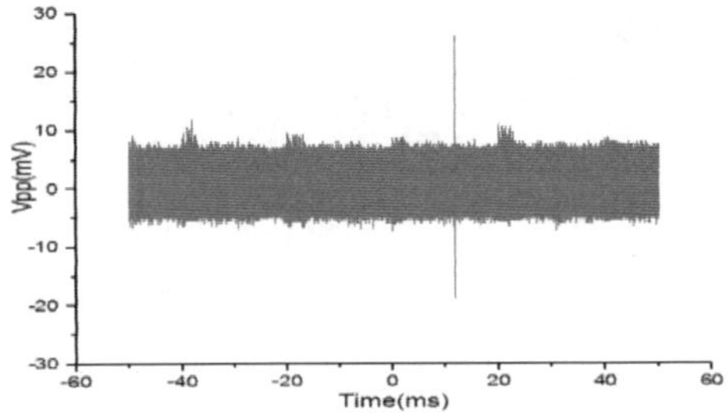

Figure 18. Positive PDIV detected by double layer bowtie antenna.

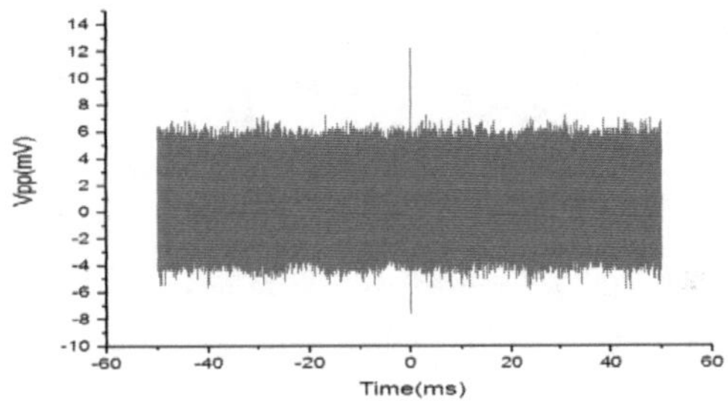

Figure 19. Positive PDIV detected by single layer bowtie antenna.

4.3. Partial discharge waveform

The PD waveform is another PD parameter that was measured during the experiment. Figure 20 and 21 show the shape of negative PD waveform while the Figure 22 and 23 show the positive PD waveform detected by the designed double layer bowtie antenna and single layer bowtie antenna. The PD measurement data was performed at 7 kV. As it can be seen on Figure 20 and 21, the Vpp magnitude for the negative PD waveform detected by double layer bowtie antenna and single layer bowtie antenna are 20.57 mV and 8.75 mV respectively while the Vpp magnitude for the positive PD waveform detected by double layer bowtie antenna and single layer bowtie antenna are 76.8 mV and 42 mV respectively as seen in Figure 22 and 23.

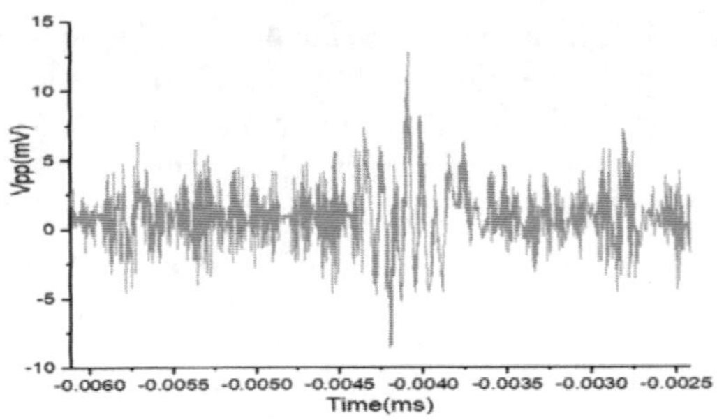

Figure 20. Negative PD waveform by double layer bowtie antenna.

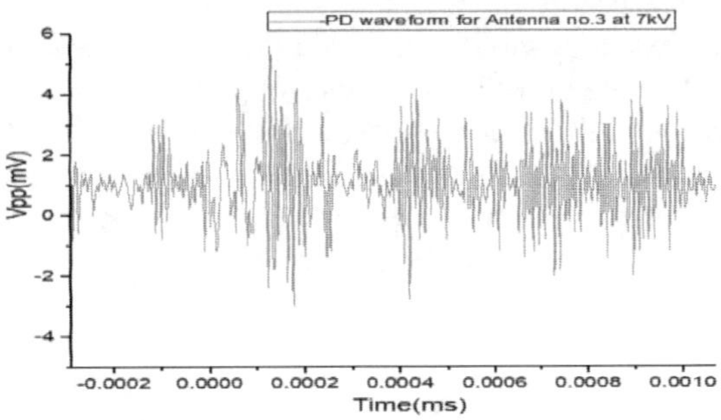

Figure 21. Negative PD waveform by single layer bowtie antenna.

IOP Conf. Series: Materials Science and Engineering **850** (2020) 012036 doi:10.1088/1757-899X/850/1/012036

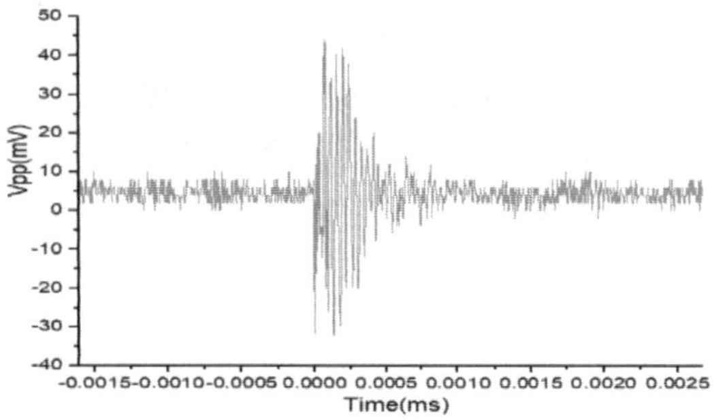

Figure 22. Positive PD waveform by double layer bowtie antenna.

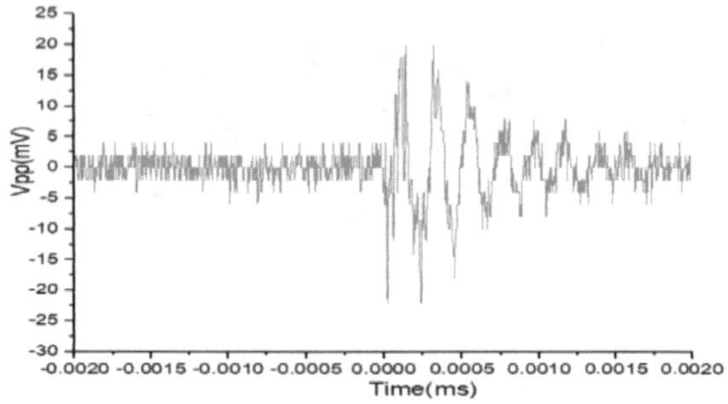

Figure 23. Positive PD waveform by single layer bowtie antenna.

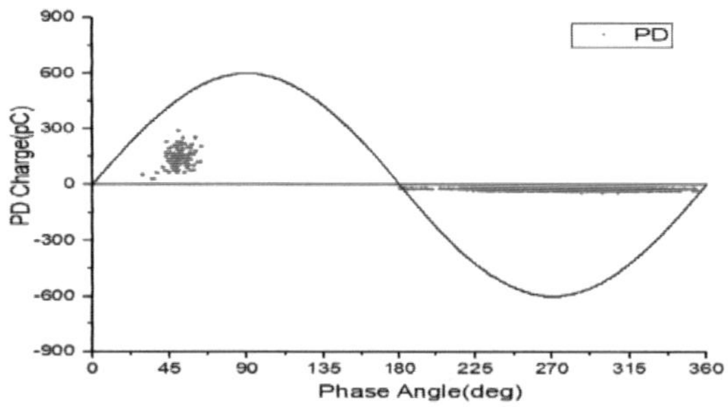

Figure 24. PD patterns by double layer bowtie antenna.

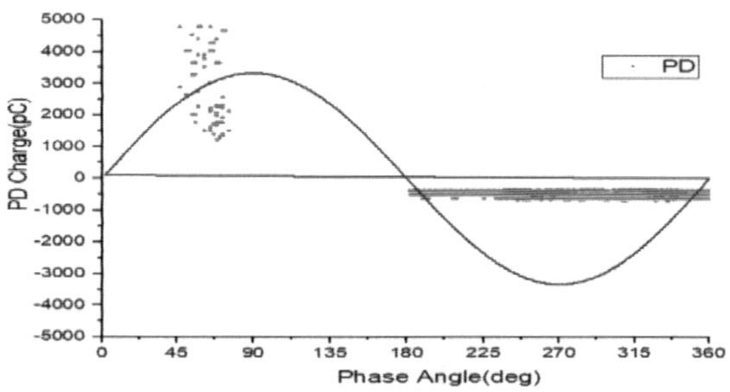

Figure 25. PD patterns by single layer bowtie antenna.

4.4. Partial discharge patterns

Figure 24 and 25 show the PD patterns detected by designed double layer bowtie antenna and single layer bowtie antenna at 7kV. The PD patterns is another PD characteristic commonly known as graph ϕ - q - n that shows the appearance of the phase angle where the PD occurs and the charge magnitude of the PD occurrence as well as the PD number.

5. Discussion and analysis

This research was about the design and application of the double layer bowtie antenna and single layer bowtie antenna as UHF sensor for partial discharge detection. The designed double layer bowtie antenna has been fabricated to operate in the UHF range of 300 MHz-3 GHz, with a minimum return loss, RL $\leq$ -10 dB and a VSWR ≤ 2 in order to be able to detect the PD signal which is the UHF signal [11]. The optimal design for the double layer bowtie antenna has a total bandwidth of 602 MHz according to the simulation results and a bandwidth of 576 MHz based on the measurement results by using the vector network analyser while the optimal design for the single layer bowtie antenna has a bandwidth of 240 MHz according to the simulation results and a bandwidth of 298 MHz based on the measurement results. As can be seen, there has been a little difference between the simulation and the measurement results. This difference may be caused by several factors which make the actual shape of the antenna to be not uniform between the design and fabrication results. The use of connectors can also be a factor that makes the results to be different. The next factor was the effect of environmental noise when testing was carried out. On the simulation side, the limitations of computer calculations can also be another factor that causes a difference between simulation results and measurement to the results obtained by using the VNA.

The experiment was conducted in air insulation by using the PD measurement circuit which include the voltage regulator, HV test transformer, limiting resistor, coupling capacitor, voltage capacitor divider, needle-plate electrode model (PD source), digital oscilloscope and the UHF sensor (bowtie antenna) which is placed near the PD source 10cm away to capture the induced EM waves emitted from the PD source. The bowtie antenna is then connected to the oscilloscope in order to measure the PD signal. Before beginning the PD measurement experiment, the background noise was measured in advance in order to know the amount of noise that is present in the laboratory and in order to distinguish noise signal from the PD signal [3]. Another PD characteristic that was measured is PDIV. During the PDIV measurement, it has been seen that the designed double layer bowtie antenna needs lower applied voltage to detect the first PD in both negative and positive half cycles compared to the single layer bowtie antenna. For example, from the PD measurement, the applied voltage for negative PDIV by double layer bowtie antenna is 3.21 mV while the needed voltage for negative PDIV by single layer bowtie antenna is 3.42 mV. This means that the double layer bowtie antenna will detect the first PD earlier than the single layer bowtie antenna. Alternatively, the peak-to-peak voltage (Vpp) magnitude for negative PDIV detected by double layer bowtie antenna is 26.13 mV while the Vpp magnitude for negative PDIV detected by single layer bowtie antenna is 14.02 mV. This indicates that the double layer bowtie antenna is more sensitive than single layer bowtie antenna in detecting PD. According to the PD waveform measured by the designed bowtie antennas at 7 kV, the Vpp magnitude for PD waveform detected by double layer bowtie antenna is higher than the Vpp magnitude for PD waveform detected by the single layer bowtie antenna. For example, the Vpp magnitude for positive PD waveform detected by double layer bowtie antenna is 76.8 mV while the Vpp magnitude for positive PD waveform detected by single layer bowtie antenna is 42 mV. This confirms again that the double layer bowtie antenna has higher sensitivity in detecting PD than the single layer bowtie antenna.

6. Conclusion

This paper has discussed the design and implementation of double layer bowtie antenna and single layer bowtie antenna as UHF sensor for PD detection in air insulation. The following conclusion has been made.

- The designed double layer bowtie antenna has a better performance than the single layer bowtie antenna based on the antenna characteristic parameters, namely return loss (RL), bandwidth and vector standing wave ratio (VSWR).
- The double layer bowtie antenna and single layer bowtie antenna were designed by using 3-D software and later on fabricated on the PCB with FR4-epoxy substrate whose relative permittivity is 4.4. After fabrication, the designed bowtie antennas were tested by using the Vector Network Analyzer (VNA) to measure some basic antenna characteristic parameters.
- According to the simulation results, the designed double layer bowtie antenna has a total bandwidth of 602 MHz while the single layer bowtie antenna has a bandwidth of 240 MHz.
- The designed double layer bowtie antenna and single layer bowtie antenna have been implemented to detect partial discharge (PD) signals in air insulation by measuring the PD characteristics such as background noise, PDIV, PD waveform and PD phase patterns.
- The designed double layer bowtie antenna is more sensitive than the single layer bowtie antenna in detecting PD in air insulation.

References

[1] Suwarno 2016 *Int. J.of Electrical Engineering and Informatics(Bandung)* **8** 147163
[2] Khayam U 2017 *4th Int. Conf. on Electrical Engineering,Computer Science and Informatics* (Yogyakarta: IEEE)
[3] Uwiringiyimana J P and Khayam U 2017 *Int. Conf. on High Voltage Engineering and Power Systems* (Bali:IEEE) pp35–39
[4] Hanalde A and Khayam U 2013 *Int.Conf. on Information Technology and Electrical Engineering* (Yogyakarta: IEEE) pp365-369
[5] Joko M, Suwarno, Khayam U and Masayuki H 2013 *Int. Conf. on Electrical Engineering and Informatics* (Malaysia: IEEE)

[6] Joko M, Suwarno, Khayam U and Masayuki H 2013 *48th Universities' Power Engineering Conference* (Dublin: IEEE)
[7] Khusnul K, Khayam U, Suwarno, Yuki T, Masahiro K and Masayuki H 2015 *5th Int. Conf.on Electrical Engineering and Informatics* (Bali: IEEE)
[8] Farrah V, Suwarno and Khayam U 2014 *Int. Conf. on Sensor, Sensor System,and Actuator* (Bnadung: IEEE)
[9] Asep A and Khayam U 2014 *Int. Conf. on Smart Green Technology in Electrical and Information Systems* (Bali: IEEE) pp. 98-101
[10] Inu S, Khayam U, Suwarno, Kiichi N, Mohamad K, Masahiro K and Masayuki H 2014 *Int. Symp on Electrical Insulating Material* (Japan: IEEE) pp. 330-333
[11] Judd M D,Yang L and Hunter I 2005 *IEEE Electr. Insul. Mag*. pp 5–14

IOP Conf. Series: Materials Science and Engineering **850** (2020) 012037 doi:10.1088/1757-899X/850/1/012037

Low interaction honeypot as the defense mechanism against Slowloris attack on the web server

N R Fitri[1], A H S Budi[1,*], I Kustiawan[1] and S E Suwono[2]

[1] Department of Electrical Engineering Education, Universitas Pendidikan Indonesia, 40154, Bandung, West Java, Indonesia
[2] Group of Faculty Member, Telkom Corporate University Center, 40152, Bandung, West Java, Indonesia

*agusheri@aol.com

Abstract. Threats and attacks on internet services are in line with developments in internet technology. One of the main risks is Distributed Denial of Service (DDoS). In this paper, we focus on the Slowloris attack which is an open-source DDoS attacker that generally attacks the Apache webserver. Apache is one of the most popular web servers in the world and is still used by many companies. This research presents another way to reduce attacks to the firewalls and load a counterweight by using a low interaction honeypot, HoneyPy. In normal conditions, attack schemes and defense mechanisms are evaluated by conducting experiments. Slowloris attacks cause high traffic because it prevents the socket from closing, so the webserver cannot handle other connections from legitimate users. When defense mechanisms are applied to the network, Slowloris attacks that enter the webserver are minimized because most packets are discarded and directed to the honeypot without the attacker's knowledge. It seems like the attacker has managed to attack the web server even when the attack becomes slower.

1. Introduction

The development of internet technology goes hand in hand with threats and attacks on internet services. One of the main risks is the Denial of Service (DoS). Denial of Service (DoS) is an attack program to prevent legitimate users from gaining access services from servers. This type of attack causes the system resources to become saturated [1]. If a DoS attack is implemented in a distributed manner and carried out by more than one machine, the attack is called Distributed Denial of Service (DDoS). The DDoS attack is launched by installing a backdoor into many computers as botnets to flood a server causes a denial of service for legitimate users. The botnets are controlled by the Command and Control (C&C) server. The attacker sends commands to the Command and Control server (C&C) then it is forwarded to the machines or computers that have been infected with malware or a backdoor to denial of service attack on the server for an undetermined time [2]. As a result, users cannot access services from the server. The most famous DDoS attack method now is to involve application-level flooding, especially on web servers. These methods can include HTTP GET Flood, HTTP Post Flood, Slowloris, DNS, etc. [3]. Slowloris is a DoS attack but can function as a DDoS because only one execution by one machine can "flood" the webserver with HTTP Flooding. Slowloris cannot be detected by Intrusion Detection System (IDS) or Intrusion Prevention System (IPS) because Slowloris does not send malformed HTTP packets so that it is detected by IDS or IPS like legitimate users [4]. Slowloris sends partial HTTP connections so that the connection remains open for a particular time so that the packets can pass IPS [5]. Apache-based web servers are vulnerable to Slowloris attacks, because in the process of handling requests with a queuing system with M/M/N so the probability of transitioning to "excessive" or saturation conditions is higher [6]. If Slowloris attacks a web server, users cannot access HTTP services or web pages as if the user does not have an internet network to access the web page. Nowadays, there are several ways used to mitigate Slowloris attacks, namely limiting the number of connections from specific hosts, determining the different timeouts for each link as a function of the number of links and delayed binding carried out by load balancers [7]. Besides, it can use Firewall and operating system security mechanisms such as Dynamic IP Restriction, mod_evasive, mod_qos, IPTables, and Fail2Ban [1].

In this paper, besides using Firewall on routers and Honeypot as a web server security system. Low to Medium Interaction Honeypot will be used as a trap server for attackers where some HTTP requests from Slowloris will be directed towards the honeypot with Port Forwarding. Honeypot is implemented on the Raspberry Pi 3. With the honeypot, it is expected to minimize the Slowloris attack on the webserver.

2. Related works

Shorey et al. explain that Slowloris uses HTTP and DNS because HTTP uses DNS as address resolution. Slowloris attacks have a high traffic rate [4]. Papadie et. al suggested the defense mechanism against Slowloris is using Firewall, Cisco and operating system load-specific tools such as Dynamic IP Restriction, mod_evasive, IPTables, and Fail2Ban [1]. There is also another proposed mitigation mechanism suggested by Arafat et al. which is a reverse web proxy to protect servers from a sudden surge of attack flows [8]. Tambunan et al. suggested a mitigation system using Glastopf Web Honeypot. But Glastopf doesn't give any reaction towards the Slowloris attack [9]. Avinash et al. proposed several ways to mitigate Slowloris attacks including increasing the number of connections that webservers allow so the legitimate users can still access the service [10]. IP-tables firewall rules can be used to limit connections from specific hosts. Then, configuring a fixed time-out connection between the web server and the client is set for how long it can stay connected and reduced the period of its connection. So that it can limit Slowloris attacks.

3. Design and testing

The network topology for this study is designed based on references in three conditions, normal conditions are shown in Figure 1, when the webserver is attacked without a security system shown in Figure 2 and with a network security system shown in Figure 3.

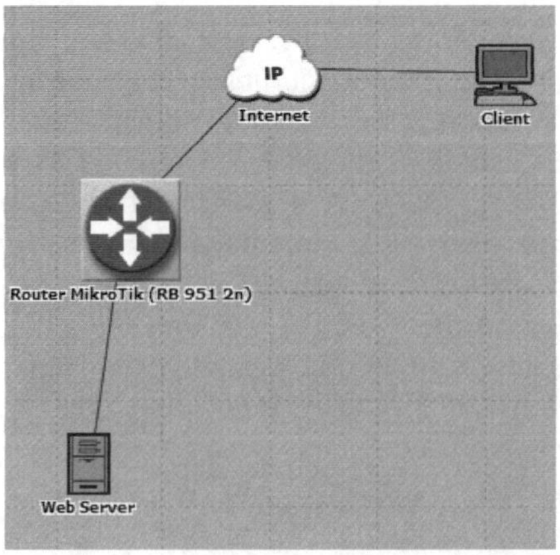

Figure 1. Normal condition network topology.

Under normal conditions, the network topology consists only of routers, web servers, and one laptop as a client. Activities on the network only browse with the browser on one laptop. There is no attack on the webserver.

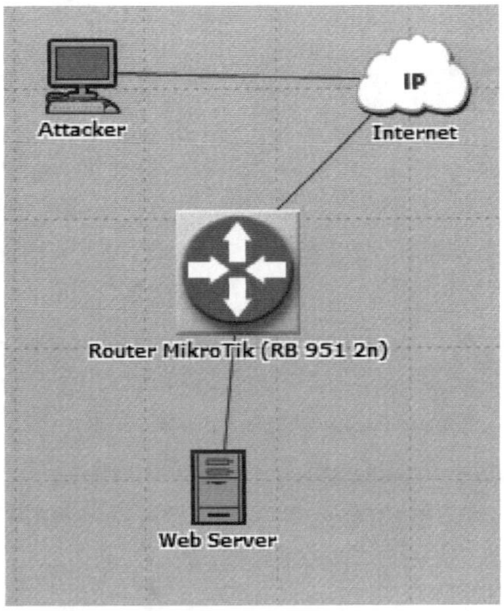

Figure 2. The attacking condition network topology.

In the attacking condition, the network topology consists of routers, web servers and one computer as an attacker without a security system. The attack was carried out for five minutes with one Ubuntu-based computer.

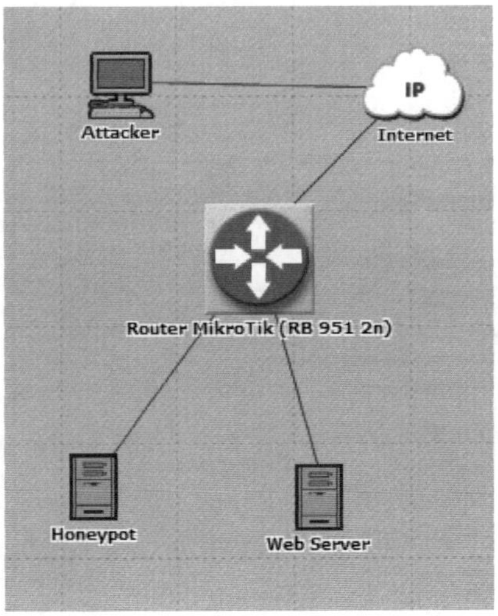

Figure 3. The attacking condition with the defense mechanism network topology.

In the attacking condition with the defense mechanism, the network topology consists of a MikroTik RouterBoard (RB 951 2n), a webserver, a computer, and a honeypot. The defense mechanism is a combination of a firewall on the router and honeypot. The attack was carried out for five minutes with one Ubuntu-based computer. This network topology is a local network consisting of a computer as the attacker, a laptop as a monitor and setting up the firewall on the router, honeypot implemented on the RaspberryPi 3 Model B +, and RaspberryPi 2 Model B as a web server. Figure 4. shows the ports on the router.

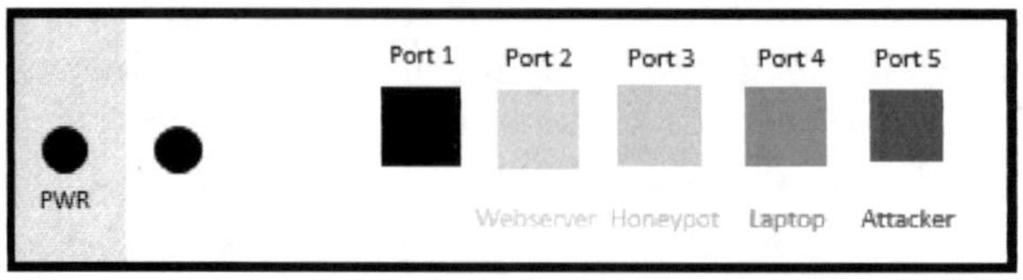

Figure 4. Ports on the router.

The router has an IP Address 192.168.100.1/24 with port 2 to port 4 as a bridge. Port 5 is removed from the bridge as a simulation of the attacker attacking through the internet with 192.168.25.1/24 as the IP Address. Port 2 is connected to a webserver, port 3 is connected to a honeypot, port 4 is connected to a laptop, while port 5 is connected to a computer as the attacker. The IP address of the webserver 192.168.100.250, the honeypot is 192.168.100.248, and the attacker is 192.168.25.252.

4. Proposed mitigation system

The security system to minimize Slowloris attacks is a combination of a firewall and a honeypot. The firewall is set up in a MikroTik router and honeypot as the fake server. We used HoneyPy as low interaction honeypot. The honeypot runs a full service, for port 80 it opens on port 10080. Figure 5 is a web server security system algorithm.

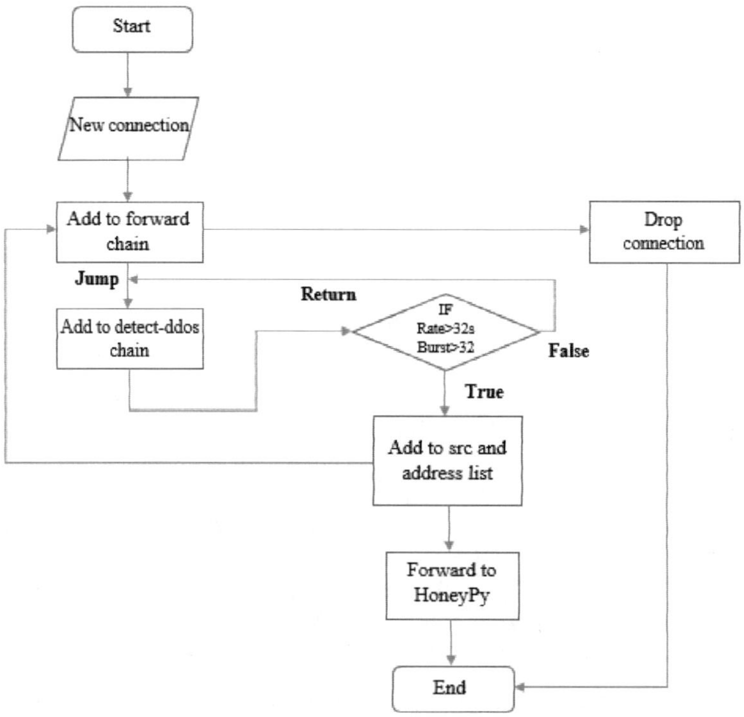

Figure 5. Defense mechanism algorithm of the webserver.

Every connection that passes through the router will be detected, and then the traffic is processed in the firewall chain. The traffic is entered and managed first in the forward chain, then the chain changes to the detect-ddos chain. If the packet data traffic detected for the source address (src-address) and destination address (dst-address) as the connection rate exceeds 32/s and the burst exceeds 32, then the sender's IP Address will be appeared on the address list as "ddoser" as well as IP The address that is the destination of the attack is appeared on the address list as "ddosed". The detected address list has a timeout of ten minutes. Then, some packet data traffic from the sender is processed to be dropped and forwarded to the honeypot without the attacker's consent.

5. Results and Discussion

The results of recording traffic data packets are recorded on the webserver using tcpdump, then the results of the data are analysed in the Wireshark. The graph displayed is an I/O Graph in TCP Connection, HTTP Connection and TCP Errors. The Y-axis indicates packet/second and the X-axis shows time. The Flow Graph displayed shows packets are sent and received by the webserver and honeypot.

5.1. Normal traffic

Figure 6 shows a TCP Connection graph and Figure 7 shows a diagram of an HTTP connection in normal condition. The webserver with the IP address 192.168.100.250 sent 200 good HTTP requests to the laptop that send HTTP GET requests seen in the Flow Graph in Figure 8.

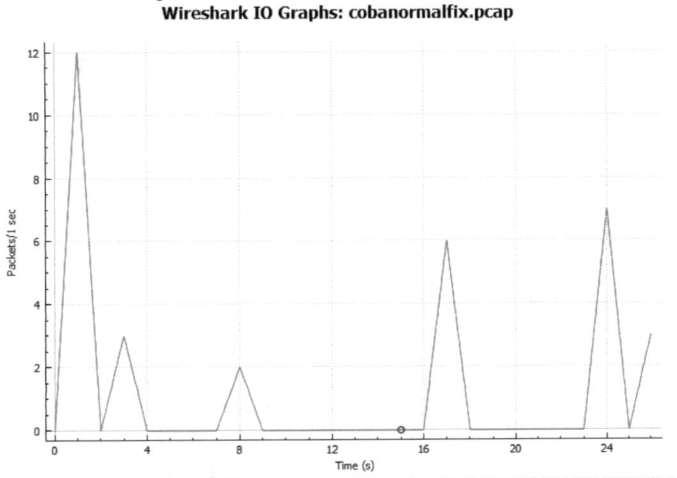

Figure 6. The graph of TCP connection in normal traffic.

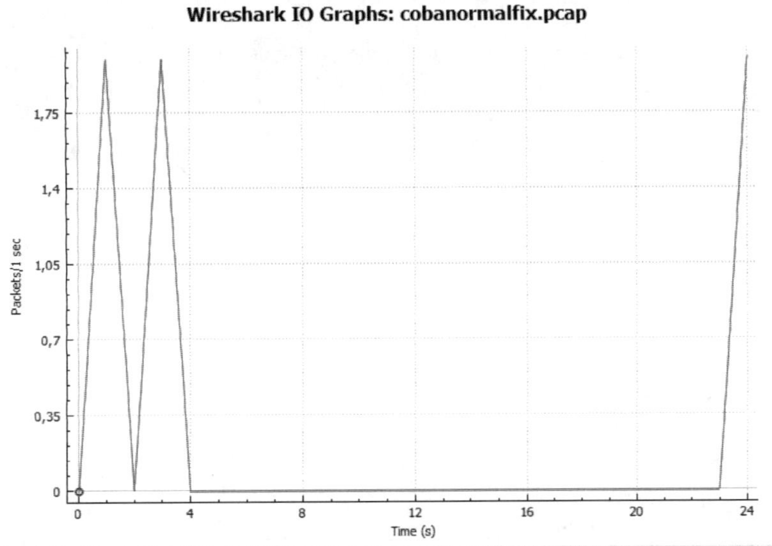

Figure 7. The graph of HTTP connection in normal traffic.

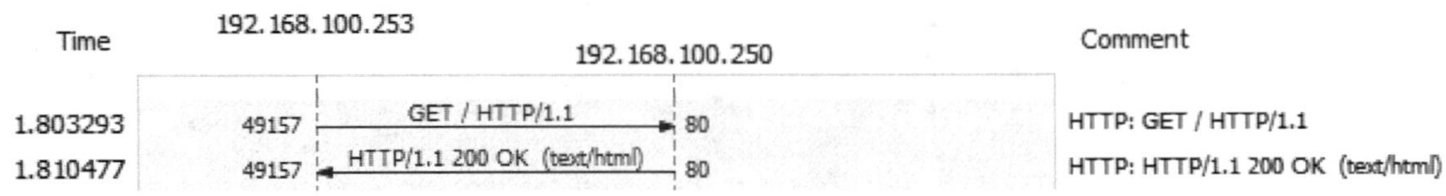

Figure 8. Flow graph HTTP connection.

5.2. Data package traffic when an attack occurs

The attack occurred for five minutes without a defense mechanism. In Figure 9. The traffic is high. In Figure 10. It shows high traffic because the Slowloris attacks are slow but numerous in a certain amount of time. The Slowloris attack takes up webserver resources, so the webserver experiences saturation as in Figure 11. Figure 12 shows the TCP errors traffic due to Slowloris attack also causes a reset connection.

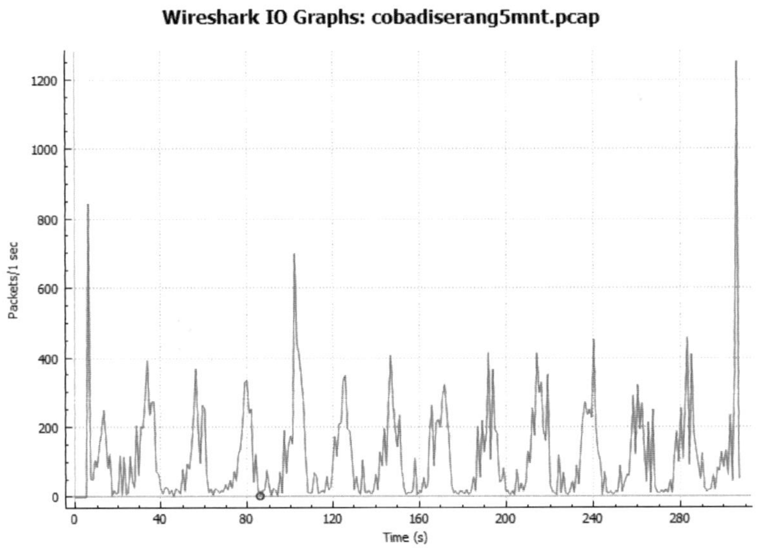

Figure 9. The graph of TCP connection for five minutes.

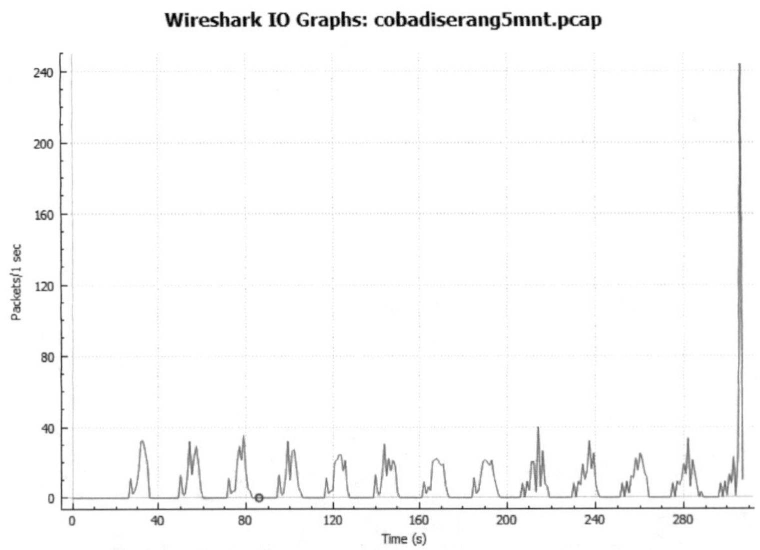

Figure 10. The graph of HTTP connection for five minutes.

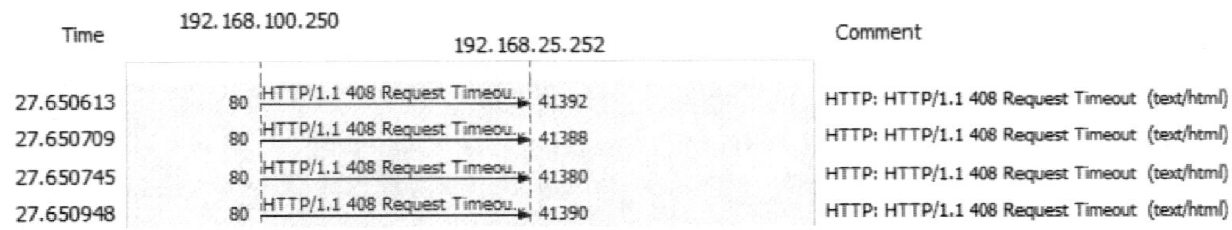

Figure 11. Flow graph HTTP request timeout.

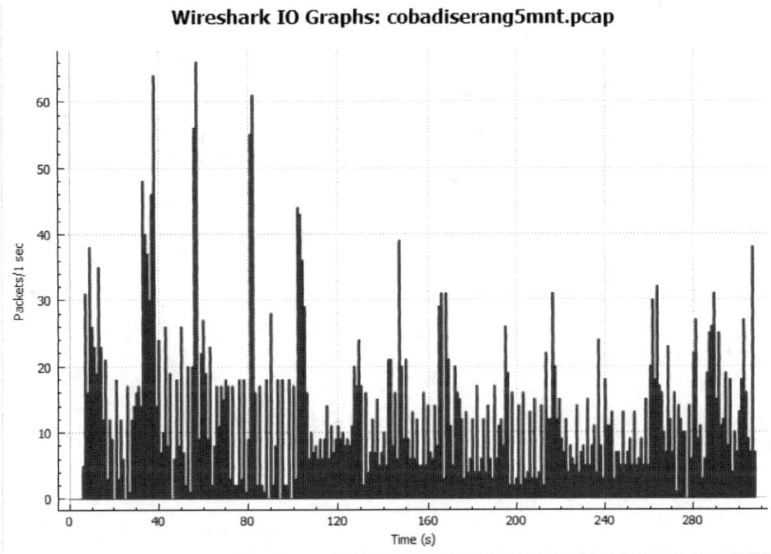

Figure 12. Graph TCP connection errors for five minutes.

5.3. The attacking condition with the defense mechanism

Slowloris is going slower when the defense mechanism is implemented the network as shown in Figure 13. The data packets are sent less than 500 packets. Figure 14 shows lower TCP traffic and high traffic only appears when the 105th second and the connection rate is reduced to 144 packets/second. HTTP traffic also becomes lower to 11 packets/second when the second is 125 as shown in Figure 15.

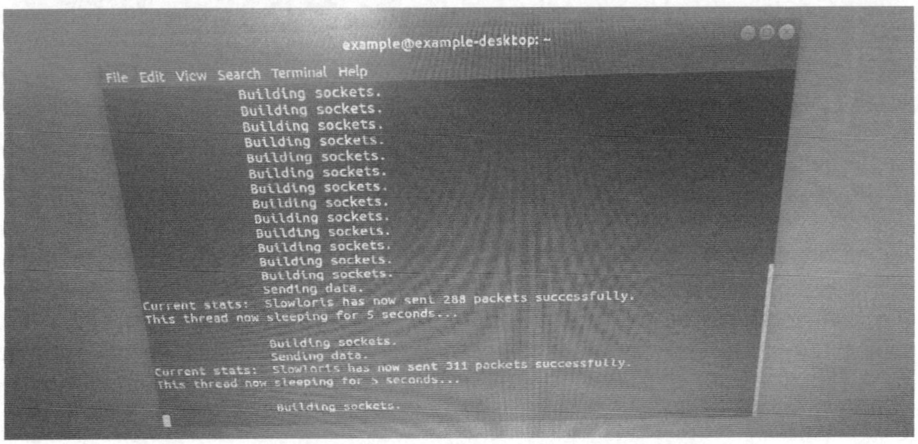

Figure 13. Slowloris attack.

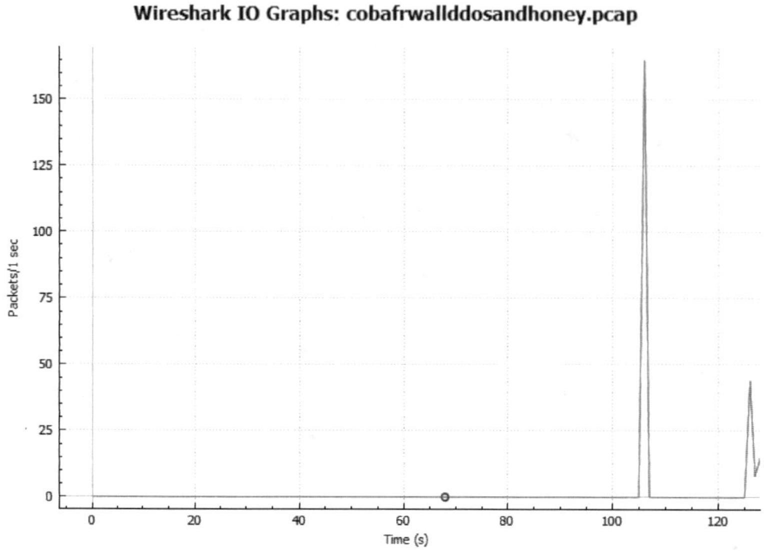

Figure 14. The graph of TCP connections for five minutes when an attack is mitigated with a defense mechanism.

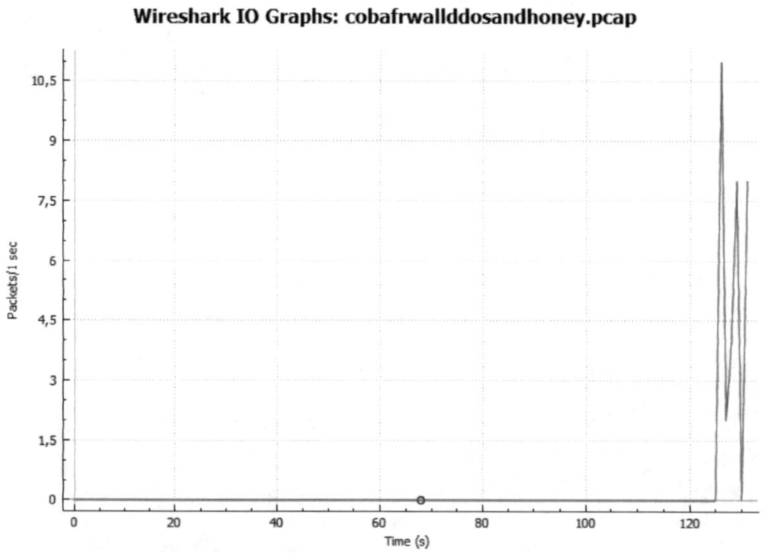

Figure 15. The graph of HTTP connections for five minutes when the attack is mitigated with a defense mechanism.

5.4. Data package traffic to the honeypot

Some Slowloris attacks are directed into the honeypot based on the firewall rule. Figure 16 shows the data traffic sent into the honeypot for 5 minutes. The traffic is high for 1000 connections every 5 seconds and the highest when the 35th second is 230 packets/s. Figure 17 shows an HTTP GET request from the attacker directed into a honeypot.

IOP Conf. Series: Materials Science and Engineering **850** (2020) 012037 doi:10.1088/1757-899X/850/1/012037

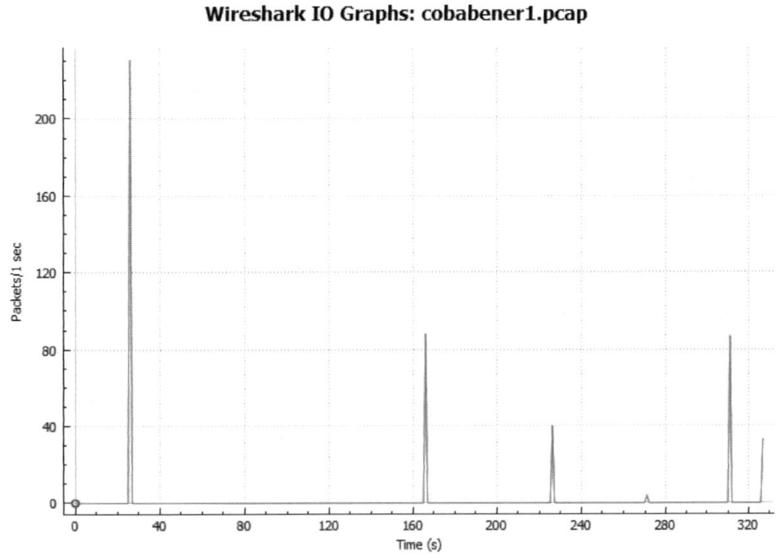

Figure 16. TCP connection graph for five minutes in honeypot.

No.	Time	Source	Destination	Protocol	Length	Info
3499	311.782171131	192.168.25.252	192.168.100.244	TCP	74	GET / HTTP/1.1 [TCP segment of a reassembled PDU]
3497	311.782166808	192.168.25.252	192.168.100.244	TCP	74	GET / HTTP/1.1 [TCP segment of a reassembled PDU]
3495	311.782164776	192.168.25.252	192.168.100.244	TCP	74	GET / HTTP/1.1 [TCP segment of a reassembled PDU]
3493	311.782162693	192.168.25.252	192.168.100.244	TCP	74	GET / HTTP/1.1 [TCP segment of a reassembled PDU]
3491	311.782160662	192.168.25.252	192.168.100.244	TCP	74	GET / HTTP/1.1 [TCP segment of a reassembled PDU]
3489	311.782158631	192.168.25.252	192.168.100.244	TCP	74	GET / HTTP/1.1 [TCP segment of a reassembled PDU]
3487	311.782156599	192.168.25.252	192.168.100.244	TCP	74	GET / HTTP/1.1 [TCP segment of a reassembled PDU]
3485	311.782154516	192.168.25.252	192.168.100.244	TCP	74	GET / HTTP/1.1 [TCP segment of a reassembled PDU]
3483	311.782152433	192.168.25.252	192.168.100.244	TCP	74	GET / HTTP/1.1 [TCP segment of a reassembled PDU]
3481	311.782148474	192.168.25.252	192.168.100.244	TCP	74	GET / HTTP/1.1 [TCP segment of a reassembled PDU]
3479	311.781157589	192.168.25.252	192.168.100.244	TCP	74	GET / HTTP/1.1 [TCP segment of a reassembled PDU]
3477	311.781155558	192.168.25.252	192.168.100.244	TCP	74	GET / HTTP/1.1 [TCP segment of a reassembled PDU]
3475	311.781153526	192.168.25.252	192.168.100.244	TCP	74	GET / HTTP/1.1 [TCP segment of a reassembled PDU]
3473	311.781151391	192.168.25.252	192.168.100.244	TCP	74	GET / HTTP/1.1 [TCP segment of a reassembled PDU]
3471	311.781146756	192.168.25.252	192.168.100.244	TCP	74	GET / HTTP/1.1 [TCP segment of a reassembled PDU]
3470	311.780852016	192.168.25.252	192.168.100.244	TCP	74	GET / HTTP/1.1 [TCP segment of a reassembled PDU]

Figure 17. HTTP GET request from attacker.

6. Conclusion

The defense mechanism consisting of a firewall and a honeypot is proven to mitigate the Slowloris attack. Most of the packets are dropped and directed into the honeypot without the attacker's knowledge. It looked like the attacker had succeeded attacked the webserver even though when the attack became slower.

References

[1] Papadie R and Apostol I 2017 Analyzing websites protection mechanisms against DDoS attacks In *2017 9th International Conference on Electronics, Computers and Artificial Intelligence (ECAI)* (pp. 1-6) IEEE

[2] Bhosale K S, Nenova M and Iliev G 2017 The distributed denial of service attacks (DDoS) prevention mechanisms on application layer *2017 13th International Conference on Advanced Technologies, Systems and Services in Telecommunications (TELSIKS)* (pp. 136-139) IEEE

[3] Choi J, Park J G, Heo S, Park N and Kim H 2016 Slowloris DoS Countermeasure over WebSocket *International Workshop on Information Security Applications* (pp. 42-53) Springer Cham.

[4] Shorey T, Subbaiah D, Goyal A, Sakxena A and Mishra A K 2018 Performance comparison and analysis of slowloris, goldeneye and xerxes ddos attack tools *2018 International Conference on Advances in Computing, Communications and Informatics (ICACCI)* (pp. 318-322) IEEE

[5] Duravkin I, Loktionova A and Carlsson A 2014 Method of slow-attack detection. In *2014 First*

International Scientific-Practical Conference Problems of Infocommunications Science and Technology (pp. 171-172) IEEE

[6] Rahmatullah D K, Nasution S M and Azmi F 2016 Implementation of low interaction web server honeypot using cubieboard *2016 International Conference on Control, Electronics, Renewable Energy and Communications (ICCEREC)* (pp. 127-131) IEEE

[7] Beard C and Stallings W 2015 *Wireless communication networks and systems* (Boston: Pearson)

[8] Arafat M Y, Alam M M and Alam M F 2015 A Practical Approach and Mitigation Techniques on Application Layer DDoS Attack in Web Server *International Journal of Computer Applications* **131**(1) 13-20

[9] Tambunan B, Raharjo W S and Purwadi J 2013 Desain dan Implementasi Honeypot dengan Fwsnort dan PSAD sebagai Intrusion Prevention System *Ultima Computing: Jurnal Sistem Komputer* **5**(1) 1-7

[10] V Avinash and K Sornalakshmi 2015 HTTP Reverse Proxy Authentication *International Journal of Advance Engineering and Research Development* **2**(03)

IOP Conf. Series: Materials Science and Engineering **850** (2020) 012038 doi:10.1088/1757-899X/850/1/012038

Characteristics of hardness at local propellers with variations of timing by using sandblasting method

T Endramawan*, A Sifa and F Dionisius

Jurusan Teknik Mesin, Politeknik Negeri Indramayu, Jl. Lohbener Lama No. 08 Lohbener, Indramayu, Indonesia

*tito@polindra.ac.id

Abstract. Sandblasting is one method of surface treatment that aims to improve the characteristics of material properties, such the hardness materials. This process was done by spraying abrasive material onto the surface of the object with the pressure. The purpose of this study was to increase the hardness of local propellers as a result of the sandblasting process. In this study, the material used aluminum alloys in the form of propellers of local fishing boats. While abrasive material used aluminum oxide with size 100 mesh. Then the pressure used 3, 4, and 5 bars with 25, 50, and 75 seconds in shooting times by angle 90. Measurement of hardness value used Universal Hardness Tester by macro-Vickers with load parameters 5 Kgf and indentation time of 10 seconds. The results showed that the longer the shooting time, the hardness of the propeller be increased 32.4% from the material that has not been carried out by the sandblasting process. The factor of time variation and shooting pressure has a significant effect on the results of the hardness in the specimen.

1. Introduction

Indramayu is an area with ecological and geographical conditions that are potential for the development of fisheries and marine businesses, then Indramayu has the largest contribution of fisheries and marine resources in West Java Province. The Indramayu Regency Government has a general policy direction that synergizes with the ideas or policies of the KKP-RI by placing increased production in the fisheries and marine sector. Increasing fisheries and marine products in Indramayu include capture fisheries, aquaculture, people's salt production and processing of fisheries and marine products [1].

Indramayu fisheries production contributes a large contribution to fisheries production in West Java, which is 61.61 percent which is the largest fish producer in West Java. Based on data from Diskanla Indramayu Regency, from January to December 2016, fish production was landed at 14 fish auction sites (TPI). The capture fisheries business is carried out by 35,929 crew members, 4,726 boat owners, 2,546 fish basket people, and 6,604 processors, while the fleet consists of 334 units of motorized boats and 5,725 outboard motors [2]. Most of the ships searching for these fish are made from the Karangsong Indramayu area which can be seen along the Prajagumiwang River, fishing boats with a length of about 28 m in length, 36 m in length, and 12 m in width, with weights up to 56 GT (gross tonnage) [3].

Ships are one of the modes of transportation whose growth is increasing because they are considered to be quite effective and relatively cheaper compared to air transportation so that it will have an impact on the use of materials, especially metals as raw materials for shipbuilding and parts such as propellers. The characteristics of the most important propellers are the high level of resistance to fatigue due to corrosion at sea, high resistance to erosion due to cavitation and high strength to load ratio. Ships consist of many parts, one of the important parts is a propeller. These boat propellers are generally made of corrosion-resistant material because they are operated directly in seawater which is a corrosion medium. The material used for making propellers is aluminum alloy and stainless steel. Other popular materials used are alloys of nickel, aluminum, and bronze which are 10 ~ 15% lighter than other materials and have higher strength. Besides, the material chosen must have good machinability, repair ability and weld ability to facilitate the production, maintenance, and repair process [4].

The application of ship propellers for the maritime industry varies. Starting from the use of fishing boats, fast boats, yachts, and commercial vessels that require a certain load and speed. For these reasons, there is a classification of propellers to determine the tolerance that the propeller must have according to their needs. The more accurate the manufacturing of ship propellers, the greater the efficiency produced by the engine [5]. The propeller has two or more blades with a snug position from the hub or boss. The propeller bar is a unified part of the hub, while the propeller hub is positioned on the shaft to be operated by the propeller [6].

The propellers in the field are mostly made of aluminum alloy made using the gravity casting technique with sand mold. The problem is that many fishermen fail on their propellers which generally occur faults on the propeller fins so that plans to find fish will be disrupted by the breakdown of the propeller. Based on the field survey carried out previously, the cause of the fracture was due to the condition of the propeller used, the condition was not good which could be seen visually in the form of many pores on the surface of the blade (figure 1) and several imperfect cracking processes. Crack initiation (figure 2) which will cause the failure of the propeller.

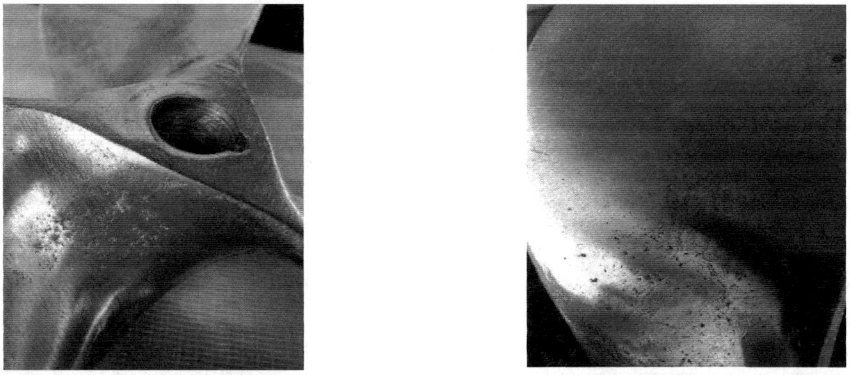

a. porosity on hub b. porosity on blade

Figure 1. Porosity on propeller surface.

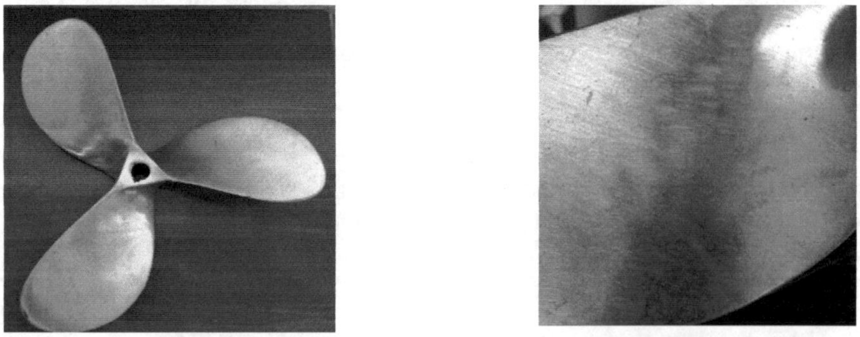

a. propeller b. crack of blade

Figure 2. Crack initiation on propeller surface.

The purpose of this study is to improve the quality of the fishing boat propellers in the form of surface hardness by repairing the surface of the propeller through sandblasting techniques so that it is expected to produce a vane surface that is free of porosity and removes scratches on the surface. Sandblasting method with the use of variations in the form of time and firing pressure to obtain optimum surface hardness results.

Sandblasting is one method that can be used to modify metal surfaces that aim to improve characteristics to improve the surface [7]. The operation of high pressure sandblasting is used to soften rough surfaces and remove sharp residual production and contaminants attached to the material [8]. The sandblasting process is carried out by firing small particles with high pressure which results in the surface of the material experiencing plastic deformation and changes in surface roughness. The farther the distance of the nozzle with the workpiece, the greater the roughness value and will decrease at a distance of 100 mm [9].

Things that determine the results of the sandblasting process include the size of the sand grains and the time of sandblasting. Sandblasting is expected to change the surface structure of the material formed into layers of nano-sized structures to improve the mechanical properties of the material. The highest hardness results reached 39.8 VHN with 40 mesh parameters and 90 seconds of shooting time [10]. The pressure and shooting angle factors also influence the results of the sandblasting process, the higher the pressure and angle used will have an impact on the decreasing corrosion rate that occurs on the surface, and the best angle is 90^0 [11].

The process of sandblasting using aluminum oxide (Al2O3) on the refinement of cobalt-chromium material concluded that the longer the firing process, the smoother the surface will be produced, the time it takes about 2-3 minutes [Endang Kusdarjanti]. The variation of spraying time will result in small changes in the value of material roughness but may affect decreasing roughness. For large particle sizes (16 mesh) has a higher roughness value compared to mesh 40, which means that the particle size becomes larger, the roughness value obtained will be larger [12].

Variable shooting angle, shooting distance, and grain size are the most influential factors on the roughness of the workpiece is the shooting distance, this refers to the statistical test that produces the highest Ra found at the shooting distance. The highest roughness value is at the shooting angle variable 600 with a firing distance of 45 cm, and the size of the sand grains is 18m with a roughness value of 7.591μm [13]. Application of methods by rotating workpieces/specimens to obtain uniform results on the surface which shows a change in the microstructure of the AISI 316L steel [14]. To obtain a roughness value of 30-85 mm on a low carbon steel plate obtained by the sandblasting method with a pressure parameter of 4 bar for 10 seconds at 45.5 mm or with 6 bars for 20 seconds which produces 76 mm [15].

2. Methods

The material used in this study is a propeller used by local fishermen in Indramayu with the brand KTM system 25 which is based on aluminum alloy. Before testing, at the initial stage material composition testing was carried out using a spectrometer test from a propeller to determine the percentage content of elements in aluminum alloy propellers.

The sandblasting process (figure 3) used aluminum oxide sand with a size of 100 mesh which is fired on a workpiece/specimen with a distance of 300 mm and by using a time variation of 25, 50 and 75 seconds. While the pressure used is 3, 4, and 5 bars. Before the sandblasting process, the object is cleaned first with a dirt-removing liquid attached to the specimen. The firing process for sandblasting uses air pressure from the compressor which will be regulated by the pressure gage regulator so that the pressure released is as desired. Particles coming out of the sandblasting torch are fired directly into the workpiece with a 90^0 angle manually using the shooting timer. When the shooting takes place it will certainly result in pressure in the box room will increase and cause the air will come out through the air filter so that the sand will not be carried out / the environment. For data collection used as many as 3 pieces for each variation both for time and pressure. After that, the specimens were tested for hardness using the Vickers method for each specimen 3 times.

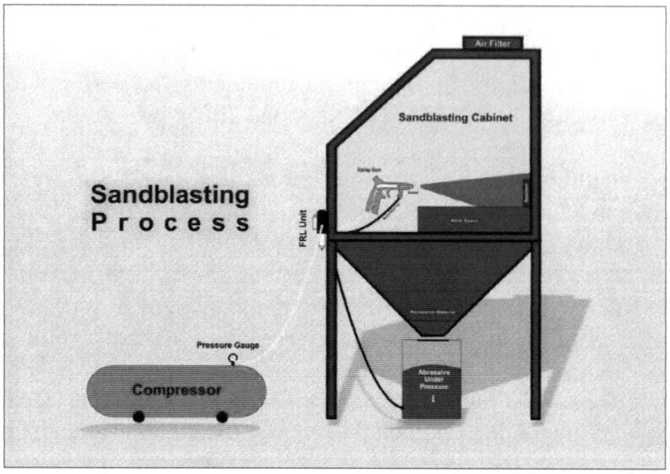

Figure 3. Schematic of sandblasting.

Hardness testing was carried out with brand macrovickers using a load of 10 Kgf and an indentation time of 15 seconds. The macrohardness test according to ASTM E-384 sets the diamond indenter with the angle between the facing surfaces is 136 ° as shown in figure 4 below.

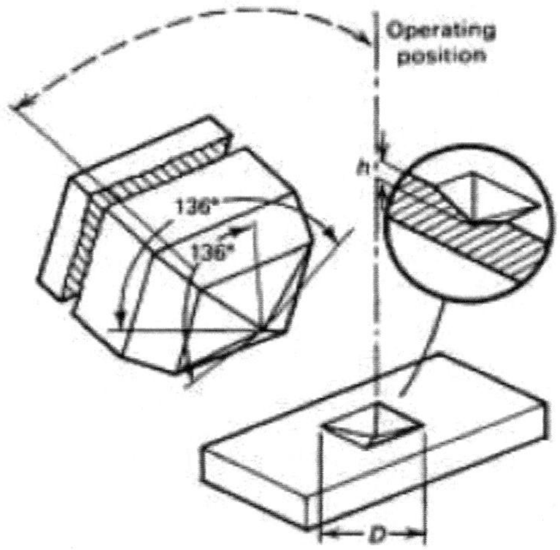

Figure 4. Hardness vickers tester [16].

After the force is removed then the diagonals are measured, so that Vickers' hardness can be formulated by the equation:

$$\mathrm{HV} = \frac{1{,}8544\ \mathrm{P}}{\mathrm{d}^2} \quad (1)$$

Where : P = load (kgf)
d = diagonal (mm)

3. Results and discussion

3.1. Material composition

The results of testing the composition of the material of a local fishing boat propeller with a spectrometer test are obtained as follows:

Table 1. Propeller composition.

No	Element	Value (%)
1	Si	5.19
2	Mn	0.128
3	Cr	0.0259
4	Ni	0.0570
5	Zn	2.30
6	Cu	1.20
7	Fe	1.44
8	Al	89.4
9	Mg	0.0579

Based on table 1. Information is obtained that the propeller is made of aluminum alloy with a content of 89.4% Al and a content of 5.19% Si so that it can be said that the material used in the manufacture of propellers is Al-Si alloy which has the advantage of being able to increase the rate flow during the casting process and has a good surface so that it is expected to produce a good propeller cast product [17]. This propeller with a content of 2.3% Zn, 1.2% Cu and 0.06% Mg, can be used for the heat treatment process to increase its hardness [18].

3.2. Hardness

Table 2. Hardness of propeller after sandblasting process.

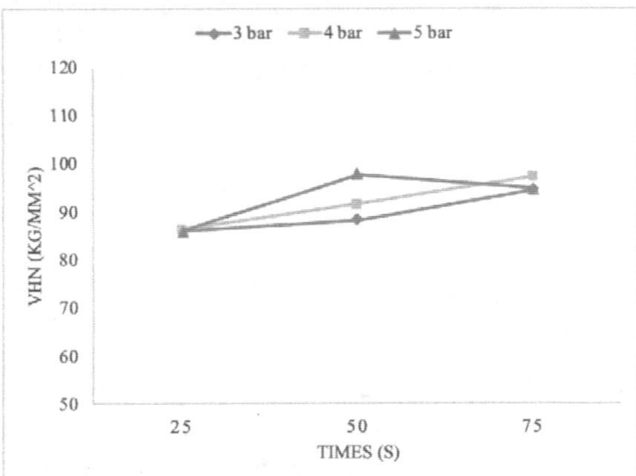

Figure 5. Hardness with variation times and pressure.

Based on figures 5, it shows that the hardness value increases with increasing sandblasting shooting time in the specimen with the highest hardness value of 94.6 VHN at 3 bar pressure and 97.1 VHN at 4 bar pressure (table 2). The existence of this sandblasting process has increased the hardness value by 31.6% of the specimens that have not been treated with sandblasting. The hardness value will increase with increasing time and the particle firing pressure to the specimen which reaches its highest value of 97.7 VHN, so that the specimen experiences an increase in hardness by 32.4% when compared to material without the sandblasting process. After the shooting time of more than 50 seconds with a pressure of 5 bars, there will be a decrease in the value of hardness as time increases, this is because the surface of the specimen has experienced saturation. This is related to changes in the characteristics of the material if the longer the surface is crushed with sandblasting, there will be an increase in mechanical properties, one of which is hardness [7].

4. Conclusion

The sandblasting process carried out on a propeller with a local fishing boat containing 89.4% Al and belonging to the type of Al-Si aluminum alloy in general can increase its hardness value by 32.4% from the material that has not been carried out by the sandblasting process. The factor of time variation and shooting pressure has a significant effect on the results of the hardness in the specimen.

Acknowledgments

Our gratitude goes to the research and service institutions of the Indramayu State Polytechnic community who have provided funding for internal research grants for the 2019 budget year.

References

[1] Dinas Kelautan dan Perikanan Kabupaten Indramayu 2017 *Profil Diskanla 2017 Kabupaten Indramayu 17 October 2017* [Online] Available: http://diskanla.indramayukab.go.id/2017/10/17/profil-diskanla-2017/ [Accessed 10 June 2019]

[2] Handayani L 2017 *Indramayu Sumbang 61,61 Persen Produksi Perikanan Jabar* [Online]. Available: https://www.republika.co.id/berita/nasional/daerah/17/01/25/okce1d365-indramayu-sumbang-6161-persen-produksi-perikanan-jabar [Accessed 12 June 2018]

[3] Pikiran Rakyat 2018 *Menelisik Sentra Kapal Nelayan di Indramayu* [Online]. Available: https://www.pikiran-rakyat.com/jawa-barat/2018/09/11/menelisik-sentra-kapal-nelayan-di-indramayu-429989 [Accessed 10 January 2019]

[4] Carlton J 2018 *Marine propellers and propulsion* (Butterworth-Heinemann)

[5] Gospodnetic D and Gospodnetic S 1996 *Integrated Propeller Manufacturing System* (Ontario: Dominis Engineering Ltd.)

[6] Harvald S A 1992 *Resistance and propulsion of ships* (Canada: John Willey & Sons, Inc.)

[7] Bouledroua O, Meliani M H, Azari Z, Sorour A, Merah N and Pluvinage G 2017 Effect of sandblasting on tensile properties, hardness and fracture resistance of a line pipe steel used in algeria for oil transport *Journal of Failure Analysis and Prevention* **17**(5) 890-904

[8] Sikka R 2019 *Sandblasting Process - Introduction, Materials Used, Pros and Cons* [Online]. Available: http://www.shotblasting.org.in/sand-blasting-materials-used-pros-and-cons.php [Accessed 10 June 2019]

[9] Widiyarta I M, Parwata I M and Lokantara I P 2014 Kekasaran Permukaan Baja Karbon Sedang Akibat Proses Sand-Blasting dengan Variasi Jarak Nosel *Jurnal Energi Dan Manufaktur* **8**(2)

[10] Ahmadi N 2018 Effects of Grains Size and Sandblasting Duration to the Surface Hardness of the Casting Product of Used Propeller *Conference SENATIK STT Adisutjipto Yogyakarta* **4** 69-74

[11] Setyarini P H 2011 Optimasi Proses Sand Blasting Terhadap Laju Korosi Hasil Pengecatan Baja Aisi 430 *Rekayasa Mesin* **2**(2) 106-109

[12] Bangun W P, Widiyarta I M and Parwata I M 2017 Pengaruh Waktu dan Ukuran partikel Dry Sandblasting Terhadap Kekasaran Permukaan Pada Baja Karbon Sedang *Jurnal Ilmiah Teknik Desain Mekanika* **6**(0) 1

[13] Kurniawan E 2013 *Analisis Kekasaran Permukaan Pada Proses Sandblasting dengan Variasi Sudut, Jarak, dan Butiran Pasir Silika Pada Pelat St 37*
[14] Mukhsen M I, Mahardika M and Suyitno S 2018 Pengaruh Sandblasting Terhadap Struktur Mikro Pada Sekrup Implan Baja Tahan Karat AISI 316L *Jurnal Mekanikal* **8**(1)
[15] Pradana R B and Kromodiharjo S 2017 Studi eksperimen pengaruh tekanan dan waktu sandblasting terhadap kekasaran permukaan, biaya, dan kebersihan pada pelat baja karbon rendah di pt. swadaya graha *Jurnal Teknik ITS* **5**(2)
[16] Kuhn H and Medlin D 2000 Asm handbook volume 8: Mechanical testing and evaluation *ASM International, Member/Customer Service Center, Materials Park, OH 44073-0002, USA, 2000 998*
[17] Suharto 1995 *Teori Bahan dan Pengaturan Teknik* (Jakarta: PT. Rineka Cipta)
[18] Saito S and Surdia T 2005 *Pengetahuan Bahan Teknik* (Jakarta: Pradnya Paramita)

IOP Conf. Series: Materials Science and Engineering **850** (2020) 012039 doi:10.1088/1757-899X/850/1/012039

Electrospinning and characterization nanofibers and nano particle of *Polyvinylpyrrolidone*

C Virginia[1,*], A Khasanah[1], J Jauhari[2,3] and I Sriyanti[1,2,3]

[1] Department of Electrical Engineering, Universitas Sriwijaya, Palembang, Indonesia
[2] Laboratory of Instrumentation and Nanotechnology Applications, Faculty of Computer Science, Universitas Sriwijaya, Palembang, Indonesia
[3] Department of Physics Education, Faculty of Education, Universitas Sriwijaya, Palembang, Indonesia

*cherinvirginia5396@gmail.com

Abstract. Electrospinning method used to fabricate fibers both micrometers and nanometers scale. Nanofibers of *Polyvinylpyrrolidone* (PVP) have been developed using an electrospinning method. It is PVP acting as matrix and ethanol analyst as a solvent. We used the same concentrate of PVP I namely 10% (w/v) for three difference molecule-weight materials namely PVP 1.300.000 kg/mol (PVP-1), PVP 360.000 kg/mol (PVP-2) and PVP 40.000 kg/mol (PVP-3). The spinning process of the three molecule-weight PVP could be produced with various parameters, 0.3 flowrate, 10 kV voltage, and 12 cm needle length. Observation using a microscope shows the morphology of each solution (PVP-1), (PVP-2), (PVP-3) will generate large fibers, medium fiber, and nanofibers. The diameter of each fiber is 2620, 1760 and 1380. From the diameter analysis, we conclude that a smaller molecule-weight of PVP will obtain smaller fibers. The FTIR results give an overview of the interactions between PVP molecules that are shown against the PVP peaks. Nanofibers and nanoparticles can be applied in medicine, high-performance air filters, military and medical protective textiles, sensors, advanced composites, photo vault cells, and separation membranes. This study will show the difference form of PVP with the same concentrate for three difference molecule-weight.

1. Introduction

Electrospinning is a simple and versatile process by which nanoscale fibers can be produced using an electrostatically driven jet of a polymer solution or polymer melt [1–3]. It has austerity and promt procedure allow production of polymeric fibers with diameter ranging from nanometer to micrometer scale [4,5]. This technique involves (1) Coulomb forces resulted from the applied electrical charge and (2) expansion of the polymer solution. Upon exposure to a very high electrical potential or voltage difference, the charged polymer is attracted to the collector and forms fiber strands [4].The morphology, size, and surface of the fibers can be customized by adjusting. processing parameters such as molecular weight, polymer concentration, electric voltage, needle-collector spacing, needle shape, polymer solution flow rate, and collector geometry; or by adjusting solution parameters such as conductivity, viscosity, and surface tension [6,7]. The structure of the nanofiber can be tailored to fulfill the desired characteristics, depending on the application. Electrospun nanofibers show great promise for developing many types of novel drug delivery systems (DDSs) owing to their high surface area, high porosity, and ability to encapsulate high drug loadings [8]. It also has function in air filtration, tissue engineering [9] and wound healing purposes [10,11].

Polyvinylpyrrolidone (PVP) along with polyvinyl alcohol (PVA) and polyvinyl acetate (PVAc) are vinyl polymers that are extensively used for the development of oral controlled-release products [12]. From the crystalline state point of view, the vinyl polymers are amorphous in nature. Then, based on the water solubility, PVP is a water soluble polymer. Among them, PVP is often used for solid dispersions containing amorphous drugs in which amorphous-state drugs show improved dissolution rates compared to their crystalline counterparts [13,14].

In this study, *Polyvinylpyrrolidone* (PVP) nanofibers were successfully prepared by the electrospinning and characterized in three different concentrations. The morphology and structure of the PVP-1, PVP-2, and PVP-3 nanofiber mats were analyze scanning electron microscope (SEM), Fourier transforms of the infrared spectrophotometer (FT-IR). The effect of increased PVP molecule weight in the PVP-1, PVP-2, and PVP-3 nanofibers on morphology.

2. Experimental

The materials used to produce PVP nanofiber and nano particle were *Polyvinypyrrolidone* (PVP) which PVP-1 (MW 1,300,000 kg mol^{-1}), PVP-2 (MW 360.000 kg mol^{-1}) and PVP-3 (MW 40.000 kg mol^{-1}) obtained from Sigma Aldrich. Ethanol (analytical grade, Merck) was used as the solvent.

The fabrication of PVP fiber started with dissolving PVP at a concentration 10% into Ethanol. A series of polymer molecule weight was prepared: 1.300.000 kg mol^{-1}, 360.000 kg mol^{-1} and 40.000 kg mol^{-1} which then labeled as PVP-1, PVP-2, and PVP-3, severally. Each precursor was stirred by using magnetic stirrer for ± 3 hours at temperature $40°C$ to produce homogeneous solution.

The polymer solution was electrospun using Electrospinning machine (Nachriebe 601) as illustrated in schematic diagram in Figure 1. The apparatus consists of a high voltage power supply, syringe with needle diameter of 0.8 mm, syringe pump, and aluminium-coated rotating collector. The apparatus was placed on a temperature and humidit-controlled chamber (25±0.5°C, RH 50%). Initially, the polymer solution was loaded into the syringe. The polymer was then dispensed by the syringe pump at a constant flowrate of 0,3 ml/hr. The needle tip was positively charged at 10 kV to attract the polymer solution out of the needle. The grounded drum collector was allowed to collect the resulted fiber at rotation speed of 200 rpm and tip collector distance of 12 cm. The Taylor Cone formation on the needle tip was monitored using a charge-coupled device (CCD) camera.

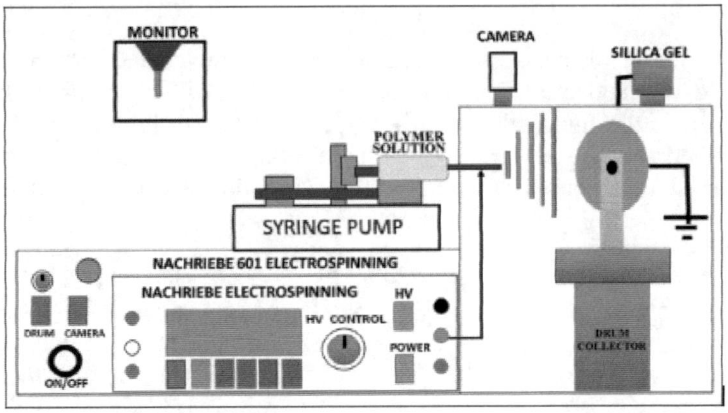

Figure 1. The schematic of nachriebe 601 electrospinning apparatus.

The morphology of PVP-1, PVP-2, and PVP-3 nanofibers were determined using a scanning electron microscope (SEM, JSM-6510; JEOL, Tokyo, Japan). The size distribution of the PVP-1, PVP-2, and PVP-3 fibers were determined using Origin ver.8 software (OriginLab Corporation, USA). The functional groups and changes occurring in PVP-1, PVP-2, and PVP-3 nanofibers were identified using a Fourier Transform Infrared Spectroscopy (FTIR) spectrum (Alpha; Bruker, Germany) at 500-4000 cm-1 spectral range. All samples were trimmed into 3 × 20 mm rectangle with similar thickness and clamped on both sides of the apparatus and then stretched at elongation rate of 20 mm/min. The functional group in each sample were determined by looking at the number of peaks in the form of transmitations. Moreover, the information about compound structure and the purity of PVP could be assessed.

3. Results and discussion

3.1. PVP nanofibers image

The physical appearance of PVP-1, PVP-2, and PVP-3 produced from the electrospinning process can be seen in Figure 2. Macroscopically. PVP which has a difference in molecular weight but dissolved by same solution concentration will obtain heterogeneous fiber morphology. The nanofiber form of PVP-1 (MW 1.300.000 kg/mol^{-1}) are shown in Figure 2a. which is shaped like fine white layer. The form of PVP-2 (MW 360.000 kg/mol^{-1}) Nanofibers are displayed in Figure 2b. whose shape is similar to PVP-1 but it's thinner and more brittle than PVP-1 layers. Afterwards, PVP-3 (MW 40.000 kg/mol^{-1}) were accumulated to form a pile of white particle which are shown in Figure 2c. The successful formation of the fibers was a result of three contributing factors. The first factor is the Coulomb force (Fc), where high electrical voltage applied on the needle tip induces charge on polymer and consequently the polymer has a tendency to move towards the grounded collector. The second factor is the loading force (Fd) from the syringe pump, which provides polymer drop available to be spun. The third factor is the surface tension force (Fγ) at the polymer-air interface, which retains the polymer droplet on the needle tip. The equilibrium of these three forces results (Eq. 1) in a cone jet formation known as Taylor's cone [15,16].

$$\mathrm{Fc} + \mathrm{Fd} - \mathrm{F}\gamma = 0 \tag{1}$$

When Fc is introduced to the polymer, the polymer chains elongates while the solvent evaporates, resulting in solidified fiber formation on the collector surface [6,17]. When the voltage was set at 10kV percusors droplets would form Taylor's cone, the equations can be written by:

$$V_c \sim \sqrt{\frac{\gamma d^2}{\varepsilon R}} \tag{2}$$

In which V_c is a critical voltage that caused precursors being attracted and forming a cone, d the distance of the needle tip to the collector, γ is a solution surface tenssion, and R is the main curvature of precursor. As shown in the figure, the electrical voltage plays an important role in the formation of nanofiber as described previous research [18]. The increased of nanofiber diameters might occur with high voltage due to the acceleration of the mass flow of the polymer solution by excessive electrostatic force [19].

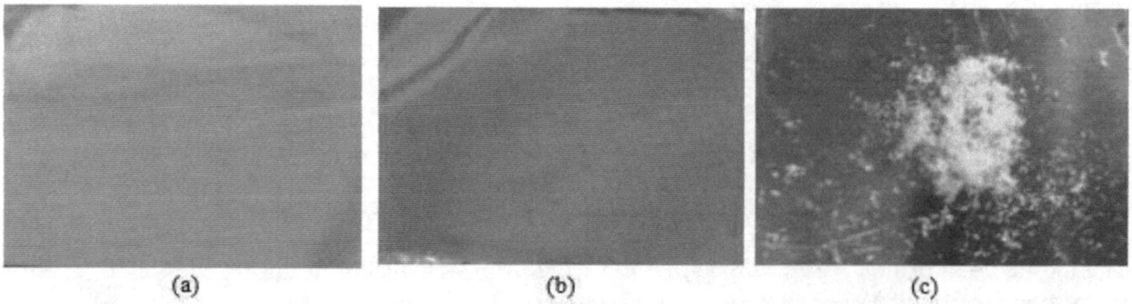

Figure 2. Electrospun nanofibers and nano particle Polyvinilpyrrolidone with 10% concentration.

3.2. Morphology and diameter of PVP nanofiber

The morphology of PVP-1, PVP-2 and PVP-3 fiber is shown in Figure 3. the level of solution viscosity was affected by solution concentration and dissolved molecular weight. It is *polyvinylpyrrolidone* with a different molecular weight that influence the precusors' viscosity. High polymer molecular weight (1.300.000 kg/mol^{-1}) produced unbeaded fibers (figure), which was attributed to the viscosity of the solution.

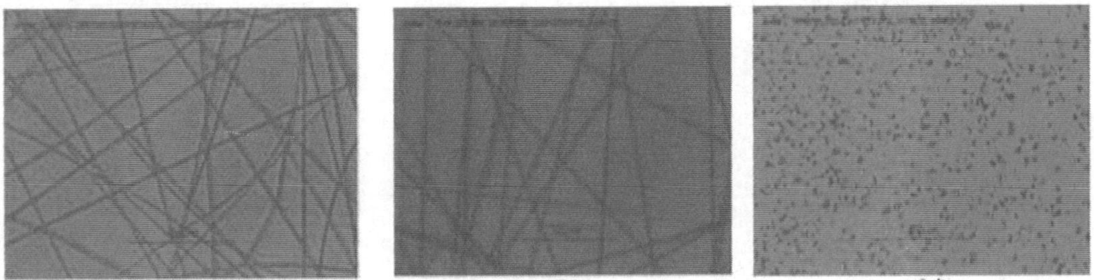

(a) (b) (c)

Figure 3. optical microscopic images of PVP nanofiber at 400× magnification and diameter distributions of (a) PVP-1, (b) PVP-2, (c) PVP-3.

At low polymer viscosity (MW 40.000 kg/mol^{-1}) has fewer inter-chain bonds, causing the polymer solution to be unable to maintain continuous elongation during the stretching of the jet [9]. This condition will result in beaded fiber. It has been established that the formation of beads is associated with viscoelasticity of the solution [11]. In electrospinning process, viscoelastic force permits continuous conversion of the jets into fibers instead of beads. In low viscosity solutions, the viscoelasticity of polymer droplet on the needle tip is unable to overcome Rayleigh instability, an axisymmetric rotation that causes the stretched jet to break up and form beads [20]. As the polymer molecular weight decreases (MW 1.300.000 kg/mol^{-1} and 40.000 kg/mol^{-1}) beaded fibers were produced (Figure 3c).

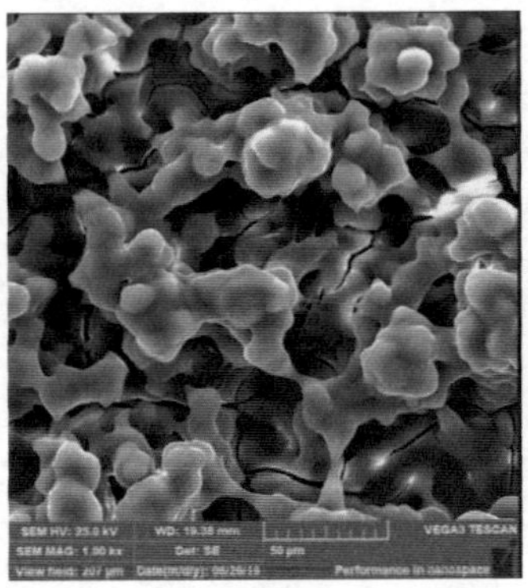

Figure 4. SEM images and diameter distributions of PVP-3.

The effect of PVP concentration on fiber diameter is displayed in Fig. 3a. It was found that higher PVP molecular weight produced fibers of greater diameter. In fact, a PVP viscosity are influenced by the weight of molecular solute. Fiber diameter is strictly controlled by the viscosity of polymer, where the degree of chain entanglement is increased with higher viscosity, leading to the formation of a larger fiber diameter [21].

In this study, the fiber diameter of *polyvinylpyrrolidone* 10% (w/v), PVP-1 (MW 1.300.000 kg/mol^{-1}) PVP-2 (MW 360.000 kg/mol^{-1}), and PVP-3 (MW 40.000 kg/mol^{-1}) are 2.620 *nm* for PVP-1 shown at figure 3a, 1.760 *nm* for PVP-2 shown at figure 3b, and 1,380*nm* for PVP-3 shown at figure 3c. It is

PVP-3 which has lowest both molecular weight and viscosity among other dissolved PVP. Therefore, from these data above, we can see that smaller polymer viscosity will obtain smaller diameter of fiber and it will contain a lot of beads.

3.3. Fourier-Transform Infrared (FTIR)

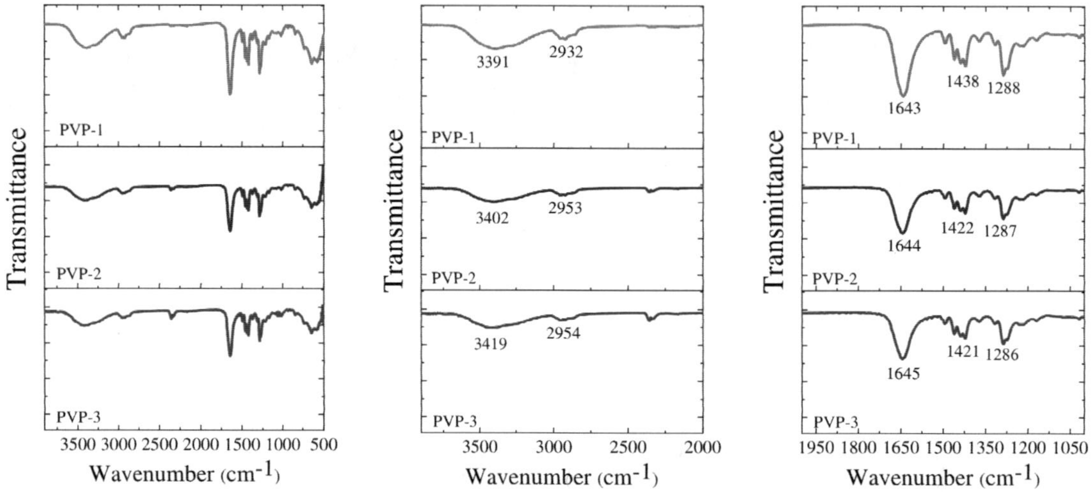

(a) (b) (c)

Figure 5. FTIR spectra of PVP, PVP-1, PVP-2, PVP-3 nanofiber.

FTIR analysis was performed to study the influence of the process on the chemical structure of the *polyvinylpyrrolidone* (PVP) 10% (w/v) nanofiber. It is also conducted to identify characteristic functional groups in PVP-1, PVP-2, and PVP-3 nanofibers. The FTIR spectrum of PVP is shown in Figure 5a. FTIR spectrum at 3000-3500cm^{-1} shows a broad peak direction at wave number 3391cm^{-1} (PVP-1), 3402cm^{-1} (PVP-2), and 3419cm^{-1} (PVP-3) which are hydroxyl groups including stretching O-H bonds [20,22]. if it is seen from the peak, hydrogen bonds from PVP-1, PVP-2 and PVP-3 peak added (up). Another peak appears at wave number 2932cm^{-1}, 2953cm^{-1}, 2954cm^{-1}, all three bonds are identified as a stretch of C-H bond. A stretch of C = O (carbonyl group) bonds regulated for PVP-1, PVP-2 and PVP-3 samples can be seen from the peaks shown at wave numbers 1643cm^{-1}, 1644cm^{-1} and 1645cm^{-1}. At peak 1288cm^{-1}, 1287cm^{-1} and 1286cm^{-1}, there is a bending of C-N bonds.While at peaks 1438cm^{-1}, 1422cm^{-1} and 1421cm^{-1}, it occurs a bending of the C-H bond. This can be seen from the peak between wave numbers 1000 - 1500cm^{-1}. For PVP 1 samples, PVP 2 and PVP 3 wave numbers move to higher values, and peaks between 500 - 1000cm^{-1} happen a bending of the C-C bond. The higher ethanol content causes the hydroxyl peak to be sharper and more visible from 3000cm^{-1} to 3500cm^{-1}, consequently the hydroxyl and carbonyl (CO) peaks move to the lower wave number direction. The stretching peak of this carbonyl appeared in 1643cm^{-1} in PVP-1, 1644cm^{-1} in PVP 2 and 1645cm^{-1} in PVP 3, the movement peak of the carbonyl stretch and other peaks was considered as a result of the comparison of PVP molecules caused by the electrospinning process.

4. Conclusion

A smooth and uniform nanofiber with diameter 2.860nm (PVP-1), 1.760nm (PVP-2), and 1.3800nm (PVP-3), was manufactured by electrospinning method using *Polyvinylpyrrolidone* with concentration 10% (w/v) in three different molecular weight PVP-1 (MW 1.300.000 kg/mol^{-1}) PVP-2 (MW 360.000 kg/mol^{-1}), and PVP-3 (MW 40.000 kg/mol^{-1}). the difference of molecular weight influenced the solution viscosity. it is also affected nanofiber structure. The PVP-3 has smallest diameter and producing lots of beads. Moreover, the polymer remained amorphous in PVP-3. This study provides an insight into enhancement of physical characteristics of hydrophilic by the right choice of polymer blends.

Acknowledgments
This research was financially supported by Universitas Sriwijaya, Republik of Indonesia under the University's Excellence Research (PUPT) Grant in the fiscal year 2019. This research was financially supported by BPDPKS, Republik of Indonesia under the Research and Development of Palm Oil in the fiscal year 2018.

References

[1] Illangakoon U E, Nazir T, Williams G R and Chatterton N P 2014 Mebeverine-loaded electrospun nanofibers: Physicochemical characterization and dissolution studies *J. Pharm. Sci.* **103** 283–92
[2] Celebioglu A and Uyar T 2011 Electrospun porous cellulose acetate fibers from volatile solvent mixture *Mater. Lett.* **65** 2291–4
[3] Vallejos M E, Peresin M S and Rojas O J 2012 All-Cellulose Composite Fibers Obtained by Electrospinning Dispersions of Cellulose Acetate and Cellulose Nanocrystals *J. Polym. Environ.* **20** 1075–83
[4] Sriyanti I, Edikresnha D, Rahma A, Munir M M, Rachmawati H and Khairurrijal K 2018 Mangosteen pericarp extract embedded in electrospun PVP nanofiber mats: Physicochemical properties and release mechanism of α-mangostin *Int. J. Nanomedicine* **13** 4927–41
[5] Jauhari J, Wiranata S, Rahma A, Nawawi Z and Sriyanti I 2019 Polyvinylpyrrolidone/cellulose acetate nanofibers synthesized using electrospinning method and their characteristics *Mater. Res. Express* **6** 0–6
[6] Rahma A, Munir M and Prasetyo A 2016 Intermolecular Interactions and the Release Pattern of Electrospun Curcumin-Polyvinyl (pyrrolidone) Fiber **39** 163–73
[7] Sriyanti I, Edikresnha D, Rahma A, Munir M M, Rachmawati H and Khairurrijal K 2017 Correlation between Structures and Antioxidant Activities of Polyvinylpyrrolidone/ Garcinia mangostana L. Extract Composite Nanofiber Mats Prepared Using Electrospinnin *J. Nanomater.* **2017**
[8] Kim C, Park S H, Lee W J and Yang K S 2004 Characteristics of supercapaitor electrodes of PBI-based carbon nanofiber web prepared by electrospinning *Electrochim. Acta* **50** 877–81
[9] Santos C, Silva C J, Büttel Z, Guimarães R, Pereira S B, Tamagnini P and Zille A 2014 Preparation and characterization of polysaccharides/PVA blend nanofibrous membranes by electrospinning method *Carbohydr. Polym.* **99** 584–92
[10] Alessandrini F, Vennemann A, Gschwendtner S, Neumann A, Rothballer M, Seher T, Wimmer M, Kublik S, Traidl-Hoffmann C, Schloter M, Wiemann M and Schmidt-Weber C 2017 Pro-Inflammatory versus Immunomodulatory Effects of Silver Nanoparticles in the Lung: The Critical Role of Dose, Size and Surface Modification *Nanomaterials* **7** 300
[11] Aruan N M, Sriyanti I, Edikresnha D, Suciati T, Munir M M and Khairurrijal K 2017 Polyvinyl Alcohol/Soursop Leaves Extract Composite Nanofibers Synthesized Using Electrospinning Technique and their Potential as Antibacterial Wound Dressing *Procedia Eng.* **170** 31–5
[12] Pusporini P, Edikresnha D, Sriyanti I, Suciati T, Munir M M and Khairurrijal K 2018 Electrospun polyvinylpyrrolidone (PVP)/green tea extract composite nanofiber mats and their antioxidant activities *Mater. Res. Express* **5**
[13] Broman E, Khoo C and Taylor L S 2001 A comparison of alternative polymer excipients and processing methods for making solid dispersions of a poorly water soluble drug **222** 139–51
[14] Ignatious F, Sun L, Lee C and Baldoni J 2010 Expert Review Electrospun Nanofibers in Oral Drug Delivery **27** 576–88
[15] Subramanian U M, Kumar S V, Nagiah N and Sivagnanam U T 2014 Fabrication of polyvinyl alcohol-polyvinylpyrrolidone blend scaffolds via electrospinning for tissue engineering applications *Int. J. Polym. Mater. Polym. Biomater.* **63** 462–70

[16] Islam M S and Karim M R 2010 Fabrication and characterization of poly(vinyl alcohol)/alginate blend nanofibers by electrospinning method *Colloids Surfaces A Physicochem. Eng. Asp.* **366** 135–40
[17] Quan J, Yu Y, Branford-White C, Williams G R, Yu D G, Nie W and Zhu L M 2011 Preparation of ultrafine fast-dissolving feruloyl-oleyl-glycerol-loaded polyvinylpyrrolidone fiber mats via electrospinning *Colloids Surfaces B Biointerfaces* **88** 304–9
[18] Jiang Y N, Mo H Y and Yu D G 2012 Electrospun drug-loaded core-sheath PVP/zein nanofibers for biphasic drug release *Int. J. Pharm.* **438** 232–9
[19] Rahma A, Munir M M, Khairurrijal and Rachmawati H 2015 The Influence of Non-Ionic Surfactant on the Physical Characteristics of Curcumin-Loaded Nanofiber Manufactured by Electrospinning Method *Adv. Mater. Res.* **1112** 429–32
[20] Konwarh R, Karak N and Misra M 2013 Electrospun cellulose acetate nanofibers: The present status and gamut of biotechnological applications *Biotechnol. Adv.* **31** 421–37
[21] Chen M, Qu H, Zhu J, Luo Z, Khasanov A, Kucknoor A S, Haldolaarachchige N, Young D P, Wei S and Guo Z 2012 Magnetic electrospun fluorescent polyvinylpyrrolidone nanocomposite fibers *Polymer (Guildf).* **53** 4501–11
[22] Ding B, Kim H Y, Lee S C, Shao C L, Lee D R, Park S J, Kwag G B and Choi K J 2002 Preparation and characterization of a nanoscale poly(vinyl alcohol) fiber aggregate produced by an electrospinning method *J. Polym. Sci. Part B Polym. Phys.* **40** 1261–8

IOP Conf. Series: Materials Science and Engineering **850** (2020) 012040 doi:10.1088/1757-899X/850/1/012040

The effect of a single quenching double tempering process on the mechanical properties of NiCrMo alloyed steel casting

B Bandanadjaja[1,*], W Purwadi[1], N Lilansa[2], D Idamayanti[1] and T R S Putra[1]

[1] Foundry engineering, Politeknik Manufaktur Bandung, Kanayakan 21, Indonesia
[2] Automation engineering, Politeknik Manufaktur Bandung, Kanayakan 21, Indonesia

*benybj@yahoo.com

Abstract. This research aimed to improve the mechanical properties of NiCrMo alloyed steel casting, especially in terms of toughness. Toughness is a combination of tensile strength, yield strength, and elongation. The method used was a multi-heat treatment process involves normalizing, tempering, double tempering, and an additional intermediate process of quenching. The results obtained through the normalizing process followed by single quenching on oil media and double tempering (single quenching double tempering) produced the best-combined result of tensile strength, yield strength, and elongation. The modulus of toughness increased by up to 745 % compared to the as-cast condition from 20 N.mm/mm3 to 149 N.mm/mm3. The best mechanical properties were obtained from tempered martensite microstructure. It is free from rest martensite and secondary carbide.

1. Introduction

There are several known microstructures of steel, such as ferrite, pearlite, bainite, martensite, and austenite. Each structure has different mechanical properties [1,2]. It is possible to get a higher strength from one of these structures. In general, quenching and tempering are effective ways to increase steel strength, which can be achieved mainly through precipitation of fine dispersion of alloy carbides during tempering [3]. The technique is well known to produce the highest level of strength in steel. Martensitic structures are rarely used without undergoing a tempering process because a large amount of internal stress is created due to the transformation process that causes the material to become more brittle [4]. Tempering at low temperatures is enough to reduce this stress significantly without fundamentally changing the basic characteristics of martensite structures. The effect of microstructure on the mechanical properties of low alloy steels has been actively studied by physical metallurgical researchers [5]. The coarse microstructure formed in the ferritic/pearlitic steel hot roll process developed conventionally results in increased ductility, toughness, and strength, which is good. The development of new steel with a better combination of strength, ductility, and toughness has led to the emergence of new types of multi-phase or ultra-fine grained high strength low alloy steel [6].

Applications in the transportation industry demand more economical high-strength steel with good ductility and impact toughness to lighten structural parts [7,8]. Mechanical properties of low alloy steel can be improved by improving austenite grains. Refinement of microstructure on martensitic steels resulting from quenching and tempering can increase steel strength and toughness. The martensitic lath structure is usually stronger, and increasing the toughness of lath martensite low alloy steel is important, especially for industrial applications used at low temperatures. To get a higher combination of strength and toughness in low alloy steel, achieved through hardening and tempering, can be done by refining austenite grains by microalloying in addition to cyclic heating, initial microstructure, austenitization temperature and holding time at austenitizing temperatures. These factors are related to each other during the heat treatment process, and the overall effect can be determined by the measurement of austenitic prior to grain size. Besides microalloying and cyclic thermal treatment, the formation of austenite grains from various initial microstructure also becomes an interesting technique adopted in the industry for microstructure refining [9].

Sarikaya et al. found that very high impact energy values could be achieved with double heat treatment, but the strength was slightly lower because of the increase in the amount of austenite held through the double quenching treatment (DQ) followed by low temperature tempering on the base structural steel Fe-Cr-C [10]. Chang et al. studied the effect of double austenitization (DA) treatment on the toughness of Ni-Cr-Mo-V steel with the same strength and found that the increase in toughness of steel treated with DA could be mainly attributed to the lower aspect ratio of carbide [11]. Therefore, steels that are processed with double quenching and subsequent double tempering (DQT) treatments can produce different properties due to different microstructure and morphology as well as the distribution of secondary carbides in steel, the main factors, which improve different mechanical properties.

This research is a continuation of previous research, proposed the use of a single quenching double tempering (SQDT) process [12,13]. The effect of double tempering on the mechanical properties after the normalizing process has been studied on Ni-Cr-Mo alloy cast steel or equivalent to AISI 4340. It was done to improve material toughness through modification of the heat treatment processes. In as-cast conditions, steel casting material has very brittle properties where elongation is only 4%, and impact value is 15 J/cm^2. Previously the heat treatment process was carried out by normalizing followed by double tempering at a temperature of 650 °C. It was found that steel casting elongation increased up to 20%, and the toughness increased to 142 J/cm^2 [12,13]. The objective of this research is to study the effects of the additional process of hardening or quenching after normalizing, then followed by a double tempering process on microstructure and mechanical properties of cast steel, primarily with increasing material toughness.

2. Experimental procedure

2.1. Material

The research involved the following steps: For experimental samples, raw materials alloy steel was made of CrNiMo cast, with the composition of the AISI 4340 standard. Steel was cast in the shape of the Y-block sample. Y-block was then cut in a 15x15x12 mm square form. Square samples were used for metallographic examination and Rockwell hardness tests in accordance with ASTM E 18 standards. In addition, the cylindrical sample was also made for tensile tests observing ASTM E 8M standards. Castings were tested for chemical composition using OES (optical emission spectroscopy). The chemical composition obtained is as per Table 1.

Table 1. Chemical composition (%).

C	Si	Mn	P	S	Cr	Ni	Mo
0,34	0,22	0,70	0,02	0,01	0,87	1,82	0,34

The raw material, NiCrMo alloy steel, in as-cast condition has a microstructure with the majority of widmanstaten ferrite, which appears like sharp needles. The other structure is the presence of small amounts of fine pearlite. This steel as-cast NiCrMo has a hardness of 25.4 HRC and a very low elongation of only 4%. Figure 1 shows the microstructure of NiCrMo alloy steel as-cast material.

Figure 1. As-Cast microstructure.

The shortage of elongation makes the steel very brittle. The brittleness comes from the existence of a sharp needle widmanstaten structure.

2.2. Heat treatment
The heat treatment process was carried out by implementing a variety of processes, as shown in Table 2.

Table 2. Heat treatment scheme.

No	Sample	Heat Treatment Process
1	AC	As-cast
2	N	Normalizing 900°C
3	NST	Normalizing 900°C - Tempering 650°C
4	NDT	Normalizing 900°C - Tempering 650°C - Tempering 705°C
5	NSQDT	Normalizing 900°C - Quenching 845° - Tempering 650°C – Tempering 705°C

The process for N sample (Normalizing) was done by heating at 900 °C for 2 hours, followed by air cooling. The NST sample (Normalizing Single Tempering) was done by heating at 900 °C for 2 hours, followed by air cooling, then tempered at 650 °C for 2 hours, followed by air-cooled. The NDT sample (Normalizing Double Tempering) was carried out by heating at 900 °C for 2 hours, followed by air cooling, then tempered twice at 650 °C and 705 °C for 2 hours and then air-cooled. The NSQDT sample (Normalizing Single Quench Double Tempering) was carried out by adding a single quench process before the double tempering process as done in the NDT sample. Single quench was done by heating the sample at 845 °C for 2 hours, followed by quench on oil media.

2.3. Mechanical testing and metallographic examination
Mechanical testing, namely tensile testing, was carried out according to E 8M standard, and Rockwell hardness testing followed E18 standard. The metallographic analysis was performed using an optical microscope; the samples etched using 3% Nital. The entire test results were analyzed for effects of the quenching process followed by double tempering treatment compared to other methods according to the various parameters determined prior to the experimental.

3. Result and discussion

3.1. Microstructures

Figure 2 shows the microstructure observation results. Figure 2 (a) N, shows the microstructure of the sample resulting from the normalizing process. There was no widmanstaten ferrite needle in the microstructure. Instead, the structure was mostly fine pearlite with higher ductility. Figure 2 (b) NST, shows a microstructure resulting from the normalizing process followed by single tempering. White ferrite structures appeared in clusters with higher ductility but lower in strength compared to the N sample. Figure 2 (c) NDT, shows the microstructure of normalized results followed by double tempering. The microstructure had finer grains, whilst the prior austenite grain boundary that appeared on the NST sample was not found, and the ferrite structure was no longer clustered but more dispersed. This makes the sample's ductility higher. Figure 2 (d) NSQDT, shows a normal microstructure followed by a single quench then continued by double tempering. The microstructure appeared as tempered martensite, which was evenly distributed and smooth. There was no longer the presence of ferrite, which made the sample stronger than the NDT sample. However, ductility may have increased because the form of temper martensite had finer grains.

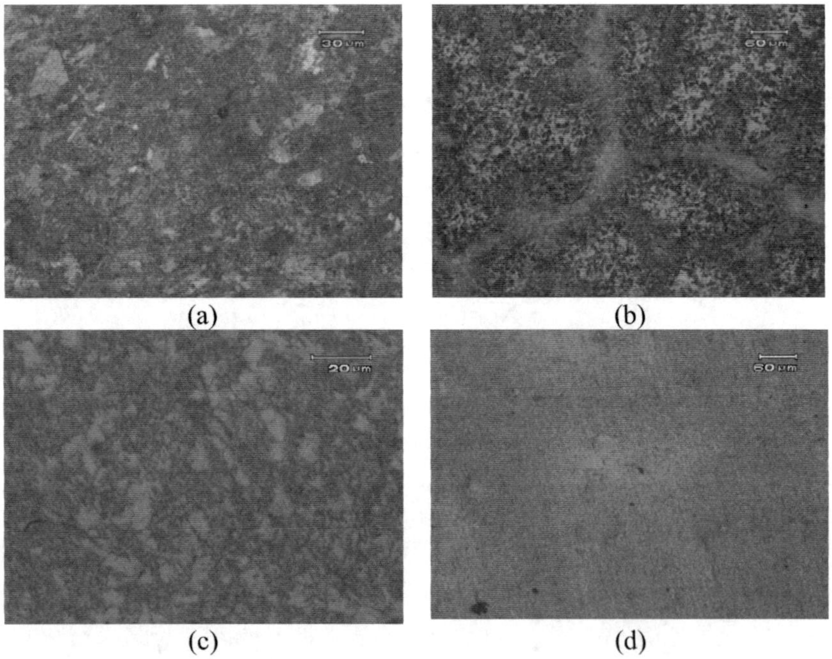

Figure 2. Microstructures (a) N (b) NST (c) NDT (d) NSQDT.

3.2. Mechanical properties

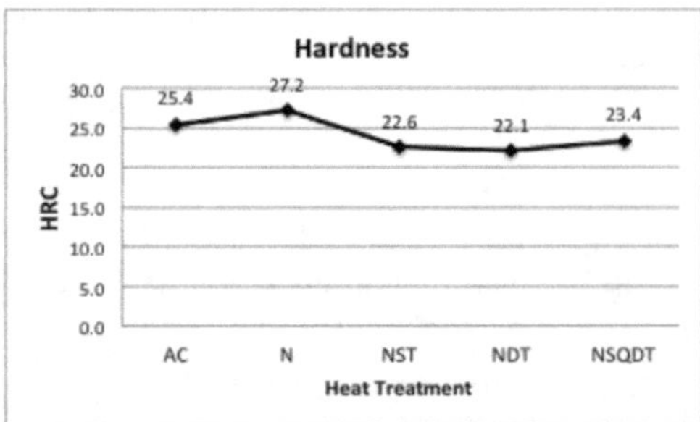

Figure 3. Hardness.

IOP Conf. Series: Materials Science and Engineering **850** (2020) 012040 doi:10.1088/1757-899X/850/1/012040

Figure 3 shows the hardness of the samples. The differences in terms of the value of hardness were insignificant. The highest value resulted from the N (Normalizing) process, with a value of 27.2 HRC, and the lowest value came from the NDT (Normalizing-Double-Tempering) process with a value of 22.1 HRC. The hardness value is ascertained based on a combination of the microstructure formed. The N sample had a microstructure of mostly fine pearlite, which was harder, whilst the presence of ferrite in the NDT sample made it softer.

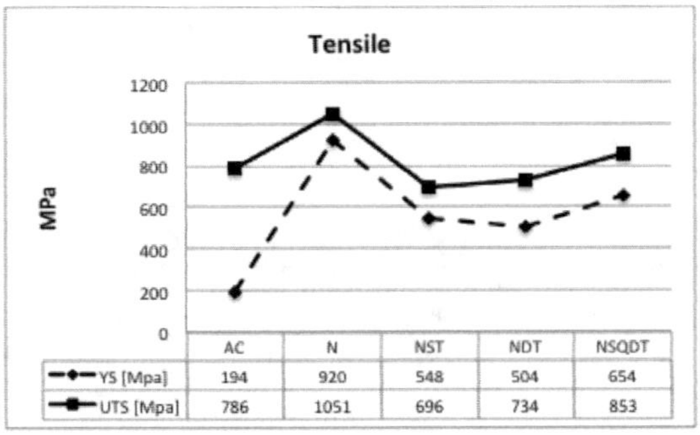

Figure 4. Tensile test.

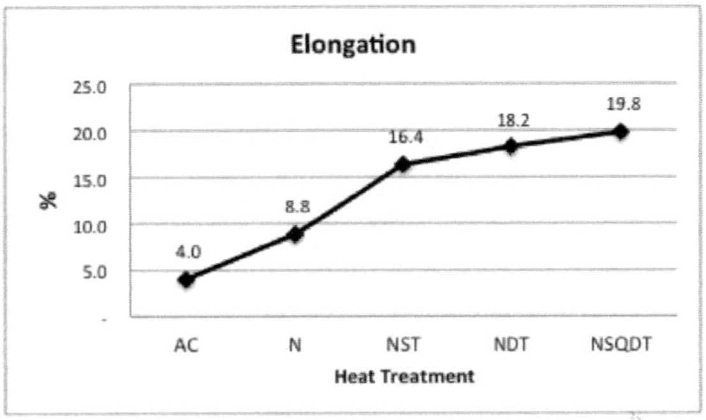

Figure 5. Elongation.

Figures 4 and 5 show the results of tensile testing data. Tensile strength in as-cast conditions was 786 MPa, with 4% elongation increasing after normalizing to 1051 MPa with elongation up to 8.8%. The result shows that the normalizing process can significantly improve the mechanical properties of cast steel by transforming the widmanstaten ferrite structure to fine pearlite [14]. This increases both elongation and strength simultaneously.

The various multiple heat treatment processes after normalizing, namely single tempering, double tempering, and quench double tempering, increase ductility for each sample. The double tempering process produced a finer grain structure [15-18]. In the double tempering process, the first tempering process reduced the hardness by transforming martensite to tempered martensite but can produce brittle secondary carbide from the steel alloy. First, tempering produced martensite from the rest austenite, resulting from the quenching process [14]. The second tempering process tempered the martensite, reducing brittleness and at the same time, producing the fine grain. The NSQDT process increased elongation up to 19.8%.

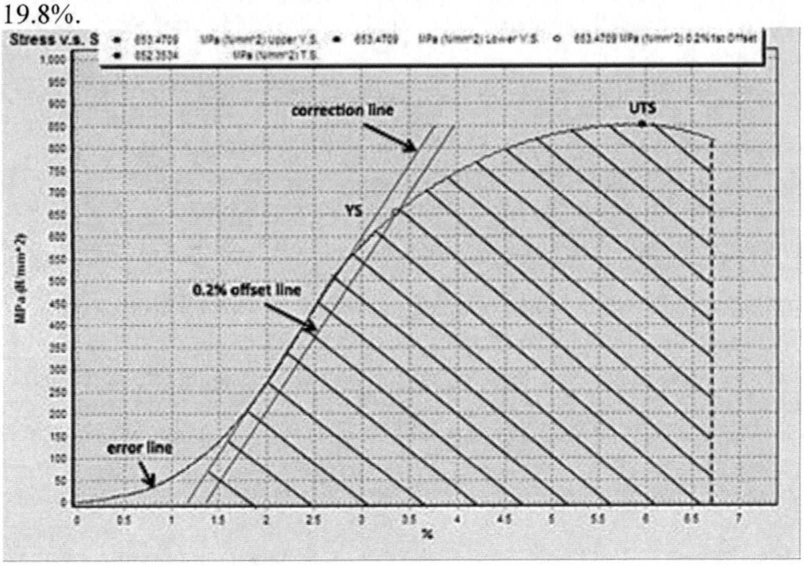

Figure 6. Tensile test curve.

Figure 6 shows the tensile test curve. The original curve from the test machine was incorrect because there was an error when recording the elongation associated with the slip in the sample gripping. Figure 6 shows the correction done. From the curve, the modulus of toughness can be calculated. This value measures the ability of the material to absorb energy, which is represented by the area under the curve. The formula to calculate the modulus of toughness is as per (1) below [19]:

$$U_T = \frac{so+su}{2} e_f \tag{1}$$

UT	Modulus of toughness (J/m^3) or (N.mm/mm^3)
S_o	Yield strength (MPa)
S_u	Ultimate tensile strength (MPa)
e_f	Elongation at fracture

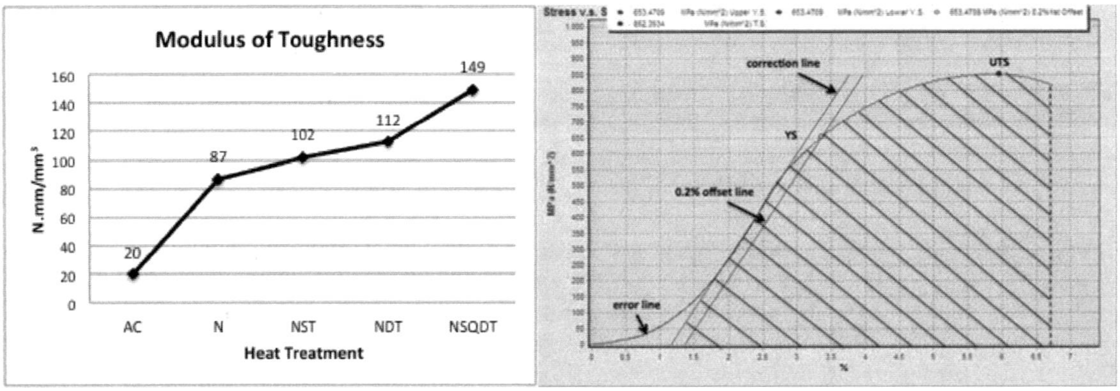

Figure 7. Modulus of toughness and tensile test curve.

Figure 7 shows the results of the calculation for the modulus of toughness for all the samples. The highest value was produced by the NSQDT sample. The NSQDT process produced CrNiMo alloy steel materials with the highest modulus of the toughness of 149 N.mm/mm^3. The toughness is a combination of the tensile strength, yield strength, and elongation. The high value was the result of the quenching process, followed by double tempering. The process formed a microstructure of tempered martensite with fine grains, free from residual "brittle" martensite and free "brittle" secondary carbide at the prior austenite grain boundary.

4. Conclusion

The results of this study showed the NSQDT (Normalizing Single Quenching Double Tempering) process had the best combination of mechanical properties with good strength, yield strength, and elongation. It contributed to the toughness of NiCrMo steel. The modulus of toughness increased by up to 745 % compared to the as-cast condition from 20 N.mm/mm^3 to 149 N.mm/mm^3.

The NSQDT process formed a microstructure of tempered martensite with fine grains, with high tensile strength and good mechanical properties. The second tempering process eliminated the martensite, after the first tempering process that made it tougher.

Acknowledgments
We would like to show our gratitude to the Kemenristekdikti and POLMAN Bandung, which have supported the research.

References

[1] Chai H F and Laird C 1987 Mechanisms of cyclic softening and cyclic creep in low carbon steel *Mater. Sci. Eng.* **93** 159–74

[2] Callister W D and Rethwisch D G 2011 *Materials science and engineering* vol 5 (John Wiley & Sons NY)
[3] Huang D H and Thomas G 1971 Structure and mechanical properties of tempered martensite and lower bainite in Fe-Ni-Mn-C steels *Metall. Trans.* **2** 1587–98
[4] Kwon H, Cha J C and Kim C H 1988 The effect of grain size on fracture behaviour in tempered martensite embrittlement for AISI 4340 steel *Mater. Sci. Eng.* **100** 121–8
[5] Meysami A H, Ghasemzadeh R, Seyedein S H and Aboutalebi M R 2010 An investigation on the microstructure and mechanical properties of direct-quenched and tempered AISI 4140 steel *Mater. Des.* **31** 1570–5
[6] Maropoulos S, Ridley N, Kechagias J and Karagiannis S 2004 Fracture toughness evaluation of a HSLA steel *Eng. Fract. Mech.* **71** 1695–704
[7] Tomita Y 2000 Development of fracture toughness of ultrahigh strength, medium carbon, low alloy steels for aerospace applications *Int. Mater. Rev.* **45** 27–37
[8] Weng Y 2009 *Ultra-fine grained steels* (Springer Science & Business Media)
[9] Sanij M H K, Banadkouki S S G, Mashreghi A R and Moshrefifar M 2012 The effect of single and double quenching and tempering heat treatments on the microstructure and mechanical properties of AISI 4140 steel *Mater. Des.* **42** 339–46
[10] Sarikaya M, Steinberg B G and Thomas G 1982 Optimization of Fe/Cr/C base structural steels for improved strength and toughness *Metall. Trans. A* **13** 2227–37
[11] Chang E, Chang C Y and Liu C D 1994 The effects of double austenitization on the mechanical properties of a 0.34 C containing low-alloy Ni-Cr-Mo-V steel *Metall. Mater. Trans. A* **25** 545–55
[12] Bandanadjaja B and Achyarsyah M 2014 Perbaikan Ketangguhan Material Baja Cor Paduan Ni-Cr-Mo Melalui Proses Tempering Ganda *Seminar Nasional Teknologi Manufaktur (STEMAN)* pp 1–4
[13] Bandanadjaja B and Idamayanti D 2017 Pengaruh Proses Tempering Ganda Terhadap Sifat Mekanik Material Baja Cor Paduan Ni-Cr-Mo *Metalurgi* **1** 29–36
[14] Thelning K-E 2013 *Steel and its heat treatment* (Butterworth-heinemann)
[15] Zhang C, Wang Q, Ren J, Li R, Wang M, Zhang F and Yan Z 2012 Effect of microstructure on the strength of 25CrMo48V martensitic steel tempered at different temperature and time *Mater. Des.* **36** 220–6
[16] Liu Y, Ye D, Yong Q, Su J, Zhao K and Jiang W 2011 Effect of heat treatment on microstructure and property of Cr13 super martensitic stainless steel *J. iron steel Res. Int.* **18** 60–6
[17] Kaijalainen A J, Suikkanen P P, Limnell T J, Karjalainen L P, Kömi J I and Porter D A 2013 Effect of austenite grain structure on the strength and toughness of direct-quenched martensite *J. Alloys Compd.* **577** S642–8
[18] Wang Q, Zhang C, Li R, Gao J, Wang M and Zhang F 2013 Characterization of the microstructures and mechanical properties of 25CrMo48V martensitic steel tempered at different times *Mater. Sci. Eng. A* **559** 130–4
[19] Davis J R 2004 Tensile testing . ASM International *Ohio, USA*

IOP Conf. Series: Materials Science and Engineering **850** (2020) 012041 doi:10.1088/1757-899X/850/1/012041

Lifetime simulation of traditional boat propellers using Ansys

T Endramawan*, A Sifa, F Dionisius and M I Syah

Politeknik Negeri Indramayu, Indramayu, Indonesia

*tito@polindra.ac.id

Abstract. In this study an age-predictive simulation of propellers used by traditional fishermen in the Indramayu region was carried out. These boats were driven by gasoline or diesel motors to drive boat propellers. This boat propeller is made of aluminum alloy with gravity casting technique. The problem is that there are many fishermen who fail on their propellers which generally occur faults on the propeller fins, the purpose of this study is to predict fatigue and age of propellers used in traditional boats, with the observation and calculation method of Finite Element Analysis (FEA) using Ansys software. The results of lifetime propeller analysis with variability in boat hull loading and boat load, with different variations between 3488N and 4788 N, obtained a lifetime for the loading did not experience a significant difference with a lifetime of 277,7 hours.

1. Introduction

Propeller is one of the driving devices used by boats, and there are still many traditional boats used in Indonesia, especially in the Indramayu area, traditional boats used by Indramayu fishermen generally use fix pitch propellers with aluminum alloy. The age of propeller use as a boat driver needs to be controlled and monitored in its use, because propeller as the main tool for boosting the boat needs to be avoided conditions of damage to the propeller due to loading on the boat, especially in the condition of fishing in the middle of the sea.

Taking into account the problem, in the loading conditions that occur, the propeller fatigue lifetime is always the main focus, because it is a model for predicting fatigue [1], it is necessary to do a lifetime simulation to predict the use of the propeller used, Lifetime prediction model Estimate the remaining cycle for the hole or space that appears. surface cracks will occur due to cyclic contact loads [1]. Finally, the cracks become large enough to produce unstable growth, need material to escape from the surface, which one cause damage [1]. According to the research bertoglio et al stated that This fatigue load analysis is very suitable for boat characterized by varied operating conditions [2]. Lifetime simulations carried out by taking into account some of the effects of boat loading and hydrostatic on propellers, lifetime simulations were carried out using the Finite Element Analysis (FEA) method using Ansys software.

IOP Conf. Series: Materials Science and Engineering **850** (2020) 012041 doi:10.1088/1757-899X/850/1/012041

2. Methods

2.1. Geometry

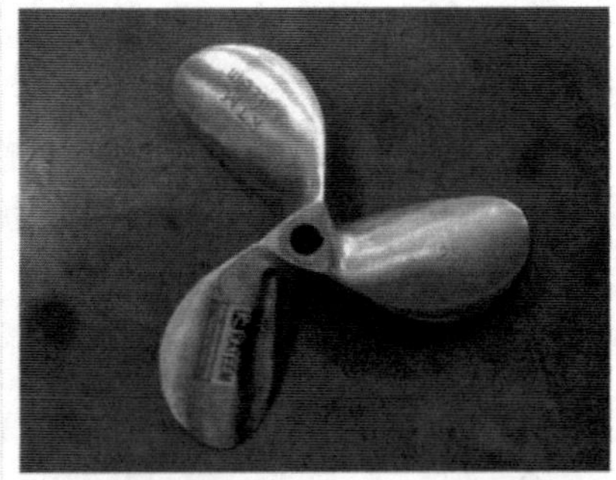

Figure 1. Fix pitch propeller.

Fix type pitch propeller depicted in figure 1 are used in the coastal area of Indramayu. Before carrying out a lifetime simulation of a traditional propeller boat, the dimensions of the propeller used by the traditional boat were carried out. The measurement process was carried out using AMATEK 3D Creator form handy scan, with the following results:

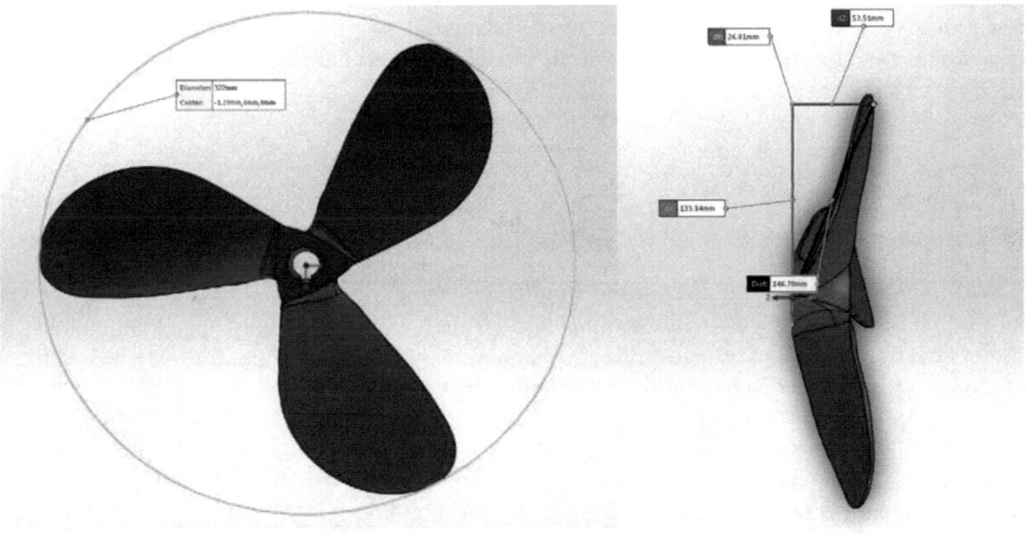

(a) (b)

Figure 2. Geometry propeller.

Figure 2 show the scan results obtained visually appear to resemble those of Figure 1. Propeller with a number of blades as much as 3, with diameter 322 mm, with a thickness of 110 mm, with a shaft diameter of 25 mm.

2.2. Material

Traditional boats use a fix pitch propeller, in this study propellers were used using aluminum alloy material with material specifications by testing compositions using the Inhouse Methode WI-4-02-22 method to determine the elemental content contained in the base metal. Metal composition testing was carried out with a Spectrometer machine with the following results:

Table 1. Chemical composition of boat propeller materials (New).

Element	Percentage (%)
Silicon (Si)	3,81
Manganese (Mn)	0,176
Chrome (Cr)	0,0338
Nickel (Ni)	0,0451
Zinc (Zn)	1,66
Copper (Cu)	1,30
Iron (Fe)	1,59
Aluminum (Al)	91,2
Magnesium (Mg)	0,0689

2.3. Parameters

2.3.1. Hydrostatic. Propeller used in traditional boats is used in water (h) 1 m, with density of sea water (ρ_h) in the amount of 1025 kg/m^3 , by using equations [3]:

$$P_H = P_0 + \rho_h . g . h \quad (1)$$

The amount of hydrostatic pressure obtained is 10045 Pa with P_o= 0 Pa, the hydrostatic pressure is used as a loading parameter in the propeller.

2.3.2. Angular velocity. The propeller speed used is rotated with the type of Dongfeng 24 HP diesel engine, with an average rotation of 2200 rpm, it can be obtained the angular velocity that occurs with the following equation:

$$\omega = \frac{2\pi}{T} \quad (2)$$

Then the large angular velocity that occurs by changing the period to be the frequency of the number of engine turns [4], obtained by the angular velocity of 230.38 rad / s.

2.3.3. Bouyancy force. As a variable in the simulation of boat loading on a propeller that serves as a driving force for traditional boats, with raw materials from teak wood weighing 700 kg without payload and with a load of 300 kg, the propeller load if it contains a load of 1000 kg, then can be calculated the size of the buoyancy force of the boat with the following equation [3];

$$F = k . \rho . g . V \quad (3)$$

The condition of the buoyancy of the boat (k) assumption 1/3, can be obtained by the amount of buoyancy force that occurs when unload is 3348 N, and the amount of buoyancy force with a load of 4788 N.

2.3.4. Initial condition

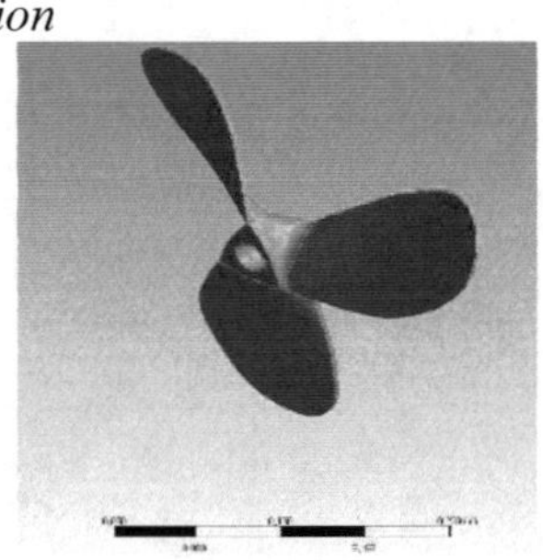

(a)

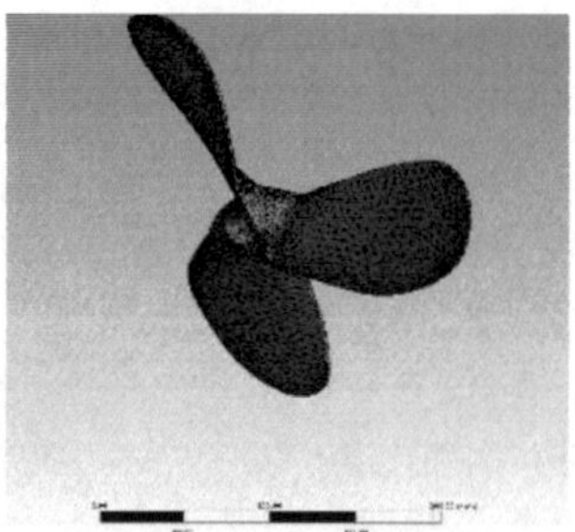

(b)

Figure 3. Initial condition (a) Propeller part, (b) Propeller mesh.

In figure 3. (a) shows the part to be simulated and in figure 3. (b) shows the part after the mesh is done with the number of nodes 25078 and the number of elements 102303 with size 2 mm, with visualization of loading as follows:

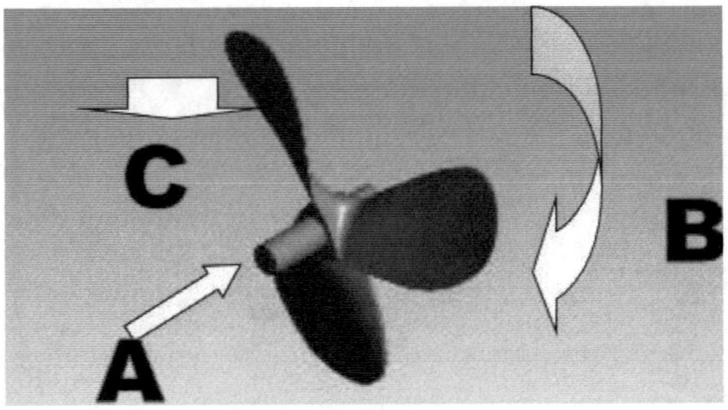

Figure 4. Propeller load.

The mechanical loads in propeller are the hydrodynamic pressure, the centrifugal force and the gravity. The non-uniform distribution of the hydrodynamic pressure on the surface of the propeller is obtained through a boundary element analysis [5]. In the figure above. Visually shows the loading on the propeller given the loading speed of sea water current (A) 3 m / s, rotation speed (B) 230.38 rad / s, and the amount of sea water pressure (C) at a depth of 1 m of 10045 Pa.

3. Results and discussion

3.1. Unload boat

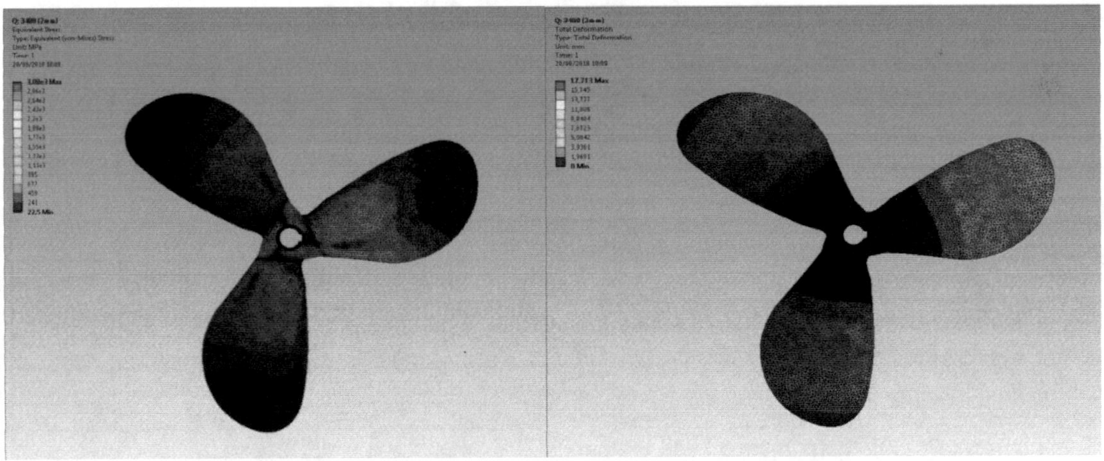

Figure 5. Results simulation (a) Stress and (b) Total deformation.

Simulation of loading on propellers with unloaded boat conditions with a calculated buoyant force of 3348 N, shown in figure 5. (a) visually shows stress with a maximum value of 3.08e3 MPa occurring on the area at the base of the propeller and stress value the minimum of 22.5 MPa, shown in the area of the propeller tip, in figure 5. (b) shows the deformation value with a minimum value of 0 mm in the propeller shaft and base area, and the maximum deformation value of 17.713 mm occurs at the end area propeller blade, according to the research bertoglio et al stated that the stress field in the blade material, varying depending on the position of the blade during one revolution [2].

IOP Conf. Series: Materials Science and Engineering **850** (2020) 012041 doi:10.1088/1757-899X/850/1/012041

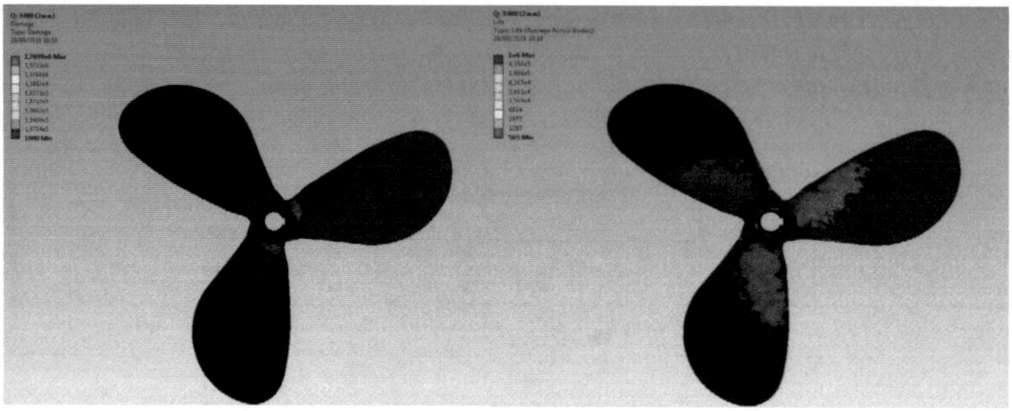

Figure 6. Result visualization (a) Damage (b) Lifetime.

Damage visualization shown in figure 6. (a) shows the area of damage at the base of the propeller blade and other areas is quite safe, while in figure 6. (b) shows a lifetime with a minimum value of 565 s, or equal to 9.4 minutes, while in maximum conditions of 1e6 s, or equivalent to 277.7 hours.

Figure 7. Safety factor.

In Figure 7. shows the value of the safety factor that occurs in the propeller shows the minimum value of 0.120 is shown in the red area and the maximum safety factor value of 15 is indicated in the blue area.

3.2. Load boat

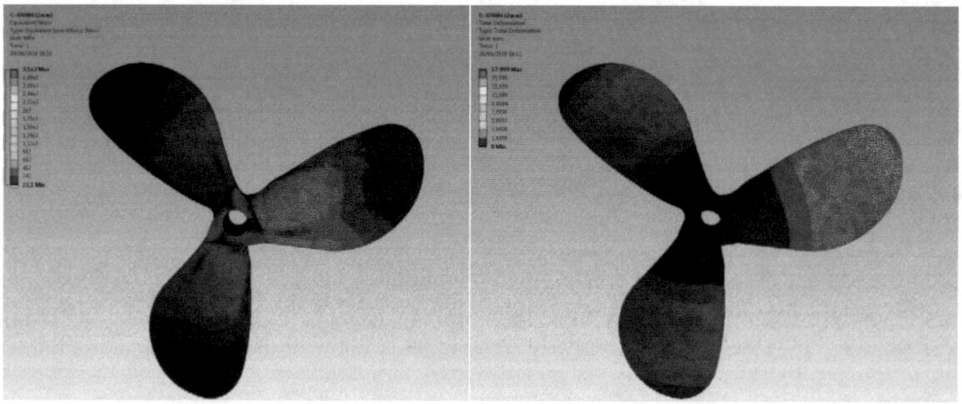

Figure 8. Results simulation (a) Stress, (b) Total deformation.

The simulation of loading on propellers with the condition of an unloaded boat with a calculated buoyant force of 4788 N, shown in figure 8. (a) visually shows stress with a maximum value of 3,1 3 MPa occurring in the area at each base of the propeller and the minimum stress value of 22.1 MPa, shown in the area of the propeller tip, in figure 8. (b) shows the deformation value with a minimum value of 0 mm in the propeller shaft and base area, and the maximum deformation value of 17.999 mm occurs in Propeller tip area.

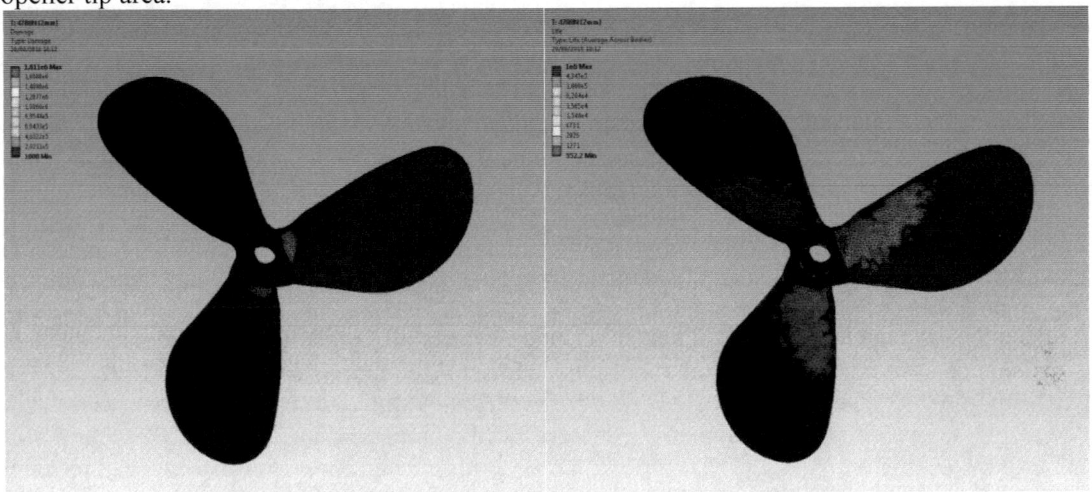

Figure 9. Result visualization (a) Damage (b) Lifetime.

Damage visualization is shown in figure 9. (a) shows the damage area at the base of the propeller blade and the other area is quite safe, the minimum damage value indicates 1000 and the maximum damage value shows 1.811e6 while in figure 9. (b) shows lifetime with a minimum value of 552, 2 s, or equal to 9.2 minutes, while at maximum conditions 1e6 s, or equal to 277.7 hours.

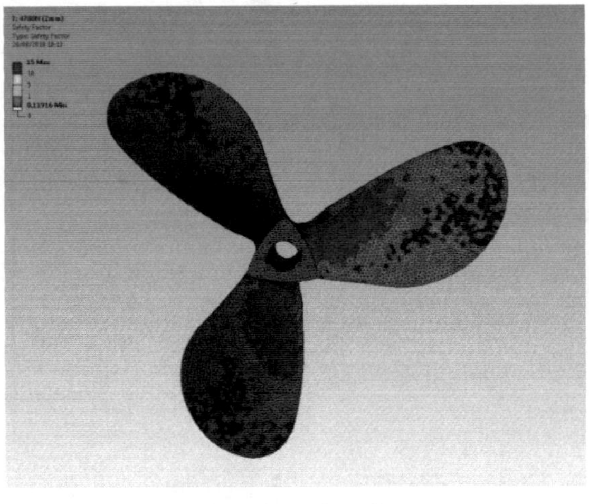

Figure 10. Safety factor.

In figure 10. shows the value of the safety factor that occurs in the propeller shows the minimum value of 0.119 is shown in the red area and the maximum safety factor value of 15 is shown in blue area.

4. Conclusion

In this study it can be concluded that in the existing loading, the cargo boat with a load weight of 3348 N has a greater buoyancy compared to a charged boat with a load weight of 4788 N, the maximum stress value on a boat without a charge of 3.03e3 MPa and a minimum stress value 22.5 MPa, the boat with a maximum stress value of 3.1e3 MPa and the minimum stress value of 22.1 MPa shows that there is an effect of charge on the stress value, and has no effect on the total deformation value, for area damage showing the same on a charged boat and not charged, and minimum lifetime on 9.4 minutes unloaded boats and on boats with 9.2 minutes, with a maximum value of 277.7 hours.

Acknowledgments
Thank you to the Head of the Indramayu State Polytechnic Research and Community Service Center for providing research funding and the Manufacturing Design Study Program - Mechanical Engineering Department - Indramayu State Polytechnic which has provided access to research facilities.

References
[1] Pritchard P J, Fox R W and McDonald A T 2010 *Introduction to fluid mechanics* (John Wiley & Sons)
[2] Yin Y, Chen Y X and Liu L 2017 Lifetime prediction for the subsurface crack propagation using three-dimensional dynamic FEA model *Mechanical Systems and Signal Processing* **87** 54-70
[3] Rao R S, Mohan S and Kumar G S 2016 Determination of Fatigue Life of Surface Propeller by Using Finite Element Analysis *International Journal of Engineering Science* **2492**
[4] Bertoglio C, Gaggero S, Rizzo C M and Viviani M 2014 Fatigue strength assessment of propellers by means of weakly coupled CFD and FEM analyses *ASME 2014 33rd International Conference on Ocean, Offshore and Arctic Engineering* American Society of Mechanical Engineers Digital Collection
[5] Yeo K B, Choong W H and Hau W Y 2014 Prediction of propeller blade stress distribution through FEA *Journal of Applied Sciences* **14**(22) 3046-3054

IOP Conf. Series: Materials Science and Engineering **850** (2020) 012042 doi:10.1088/1757-899X/850/1/012042

Traffic sign detection and recognition by improving Region of Interest (ROI) division to support driver assistance system

A Suheryadi*, A Sumarudin, A Puspaningrum and E Prastyo

Informatic Department, Politeknik Negeri Indramayu, Indramayu, Indonesia

*adisuheryadi@polindra.ac.id

Abstract. Traffic signs detection and recognition is a technology that allows automatic identification of traffic signs. In the implementation, it mostly uses a camera to capture video or real-time conditions. The Traffic detection and recognition system can give benefits to improve safety for drivers or passengers. Therefore, traffic signs detection and recognition system are one of the important parts in Driver Assistance Systems (DAS), besides that it is also used in video surveillance to improve traffic safety in the field of Intelligent Transportation System (ITS). In this paper, we propose a method for detecting and recognizing traffic signs by combining color and shape detection and divide the interest object to several regions for object recognition. The first step is image transformation, which aims to determine interest objects in the camera view. Such a process was using hue, saturation, value (HSV) color space, grey level, and black and white (BW) level. The next step is the filter and the segmentation process. The result of the filter and segmentation process then detects area objects by using region determination. The last step is the region of interest (ROI), which is used to recognizing sign traffic by the sum of absolute differences (SAD), and we try to improve area selection of ROI around nine parts. The results of the experiment proved that the proposed method was relatively good with the success rate for detecting objects about 97,38 %, and the true positive of recognition result about 81,67 %.

1. Introduction

The aim of the intelligent transportation system (ITS) development to provide convenience, comfort, and safety of road users, i.e., vehicle drivers. Safety driving is a priority that must be considered by the driver. However, one of the safety driving keys is that the driver should focus on driving a vehicle [1]. About 90% of accidents occur due to negligence in driving. One of the biggest causes of traffic accidents is driver fatigue and unable to focus on driving. In general, accidents occur due to the driver losing vigilance. Furthermore, some factors that cause accidents could be categorized, such as drivers (errors & disturbances), environment, and vehicle damage factors [2]. A driver warning system is needed to help drivers stay alert. The driver assistant system (DAS), a part of ITS, will give a warning to the driver. DAS is a technology applied to vehicles that aim to reduce driver error.

Most of the studies on DAS lead to computer vision. The study consists of object detection and conditions around the traffic lane that the driver passes, such as lane detection [3], pedestrian detection [4-6], distribution of vehicles on the road by detecting vehicles [7,8], traffic signs detection and recognition [9], etc. One of the popular topics in this study is traffic sign detection and recognition (TSDR). The traffic signal detection (TSD) process is the first process that greatly determines the success of the algorithm because the accuracy of TSD to produce the position of traffic signs has a major influence on the tracking or classification (TSR) process.

IOP Conf. Series: Materials Science and Engineering **850** (2020) 012042 doi:10.1088/1757-899X/850/1/012042

The TSD process aims to identify the Region of Interest (ROI) and find the location of traffic signs in the observation area. The fewer errors in detecting traffic signs, the better the TSD algorithm used. The challenge faced in detecting traffic signs is finding objects in complex conditions, unclear images, and speed of detection. The results of the detection process are input to the TSR process. TSR aims to classify traffic signs so that traffic signs can be identified. The thing that is often faced in recognizing traffic signs is damage to the image both due to noise, lighting, and environmental disturbances.

TSDR-related studies are using a method by colors [10-12], shape [13,14], and combinations of colors and shapes [15]. Salti et al. [7] referring to a study using color segmentation with red, green, and blue (RGB) color structures. Selcan et al. [9] proposed a TSDR approach using color and loop detection algorithms, while the classification used local and Gabor binary patterns. Huang et al. [12] proposed sign detection using the YCbCr color segmentation and Gaussian color model for segmentation. On the other hand, TSDR research with a focus on the shape of an object has come from Liang et al. [14], They proposed the TSD method by using ROI and matching template forms, with learning using SVM. The most previous study methods above proposed focus on the possibility of detecting traffic signs without considering the area of space observed.

In this paper, the stages of method that are possible to reduce the space of observation area to detecting and recognizing objects are proposed. The discussion in our paper uses an HSV color structure to segmentation processes and threshold of interest objects to reduce the observation area. Then, the result of the interest object area is processed using the development of ROI distribution and the calculate of SAD to recognizing objects.

2. Methods

The proposed method consists of two main processes that were detection and classification. The traffic signs detection process is done gradually by reducing the ROI. This procedure could reduce the observation area to the detecting process. Whereas the method of recognizing traffic sign uses SAD of traffic sign template. The process stages were image conversion, image filtering, region determination, and traffic sign recognition. The process diagram of this study is shown in Figure 1.

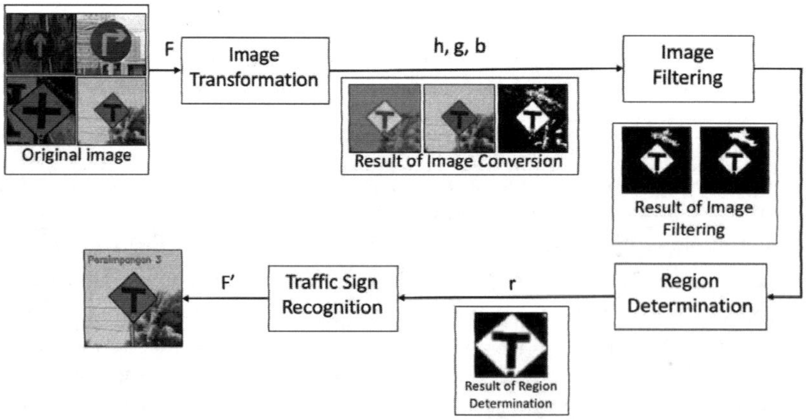

Figure 1. The stages of image processing.

2.1. Image transformation

The image transformation stage aims to convert the original image to detect candidate traffic sign areas easily. This image transformation stage is the conversion of RGB to HSV, HSV to gray, and gray to black and white conversion. The input of image transformation is the original color image (F). The input is processed in the RGB to HSV conversion process. This process will produce a new image on the HSV level (h). The RGB to HSV conversion is processed because, in the HSV color level, the traffic sign is more distinguishable from the surrounding background. The results of the RGB to HSV conversion process are then converted to a gray level (g). Finally, g is converted to a black and white level to reduce existing noise. These stages can be seen in Figure 2.

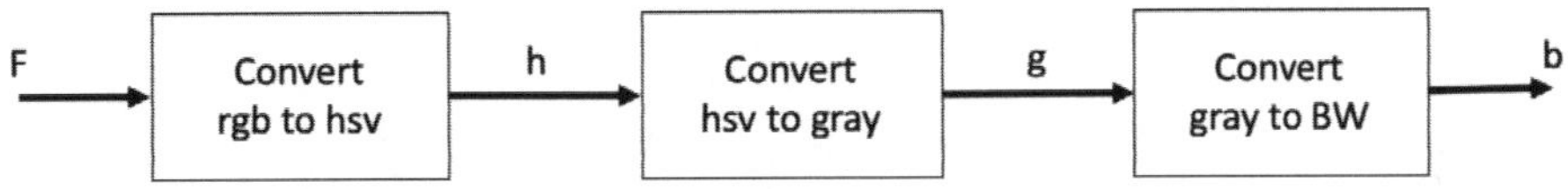

Figure 2. Proses of image transformation.

2.2. *Image filtering*

At this stage, there are two processes, i.e., selection of blob limiting the member and morphological filter. Blob selection was operated to get blob candidates, which is part of the traffic sign. The process of selection of blob is explained in equation 1.

$$m = \begin{cases} m, \sum_{i=0}^{n} m_i > th \\ 0 \end{cases} \quad (1)$$

Where m is the number of blob member candidates, and th is the threshold used to determine the limit of the number of blob members. Next is the morphological filter; this process is used to reduce noise and improve blob conditions. The result of image filtering is shown in figure 3.

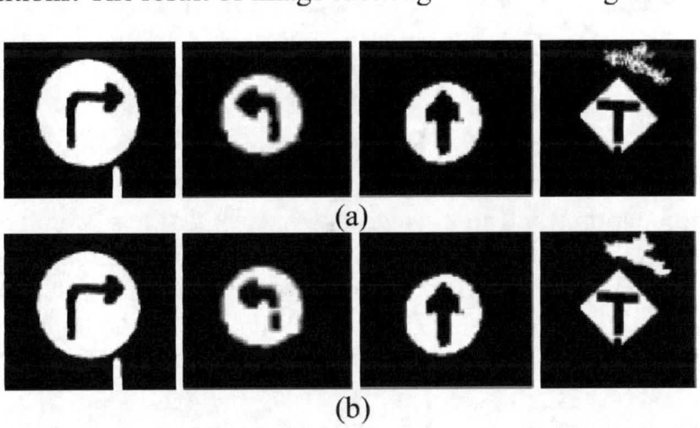

Figure 3. Image filtering: (a) limited member result; (b) morphological filter result.

2.3. *Region determination*

After obtaining a candidate blob of a traffic sign, the next process was to determine the region. It aimed to reduce the object's observation area for easing the process of matching traffic signs on the template. The determination of the observation area was done by comparing the area of the candidate of the blob. Then the candidate blob with the largest regional candidate is chosen as the observation area. The reduction of the observation area was made by looking at the corner, i.e., the width and height of the selected blob candidate. The matrix of the region determination process is shown in equations 3 and 4, and the algorithm is shown in Figure 4.

$$cO = [x, y, w, h] \quad (2)$$

$$f' = \begin{bmatrix} (x, y) & (x, x + w) \\ (y, y + h) & (x + w, y + h) \end{bmatrix} \quad (3)$$

The cO is a candidate object, and (x, y) is the position of the corner blob. (w, h) is the width and height of the blob and f' is a new observation area as a result of region determination, as shown in Figure 5.

```
Procedure contourFiltering (in m, img; out F')
Algorithm
        get contour each of blob
        mx, max_area ← 0
        for each contours
                get corner of blob (x, y, w, h)
                area ← w*h
                if (area > max_area) then
                        mx ← x, y, w, h
                        max_area ← area
        x, y, w, h = mx
        f' ← img[y:y+h, x:x+w]
```

Figure 4. The algorithm of filtering.

Figure 5. Region determination result

2.4. *Traffic sign recognition*

The traffic sign recognition process seeks to provide alternative methods that are simple but still reliable to identify traffic signs. The basic concept of traffic sign recognition is to mapping candidate observation areas of the region determination result to the template dataset that has been labeled for each subclass. The stages of the process in this TSR have divided the area into nine parts symmetrically through ROI and then determining the potential area owned to do the matching process. The simple matching method was SAD. The SAD process used equation 5 that have three-level iteration because the process should be evaluating each of the subclass template data.

$$\sum_{i=0}^{n}\sum_{j=0}^{p}\sum_{k=0}^{q}\left|tM_{i,j,k} - f'_{j,k}\right| \tag{4}$$

Where i, j, k are the index of iteration and n, o, p are the iteration's maximum limits. tM is a template dataset for each subclass, and $f\,'$ is the result of the region determination process.

3. Results and discussion

In this section, we will explain the results of experiments that have been carried out. We conduct our experiments by data form German (GTSD) and combine with traffic signs in our country to show the method performances. First, we give a brief introduction to data kinds. Second, we explain the results of our experiment.

We use eight types of traffic signs, which consist of three categories, namely three types of command signs, four types of warnings, and one type of prohibition. These traffic signs are familiar signs used in our country. We provide one template for each type of traffic signs and fifteen pieces for testing so that the total signs we test are around 120 traffic signs. The dimensions of the images used are around 150 x 150 to 300 x 300 pixels. Examples of tested traffic signs are shown in Figure 6.

IOP Conf. Series: Materials Science and Engineering **850** (2020) 012042 doi:10.1088/1757-899X/850/1/012042

Figure 6. Sample of traffic sign kinds.

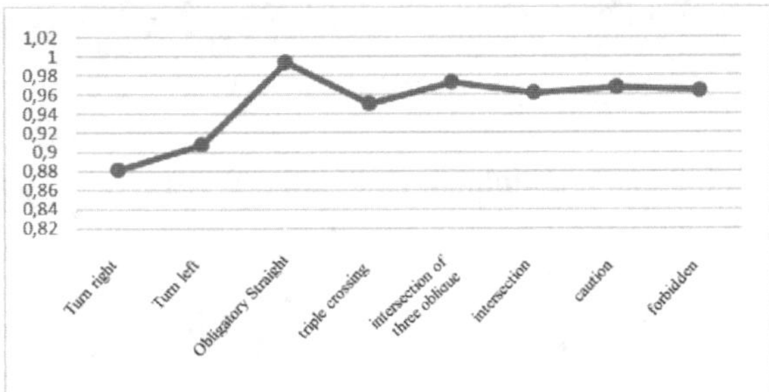

Figure 7. Graph of recall result.

In measurement testing, we use two-parameter, that are recall and accuracy. The recall is the success rate of methods for detecting objects. The recall is generated from true positive, represented by the number of true positives and false negative [16]. However, the accuracy in this test is the success rate of the method for identifying traffic signs. Our accuracy is to compare the method output with ground truth. The results of the experiment show that the average recall obtained from 120 traffic signs is 97.38%. This shows that the detection rate of this method is good enough, while the average accuracy obtained from this experiment was 81.67%.

4. Conclusion

This paper presented alternative methods for detecting and recognizing traffic signs. There were four main processes in the method, i.e. image transformation, image filtering, region determination, and traffic sign recognition. The main technique to detect traffic signs focuses on reducing the ROI and selecting areas that have potential as traffic signs. The experimental result shows that the proposed method can detecting the traffic signs as shown in 97,38 % for true positive of interest object (recall) and 81,67 % for the accuracy reached to recognize. For future work, the proposed method can be improved by machine learning to optimize traffic sign detection and recognition.

Acknowledgments

We want to thank Politeknik Negeri Indramayu for funding the research.

References

[1] Tideman M, van der Voort M C, van Arem B and Tillema F 2007 A review of lateral driver support systems *2007 IEEE Intelligent Transportation Systems Conference* pp 992–9

[2] Brookhuis K A, De Waard D and Janssen W H 2019 Behavioural impacts of advanced driver assistance systems--an overview *Eur. J. Transp. Infrastruct. Res.* **1**

[3] Neven D, De Brabandere B, Georgoulis S, Proesmans M and Van Gool L 2018 Towards end-to-end lane detection: an instance segmentation approach *2018 IEEE Intelligent Vehicles Symposium (IV)* pp 286–91

[4] Geronimo D, Lopez A M, Sappa A D and Graf T 2009 Survey of pedestrian detection for advanced driver assistance systems *IEEE Trans. Pattern Anal. Mach. Intell.* 1239–58

[5] Bertozzi M, Broggi A, Fascioli A, Graf T and Meinecke M-M 2004 Pedestrian detection for driver assistance using multiresolution infrared vision *IEEE Trans. Veh. Technol.* **53** 1666–78

[6] Suheryadi A and Nugroho H 2016 Spatio-temporal analysis for moving object detection under complex environment *2016 International Conference on Advanced Computer Science and Information Systems (ICACSIS)* pp 498–505

[7] Mukhtar A, Xia L and Tang T B 2015 Vehicle detection techniques for collision avoidance systems: A review *IEEE Trans. Intell. Transp. Syst.* **16** 2318–38

[8] Sumarudin A, Darsih D, Iryanto I and Suheryadi A 2019 Aplikasi Penghitung Kendaraan Pada Jalur Pantura Menggunakan Blob Deteksi Dan Kalman Filter *J. Appl. INFORMATICS Comput.* 8–11

[9] Liu C, Li S, Chang F and Wang Y 2019 Machine Vision based Traffic Sign Detection Methods: Review, Analyses and Perspectives *IEEE Access*

[10] Salti S, Petrelli A, Tombari F, Fioraio N and Di Stefano L 2015 Traffic sign detection via interest region extraction *Pattern Recognit.* **48** 1039–49

[11] Berkaya S K, Gunduz H, Ozsen O, Akinlar C and Gunal S 2016 On circular traffic sign detection and recognition *Expert Syst. Appl.* **48** 67–75

[12] Huang H and Hou L Y 2017 Speed limit sign detection based on Gaussian color model and template matching *Proceedings - 2017 International Conference on Vision, Image and Signal Processing, ICVISP 2017* vol 2017-Novem pp 118–22

[13] Boumediene M, Lauffenburger J-P, Daniel J, Cudel C and Ouamri A 2014 Multi-ROI association and tracking with belief functions: application to traffic sign recognition *IEEE Trans. Intell. Transp. Syst.* **15** 2470–9

[14] Liang M, Yuan M, Hu X, Li J and Liu H 2013 Traffic sign detection by ROI extraction and histogram features-based recognition *The 2013 international joint conference on Neural networks (IJCNN)* pp 1–8

[15] Yuan X, Guo J, Hao X and Chen H 2015 Traffic sign detection via graph-based ranking and segmentation algorithms *IEEE Trans. Syst. Man, Cybern. Syst.* **45** 1509–21

[16] Junker M, Hoch R and Dengel A 1999 On the evaluation of document analysis components by recall, precision, and accuracy *Proceedings of the Fifth International Conference on Document Analysis and Recognition. ICDAR'99 (Cat. No. PR00318)* pp 713–6

IOP Conf. Series: Materials Science and Engineering **850** (2020) 012043 doi:10.1088/1757-899X/850/1/012043

Performance analysis of Pico Hydro-Solar Photovoltaic Hybrid System

I W Ratnata*, S Sumarto and W S Saputra

Faculty of Technology and Vocational Education, Universitas Pendidikan Indonesia, Bandung, Indonesia

*wratnata@gmail.com

Abstract. This paper presents experimental results from the operation of the prototype of Pico Hydro-Solar Photovoltaic Hybrid System. This device includes a pico hydro power with a capacity of 600 VA, a 3x200 WP solar panel, 4x60Ah energy storage, solar charge controller, a 1500 W inverter, and a 15-200 watt variable load. In this experiment, measurements of the current flow and terminal voltage were carried out at the output of photovoltaic, picohydro, battery, and inverter at various different load values, ranging from 15 watts to 200 watts. Measurements are made during the day when the intensity of solar radiation is quite high. The measurement results show that the output voltage on each component is relatively stable at various load values, and the load energy requirements are supplied more by solar photovoltaic components in sunny weather and more by pico-hydro in cloudy weather.

1. Introduction

Indonesia is one of the countries with the highest growth rate in energy consumption in the world. The total final energy consumption (without traditional biomass) in 2018 was around 114 MTOE derived from 40% transportation, 36% industry, 16% household, 6% commercial sector and 2% other sectors [1]. This total energy consumption has increased by 3.17% from 2017 which was 110.5 MTOE [2].

The declining fossil energy production especially crude oil and the global commitment in reducing greenhouse gas emission have encouraged the government to increase the role of new and renewable energy continuously to maintain energy security and independence. As stated in Government Regulation No. 79 of 2014 on National Energy Policy, new and renewable energy mix target is at least 23% by 2025 and 31% by 2050. Indonesia has respectable potential of new and renewable energy to meet the primary energy mix target as seen in Table 1 below [1].

Table 1. Renewable energy potential [1].

Energy Source	Potential
Hydro	94.3 GW
Geothermal	28.5 GW
Bioenergy	Bio PP: 32.6 GW and Biofuel: 200 Thousand bpd
Solar energy	207.8 GWp
Wind	60.6 GW
Ocean energy	17.9 GW

The total renewable energy potential is equivalent to 442 GW for power plant, while 200 thousand bpd (barrel per day) of biofuel and biogas are used as fuel in transportation, household, commercial sector and industry. New and renewable energy (NRE) for power plant in 2018 was 8.8 GW or 14% from the total 64.5 GW of power plant capacity (fossil and non-fossil).

NRE and alternative clean energy generation technologies have attracted a lot of concern because they have several advantages such as being able to reduce dependence on fossil energy use, especially coal and oil; can provide electricity on a regionally local scale; able to utilize the potential of local energy resources, and environmentally friendly. Although Indonesia's geographical condition has the potential for renewable energy development, the transition to this energy is still a major challenge. The low NRE utilization for electricity generation is due to the high NRE power plant production cost. Thus, it is difficult to compete with fossil fuel power plant especially coal. Furthermore, the lack of support to domestic industry concerning renewable energy power plant components and the difficulty in obtaining low interest-financing have contributed to the slow development of renewable energy.

Another thing that causes NRE to be less competitive compared to conventional energy sources is uncertainty and intermittence caused by dependence on weather, and high initial costs [3]. Therefore, extensive research on renewable energy technology has been carried out throughout the world that has resulted in significant developments in renewable energy materials, reduced costs of renewable energy technologies, and increased efficiency. The results of these studies are the development of a power generation system that integrates two or more energy sources and operates them in hybrid system mode. Hybrid Renewable Energy Systems (HRES) can be a combination of one renewable energy source and one conventional energy source, or more than one renewable energy source with or without conventional energy sources, working in a standalone mode or connected to a pre-existing network [3-5]. With hybrid systems, several alternative energy sources are expected to complement each other to a certain extent and achieve higher total energy efficiency than can be obtained from a single renewable energy source.

Related to renewable energy sources, just behind the Electric Power Engineering Laboratory, Department of Electrical Engineering Education, Indonesia University of Education (UPI), there is a river flow based on measurement results having an average discharge of 0.23 m^3/s with a water fall height of 2.8 m (*head*). At 50% system efficiency [4], the potential power generated is around 3.16 kW. Based on this potential, this research will design and build a power plant in the form of a Pico Hydro Power Plant (PHP). Furthermore, this PHP will be combined with a Solar Photovoltaic (SPV) which will also be developed at the site to form a prototype of Hybrid Power Plants (HPP). Based on data from Surface meteorology and Solar Energy (SSE), NASA, in 2017, the average solar radiation energy at that location is 10.02 $kWh/m^2/day$, which is very potential to produce solar electricity.

2. Hybrid Power Plants (HPP)

Hybrid Power Plants (HPP) is a system that combines several types of power plants, in general between fossil fuel-based power plants and renewable energy-based power plants. Generally, power generating systems that are widely used for HPP are diesel generators, solar photovoltaic (SPV), microhydro, and wind power plant. HPP is a solution to overcome the fuel crisis and the lack of electricity in remote areas, small islands and also in urban areas.

The HPP can operate parallel connected to the main grid (on-grid operation), stand alone (off-grid operation) or connected to a micro grid, with other generators working in parallel (grid supporting mode). The configuration of stand-alone hybrid systems can be divided into three types consisting of serial, parallel and switched systems; while the grid-connected hybrid system consists of centralized dc-bus, centralized ac-bus and distributed ac-bus. This configuration can be seen in figure 1 and 2 [3,6].

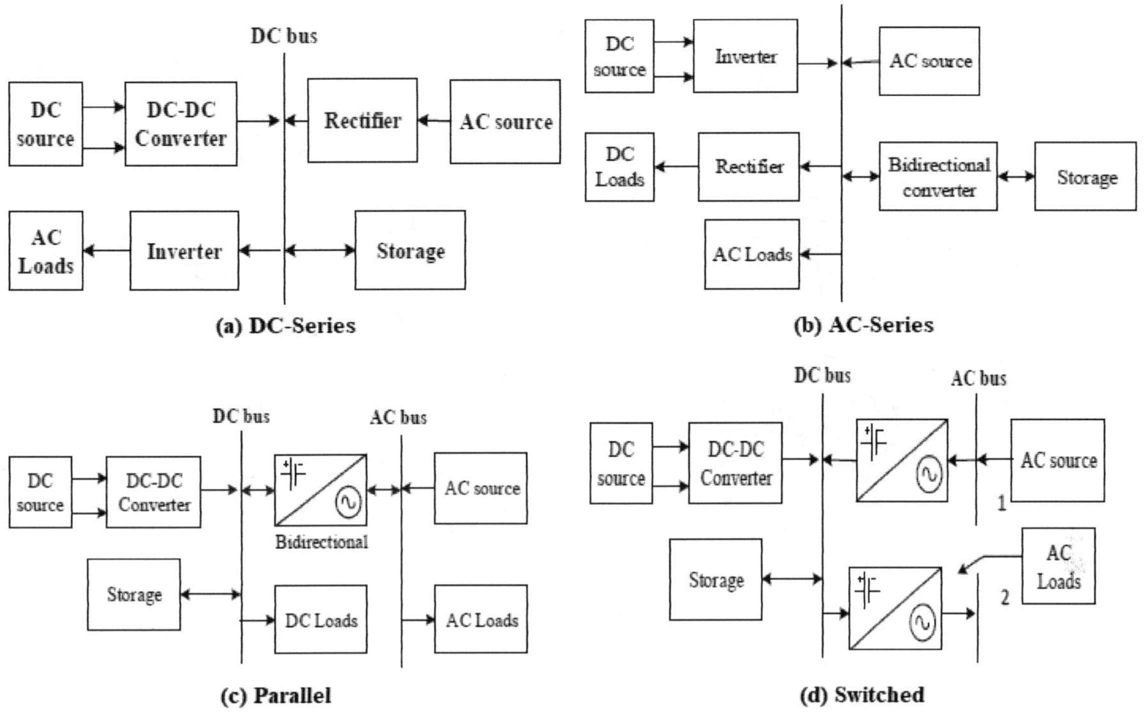

Figure 1. Various stand-alone hybrid system configurations.

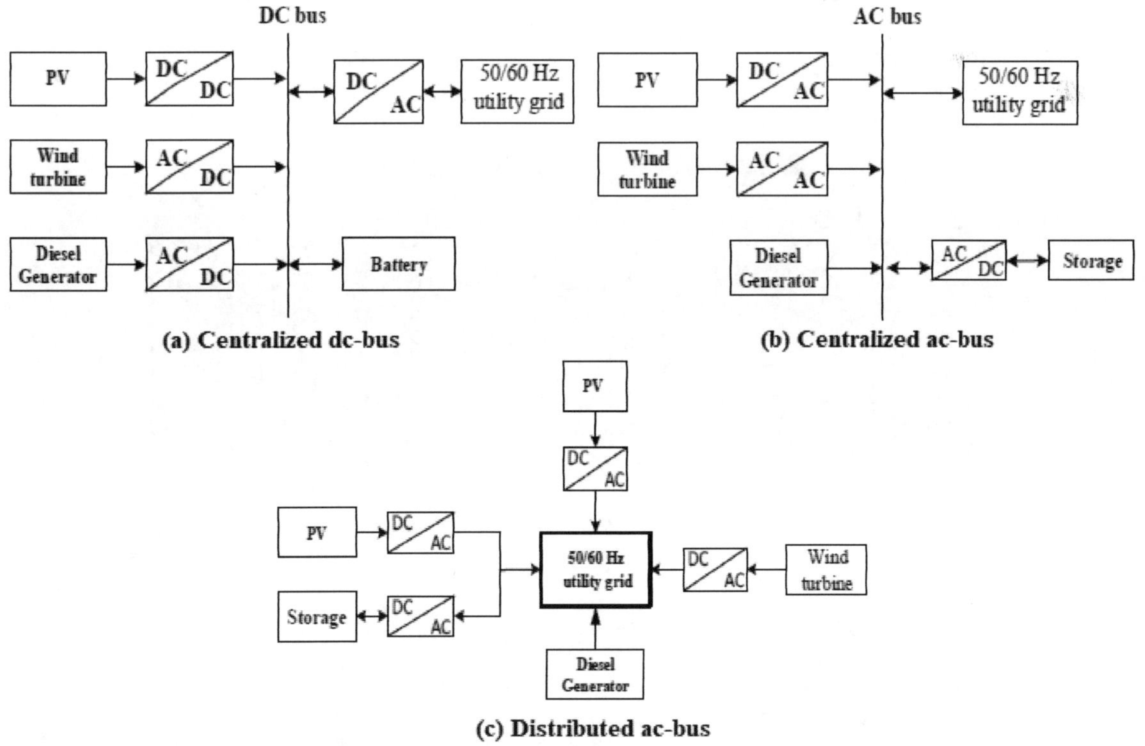

Figure 2. Various grid-connected hybrid system configurations.

3. Systems design

The system developed in this study has the configuration of stand-alone hybrid switched systems, as shown in figure 3.

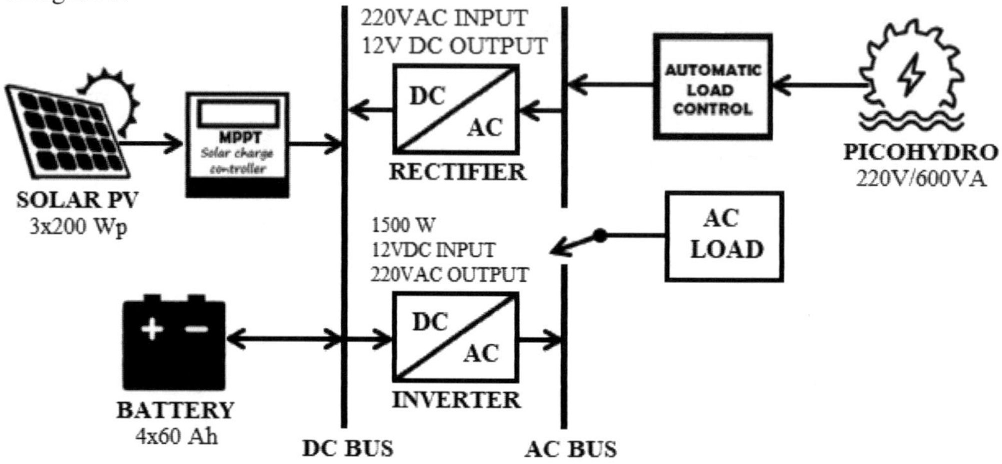

Figure 3. Pico Hydro-Solar Photovoltaic Hybrid System scheme.

The hybrid renewable energy system developed consists of the following components:

- The pico hydro power plant consists of an induction generator with a rating of 220V / 600VA, 50 Hz which is driven by Propeller Open Flume Turbine. Water flow used to drive turbines has an average flow of 0.23 m^3/s and a head of 2.8 m. This generator is equipped with an Automatic Load Controller to adjust the load balance.
- Solar panels with 3x200 Wp capacity equipped with MPPT Solar Charge Controller. This solar panel serves to convert the sun's radiant energy into electrical energy. The average solar radiation energy at the study site during 2017 can be seen in figure 4.

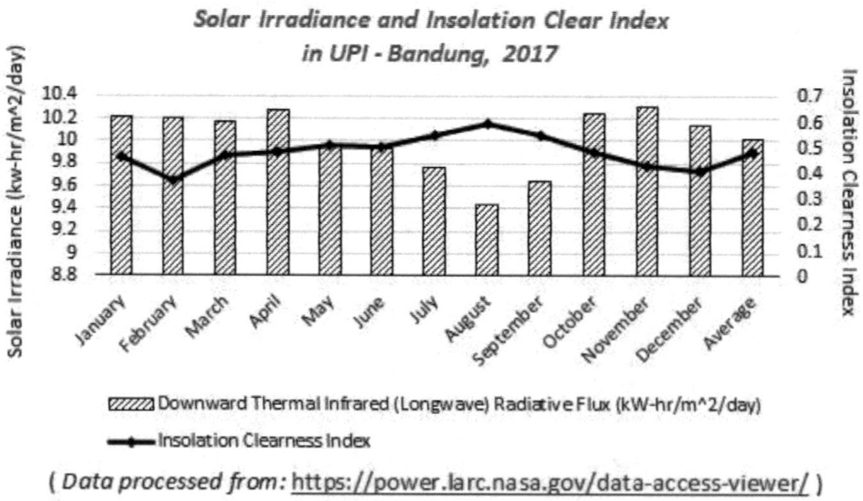

Figure 4. Solar irradiance and insolation clear index in UPI-Bandung, 2017.

- A battery with a capacity of 4x60 Ah which functions to store direct current energy.
- Rectifiers that convert AC electrical energy from generators into DC electric energy. This rectifier functions to change the voltage 220VAC to 12VDC.

- Inverter which functions to change the voltage 12VDC to 220VAC, 50 Hz, with a capacity of 1500 watts.

In this system, AC loads can be supplied in two ways: (1) directly from the AC generator; or (2) from a DC source that is converted into an AC source by an inverter. This DC source can come from batteries, from solar cells, or from AC generators that have been converted into DC by rectifiers. This research is focused on the performance of hybrid energy systems that supply AC loads through (2).

4. System performance analysis

After the installation process of the developed HPP is carried out, the next step is the testing process to find out its performance. This testing is done by measuring current and voltage in parts as shown in figure 5. In this test, measurements of current and / or voltage on solar cell output <1>, MPPT solar charge controller <2>, rectifier <4>, battery <5>, inverter input <6>, and inverter output/load input <7>.

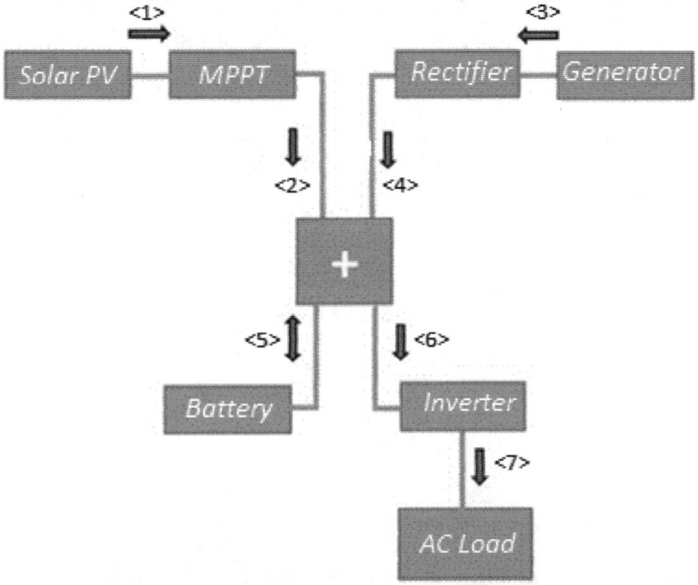

Figure 5. Measurement scheme.

After taking measurements during September and October 2019 in the time range of 10:00 to 14:00 WIB, the average value of the measured quantities was obtained. The results of these measurements are shown in table 2 for relatively sunny weather conditions and table 3 for relatively cloudy weather. Referring to Figure 5, according to Kirchoff's Law I, the DC bus applies that $I_2 + I_4 = I_5 + I_6$. By considering the electric current data in Table 1, it can be seen that in general, the value of $I_2 + I_4$ is almost the same or close to the value of $I_5 + I_6$, which means that the performance of the current divider is appropriate. However, at certain loads, this has not been fulfilled, for example at 100, 175 and 150 watt installed loads. The greatest deviation in this condition (sunny weather) occurs at a 175 watt installed load. Data on relatively cloudy weather conditions (table 3) also shows the same results, with deviations occurring at installed loads of 100, 125, and 175 watts and the deviation at 125 watts load has the greatest value.

Ideally, the difference in current flow should be zero. In the measurement data table, it can be seen that there is no zero current difference; some are close to zero, and some are greater than zero. This is possible because the reading of the measuring instrument is not done simultaneously and a sudden decline in the output of the solar cell due to changes in the intensity of solar radiation.

IOP Conf. Series: Materials Science and Engineering **850** (2020) 012043 doi:10.1088/1757-899X/850/1/012043

Table 2. Current and voltage measurement results in Pico Hydro-Solar Photovoltaic Hybrid System in relatively sunny weather conditions.

	Solar PV Output	MPPT Output		Rectifier Output		Battery		Inverter Input		Inverter Output		
Installed Load	V_1 (V)	I_2 (A)	V_2 (V)	I_4 (A)	V_4 (V)	I_5 (A)	V_5 (V)	I_6 (A)	V_6 (V)	I_7 (A)	V_7 (V)	P-load
0	37	2.45	13.27	0.03	13.83	2.42	13.30	0.47	13.21	0	239.8	0
15 W	36	2.56	13.24	0.05	13.83	0.82	13.23	1.96	13.07	0.07	239.7	18 W
25 W	36	2.98	13.21	0.06	13.82	0.41	13.19	2.71	12.98	0.11	239.6	26 W
50 W	36	4.36	13.08	0.11	13.81	-0.82	13.03	5.12	12.71	0.22	239.3	53 W
75 W	35	5.99	12.91	0.22	13.78	-1.53	12.84	7.52	12.39	0.33	239.0	78 W
100 W	34	7.86	12.73	0.32	13.75	-3.04	12.70	9.93	12.08	0.43	238.8	103 W
125 W	33	9.30	12.58	0.88	13.65	-3.35	12.46	12.66	11.75	0.55	238.3	129 W
150 W	32	10.53	12.45	1.94	13.42	-4.37	12.35	15.60	11.43	0.65	237.6	156 W
175 W	30	11.56	12.31	2.64	13.29	-6.18	12.18	18.71	11.12	0.76	237.3	179 W
200 W	32	11.67	12.32	2.32	13.36	-8.55	12.10	21.40	10.90	0.87	237.8	204 W

Note: The negative sign (-) on I5 indicates the battery is discharging, and the positive (+) sign indicates the battery is charging.

Table 3. Current and voltage measurement results in Pico Hydro-Solar Photovoltaic Hybrid System in relatively cloudy weather conditions.

	Solar PV Output	MPPT Output		Rectifier Output		Battery		Inverter Input		Inverter Output		
Installed Load	V_1 (V)	I_2 (A)	V_2 (V)	I_4 (A)	V_4 (V)	I_5 (A)	V_5 (V)	I_6 (A)	V_6 (V)	I_7 (A)	V_7 (V)	P-load
0	30	3.30	13.29	0.02	13.83	2.94	12.88	0.47	13.28	0	239.4	0
15 W	30	3.44	13.33	0.02	13.83	1.66	12.83	1.95	13.15	0.08	239.2	18 W
25 W	30	3.29	13.22	0.03	13.83	0.75	12.76	2.71	13.03	0.11	238.6	26 W
50 W	30	3.32	13.71	1.28	13.84	-0.80	12.24	5.23	12.48	0.22	238.4	53 W
75 W	30	3.32	12.51	2.47	13.36	-2.63	12.10	7.76	12.01	0.33	238.2	79 W
100 W	30	3.46	12.27	3.65	13.17	-4.48	11.79	10.29	11.64	0.44	237.7	103 W
125 W	28	3.36	12.04	5.03	12.95	-7.15	11.60	13.16	11.29	0.55	237.1	129 W
150 W	30	3.38	12.20	9.95	13.20	-3.03	11.79	15.78	11.33	0.66	236.8	156 W
175 W	28	3.90	12.12	11.15	13.01	-5.36	11.66	18.83	11.04	0.77	236.7	180 W
200 W		3.30	13.29	0.02	13.83	2.94	12.88	0.47	13.28	0	239.4	

Note: The negative sign (-) on I5 indicates the battery is discharging, and the positive (+) sign indicates the battery is charging.

Considering the output voltage, this prototype also showed quite stable performance, especially the MPPT solar charge controller (V_2) output voltage, picohydro output voltage after passing the rectifier (V_4), battery voltage (V_5), and inverter output voltage (V_7). There is a decrease in voltage on these components with an increase in load, but relatively insignificant. While the input voltage of the inverter (V_6), its value tends to decrease with the increase in installed load. The performance of this voltage can be seen in figure 6.

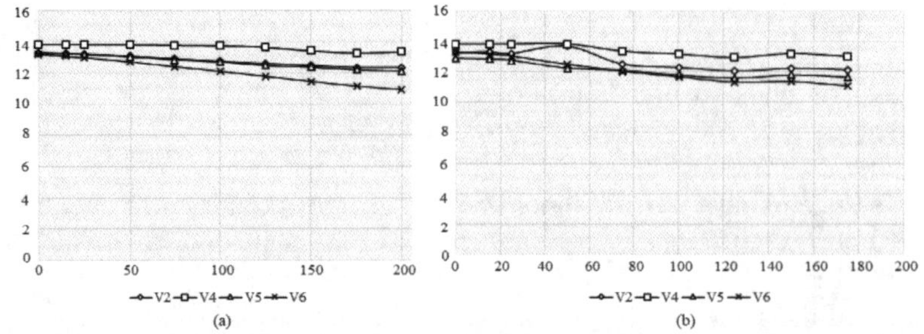

Figure 6. The output voltage as a function of the installed load for the data in: (a) Table 1 (sunny weather); (b) Table 2 (cloudy weather).

The data in Tables 2 also show that with the increase in installed load, the output current generated by solar photovoltaic and generators both increases. However, the magnitude of the current in the solar photovoltaic (I_2) is much greater than the magnitude of the current in the generator (I_4) for a relatively equal voltage ($V_2 \approx V_4$). This shows that the electrical energy needs of the load is more supplied by solar photovoltaic. For more details, consider the graph in figure 7 (a) below. A different phenomenon is shown by the data in table 2. In cloudy weather conditions, an increase in installed load will increase the generator output current, but does not increase the solar photovoltaic output. This can be seen in Figure 7 (b).

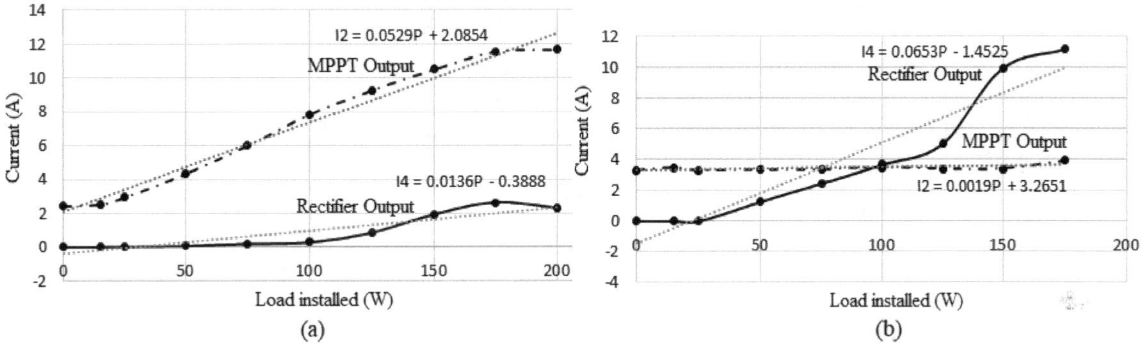

Figure 7. The output current of solar photovoltaic and generators as a function of the installed load for the data in: (a) Table 1 (sunny weather); (b) Table 2 (cloudy weather).

5. Conclusion

Based on the experimental results on the hybrid energy system the following conclusions are obtained:

- The system has a relatively stable output voltage, especially at the MPPT solar charge controller output voltage, the Pikohidro output voltage after passing through the rectifier, and the battery voltage. While the input voltage of the inverter, its value tends to decrease with the increase in installed load.
- In supplying load energy demand, in sunny weather conditions, electrical energy derived from solar photovoltaic is more dominant than electrical energy from picohydro, and in cloudy weather conditions, electrical energy derived from pikohidro is more dominant.

Acknowledgments

The author would like to thank the Head of the Electrical Power Laboratory, the Electrical Engineering Education Department, and the staff who have provided laboratory facilities for this research activity.

References

[1] Secretariat General of the National Energy Council 2019 *Indonesia Energy Outlook (IEO)* 2019 pp 1-6
[2] Secretariat General of the National Energy Council 2018 *Indonesia Energy Outlook (IEO)* 2018 p 1
[3] Ibrahim Md., Khair A and Ansari S 2015 *Int. Journal of Engineering Research and Applications.* **5** 42-48
[4] Maher P and Smith N 2001 *Pico Hydro for Village Power A Practical Manual for Schemes up to 5 kW in Hilly Areas.* Project UK Departement for International Development (DfID)
[5] Lazarov V D, Notton G, Zarkov Z, and Bochev I 2005 Hybrid power systems with renewable energy sources types, structures, trends for research and development *Proc of International Conference ELMA* 515-20
[6] Bajpai P and Dash V 2012 Hybrid renewable energy systems for power generation in stand-alone applications: A review *Renewable and Sustainable Energy Reviews* **16** 2926-2939

Oyster mushroom cultivation solution based IoT for Nova Farmers

R M Fauzi[1,*], F S Waranti[2], I K Rais[3] and Y Somantri[1]

[1] Department of Electrical Engineering Education, Universitas Pendidikan Indonesia, Jl Dr. Setiabudhi No.229, Bandung, Indonesia
[2] Department of Physics Education, Universitas Pendidikan Indonesia, Jl Dr. Setiabudhi No.229, Bandung, Indonesia
[3] Department of Mechanical Engineering Education, Universitas Pendidikan Indonesia, Jl Dr. Setiabudhi No.229, Bandung, Indonesia

*mochamadrizki@student.upi.edu

Abstract. Nova Farmers is a group of oyster mushroom farmers in Pasirhalang Village, Cikarangmulya, West Bandung, which has a vertical mushroom baglog. Farmers who are dominated by women need to climb Baglog to water Baglog at every level. The purpose of this study is to design a tool that can maximize the growth of oyster mushrooms, temperature, and humidity of the kumbung to be maintained at temperatures of 28 ° C and 80% through watering. To improve watering techniques by farmers, oyster mushroom sprinklers are automatically made based on the Internet of Things that takes into account temperature and humidity factors that can be monitored and controlled through the ABUJAT android application using Android Studio software. The sprinkler section can move vertically up and down to be able to sprinkle all baglogs at each level. This system is made using a DC PG45 motor to drive the sprinkler section which is controlled by DOIT ESP32 DevKit V1 as a microcontroller and combined with a DHT22 sensor as a temperature and humidity sensor, DS1307 RTC module, water sensor, buzzer module for watering the alarm notification, 12VDC pump for pumping water, 16x2 LCD to display sensor values, limit switches, and BTS7960 motor drivers. Based on the results of the experiment, our application can function properly and can be used by Nova Farmers.

1. Introduction

Oyster mushroom or in Latin called Pleurotus sp. is one of the high value consumption mushrooms. In addition to consumption, oyster mushrooms can be used for the treatment and cure of diseases. Oyster mushrooms have high economic value and are prospective for farmers [1-3].

In Indonesia, white oyster mushrooms are the most widely cultivated mushrooms. But the market demand for white oyster mushrooms is still not being met every day. This is due to the availability of oyster mushrooms whose numbers are below market demand [4,5]. This condition can be an opportunity for farmers or the community to open an oyster mushroom business, as did Nova Farmers, a group of women farmers in Pasirhalang Village, Cikarangmulya, West Bandung Regency. This farmer group utilizes their yard to make kumbung measuring ± 10m x 5 m.

To maximize the amount of oyster mushroom production with the width of the kumbung, the farmers took the initiative to arrange baglog on the shelf vertically so that the number of baglogs loaded in the kumbung was higher when compared horizontally. During the watering process, farmers who are

IOP Conf. Series: Materials Science and Engineering **850** (2020) 012044 doi:10.1088/1757-899X/850/1/012044

dominated by middle-aged women, water the baglog by climbing the baglog rack level by level. In this paper, we propose watering techniques carried out by farmers using ABUJAT technology with the aim of simplifying the work of oyster mushroom farmers.

2. Method

2.1. Material

In making ABUJAT technology, some electronic components and modules used consist of; DOIT ESP32 DevKit V1 as an integrated microcontroller with WiFi module for data transmission using the internet, DHT22 for air temperature and humidity sensors, RTC DS1307 module as a time data provider, water sensor as a sensor to detect watering, buzzer as a sound indicator, control relay to control 12 V water pump so that water can be sucked in and out through the sprayer, 16x4 LCD as display media for display time values, temperature and humidity and watering status, limit switches as electromechanical devices, PG45 DC motors as sprinkler actuators, and BTS7960 motor drivers to control speed and the direction of the DC motor.

2.2. Method

The implementation of this research refers to the stages of the research and development approach. This approach is used to produce certain products and test the effectiveness of a product [6]. Figure 1 shows the steps taken.

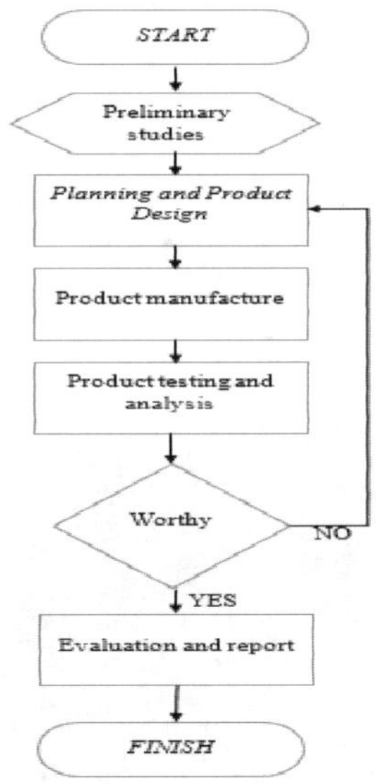

Figure 1. Implementation stages flowchart.

2.2.1. Preliminary studies. This activity is the process of finding basic knowledge of making tools. This activity is divided into literature studies and field studies. Literature study aims to find information that is relevant to the research topic as a basis for making theories sourced from the internet, books, journals, and others. While the field study aims to collect data found in the field as a reference basis for product design. In this case, we visited the Nova Kumbung Farmer Group.

2.2.2. Planning and product design. At the product planning and design stage, planning and design tools consist of mechanical system design, hardware, software, and user interfaces of the ABUJAT android application. Watering system with ABUJAT is made by compiling the work instructions of the tool through a programming language. The ABUJAT watering system consists of a manual system and an automatic system. Manual watering system is controlled through the buttons on the hardware and the

ABUJAT android application. While the automatic control of the watering system is consistent every day when the time is 8:00 a.m. and 3:00 p.m. while detecting the temperature and humidity of the kumbung. Table 1 shows the performance of our system regarding the temperature (T) and humidity (H) of the kumbung.

Table 1. References value of temperature and humidity.

Humidity	Temperature < 26°C	Temperature 26°C-28°C	Temperature >28°C
< 80%	Not Watering	Not Watering	Watering
80-90%	Not Watering	Not Watering	Not Watering
≥ 80%	Not Watering	Not Watering	Not Watering

Based on the system requirements that have been designed, then other system designs are carried out such as hardware, mechanics, and UI design of the ABUJAT android application.

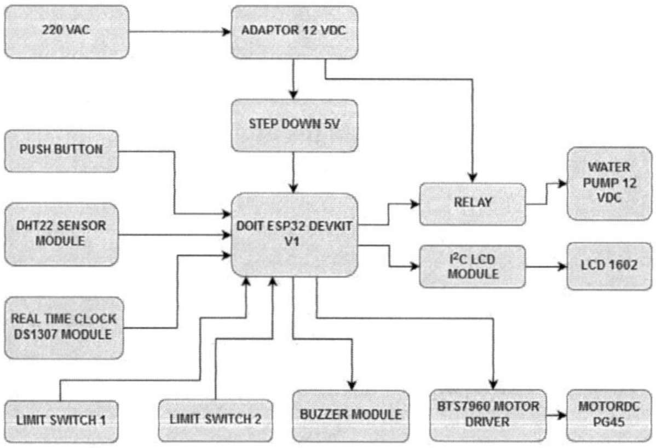

Figure 2. Block diagram system of ABUJAT.

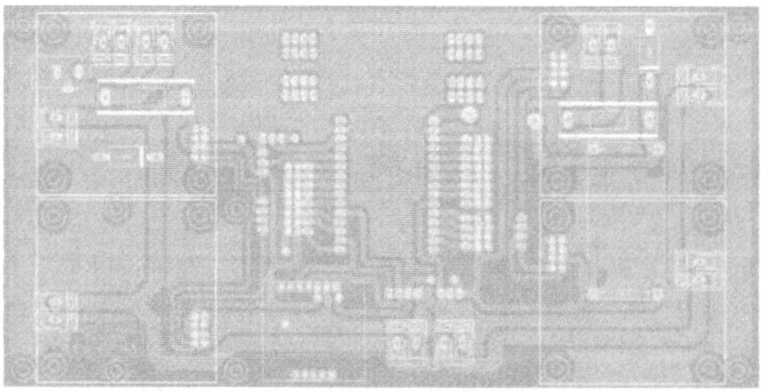

Figure 3. ABUJAT PCB layout design on EAGLE 6.4.0 application.

Figure 4. Design of ABUJAT on Solidwork application.

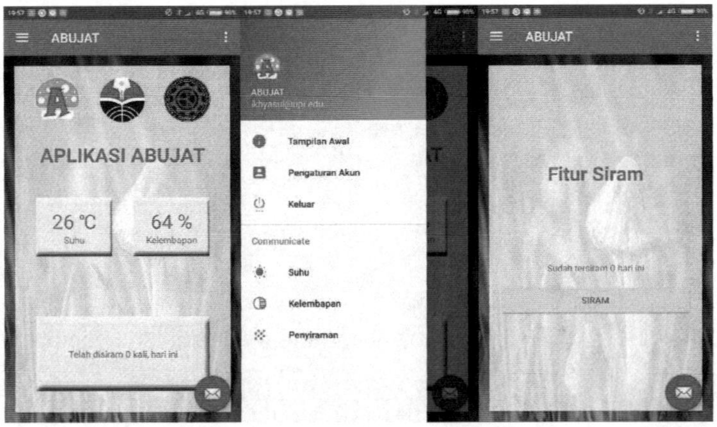

Figure 5. User Interface (UI) design of the ABUJAT android application.

2.2.3. Product manufacture. The process of making a product is based on the design that has been made and has been revised. Revision of the design is based on the evaluation results from the previous design to maximize the functionalities of the tool.

2.2.4. Product testing and analysis. Products are tested according to the automatic and manual work systems that have been designed. If the work of the tool is appropriate, the process continues to the evaluation and report stages. If it is not appropriate, the process will be reviewed back to the product design and design stage to be repaired until finally the tool works properly according to the system designed.

2.2.5. Evaluation and report. Evaluation is carried out from the beginning of the preliminary study until the product is finished. The conditions and working system of members are also evaluated to improve work efficiency. Then after that report is made.

2.3. Place and time of implementation
The process of creating and function testing of ABUJAT held in UKM Komunitas Mahasiswa Penggemar Otomasi dan Robotika Universitas Pendidikan Indonesia (KOMPOR UPI) in April-June 2019.

3. Results and discussion
Testing is carried out on the performance of the ABUJAT tool both manual and automatic systems and the performance of the ABUJAT android application. With T representing temperature and H is humidity, the test result data are shown in Table 2, Table 3, and Table 4.

Table 2. Data from the ABUJAT automatic system test results.

Day-	Automatic System			
	8 AM.	3 P.M.		
	Watering status	$T_{average}$ (°C)	$H_{average}$ (%)	Watering status
1	√	26	77	×
2	√	27	79	×
3	√	27	78	×
4	√	26	77	×
5	√	26	77	×

Table 3. Data from the manual system testing of ABUJAT with push button hardware (Hw) and android application ABUJAT (App).

Day-	Watering Status	
	Hw	App
1	√	√
2	√	√
3	√	√
4	√	√
5	√	√

Table 4. Data from the ABUJAT android application testing results.

Day-	Display LCD		Display Apl.	
	T (°C)	H (%)	T (°C)	H (%)
1	22	78	22	78
2	23	79	23	79
3	23	78	23	78
4	21	78	21	78
5	22	78	22	78

In our design, there are two operating systems namely automatic and manual systems. In an automatic system, this tool will do watering every day scheduled at 08.00 a.m. and will do the watering at 3:00 p.m. If the average temperature and humidity conditions in the mushroom house are from 08.00 a.m. obtained temperature> 28 ° C and humidity <80%, thus indicating unfavorable conditions.

The results of the automatic functioning of the ABUJAT system are shown in Table 2 for 5 days of testing. Based on trial results, the tool is functioning properly. Table 3 is a table of results of a 5 day manual trial system. This experiment was carried out through a manual watering system button found on the ABUJAT tool and on the ABUJAT android application features. In every manual system trial, ABUJAT flushes the well when the flush button on the device or application is pressed. In addition, communication testing between DOIT ESP32 DevKit V1 and the ABUJAT Android application was tested by comparing the display of air temperature and humidity values on the LCD and the ABUJAT Android application. The results of this communication test can be seen in table 4. Displays the temperature and humidity values on the LCD and the ABUJAT android application always displays the same value. Every time there is an update of the temperature and humidity value displayed on the LCD, the display of the temperature and humidity value of the Android application ABUJAT changes immediately.

4. Conclusion

Automatic and manual watering systems in our application can work well according to the planned system. Functionally this ABUJAT technology can be applied in the cultivation of oyster mushrooms. One thing we need to think about in the future is how the impact of this system on oyster mushroom production.

References

[1] Zhang R, Li X and Fadel J G 2002 Oyster mushroom cultivation with rice and wheat straw *Bioresource technology* **82**(3) 277-84
[2] Shah Z A, Ashraf M and Ishtiaq C M 2004 Comparative study on cultivation and yield performance of oyster mushroom (Pleurotus ostreatus) on different substrates (wheat straw, leaves, saw dust) *Pakistan Journal of Nutrition* **3**(3) 158-60
[3] Das N and Mukherjee M 2007 Cultivation of Pleurotus ostreatus on weed plants *Bioresource technology* **98**(14) 2723-6
[4] Djajanegara I and Masduki A 2013 Protoplast fusion between white and brown oyster mushrooms

Indonesian Journal of Agricultural Science **11**(1)16-23
[5] Sucipto E, Oktaviani R and Rizal R 2015 The effects of partnership and entrepreneurship toward business performance of oyster mushroom (pleurotusostreatus) *Indonesian Journal of Business and Entrepreneurship (IJBE)* **1**(1) 32
[6] Yulian V N 2018 Developing Teaching Materials Using Comic Media to Enhance Students' Mathematical Communication *IOP Conference Series: Materials Science and Engineering* **335** 1 012110

IOP Conf. Series: Materials Science and Engineering **850** (2020) 012045 doi:10.1088/1757-899X/850/1/012045

Synthesis of ZnO/C dots as antibacterial dental bracket

H R Taufiq*, M Mirza, S A Paramita and H Sutanto

Physic Department, Faculty Science dan Mathematics, Diponegoro University, Semarang, Indonesia

*heydarruffa@st.fisika.undip.ac.id

Abstract. Demineralization brackets has caused serious problems in orthodontic techniques effect by increasing streptococcus mutans bacteria in the mouth. To overcome this problem, antibacterial brackets based on ZnO and carbon dots materials were made using spray coating-microwave irradiation methods which have environmentally friendly, non-toxic and low-cost. C-dots with variations power of microwave are 450 W and 540 W were successfully made with microwave time of 15, 30 and 45 minutes. Optical properties showed the highest absorbance owned by ZnO/C-dots at 15 minutes and power of 450 W. Data obtained in the form of absorbance graphs were then processed using the Tauch Plot method to determine the value of the energy band gap. The shift of light absorption of ZnO/C-dots towards the higher wavelength causes a decrease in the energy band gap of 3.16 eV from before ZnO value is 3.26 eV. Therefore, ZnO/C-dots is effectively used under visible light radiation. After modification with C-dots, the surface of the bracket shows a more rugged morphology, covered by a dense and uniform ZnO/C-dots layer which is useful for protecting brackets.

1. Introduction

The second largest dental disease after dental caries is a malocclusion disease that is proven to often interfere with patients in their daily lives. This malocclusion disease occurs in patients due to normal teeth deviating between the other teeth in one arch and the teeth in the arch's jaw [1]. The most effective technique in dental practice is orthodontic techniques. The orthodontic braces technique has been commonly used in terms of health and smoothing the position of teeth. However, it was reported that this technique has a disadvantage where there have been 14 cases of mineral erosion (demineralization) from 37 cases of orthodontic treatment for 6 months (38% erosion ratio). The erosion of these minerals is caused by the bacteria *Streptococcus mutans* in the mouth that produce acidic substances [2].

Mineral erosion of dental brackets causes serious problems such as difficulty cleaning the surface of the teeth and maintaining oral hygiene. Besides, the complex form of brackets that will be a breeding ground for pathogenic bacteria (S. Mutans) causes an increase in the number of Streptococcus Mutans bacteria [3]. To overcome this, an antibacterial based photocatalyst material for dental brackets that has the advantages of non-toxic, safe and low cost is needed so that in the future it can be useful for the treatment of braces in the future.

Photocatalyst material is a material that is widely used for antibacterial applications such as ZnO and TiO_2. ZnO is one of the alternative solutions as an antibacterial material currently [4]. The antibacterial properties of ZnO in an environment in the form of water occur through the activity of photocatalyst materials that cut off the water bonds (H_2O) and produce high-energy hydroxyl (OH^+) radicals which can kill and inhibit bacterial growth [5].

Besides that, research on C-dots material has been carried out because of its low cost, environmentally friendly, strong fluorescence and high biocompatibility. C-dots material can be applied to solar cells, photocatalyst bioimaging, bacterial detection, imaging and antibacterial [6].

According to Zhang's research (2018), the modification of ZnO composites with Carbon Dots (C-dots) can improve photocatalyst efficiency and provide high antibacterial properties. C-dots in fluorescence up conversion conditions in photocatalysts are able to convert light absorption of ZnO material from UV light absorption to visible light so that modification of ZnO with C-dots is suitable for photocatalyst applications.

From the results of this study, ZnO/C-dots synthesis and deposition has been carried out on braces brackets and their application to degrade Streptococcus mutans bacteria. The manufacture of antibacterial brackets coated with ZnO/C-dots composites was carried out by microwave irradiation and spray coating methods. The ZnO/C-dots thin layer formed on the surface of the bracket will be tested for its degradation efficiency against streptococcus mutans in visible light.

2. Experimental methods

ZnO doping Cdots (ZnO/C-dots) with a variation of microwave power 50% and 60% synthesized by spray-coating technique sol-gel method on a bracket substrate. In general there are four stages, the first stage synthesis C-dots process, the second stage of made ZnO/C-dots sol-gel, the third stage of ZnO/C-dots layer deposition on the bracket substrate with spray-coating technique and the fourth stage characterization and testing photodegradation of S. mutans bacteria.

2.1. C-dots synthesis

C-dots synthesis begins by mixing 2 grams of citric acid, 4 grams of urea and 60 ml of distilled water. Then, homogenizing sample using a magnetic stirrer at 70°C for 15 minutes. The next ultrasonic process was carried out for 20 minutes. Then C-dots were synthesized using microwave at sample time 15, 30 and 45 minutes with power variations of 50% (450 W) and 60% (540 W).

2.2. Preparation of ZnO

Sol-gel ZnO preparing by Zinc acetate (Zn $(CH3COO)_2.2H_2O$) 0.5M put into isopropanol solution stirred with magnetic stirred at 60°C for 15 minutes until the solution is homogeneous, then drops Monoethanolamine (MEA) for 15 minutes at 60°C until the solution is colorless or transparent. After that, enter the C-dots doping then stir for 15 minutes at 60°C.

2.3. ZnO/C-dots coating on the bracket surface

First, C-dots were dissolved using distilled water with a concentration of 30 mg/20 ml for preparation of spray-coating techniques, before the bracket substrate was cleaned using ultrasonic for 30 minutes with DI water (deionized water). Then put on the hot plate by adjusting the temperature of 100 ^{o}C, spray-coating is done until the solution runs out deposition. After the coating is deposited on the substrate annealed for 30 minutes at atmospheric pressure. Second, coating ZnO at a bracket temperature of 450°C was carried out until the solution was depleted. Then, the sample is annealed at 350°C for 1 hour.

2.4. Characterization of ZnO/C-dots

The optical properties of ZnO and ZnO/C-dots thin films was carried out using UV-Vis Spectroscopy (UV-VIS 1240 SA) to obtain absorbance values at a wavelength of 300 nm-800 nm. To determine the surface morphology of the ZnO/C-dots layer deposition results on the bracket substrate were carried out by Scanning Electron Microscopy (SEM). Then, to test the phototalytic ability of ZnO/C-dots the degradation of Streptococcus Mutans bacteria under sunlight.

2.5. Bacterial degradation

S. mutans bacteria were prepared with a concentration of 1.0 x 10^6 CFU / mL of 3 mL. Then added ZnO/C-dots composites which have been deposited on the dental bracket. Samples exposed with visible light were optimized for 60 minutes with stirring treatment during the reaction process. Then calculated the number of S. mutans bacteria that die with the pour plate method (pouring method). Determination of antibacterial activity quantitatively by calculating the percentage of bacterial reduction.

$$\text{Antibacterial rate (\%)} = \frac{Control\ group - Experiment}{Control\ group}\ x\ 100\% \quad (1)$$

3. Results and discussion

C-Dots synthesized have brownish-yellow physical properties and light green if expose to UV laser. The synthesis results of C-Dots are shown in Figure 1. The success of the C-dots synthesis can be seen through UV-Vis spectrophotometer tests, and analysis of fluorescent ability when illuminated by UV light is shown in figure 1.

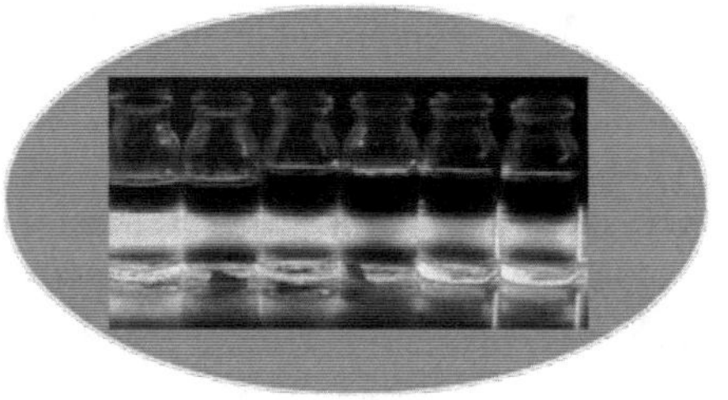

Figure 1. C-dots solution when exposed to UV light.

3.1. Optical properties

The optical properties of C-dots were tested using a UV-Vis spectrophotometer. Testing of optical Absorbance spectrum (% T) thin layer of C-dots in the wavelength range of 300-800 nm microwave power variations namely 450 W and 540 W are shown in figures 1a and 1b with a manufacturing time of 15, 30 and 45 minutes. At power 450 W the absorbance spectrum decreases due to the longer duration of the C-dots synthesis. Whereas, at 540 W the absorbance spectrum increases as the synthesis time of C-dots increases. Where C-dots have absorption edges in areas of ultraviolet light from 360-500 nm. It is seen that the C-dots that have the highest absorbance are C-dots with a power of 450 W and a time of 15 minutes. This shows that C-dots 450 W/15 minutes have good optimum absorption ability in visible light areas.

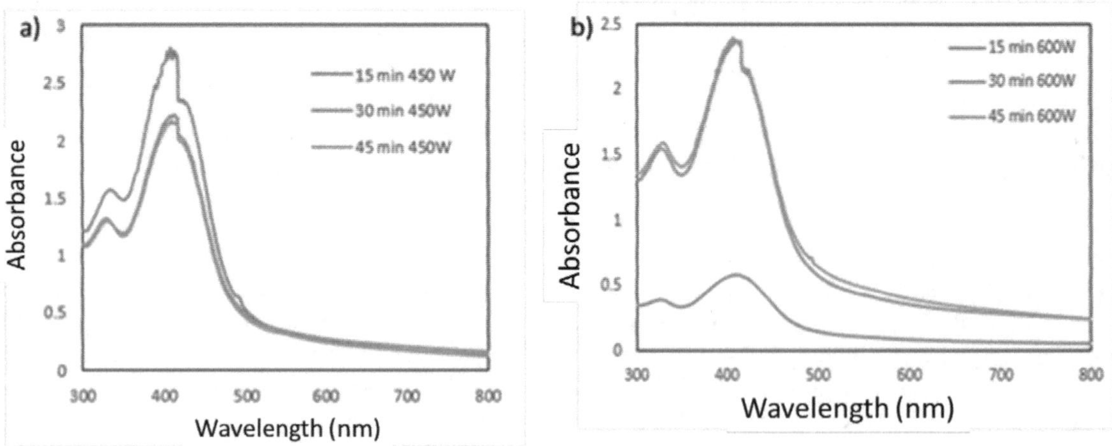

Figure 2. Absorbance spectrum of C-dots with power a) 450W and b) as a wavelength function.

The absorbance values of C-dots used to increase the ZnO photocatalyst activity. The Absorbance ZnO/C-dots spectrum is shown in Figure 2.

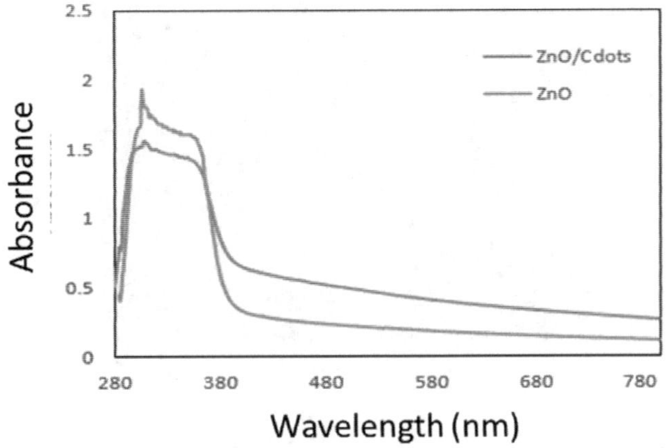

Figure 3. Spectrum absorbance ZnO/C-dots dan ZnO as a wavelength function.

All thin layers have absorption edge regions in the range of 300 nm - 380 nm which have absorption in the UV light region. In the ability to absorb the light spectrum, the absorbance of the ZnO thin layer is higher than ZnO/C-dots.

To find out the light absorption of ZnO/C-dots thin layer so that it can work in visible light or not, it is processed using the Tauch's plot method to obtain the energy band gap and its absorption area. Ploting with a tauch plot connecting between hv and $(\alpha hv)^2$ can be seen in Figure 3, then extrapolating the linear portion of the curve to the zero absorption line gives the energy band gap value for the direct transition.

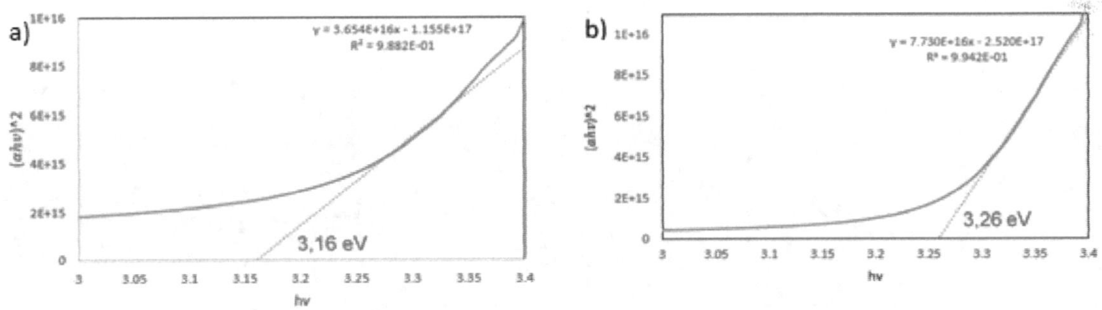

Figure 4. Determination of energy bandgap using the Tauch Plot method.

Figure 4 shows the results of the energy band gap slitting directly from the ZnO/C-dots and ZnO layers. Shifts The optical absorbance of ZnO/Cdots thin film toward higher wavelengths can be indicated by a decrease in energy band gap from 3.26 eV to 3.16 eV.

As reported (Huang et al., 2008), narrower energy bandgap can absorb more photons for electronic excitation from the valence band to the conduction band. Therefore, the effect of the C-dots on ZnO is able to increase the absorption of visible light for bacterial degradation.

3.2. Morphology ZnO/Cdots on bracket

Analysis SEM-EDX was used to determine the physical properties of the material including the microstructure of the coating and the surface morphology of the material. Meanwhile, to determine the chemical properties including the composition of the ingredients used EDX. The results of SEM ZnO/C-dots analysis on the bracket are shown in Figure 5.

Figure 5. SEM image processing 5000x magnification of ZnO/C-dots on the bracket surface.

The bracket modified with the ZnO/C-dots thin film has a yellowish surface. It is relevant that ZnO/C-dots attach tightly to the bracket surface. Besides that, before modification, the surface of the bracket looks smooth. However, after coating ZnO/C-dots, the bracket surface shows a relatively more rugged morphology, covered by uniform and dense particles that attach to the surface so that it is useful for protecting the bracket surface.

The results of EDX characterization obtained data in the form of chemical compositions of ZnO/C-dots brackets are shown in table 1.

Table 1. Composition of chemical elements contained in ZnO/C-dots.

Compound	Mass (%)	Atom (%)
C	32.83	57.04
N	5.49	8.18
O	12.57	16.40
Al	0.49	0.38
Si	0.38	0.28
Cr	7.45	0.99
Mn	1.62	0.61
Fe	28.22	10.55
Ni	2.06	0.73
Zn	8.88	2.83

3.3. Phase analysis of Zno/Cdots

X-ray diffraction method (XRD) were used for analysis phase of bracket. PanAnalytical, Type: E'xpert Pro of XRD used to get result of crystal structre of Zno and Cdots on the surface of bracket.

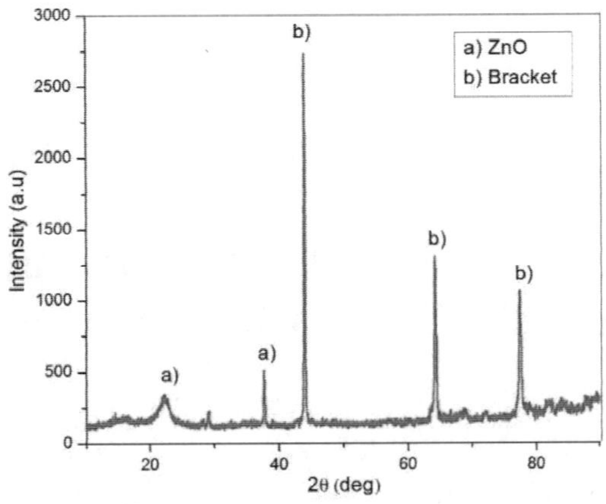

Figure 6. XRD result of bracket.

The results of XRD brackets that have been coated with ZnO/C-dots show that the diffraction peaks of the Crystal structure of Zn (002) is at 22.13° and 37.73° which matches the type of wurtzite *zinc oxide* structure. This shown that the surface of the bracket is covered by Zno crystals. On the other side the characterization of the C-dots Crystal structure is not found in the graph. XRD because there is no C-dots Crystal and low number of C-dots crystals on the bracket.

4. Conclusion

The synthesis and deposition of ZnO/C-dots on brackets was successfully made by spray coating method - microwave irradiation. The ZnO/C-dots layer has an absorption area of 300 nm - 380 nm. The effect of C-dots on ZnO will cause a shift in the absorption edge towards a higher wavelength (redshift). The shift of light absorption of ZnO/C-dots towards the high wavelength causes of a decrease in the energy band of 3.16 eV from before ZnO value is 3.26 eV. Therefore, ZnO/C-dots are effectively used under visible light radiation. After modification with C-dots, the surface of the bracket shows a more rugged morphology, covered by a uniform and uniform ZnO/C-dots is useful for protecting brackets. ZnO/C-dots brackets can be used to degrade s. mutans.

References

[1] Dentalogy 2014 *Kawat Gigi (Orthodontic Treatment)* [Online] Retrieved from: http://dentalogy.net/kawat-gigi-orthodontic-treatment/ [Accessed on: 20 November 2018]

[2] Zhang J, An X, Li X, Liao X, Nie Y and Fan Z 2018 Enhanced antibacterial properties of the bracket under natural light via decoration with ZnO/carbon quantum dots composite coating *Chemical Physics Letters* **706** 702-707

[3] Shukla C, Maurya R K, Singh V and Tijare M 2016 Evaluation of changes in Streptococcus mutanscolonies in microflora of the Indian population with fixed orthodontics appliances *Dental Research Journal* **13**(4) 309-314

[4] Tufekci E, Dixon J S, Gunsolley J C and Lindauer S J 2011 Prevalence of white spot lesions during orthodontic treatment with fixed appliances *Angle Orthod* **81** 206-210

[5] Taftiari M F, Subagio A and Nurhasanah I 2012 Antibakteri Fotokatalis TiO2: KA untuk Sterilisisasi Air yang Tercemar Bakteri Escherichia Coli *Jurnal Sains dan Matematika* **20**(1) 1-4

[6] Huang Y, Xie Y, Fan L, Li Y, Wei Y, Lin J and Wu J 2008 Synthesis and photochemical properties of La-doped HCa2Nb3O10 *International Journal of Hydrogen Energy* **33**(22) 6432–6438

IOP Conf. Series: Materials Science and Engineering **850** (2020) 012046 doi:10.1088/1757-899X/850/1/012046

Photocatalytic degradation of heavy metal in dairy milk by N-doped ZnO thin film using spray coating technique

N I Pratiwi[1,*], M I Ilsatoham[1], M Z Asror[2] and H Sutanto[1]

[1] Physics Department, Faculty of Science and Mathematics, Universitas Diponegoro, Indonesia
[2] Animal Science Department, Faculty of Animal and Agricultural Science, Universitas Diponegoro, Indonesia

*novita.i.pratiwi@gmail.com

Abstract. Zinc Oxide (ZnO) is a photocatalytic material. N-doped ZnO thin film can reduce heavy metals in milk. In the process, N-doped ZnO thin film photocatalyst is deposited on the glass by spray coating technique. N-doped ZnO thin film was characterized using UV-Vis (Ultraviolet/Visible Spectroscopy), X-Ray Diffraction (XRD) and FTIR (Fourier Transform Infrared). This study aims to investigate the reduction of heavy metals in milk due to N-doped ZnO thin film. In this study, dairy milk was obtained from farms in Central Java. Characterization results show that the band gap of N-doped ZnO thin film is 3.2 eV that N- doped ZnO thin film can function under visible light. Photocatalytic activity of prepared N-doped ZnO thin film was evaluated by the degradation of heavy metals in dairy milk. Duration of N-doped ZnO thin film degradation was found to vary from 0, 60, and 120 to 180 minutes. Heavy metal content in dairy milk is 0.5775 ppm. Addition of N-doped ZnO thin film to dairy milk reduced heavy metal content to 0.0053 ppm. In can then be inferred that heavy metal degradation using N-doped ZnO thin film is by as much as 89-99 %.

1. Introduction

Researchers are interested in the photocatalytic nature of semiconductors with promising technological applications [1]. Semiconductors are even more appealing with their superior properties, namely conductivity, magnetic and optical properties, high compatibility and sensitivity to chemical species [2]. One of the semiconductor photocatalysts is Zinc Oxide (ZnO) [3]. ZnO semiconductors have direct band gap which is useful in various applications such as photocatalysis and heavy metal degradation [4].

Zinc Oxide (ZnO) has a direct gap and width of 3.37 eV that it only works in UV light [5]. It is expected that ZnO can also work in visible light. As solar radiation is 45% visible light and 10% UV light, it is necessary to add cations (transition metals) and anions (carbon, sulphur, boron, and nitrogen) into semiconductor materials [6-7]. They are meant to improve photocatalytic activities of semiconductors.

Performance of ZnO in its application can be increased by adding metal to it. In this case, nitrogen (N) can be added to ZnO to increase its efficiency. N ion scanning acts as an acceptor in ZnO and can alter ZnO ions. Other than that, the addition of N ions can increase the absorption of light wavelengths. This is because the addition of N ions reduces energy gap in semiconductors [8-10].

Several methods for synthesizing N-doped ZnO have been developed. These include pyrolysis spray [11], pulsed laser deposition (PLD) [12], metal oxide chemical vapour deposition (MOCVD) [13], hydrothermal processes [14], wet chemical processes [15], reactive magnetron sputtering [16], high energy milling [17], pulsed laser deposition [18], and sol-gel [19,20]. The most widely used method is the sol-gel method because the material is easily prepared and it is of low cost. This research studies N-doped ZnO thin films with sol-gel method using a spray coating technique.

In its application, N-doped ZnO photocatalysts can be used to reduce heavy metal contamination. It is that known that heavy metal contamination also affects food products such as milk. Heavy metal contamination in dairy milk involves Lead (Pb), Arsenic (As), Cadmium (Cd) and Mercury (Hg) which pose threats to human health upon consumption [21]. Earlier studies have shown that heavy metal contamination in milk exceeds the Indonesian National Standard level as it is detected to contain 0.53-1.31 ppm [22] and 0.1-1.4 ppm of Pb [23], which is way above the standard level of 0.02 ppm for Pb, 0.03 ppm for Hg, and 0.1 ppm for As [24]. Dairy farmers must ensure the safety of their produce. Therefore there is a need for a method to reduce heavy metal contamination and one of the solutions is by using N-doped ZnO photocatalysts.

2. Materials and methods

2.1. Materials

The materials used in this research were all chemicals in the reagent class and were used without further purification. The precursor was Zinc acetate dihydrate (Zn $(COOCH_3)$. $2H_2O$, Isopropanol $((CH_3)_2CHOH)$. The solvent and stabilizer were monoethanolamine (C_2H_7NO). Urea (CH_4N_2O) was used as source of nitrogen. Glass was sterilized using methanol (CH_3OH), acetone (C_3H_6O) and distilled water. Meanwhile, the N-doped ZnO was deposited on the glass.

2.2. Deposition of N-doped ZnO thin film

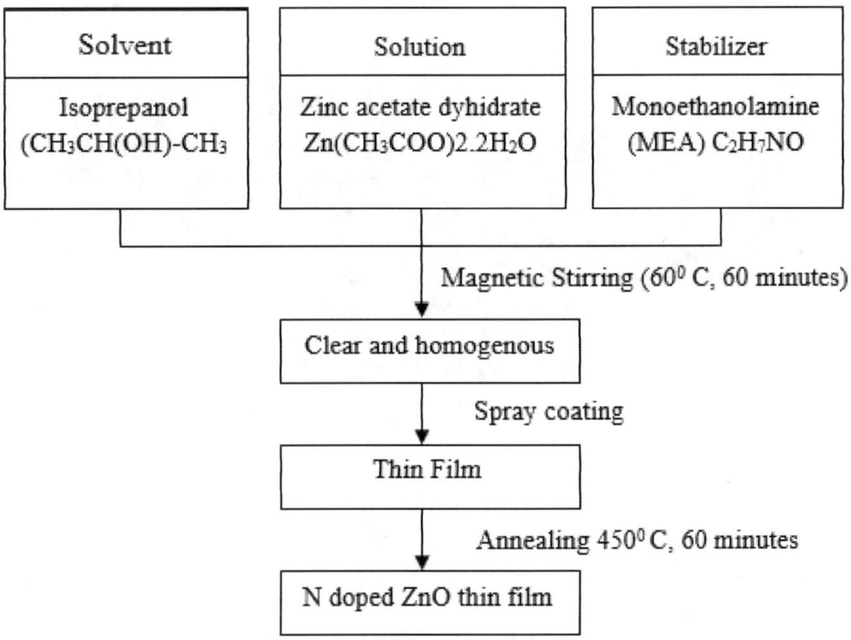

Figure 1. Schematic diagram of N-doped ZnO thin film deposition.

Deposition of N-doped ZnO thin film was performed by sol-gel method using spray coating technique. The schematic for the formation of deposition is shown in **figure 1**. The sol-gel method was started with dissolving Zinc acetate dehydrate, Zn $(COOCH_3)_2.2H_2O$ in 2-propanol (CH_3CH (OH) CH3) at 60° C to obtain a 0.3 M solution. The solution was stirred for 15 minutes. Then, monoethanolamine (MEA) was added dropped-wise into the solution while stirring for 15 minutes to obtain a clear solution of ZnO at 60° C. In case of making N-doped ZnO thin films and then nitrogen from urea (CH_4N_2O) was mixed into a clear solution Zinc acetate dehydrate to produce a weight ratio of ZnO-40% N. The solution was stirred for 30 minutes.

Glass substrate sterilization uses the Radio Corporation of America (RCA) method. This was conducted before deposition the N-doped ZnO solution on the glass substrate. Glass substrate was cleaned by acetone, and methanol for 10 min. Glass was placed in an ultrasonic cleaner because it was useful for washing dust and oil. Then, the glass was soaked in distilled water for 8 minutes and substrates were dried using a compressor gun. Before being deposited, the glass was heated at 450° C for 10 minutes. The N-doped ZnO solution which was successfully made was then sprayed onto the glass substrate for 1 hour at 450° C using spray coating technique. Furthermore, N-doped ZnO thin film was annealed at 450° C for 1 hour.

2.3. Degradation process

Experiments were carried out by comparing dairy milk without degradation of N doped ZnO thin films and milk with degradation of N-doped ZnO thin films. The schematic diagram of degradation process is shown in **figure 2**. N-doped ZnO thin films were activated in visible light. The degradation process was carried out on a laboratory scale. 200 ml of fresh milk was put into a cup and then illuminated by visible light.

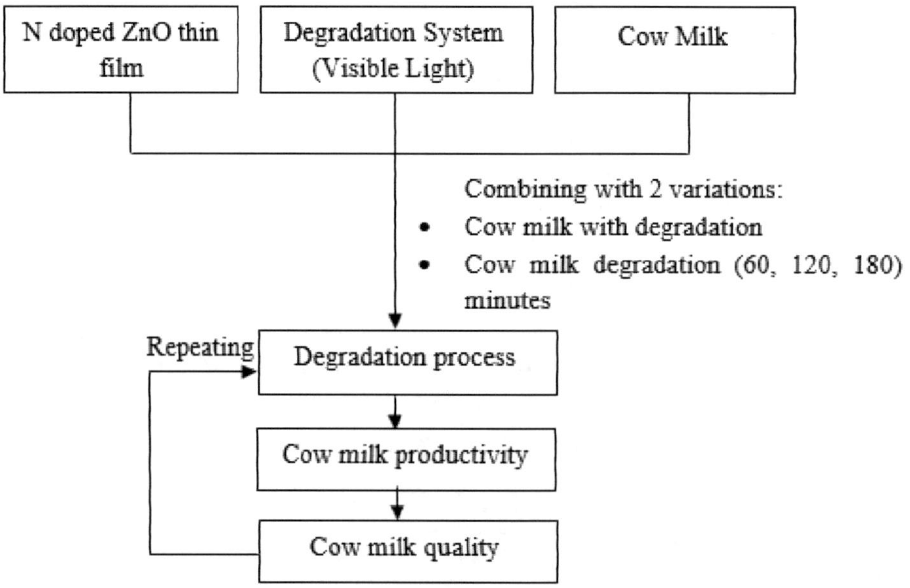

Figure 2. Schematic diagram of degradation process.

Degradation efficiency of heavy metal in milk (η) was calculated using:

$$\eta = \frac{(C_0 - C)}{C_0} x100 \qquad (1)$$

Where $C_0 C_0$ is concentration and CC is degradation concentration by photocatalysis after t minutes.

IOP Conf. Series: Materials Science and Engineering **850** (2020) 012046 doi:10.1088/1757-899X/850/1/012046

2.4. Optical characterisations of prepared thin film
X-Ray Diffraction (XRD) analysis of the samples to determine structural characterization was carried out at room temperature with a diffractometer (Philips type 7000 diffractometer meter (40 kV, 30 mA) using nickel filters and copper radiation). Optical absorbance of N doped ZnO thin films was carried out using UV-Vis spectrophotometer (Shimadzu 1240 SA). FTIR (Fourier Transform Infra-Red) analysis of N doped ZnO was carried out by PerkinElmer Spectrum IR Version 10.6.1 to see the functional groups of N-doped ZnO thin films that is, showing the relationship of N and ZnO bonds in the sample.

3. Results and discussion

3.1. UV-visible spectrophotometer analysis

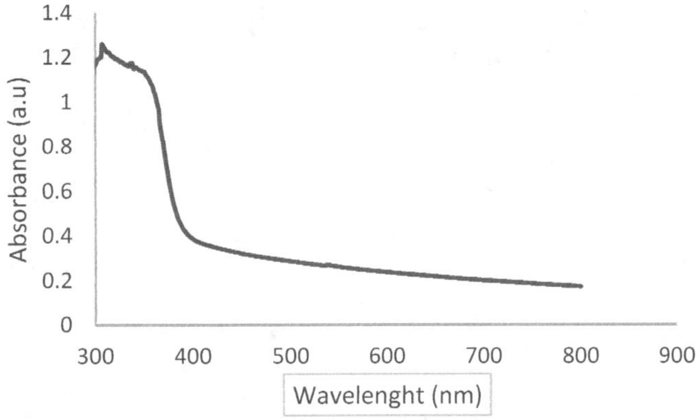

Figure 3. UV-Visible light absorption of the N-doped ZnO.

The UV-Vis (Visible Spectrophotometer) spectrophotometer analysis aimed to determine the optical properties of N-doped ZnO thin films (ZnO-40% N). Result were data in the form of a relationship between absorbance and transmission of light to materials in the absorption spectrum of wavelengths between 300-800 nm. Absorbance is the percentage of light intensity absorbed by a thin layer, while transmittance is the percentage of the intensity of light transmitted by a thin layer. Data from this sample uses transmittance, as shown in **figure 3**. Transmission in doped material (ZnO-40% N) shifts from the material value without doping transmittance (ZnO). ZnO shows UV absorption band at 310 nm. The presence of N dopant on a photocatalyst may affect optical properties of ZnO. The absorption peak shifts towards the visible light spectrum.

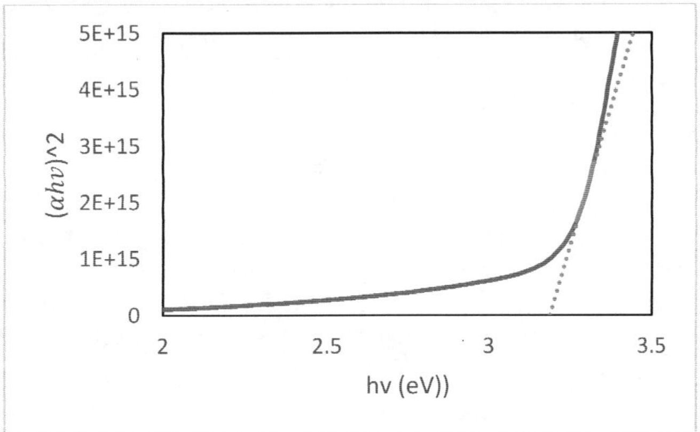

Figure 4. Bandgap of N-doped ZnO thin film.

Absorption of the ZnO region is extended to the visible light region. This is caused by the resonance absorption of plasmon N surface nanoparticles. Absorption peak of N-doped ZnO thin films at around 380 nm was observed, which must be associated with the absorption characteristics of N nanoparticles [25]. This shift shows the movement of electrons from the level of metal N to ZnO nanoparticles. Graphs generated from UV-Vis testing can be used to find thin layer energy gap values.

It can be seen in the graph of UV-Vis in **figure 3** that the peak is formed in the range of 366 nm. If plotting is done using the tauch plot method shown in **figure 4**, N-doped ZnO band gap of 3.2 eV is obtained, which can work in visible light.

3.2. XRD (X-Ray Diffractogram) analysis

XRD pattern of N-doped ZnO thin films have been obtained from the approved sol-gel method, as shown in **figure. 1**. Results of XRD pattern for N-doped ZnO thin film were confirmed using JCPDS 36-1451 for the analysis of ZnO and JCDPS No. 23-1294 for the analysis of N. In the thin films, ZnO and N were found at all peaks. Bare diffraction peaks were found to be (1 0 0), (0 0 2), (1 0 1), (1 0 2), (1 0 0), and (2 0 0). Other researchers suggest that ZnO peak patterns are similar to all samples. They conclude that the addition of both nitrogen and various syntheses did not alter the crystal structure of ZnO [20].

In a study reported by another group, it was also explained that thin films show evidence of polycrystalline structure that is hexagonal [26,27]. On this occasion, diffraction peaks (1 0 0) and (0 0 2) will be analysed for samples. The catalyst is of single-phase because no additional peaks were observed. It is observed that with N doping, the peak is expanded and the intensity is lowered. This is most likely due to a decrease in crystallinity as a result of crystal growth inhibition by urea [20] or because of O replacement with N [28]. The presence of these peaks shows that N-doped ZnO thin films have good crystal structure [29].

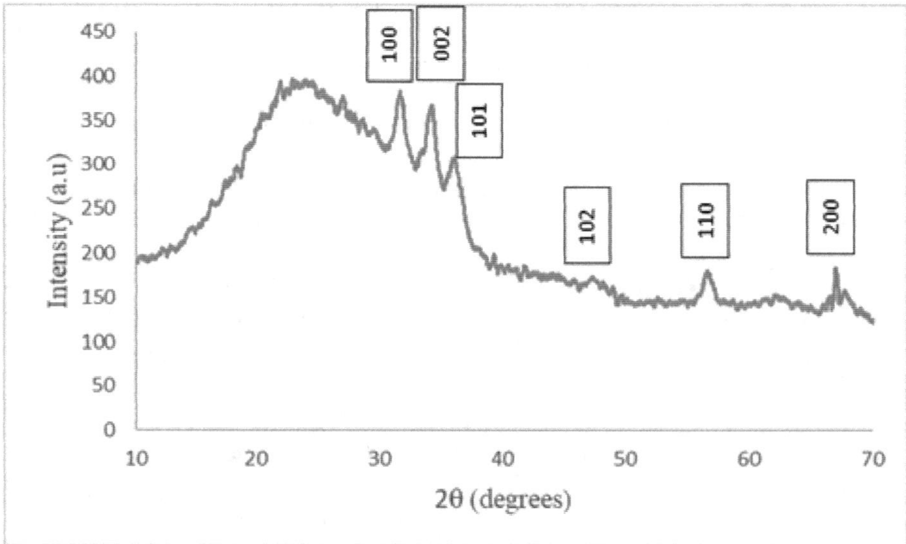

Figure 5. XRD spectrum of N-doped ZnO thin film.

In **figure 5**, 2 main peaks showing the structure of N-doped ZnO crystal was analysed and it is in line with results from an earlier study [2]. The first peak is located at 31.879 of 2 theta, which is the characteristics Wurtzite hexagonal structure of ZnO (JCDPS file No. 36-1451). The second peak is located at 33.941 of 2 theta, which is the characteristic hexagonal structure of N (JCDPS file No. 23-1294).

3.3. FTIR (Fourier Transform Infrared) analysis

FTIR analysis aims to find the functional groups produced. In this research, FTIR analysis for was performed on N-doped ZnO thin film synthesized by sol-gel method using zinc acetate dehydration and nitrogen.

.

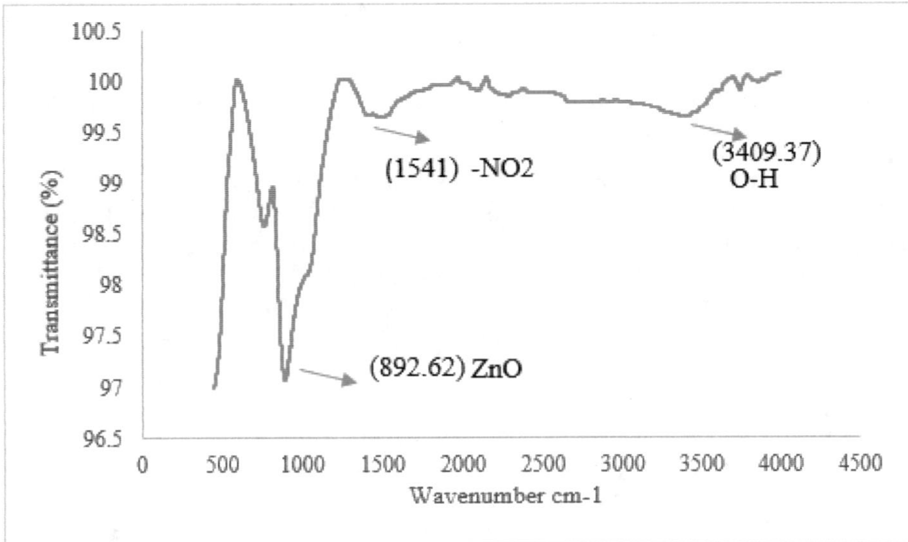

Figure 6. FTIR curve of N-doped ZnO thin film.

Resulting curve from FTIR analysis on N-doped ZnO film can is given in **figure 6.** The peak at 892.62 cm^{-1} shows ZnO [30]. In general, all metals and metal oxides have peaks at a voltage of 400-800 cm^{-1}. The peak at 1541 cm^{-1} shows N = N (nitrogen) [31]. The vibration at 3409.37 cm^{-1} shows the O-H voltage vibration [18]. Therefore, it can be concluded that N-doped ZnO thin films was successfully made as proven in the presence of ZnO and N functional groups.

3.4. Degradation of heavy metal in dairy milk using N-doped ZnO thin film

Table 1 show that the use of N-doped ZnO thin films can reduce heavy metal content in dairy milk. Decrease in the content of Pb, Hg and As in dairy milk is shown here.

Table 1. Effects of addition of N-doped ZnO thin films on dairy milk during the degradation process.

Test	SNI (2011)	Before Storage	After Storage	Unit	Method
Pb	0.02	0.7	0.01	ppm	
Hg	0.03	0.2	0.02	ppm	AAS
As	0.1	0.5	0.1	ppm	

Table 1 shows test results for the heavy metal content of dairy milk using AAS method. It can be seen that dairy milk contents before storage are lead (Pb) 0.7 ppm, mercury (Hg) 0.2 ppm and arsenic (As) 0.5 ppm, while after storage the contents are Pb 0.01 ppm, Hg 0.02 ppm and As 0.1 ppm, respectively. N doped ZnO thin films reduce the content of heavy metals Pb, Hg, As in dairy milk, hence, the test was successful. These results indicate heavy metal contents are now in agreement with SNI (Indonesian National Standard) requirements. According to SNI, fresh milk should only contain Pb of 0.02 ppm, As of 0.1 ppm and Hg of 0.03 ppm, respectively.

Table 2. Effect of varied addition time of N-doped ZnO thin films on the degradation of Pb in dairy milk.

Time (minutes)	Pb content (ppm)	Decrease Percentage (%)	SNI (2011)
0	0.5775	0	0.02
60	0.0273	91.81	0.02
120	0.0113	96.61	0.02
180	0.0053	99.08	0.02

It can be seen in Table 2 that longer addition time of N-doped ZnO thin films on Pb degradation in dairy milk shows significant results. Pb degradation at minute 0 is 0.5775 ppm. Meanwhile, Pb degradation at minute 60 is 91.81%, that is 0.0273 ppm. Pb degradation at 120 minutes is 0.0113 ppm with a reduction percentage of 96.61%. The best results were obtained at minute 180 with 0.0053 ppm, at a reduction percentage of 99.08%. This means that Pb degradation was successful. Pb level is now in agreement with SNI requirements. According to SNI, fresh milk should only contain Pb of 0.02 ppm.

3.5. Analysis of dairy milk quality

Photocatalytic activity in the degradation process is shown in **Tables 1 and 2**. Data there show reduction in metal (Pb, Hg, and As) content due to photocatalytic activity. Photocatalytic process occurs when N-doped ZnO materials are irradiated by light so that the energy is greater than the ZnO gap energy. A schematic diagram of the mechanism of heavy metal degradation reactions on the surface of N-doped ZnO nanoparticles is shown in **Figure 7.**

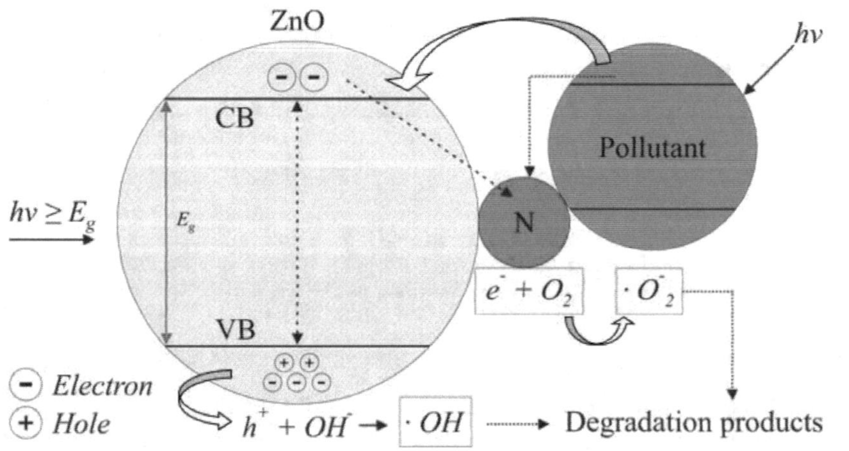

Figure 7. Schematic diagram of the mechanism of heavy metal degradation reactions on the surface of N-doped ZnO nanoparticles.

The working principle of a photocatalyst is the activation of semiconductors by the energy of photons that produce electron-hole pairs (e^- / h^+) followed by the formation of radical ions. Electrons and holes then react with dissolved oxygen and OH superoxide (O_2*-) and hydroxyl radicals (OH *) [32]. Several recent studies have proven that ZnO has high efficiency because of its ability to produce H_2O_2 more efficiently than TiO_2. [33].

ZnO is a p-type semiconductor material. P-type semiconductors have more whole concentrations than n-type semiconductors. Thus, ZnO holes are denser and hydroxyl formation is more radical (OH*). Viewed from the number of valence electrons, the OH reactivity is higher than O_2*-. This explains that

OH* lacks one electron while O_2*- has one extra electron and this makes OH* more likely to reach equilibrium [34].

The addition of N to ZnO results in better load separation. The presence of N ions increases photocatalyst activity because it traps photoelectrons that react with the surface of absorbed oxygen molecules [35]. Radical hydroxyll that has been produced reacts with organic molecules and breaks down organic compounds into other intermediate compounds which in turn, undergo further reactions. Excited electrons react with oxygen to form superoxide anion species. Superoxide anions react with compounds that break down organic molecules. Interactions between dairy milk and compounds produced during degradation process degrade heavy metal content in dairy milk.

4. Conclusion

Photocatalytic degradation of heavy metals (Pb, Hg and As) in dairy milk by N-doped ZnO thin film using spray coating technique was successfully carried out. A systematic study of the effect of N-doped ZnO content on the optical properties of thin films, the morphological of thin films, functional groups in thin films, time variations in the duration of Pb heavy metal degradation and photocatalytic activity of thin films were successfully carried out. In principle, ZnO can be activated in UV light without doping. The band gap of N-doped ZnO thin film becomes small so that it can be activated using visible light. Addition of N to ZnO also increases photocatalyst activity in degrading heavy metals (Pb, Hg, As) content in fresh dairy milk. After the addition of N-doped ZnO thin film to dairy milk, there is a decrease in the content of heavy metals by 0.0053 ppm, which is an 89-99 % reduction. These experimental results of Pb heavy metal degradation in dairy milk are satisfactory they are in agreement with SNI requirements.

Acknowledgments

The authors would like to thank the Ministry of Research, Technology and Higher Education of the Republic of Indonesia for funding this research.

References

[1] Serpone N and Emeline A V 2012 Semiconductor photocatalysis past, present and future outlook *J. Phys. Chem. Lett.* **3**(2012) 673-677

[2] Gionco C, Debora F, Calza P and Paganini M C 2016 Synthesis, Characterization, and Photocatalytic Tests of N-Doped Zinc Oxide: A New Interesting Photocatalyst *Journal of Nanomaterials* **2016** 1-7

[3] Sutanto H, Nurhasanah I and Hidayanto E 2015 Deposition of Ag2~6mol%-doped ZnO photocatalyst thin films by thermal spray coating method for E. colibacteria degradation *Materials Science Forum* **827** 3–6

[4] Poongodi G, Kumar R M and Jayavel R 2015 Structural, optical and visible light photocatalytic properties of nanocrystalline Nd doped ZnO thin films prepared by spin coating method *Ceramics International* **41**(3) 4169–4175

[5] Lamba R, Umar A, Mehta S K and Kansal S K 2015 Well-crystalline porous ZnO-SnO_2 nanosheets: an effective visiblelight driven photocatalyst and highly smart sensor material *Talanta* **131**(2015) 490-498

[6] Anandan S, Ohashi N and Miyauchi M 2010 ZnO-basedvisiblelight photocatalyst: band-gap engineering and multi-electron reduction by co-catalyst *Applied Catalysis B: Environmental* **100**(3-4) 502–509

[7] Nair M G, Nirmala M, Rekha K and Anukaliani A 2011 Structural, optical, photocatalytic and antibacterial activity of ZnO and Co-doped ZnO nanoparticles *Materials Letters* **65**(12) 1797–1800

[8] Bian H Q, Ma S Y, Zhang Z M, Gao J M and Zhu H B 2014 Microstructure and Raman scattering of Ag-doping ZnO films deposited on buffer layers *Journal of Crystal Growth* **394** 132–136

IOP Conf. Series: Materials Science and Engineering **850** (2020) 012046 doi:10.1088/1757-899X/850/1/012046

[9] Zhang Y, Zhang Z, Lin B, Fu Z and Xu J 2005 Effects of Ag doping on the photoluminescence of ZnO films grown on Si substrates *The Journal of Physical Chemistry B* **109**(41) 19200–19203
[10] Fan J and Freer R 1995 The roles played by Ag and Al dopants in controlling the electrical properties of ZnO varistors *Journal of Applied Physics* **77**(9) 4795–4800
[11] Zhao J L, Li X M, Bian J M, Yu W D and Zhang C Y 2005 Growth of nitrogen-doped p-type ZnO films by spray pyrolysis and their electrical and optical properties *J. Cryst. Growth* **280**(2005) 495–501
[12] Li Q, Wang Y, Liu J, Kong W and Ye B 2014 Structural and magnetic properties in Mn-doped ZnO films prepared by pulsed-laser deposition *Applied Surface Science* **289** 42–46
[13] Biethan J P, Sirkeli V P, Considine L, Nedeoglo D D, Pavlidis D and Hartnagel H L 2012 Photoluminescence study of ZnO nanostructures grown on silicon by MOCVD *Materials Science and Engineering: B* **177**(8) 594-599
[14] Gautam U K, Panchakarla L S, Dierre B, Fang X, Bando Y, Sekiguchi T ... and Rao C N R 2009 Solvothermal Synthesis, Cathodoluminescence, and Field-Emission Properties of Pure and N-Doped ZnO Nanobullets *Advanced Functional Materials* **19**(1) 131-140
[15] Bhirud A P, Sathaye S D, Waichal R P, Nikam L K and Kale B B 2012 An eco-friendly, highly stable and efficient nanostructured p-type N-doped ZnO photocatalyst for environmentally benign solar hydrogen production *Green Chemistry* **14**(10) 2790-2798
[16] Nakano Y, Morikawa T, Ohwaki T and Taga Y 2005 Deep-level characterization of N-doped ZnO films prepared by reactive magnetron sputtering *Applied Physics Letters* **87**(23) 232104
[17] Lu J, Zhang Q, Wang J, Saito F and Uchida M 2006 Synthesis of N-Doped ZnO by grinding and subsequent heating ZnO-urea mixture *Powder technology* **162**(1) 33-37
[18] Duclère J R, Novotny M, Meaney A, O'Haire R, McGlynn E, Henry M O and Mosnier J P 2005 Properties of Li-, P-and N-doped ZnO thin films prepared by pulsed laser deposition *Superlattices and Microstructures* **38**(4-6) 397-405
[19] Sutanto H, Nurhasanah I and Hadiyanto H 2014 Effect of Mn concentration on magnetic and structural properties of GaN: Mn deposited on silicon substrate using chemical solution deposition method *Revista Romana De Materiale-Romanian Journal Of Materials* **44**(3) 298-303
[20] Qin H, Li W, Xia Y and He T 2011 Photocatalytic activity of heterostructures based on ZnO and N-doped ZnO *ACS applied materials & interfaces* **3**(8) 3152-3156
[21] Brown G E and Parks G A 2001 Sorption of trace elements on mineral surfaces: Modern perspectives from spectroscopic studies, and comments on sorption in the marine environment *Int. Geol. Rev.* **43** 963–1073
[22] Salundik S, Suryahadi S, Mansjoer S S, Sopandie D and Ridwan W 2012 Cemaran timbal (Pb) dan arsen (As) pada susu sapi perah yang diberi pakan limbah organik pasar di peternakan sapi perah Kebon Pedes Bogor *Jurnal Peternakan Indonesia (Indonesian Journal of Animal Science)* **14**(1) 308-318
[23] Harlia E, Rahmah K N and Suryanto D 2018 Food safety of milk and dairy product of dairy cattle from heavy metal contamination *IOP Conference Series: Earth and Environmental Science* **102**(1) 012050)
[24] National Standardization Agency 2011 *Indonesian National Standard (SNI): SNI Number 3141.1: 2011. Fresh Milk - Part 1: Cow* (Jakarta: National Standardization Council)
[25] Wang T, Jiao Z, Chen T, Li Y, Ren W, Lin S, ... and Bi Y 2013 Vertically aligned ZnO nanowire arrays tip-grafted with silver nanoparticles for photoelectrochemical applications *Nanoscale*, **5**(16) 7552-7557
[26] File P D 1967 Joint committee on powder diffraction standards (Philadelphia, Pa: ASTM) 9-185
[27] Chen X and Mao S S 2007 Titanium dioxide nanomaterials: synthesis, properties, modifications, and applications *Chemical Reviews* **207**(7) 2891–2959

[28] Patil A B, Patil K R and Pardeshi S K 2011 Enhancement of oxygen vacancies and solar photocatalytic activity of zinc oxide by incorporation of nonmetal *J. Solid State Chem.* **184**(2011) 3273–3279
[29] Saboor A, Shah S M and Hussain H 2019 Band gap tuning and applications of ZnO nanorods in hybrid solar cell: Ag-doped verses Nd-doped ZnO nanorods *Materials Science in Semiconductor Processing* **93** 215-225
[30] Ningsih S K W and Khair M 2015 *Sintesis dan Karakterisasi Nanopartikel ZnO Melalui Proses Sol-Gel untuk Bahan Solar-Cell* (Padang: Universitas Negeri Padang)
[31] Chang M C, Shu H Y, Tseng T H and Hsu H W 2013 Supported zinc oxide photocatalyst for decolorization and mineralization of orange G dye wastewater under UV365 irradiation *International Journal of Photoenergy* **2013** 1-12
[32] Robertson P K, Robertson J M and Bahnemann D W 2012 Removal of microorganisms and their chemical metabolites from water using semiconductor photocatalysis *Journal of hazardous materials* **211** 161-171
[33] Bizarro M 2010 High photocatalytic activity of ZnO and ZnO: Al nanostructured films deposited by spray pyrolysis *Applied Catalysis B: Environmental* **97**(1-2) 198-203
[34] Habibi M H and Khaledi Sardashti M 2008 Preparation and proposed mechanism of ZnO Nanostructure Thin Film on Glass with Highest c-axis Orientation *International Journal of Nanoscience and Nanotechnology* **4**(1) 13-16
[35] Xu Y C and You H 2014 TiO2 modified with Ag nanoparticles synthesized via ultrasonic atomization-UV reduction and the use of kinetic models to determine the acetic acid photocatalytic degradation *Applied Surface Science* **321** 481-487

IOP Conf. Series: Materials Science and Engineering **850** (2020) 012047 doi:10.1088/1757-899X/850/1/012047

Green Chemistry principle: Ionic Liquids as a material teaching for vocational students

I Yulianti*, I Hamidah and M Komaro

Technology and Vocational Education Study Program, Postgraduate School, Universitas Pendidikan Indonesia, Bandung, Indonesia

*yuliantiina@gmail.com

Abstract. The development of human civilization is characterized by the type of material used by society. The rapid development of material engineering and technology produces a variety of advanced materials such as smart materials, natural materials, chemical materials, nanomaterials, advanced materials, and environmentally friendly materials. Ionic Liquids (ILs) have criteria that are expected to be environmentally friendly because of their Green Chemistry (GC). The vocational syllabus needs to be reconstructed to be able to adapt this material technological progress. The purpose of this study was to determine the opinions of lecturers on the adoption of Ionic Liquid material teaching materials in vocational education in an effort to deal with the use of environmentally friendly materials that are indispensable in the industrial field. The subject of the study was taken through a purposive sampling method. There were 5 lecturers in one of the private polytechnics. Data collection through questionnaires is intended to determine the views of lecturers on ionic fluid material teaching materials and analyze the material requirements to be adopted into the vocational curriculum. The investigation results show that the interest and willingness of the lecturer to study and disseminate reached a score of 80%, this means that the lecturer needed the Ionic Liquid material. However, the teacher's understanding score still needs to be improved, so that it is necessary to design an Ionic Liquid teaching material that is interesting and easy to understand by both lecturers and students.The implications of this study can be used as recommendations in learning to be able to enrich the syllabus by adding Ionic Liquid material teaching material as an indispensable teaching material along with technological advances in industry 4.0.

1. Introduction

The results of the direct survey analysis on the content of vocational teaching materials in one of the private polytechnics proved that teaching materials still contained old content with constant material in recent years. While the rapid development of science and technology changes are evidenced by the discovery of several advanced materials such as smart materials, natural materials, chemical materials, chemical materials, nano materials, advanced materials and environmentally friendly materials such as "Ionic Liquids (ILs)" [1]. Green, because the material is non-flammable, vapor pressure is close to zero, a wide temperature range that allows the use of ionic liquids at low temperatures or room temperatures in the liquid phase. Its non-volatile and non-flammable nature allows this Ionic Liquid to be used as a reliable solvent replacing volatile and flammable organic solvents which have been used in the synthesis of chemical compounds. Flammability and volatility of organic solvents can cause disturbances in the natural balance, and can cause exposure to toxicity to users [2]. This is very contrary to the 12 principles of Green Chemistry which suggest not to use ingredients that are toxic and can damage the surrounding environment [3]. There is a gap between the progress of this technology and the content of the teaching materials of the existing vocational curriculum. Syllabus / chemical engineering curriculum must be reconstructed to be able to insert Ionic Liquid material so that it can be a reference for vocational students to be able to have the competency to choose safe and environmentally friendly solvents, because this competency is very much needed in the planning and processing of chemicals as a standard chemical engineering specialist implied in the vocational curriculum in Indonesia. This competency is very important to convey in the vocational curriculum so that efforts are needed to be able to socialize the material to students and lecturers through teaching materials. The long term angle of Ionic Liquid material is discussed in the fields of science and technology, which was initiated from the discovery of the first Ionic Liquid in 1914 [4], but in the field of education and teaching it still requires considerable time to be able to socialize it.

Therefore Ionic Liquid teaching materials will be created with the orientation of Green Chemistry specifically for vocational students, but beforehand there will be an analysis of the needs of first Ionic Liquid teaching materials for five respondents of vocational lecturers (polytechnic).

2. Material and methods

2.1. Ionic Liquids

Ionic Liquid teaching materials need to be inserted into vocational curriculum / syllabus with the aim of enriching the knowledge of students and lecturers in the selection of Green Chemistry. Ionic liquids are a high-tech material that has some extraordinary characteristics, which make it more preferred than traditional types of solvents that have ever existed before. This is because (1) there is a solvating ability: ionic liquids are used as solvents, there are only ions, the reaction takes place in very different conditions compared to using water or ordinary organic solvents which are neutral molecules; (2) no / little vapour pressure: Ionic Liquid vapour pressure can be neglected which makes it important to minimize exposure to toxic volatile vapours such as in the case of organic solvents; (3) non-flammable; (4) high thermal, chemical and electrochemical stability; (5) large fluid range: wide temperature range between liquid and gas; (6) easy to recycle; (7) are very polar; (8) not coordinated; (9) incompatibility with water and other organic solvents; (10) chiral ionic liquids can control stereo selectivity; (11) which has a safer period (not decomposed); (12) has conductivity. Ionic liquids are materials that are only formed from anions and cations while solvents such as benzene, chloroform, methanol, water and others consist of neutral species [5].

Ionic liquids are very different from ionic salts [6]. Ionic salts are solid at room temperature. Ionic salt is an ionic compound formed because of the electrostatic forces between cations which are generally metals with non-metallic anions. In general, ionic materials can be liquid if heated at high temperatures, such as NaCl melts at a temperature of around 800°C, this makes the conditions very impractical especially when organic synthesis. Ionic liquids are a class of liquid salts with a melting point below 100 °C, even below room temperature. Ionic liquids formed from large and asymmetric organic cations as shown in Figure 1, are derivatives of imidazolium, pyridinium, pyrrolidinium, ammonium, phosphonium and sulfonium. Common anions include; halides, tetrachloroaluminate, hexafluorophosphate, tetrafluoroborate and bis (trifluoromethylsulfonyl) imides - which are inorganic and organic anions such as alkylsulfate, alkylsulfonate, p-toluenesulfonate (tosylate) and trifluoroacetate.

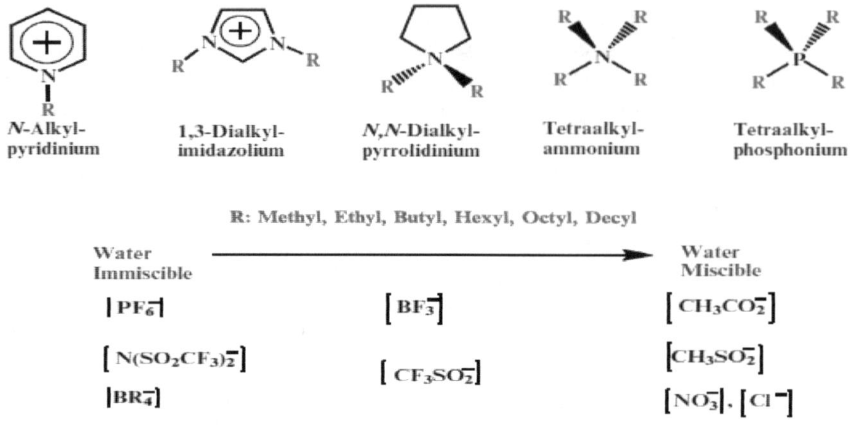

Figure 1. Some cations and ionic anions [7].

The properties of ionic liquids depend on the cations and their constituent anions. This provides great potential in designing Ionic Liquids for certain purposes [8]. Table 1 shows the types of Ionic Liquids.

Table 1. Types of Ionic Liquids with physical-chemical properties (BASF).

Type of Cation	Ionic Liquids	Melting point (oC)	Viskosity Room T (mPa)	Densitas Room T (gr/cm3)	Flash point (oC)	Solubility in Water
Imidazolium	1-Ethyl-3-methylimidazolium tetrachloroaluminate (EMIM AlCl4)	9	26	1.3040	218	react with water
	1-Butyl-3-methylimidazolium tetrachloroaluminate (BMIM AlCl4)	-10	32	1.234	198.5	react with water
	1-Butyl-3-methylimidazolium Hydrogen sulphate (BMIM HSO4)	28	4320	1.2770	>249	
	1-Ethyl-3-methylimidazolium acetate (EMIM Asetat)	-10	93	1.027	164	infinity
	1-ethyl-3-methylimidazolium thiocyanate (EMIMSCN)	< -20	22	1.1140	162	infinity
Tetraalkil ammonium	Methyl-tri-n-butylammonium methylsulfate (MTBS)	62			179.5	Infinity

The advantages of ionic liquids are; non-volatile, stable to heat up to 300oC, soluble in various organic-inorganic and organometallic compounds, properties such as polarity, hydrophilic or lipophilic can be adjusted and some physical and chemical properties. Ionic liquids are also classified as salts which have a melting point <100oC [9].

Based on the melting point data in Table 1, it is known that there is room temperature Ionic Liquid (RTIL) [10]. RTIL is a type of Ionic Liquid that is the focus of development because it is liquid at room temperature. The type of cation determines the density, viscosity, and lipophilic or hydrophilic properties of an ionic liquid. The longer the alkyl chain, the higher the viscosity. The hydrophobic / lipophilic nature of ionic liquids also increases with the length of the alkyl chain [11]. Meanwhile, the type of anion determines the acidic, basic or neutral nature of an ionic liquid. Anions such as Cl-, AlCl4-, HSO4-, BF4-, PF6- produce acidic ionic liquids while anions such as acetate provide alkaline properties [12].

Providing solvation media for chemical reactions and creating environmentally friendly reaction processes is the latest effort to replace the old provocative. For conduction of chemical processes and synthesis of materials, the aqueous and organic reaction media are always used. The availability of water as a solvent is very easy and inexpensive, non-flammable, non-toxic, and environmentally friendly [13]. However, the high or low temperature conditions of the reactions associated with organic solutes with water as solvents are a challenge to the temperature of the liquid range (0-100°C) and the limitations in dissolving each ingredient. Some organic solvents are explosive, flammable and toxic while ionic liquids are green solvents [14].

To improve work competence in the field of chemical engineering in the planning, preparation and processing of chemicals for production, it is necessary to add new knowledge about ionic liquids as green materials that can be a reference in the selection of environmentally friendly chemicals.

2.2. Green Chemistry (GC)

The rapid growth of the chemical industry has caused environmental problems. The loss of biological species in forests and in the waters, ozone depletion, agricultural pollution, freshwater and marine pollution, persistent pollutants into ecosystems, climate change, droughts, sea level changes and the spread of disease are the world's major problems today. To overcome this, the Meeting of the American Chemical Society (ACS) in August 2000 approved an alliance between ACS and Green Chemistry Institute (GCI) to build Green Chemistry as a national research priority to harmonize the interests of policy makers, business leaders and the scientific community in new initiative. ACS and GCI in partnership with the International Association of Pure and Applied Chemistry (IUPAC) sponsored the CHEMRAWN XIV World Conference on Green Chemistry, "Towards Environmentally Friendly Processes and Products". The conference brought together leaders of government, industry and academia with the aim of formulating a sustainable development pathway where pollution prevention is used to create a balance between economic development and environmental protection.

Green Chemistry (Green Chemistry / GC) is also called sustainable chemistry , is a philosophy of research and engineering / chemical engineering that advocates the design of products and processes that minimize the use and creation of hazardous compounds [15]. Green Chemistry is the use of chemicals for the prevention of pollution, is the design of chemical products and processes that are environmentally friendly [16]. GC is an approach to overcome environmental problems in terms of the chemicals produced, the process or stages of the reaction used. This concept emphasizes the reduction and use of hazardous chemicals both in terms of design and process. The dangers of chemicals intended in this GC concept include various threats to human health and the environment, including toxicity, physical hazards, global climate change, and depletion of natural resources [17]. Green Chemistry covers all aspects and types of chemical processes to reduce negative impacts on human health and the environment.

The United States Environmental Protection Agency (US-EPA) explains several principles in Green Chemistry, so that various development activities, including industries, still pay attention to "cleanliness" and "health" in pre-production, production and post-production activities. Green Chemistry technology provides a number of benefits including reducing waste, reducing the cost of expensive pipeline maintenance, safer products, reducing energy and natural resource use, and increasing the competitiveness of chemical plants [18]. The development of Green Chemistry will have many positive impacts on living things, the environment, and the increase in the economic sector along with the increasing development of industry in Indonesia [19]. Green Chemistry has evolved into a major practice supported by academics, industry, and government GC includes human and environmental health, which is guided by very specific principles of chemical practice. This principle is summarized in the Twelve Principles of Green Chemistry, namely: (1) preventing the emergence of waste; (2) atomic efficiency (efficiency of materials used in the process); (3) design of a safe synthesis process (little toxicity); (4) design of safe chemical products; (5) creating renewable raw materials; (6) catalyst (increased selectivity of reagents and energy use); (7) reducing the derivatization process (minimizing the use of additional reagents); (8) solvents and additives are safe; (9) energy efficiency; (10) design for degradation (chemical products must be designed with consideration of environmental aspects, must be easily degraded and not accumulate in the environment); (11) direct analysis to reduce pollution (analytical methods carried out in real-time can reduce the formation of unwanted by-products, focusing on developing analytical methods and technologies that can reduce the use of hazardous chemicals in the process). (12) minimize the potential for accidents (chemicals used in chemical processes must be chosen in such a way that the potential for accidents can lead to the entry of chemicals into the environment, including explosions, and fires can be avoided). The realization of the 12 basic principles can be done through the use of a curriculum with a Green Chemistry approach [20].

2.3. Development of Ionic Liquid teaching materials
To be able to apply Ionic Liquid material into the vocational curriculum, teaching materials must be made first by adopting the material from science and technology books [21]. The books are adopted so that they can be easily understood by students. It is planned that the adoption of Ionic Liquid teaching materials will be carried out with reference to the Model of Educational Reconstruction (MER) [22], as shown in **Figure 2.**

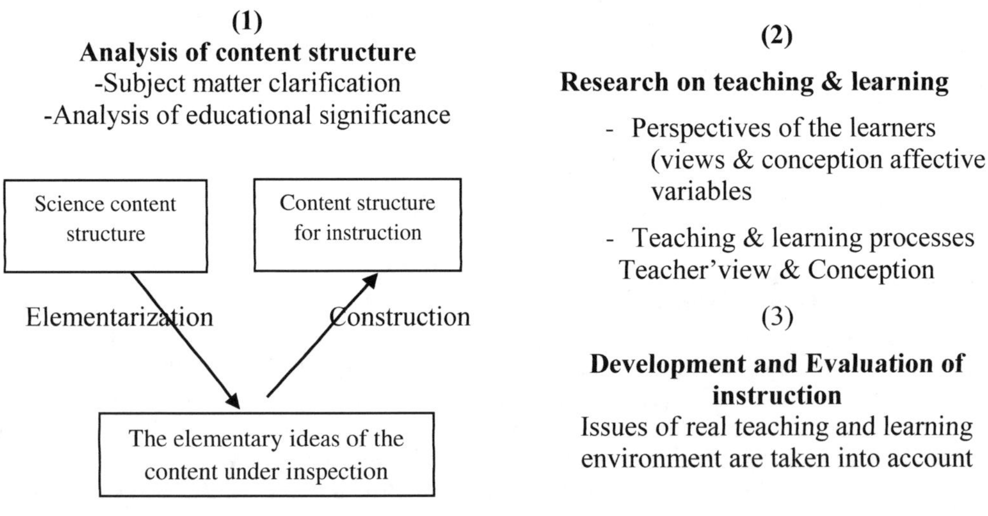

Figure 2. The model of educational reconstruction [23].

This model is designed to standardize the structure of science content into the structure of learning content. The basic ideas on the structure of content are taken directly through the elementarization process. These basic ideas are then constructed according to cognitive and affective learning goals. This model of Educational Reconstruction (MER) consists of three components, namely: 1) clarification and analysis of science content; 2) research on the teaching and learning process; and 3) Design and evaluation of the teaching and learning environment. Each component affects the other components back and forth. In developing this Ionic Liquid teaching material, only the first step of MER will be used, namely clarification and analysis of science content that has the purpose of clarifying specific

IOP Conf. Series: Materials Science and Engineering **850** (2020) 012047 doi:10.1088/1757-899X/850/1/012047

science conceptions and content structures from an educational point of view. This first component is described as clarifying subject matter and analysis of the significance of education. Clarification of subject matter illustrates qualitative analysis of content of textbooks and several journals. Analysis of the significance of education illustrates certain rules for pedagogical standards and learning objectives [23]. Analysis of science content needs to be done because university textbooks are generally consumed only by experts (scientists), while for students, presentation of content directly from university textbooks is difficult, so the teaching materials are constructed to be simpler [22].

2.4. Data collection

This study emphasizes the analysis of material requirements for Ionic Liquid teaching materials oriented to Green Chemistry to be delivered to vocational students as the latest material developments that are very useful to be able to be applied through the insertion of these materials into chemical engineering curriculum / syllabus.

Table 2. The questionnaire instrument covered positively aspects of attitude, knowledge and skills.

No.	POSITIVE STATEMENTS	SCORE				
		VA	A	N	DA	VDA
		5	4	3	2	1
A	**ATTITUDE**					
1	I've heard the term Ionic Liquid (Ionic Liquids / ILs)					
2	Ionic Liquid learning is needed at the polytechnic level as enrichment material					
3	Ionic liquid material has actually been around for a long time, but just heard the term now					
4	I am interested in new ionic liquids					
5	I am willing to study and disseminate in learning this new material as an advance in science & technology					
B 1	**KNOWLEDGE / SKILLS**					
6	Ionic liquids are ionic salts					
7	Ionic liquids are liquid at room temperature					
8	Ionic liquids are green solvents					
9	One of the properties of ionic liquids is high steam pressure					
10	Ionic liquids are high-tech materials					
11	We can synthesize ionic liquids					
12	Ionic liquids are electrolyte					
13	Ionic liquids are ionic compounds					
14	Ionic liquids do not contain solvents					
15	The use of ionic liquids will not damage the balance of nature					
16	The function of ionic salts can be replaced by ionic liquids					
17	Ionic liquids are organic compounds					
18	There are organic elements in ionic liquids					

Table 3. The questionnaire instrument was covered by negative aspects of knowledge.

No.	NEGATIVE STATEMENTS	SCORE				
		VA	A	N	DA	VDA
		1	2	3	4	5
B 2	**KNOWLEDGE**					
19	Ionic liquids are the same as ionic solutions					
20	Ionic liquids are formed from metal cations with non-metallic anions					

The research sample was taken through purpose sampling technique, namely five lecturers of chemical engineering as respondents in one of the private polytechnics [24] . Two respondents were graduates of chemical engineering master and the other three were master of science graduates who all came from well-known state universities. The research instrument was a questionnaire containing 20 statements with 18 items positive statement items and 2 negative statement items as shown in **Table 2** and **Table 3**. The statement is also divided into two categories, namely 5 items in the form of attitude aspects (statement no. 1-5) and 15 items in the form of knowledge and skills aspects (statement no. 6-20), as shown in Figure 2. The plan is material for Ionic-oriented Green Liquid Chemistry will be given to vocational students especially chemical engineering majors with the aim of increasing the competency of students' knowledge, skills and attitudes related to the planning and processing of chemicals [25].

Data from respondents were collected through questionnaires and measured using a Likert scale with five choices [26] . For a positive statement, a score 5 = SA (very agree/very good/very like), score = 4 A (agree /good/like), score 3 = N (neutral), score 2 = DA (disagree / not good / dislike), score 1 = VDA (very disagree /not very good/strongly disagree). As for negative statements, scores 5 = VDA (very disagree/not very good/strongly disagree), score 4=DA (disagree /not good/dislike), score 3=N (neutral), score 2 = A (agree/good/like), score = 1 VA (very agree/very good/very like). The score of the measurement results, then interpreted so that it becomes information that is needed by researchers for the next treatment [27].

3. Results and discussion

The results of filling out and processing positive response questionnaires and negative responses both for attitude, knowledge and cognitive sciences from the five respondents are shown in **Table 4**. Interpretation of the results of Likert scale questionnaire processing on the five respondents, for five attitude statements, all respondents agreed that ionic fluid learning was needed at the Polytechnic as enrichment material, interest and willingness to study and disseminate the material to vocational students reaching 80% with the fifth category of respondents very agree, (very good / very fond). The results of the questionnaire for the statement "have heard the term Ionic Liquid and actually Ionic Liquid material has already existed but have only now heard the term" only reached 76% with the agreed category (good / like), this is possible because respondents experienced a bit of confusion in terms of ionic liquids with ionic solutions, but in general the average attitude score reached 78.4%, which proves that Ionic Liquid material agrees to be delivered to vocational students [28].

For the results of questionnaire processing on the cognitive and skills aspects of 13 positive statements, the average response score of respondents reached 69.85% with the agree/good/like category on the Ionic Liquid material to be developed. The average score is in the middle level range of 60% - 79.99% with the agree/good/like category. This proves that lecturers' understanding of Ionic Liquid material still needs to be improved.

For negative statements, the average score only reaches 42% with a sufficient / neutral category even approaching the category of less / less agree. This interprets that respondents are still hesitant in understanding Ionic Liquid teaching materials. This is understandable because the teaching materials for ionic liquids are new material that previously did not exist in the vocational syllabus. To be able to increase the understanding of respondents, it is necessary to conduct training or technical guidance on Ionic Liquid teaching materials so that respondents can transform their knowledge well to students.

Table 4. Respondents' positive response processing.

No.	**SCORE**					**∑Score**	**Score (%)**	**DESCRIPTION**
	VA	**A**	**N**	**DA**	**VDA**			
	5	**4**	**3**	**2**	**1**			
A	**ATTITUDE**							
1	1	2	2			19	76	agree /good / like
2		5				20	80	very agree /very good / very like
3		4	1			19	76	agree /good / like
4	1	3	1			20	80	agree /good / like
5	1	3	1			20	80	very agree /very good / very like
	AVERAGE SCORES **(%)**						**78,4**	**agree /good / like**
B1								
6		5				20	80	very agree /very good / very like
7	1	3			1	18	72	agree /good / like
8	1		4			17	68	agree /good / like
9		3	2			18	72	agree /good / like
10		1	3	1		15	60	agree /good / like
11	1	1	3			18	72	agree /good / like
12		4	1			19	76	agree /good / like
13		1	3	1		15	60	agree /good / like
14		3	1	1		17	68	agree /good / like
15		1	4			16	64	agree /good / like
16		4	1			19	76	agree /good / like
17		2	3			17	68	agree /good / like
18	1	2	1	1		18	72	agree /good / like
	AVERAGE SCORES **(%)**						**69,85**	**agree /good / like**

Table 5. Processing of respondents' negative responses.

No.	**SCORE**					**∑Score**	**% Score**	**DESCRIPTION**
	VA	**A**	**N**	**DA**	**SDA**			
	1	**2**	**3**	**4**	**5**			
B 2	**KNOWLEDGE**							
19		5				10	40	neutral
20		4	1			11	44	neutral
	AVERAGE SCORES **(%)**						42	

All respondents agree that ionic liquids are ionic salts (response no. 6), but respondents have difficulty responding to negative statements "ionic liquids are the same as ionic solutions" and ionic liquids are formed from metal cations and non-metallic anions, the average response score of respondents is only 42% approached the category of minimum / disagreeing. Respondents answered agreeing to the negative statement, whereas in fact ionic liquids differ from ionic solutions because in ionic solutions there are solvents whereas in ionic liquids there are no solvents but only organic cations and organic anions. Difficulty answering questionnaires is possible because of forgetting to distinguish ionic liquids from ionic solutions. Therefore it is necessary to disseminate Ionic Liquid teaching materials to lecturers and students in the polytechnic environment. Statement no. 12 ionic liquids are electrolyte and ionic salt functions can be replaced by ionic liquids (statement no.16), the average score reaches 76% with the agree / good / like category, this gives an understanding that ionic liquids are electrolyte and can replacing ionic salt functions because they use the term ionic which means identical to the electrolyte. For the statement "there is an organic element in ionic liquids, the respondent answers that there are some who answer strongly agree, agree, neutral and some answer do not agree". This interprets that the respondent's answer is only guessing. But if the scores of the three aspects are averaged, including attitudes (78.4%), positive statements of knowledge and skills (69.85%) and negative statements of knowledge and skills (42%), results obtained at 63.42% with categories agree / good / like with Ionic Liquid teaching materials. This data provides an understanding that respondents are quite interested in ionic fluid teaching materials, willing to study and disseminate vocational students.

4. Conclusion

Green Ionic Liquid teaching materials are new material in the field of education, therefore socialization is needed. The interest and willingness of the lecturer to study and disseminate reached a score of 80%, this means that the lecturer needed the Ionic Liquid material. However, the teacher's understanding score still needs to be improved, so that it is necessary to design an Ionic Liquid teaching material that is interesting and easy to understand by both lecturers and students. Through expected curriculum reconstruction, the teaching materials can be easily conveyed to vocational students as a contribution to improve the competency of planning and processing chemicals.

References

[1] Prado R and Weber C C 2016 Applications of Ionic Liquids *Application, Purification, and Recovery of Ionic Liquids* 1-58
[2] Wang X, Chi Y and Mu T 2014 A review on the transport properties of ionic liquids *Journal of Molecular Liquids* **193** 262-266
[3] Ivanković A, Dronjić A, Bevanda A M and Talić S 2017 Review of 12 principles of Green chemistry in practice *International Journal of Sustainable and Green Energy* **6**(3) 39
[4] Wasserscheid P and Keim W 2000 Ionic liquids—new "solutions" for transition metal catalysis *Angewandte Chemie International Edition* **39**(21) 3772-3789
[5] Irge D D 2016 Ionic liquids: a review on greener chemistry applications, quality ionic liquid synthesis and economical viability in a chemical processes *American Journal of Physical Chemistry* **5**(3) 74-79
[6] Dupont J 2011 From molten salts to ionic liquids: a "nano" journey *Accounts of chemical research* **44**(11) 1223-1231
[7] Mallakpour S and Dinari M 2012 Ionic liquids as green solvents: progress and prospects *Green Solvents II* 1-32
[8] Ventura S P, Gonçalves A M, Sintra T, Pereira J L, Gonçalves F and Coutinho J A 2013 Designing ionic liquids: the chemical structure role in the toxicity *Ecotoxicology* **22**(1) 1-12
[9] Hough-Troutman W L, Smiglak M, Griffin S, Reichert W M, Mirska I, Jodynis-Liebert J ... and Pernak J 2009 Ionic liquids with dual biological function: sweet and anti-microbial, hydrophobic quaternary ammonium-based salts *New Journal of Chemistry* **33**(1) 26-33

[10] Hasanzadeh M, Shadjou N, Eskandani M and de la Guardia M 2012 Room-temperature ionic liquid-based electrochemical nanobiosensors *TrAC Trends in Analytical Chemistry* **41** 58-74
[11] Papović S, Bešter-Rogač M, Vraneš M and Gadžurić S 2016 The effect of the alkyl chain length on physicochemical features of (ionic liquids+ γ-butyrolactone) binary mixtures *The Journal of Chemical Thermodynamics* **99** 1-10
[12] Luqman M (Ed.) 2012 *Ion Exchange Technology I: Theory and Materials* Vol 1 (Germany: Springer Science & Business Media)
[13] Mannekote J K, Kailas S V, Venkatesh K and Kathyayini N 2018 Environmentally friendly functional fluids from renewable and sustainable sources-A review *Renewable and Sustainable Energy Reviews* **81** 1787-1801
[14] Khandelwal S, Tailor Y K and Kumar M 2016 Deep eutectic solvents (DESs) as eco-friendly and sustainable solvent/catalyst systems in organic transformations *Journal of Molecular Liquids* **215** 345-386
[15] Welton T 2015 Solvents and sustainable chemistry *Proceedings of the Royal Society A: Mathematical, Physical and Engineering Sciences* **471**(2183) 20150502
[16] Płotka-Wasylka J, Rutkowska M, Owczarek K, Tobiszewski M and Namieśnik J 2017 Extraction with environmentally friendly solvents *TrAC Trends in Analytical Chemistry* **91** 12-25
[17] Ibanez J G, Hernandez-Esparza M, Doria-Serrano C, Fregoso-Infante A and Singh M M 2007 The Minimization and Prevention of Pollution; Green Chemistry *Environmental Chemistry: Fundamentals* 296-323
[18] Mulvihill M J, Beach E S, Zimmerman J B and Anastas P T 2011 Green chemistry and green engineering: a framework for sustainable technology development *Annual review of environment and resources* **36** 271-293
[19] Schneider F, Kallis G and Martinez-Alier J 2010 Crisis or opportunity? Economic degrowth for social equity and ecological sustainability. Introduction to this special issue *Journal of cleaner production* **18**(6) 511-518
[20] Anastas N D and Maertens A 2018 Integrating the Principles of Toxicology into a Chemistry Curriculum *Green Chemistry* 91-108
[21] Kuhnell M 3020 Science & Technology *Chem. Eng. News* **83**(45) 29
[22] Duit R and Treagust D F 2012 How can conceptual change contribute to theory and practice in science education? *Second international handbook of science education* 107-118
[23] Duit R 2007 Science Education Research Internationally: Conceptions, Research Methods, Domains of Research *Eurasia Journal of Mathematics, Science & Technology Education* **3**(1)
[24] Wong S H 2017 *Therapeutic drug monitoring and toxicology by liquid chromatography* (London,UK: Routledge)
[25] Hartley R, Kinshuk, Koper R, Okamoto T and Spector J M 2010 The education and training of learning technologists: A competences approach *Journal of Educational Technology & Society* **13**(2) 206-216
[26] Wyse D, Selwyn N, Smith E and Suter L E (Eds.) 2016 *The BERA/SAGE handbook of educational research* (US: Sage)
[27] Budiaji W 2013 LIKERT (The Measurement Scale and The Number of Responses in Likert Scale) *Jurnal Ilmu Pertanian* **2**(2) 127-133
[28] Kossewska J M 2016 Vocational education-new generation teaching material for people with ASD *Social Welfare: Interdisciplinary Approach* **2**(5) 101-109

IOP Conf. Series: Materials Science and Engineering **850** (2020) 012048 doi:10.1088/1757-899X/850/1/012048

The sensitivity of sensor based on Microring Resonator (MRR) and Surface Plasmon Resonance (SPR) for diabetes monitoring application

H S Nugroho[1], L Hasanah[1], C Wulandari[1], R E Pawinanto[2], M H Haron[3], P S Menon[3], A R M Zain[3,4], D D Berhanuddin[3,*] and B Mulyanti[2]

[1] Department of Physics Education, Universitas Pendidikan Indonesia (UPI), Bandung 40154 Jawa Barat – Indonesia
[2] Department of Electrical Engineering Education, Universitas Pendidikan Indonesia (UPI), Bandung 40154 Jawa Barat – Indonesia
[3] Institute of Microengineering and Nanoelectronics (IMEN), Universiti Kebangsaan Malaysia (UKM), 43600 Bangi, Selangor, Malaysia
[4] 213 John McKay Lab, John A.Paulson Building, 31 Oxford St, Harvard University, Cambridge MA02138, U.S.A

*dduryha@ukm.edu.my

Abstract. Diabetes mellitus is a disease of metabolic disorders where human blood contains a high amount of sugar level for a prolonged period. This study aims to simulate how effective sensor based on MRR and SPR for the purpose of diabetes monitoring. The results show that sensor based on MRR and SPR provides the ability to detect low level of glucose concentration as low as 7 mmol/L (0.13%) and the ability to detect resonance shift for very small glucose concentration change. Therefore, the sensor based on MRR and SPR can be used for diabetes monitoring. The sensitivity of the sensor based on MRR and SPR is 85.84 nm/RIU and 116.69 °/RIU respectively. These findings are important for the development of diabetes monitoring based on MRR or SPR.

1. Introduction

Diabetes mellitus or famously known as diabetes is a disease of metabolic disorders. Diabetes is a condition where human blood contains a high amount of sugar level for a prolonged period [1]. Diabetes can lead to many horrible conditions such as diabetic ketoacidosis, cardiovascular disease, stroke, chronic kidney disease, and even death [2,3]. Diabetes can be diagnosed by the amount of glucose concentration in blood. The category of diabetes diagnosis can be seen in Table 1.

Table 1. Diabetes diagnosis category [1].

Category	Glucose concentration in blood post-meal
Healthy	< 8 mmol/L
Pre-diabetes	7.8 – 11.11 mmol/L
Diabetes	> 11.11 mmol/L

IOP Conf. Series: Materials Science and Engineering **850** (2020) 012048 doi:10.1088/1757-899X/850/1/012048

From Table 1 we can confirm how much concentration of glucose in human blood. For example, 10 mmol/L glucose concentration in the blood means that there is 10 mmol or 0.01 mol glucose in 1 L of blood. By multiplying 0.01 mol glucose with the relative mass of glucose we can confirm the mass of glucose in gram scale. By dividing this mass glucose in gram scale with 1L or 1000 ml and then multiplying with 100 we could get the percentage of glucose in the blood [4]. So, it means that 10 mmol/L is equal to 0.18016%. For sensing such an ultra-low-level concentration of biomolecule we need a sensor with very high sensitivity. Sensor-based on MRR or SPR could be used to achieve these goals [5–7]. Therefore, this study is held to simulate how effective sensor based on MRR and SPR for the purpose of diabetes detection.

1.1. Microring resonator

The geometry of MRR consisted of two different types of the waveguide, straight waveguide, and ring waveguide, which interact with each other at a very small distance called gap [8,9]. Figure 1 shows the geometry MRR. Free Spectral Range (FSR) and Quality Factor (Q-Factor) are the main characteristics of MRR which have an important role to know the sensitivity and limits of sensor detection. MRR-based sensors must have large FSR to improve its performance. The greater value of Q-factor the better detection limit resulted and in turn will cause the higher sensor sensitivity [10]. Based on how it works as a sensing device, a light comes through the waveguide then get coupled by the ring and finally produce a resonance [11,12]. Interaction of the MRR with analyte caused a change of MRR waveguide refractive index [13–15]. Thus, the resonance shift occurs. The resonance shift is indicated a change of MRR environment [16].

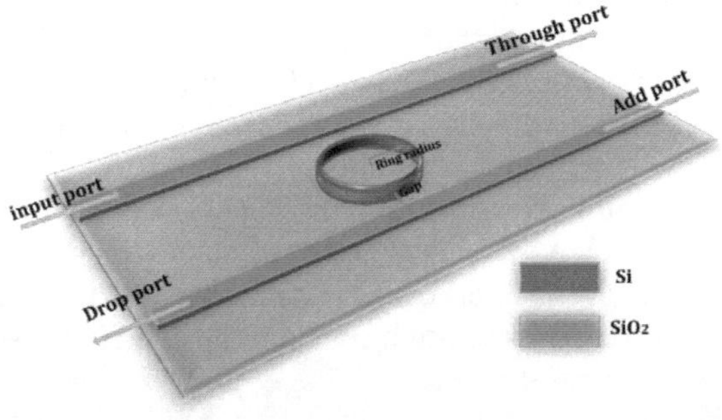

Figure 1. The geometry of MRR which consisted of two straight waveguides and one ring waveguide.

1.2. Surface plasmon resonance

In the previous study by Miyazaki, surface plasma wave is produced at a metal interface due to the oscillations of charge density at the interface of metal-dielectric, called surface plasmon, and travels along with it, while in both the media a field associated to these waves decays exponentially [17]. Surface plasma waves die out due to non-radiative decay after propagating at a certain length and the dissipated energy has turned the form into heat. As a transverse magnetically (TM) polarized waves, therefore, surfaces plasma waves get excited when the surface plasmons are matches with the wave vector of a beam of TM polarized light incident on the metal-dielectric interface. This phenomenon is what we called as surface plasmon resonance [6]. Figure 2 shows the conventional configuration of Kretschman based surface plasmon resonance. Surface plasmon resonance for the sensor is based on the detection of changes in the metal (gold) refractive index which converted to resonance angle shift [18]. The resonance angle shift is caused by the binding of analytes and ligands in the gold surface [19].

IOP Conf. Series: Materials Science and Engineering **850** (2020) 012048 doi:10.1088/1757-899X/850/1/012048

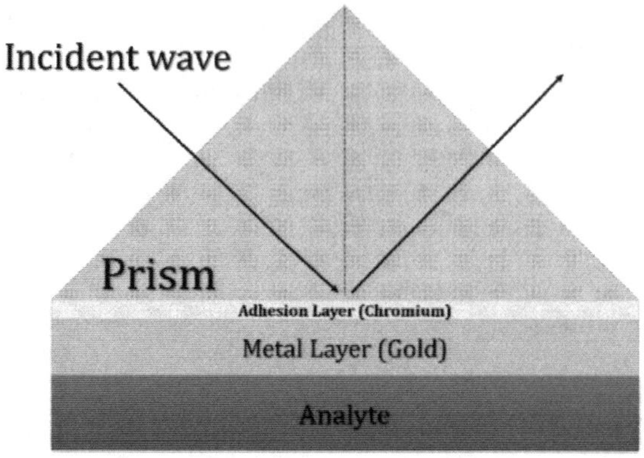

Figure 2. The conventional configuration of Kretschman based SPR.

2. Methodology

The finite difference time domain (FDTD) method is generally used for modeling photonic devices. FDTD method provides the ability to calculate both the spatial and temporal properties in a single run, hence these method is very suitable to analyze many structures [20]. FDTD method is used to solves numerical scattering problem and electromagnetic absorption on the basis of Maxwell's equation based on the Yee Mesh [21]. During this study, Lumerical mode solutions® is used to simulate MRR while Lumerical FDTD solutions® is used to simulate SPR. The initial design of MRR in this study consist of MRR add-drop system based on silicon-on-insulator (SOI) material with waveguide cross-section and ring radius of 500 nm x 220 and 4.5 μm respectively. In the other hand, the initial design of SPR in this study works at 670 nm wavelength and is consist of 3 layers as we can see in figure 2. As for the dimension for each layer are consecutively 50 nm gold layer, 2 nm chromium layer, and 7.95 μm prism (BK7) layer.

We study the response of sensor based on MRR and sensor based on SPR for diabetes monitoring. Therefore, we determine the variation of glucose concentration based on the data of human with diabetes disease in table 1. The glucose concentration of 0 mmol/L (0 %) is used as the baseline point for the simulation. Meanwhile, the glucose concentration of 7 mmol/L (0.13 %) serves as the reference point of healthy human and glucose concentration of 12 mmol/L (0.22 %) and 17 mmol/L (0.31 %) serves as the reference point of human with diabetes disease. In order to check the stability of sensors, we improve the variation for higher glucose concentration, namely 22 mmol/L (0.40 %), 55.5 mmol/L (1 %), 111 mmol/L (2 %), 166.6 mmol/L (3 %), 222 mmol/L (4 %), and 277.5 mmol/L (5 %).

3. Results and discussion

3.1. The response of sensor based on microring resonator for diabetes monitoring

The resonance of MMR occurs when the roundtrip length is proportional to an integer of guided wavelength, hence a constructive interference of light occurs then gives sharp resonance with large intensity [8]. Figure 3 shows the response of a sensor based on MRR for index background of 0 mmol/L (0 %) glucose concentration. The maximum point at the curve provides the resonance peak at the through the port transmission. This curve also serves as the baseline reference to observe the ability of sensor based on MRR for diabetes monitoring.

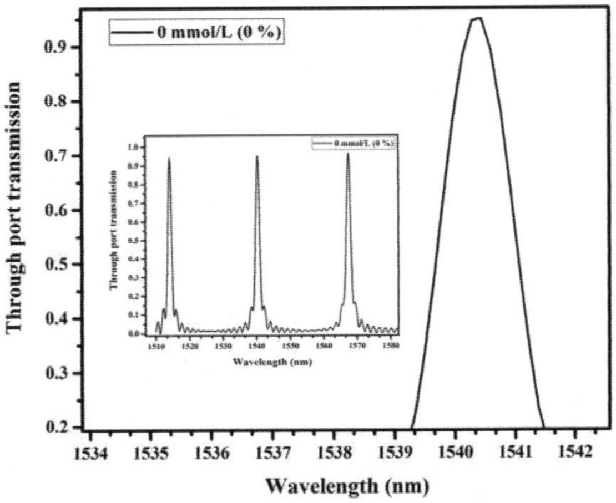

Figure 3. The transmission of MRR through port point for index background of 0 mmol/L (0 %) glucose concentration.

The presence of glucose analyte in the cladding layer affects the light propagation mode inside MRR waveguides to change [22]. The analyte will absorb the energy from the evanescent field that leaked out from MRR waveguides which causes to decreasing on the phase velocity of propagated light in MRR waveguides. The decreasing on the phase velocity of MRR waveguides induces the value of waveguides effective refractive index to increase [23]. The glucose concentration we used in this simulation resembles glucose levels in human blood. Therefore, we simulated the MRR based-sensor for each variation of the glucose concentration using the finite difference time domain (FDTD) method. Figure 4.a shows that the increase in glucose concentration induces the resonance shift of MRR. The resonance shifts occur for very low glucose concentration change from the baseline point of 0 mmol/L (0 %) to 7 mmol/L (0.13 %), 12 mmol/L (0.22 %), 17 mmol/L (0.31 %), and 22 mmol/L (0.40 %) in figure 4.a has proven the ability of MRR based-sensor for diabetes monitoring. Hence, the resonance shift to larger wavelength occurs. Furthermore, we observed the linearity between the resonance shift and the variation of the cladding layer refractive index in figure 4.b. The sensitivity of MRR based-sensor then can be known by calculating the slope of the linear curve in figure 4.b which proportional to the ratio between the resonance shifts and the refractive index variations of the cladding layer. Hence, the sensitivity of MRR based-sensor in this simulation is 85.84 nm/RIU. The sensitivity of MRR based-sensor is used to define its performance for diabetes monitoring application. The sensitivity is defined as the ratio between resonance shifts and refractive index variations of the cladding layer due to the change of glucose concentration.

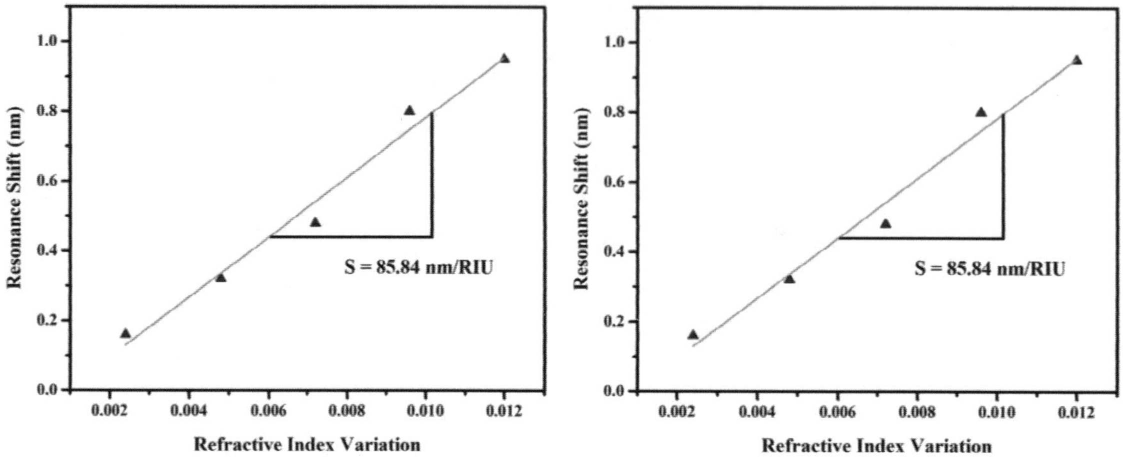

(a) (b)

Figure 4. FDTD approximation for (a) through port transmission spectra of the MRR based-sensor for various glucose concentration and (b) resonance shift response to refractive index variation.

3.2. The response of sensor based on surface plasmon resonance for diabetes monitoring

Incident wave 670 nm wavelength is emitted to the surface of the chromium adhesive layer from a light source. The interaction between the incident wave and the surface of the chromium adhesive layer induces an evanescent field on the gold layer. This evanescent field then interacts with plasmon of gold which produces by the interaction between the gold layer and dielectric medium [24]. The surface plasmon wave then generated at the boundary of gold and dielectric layer by the interaction between the evanescent field and the plasmon of gold [19]. The incident wave is swept at the surface of the chromium layer with incident angle 36° to 80°. The resonance of the surface plasmon then occurs at a certain incident angle when most of the incident wave energy is absorbed by the plasmon in the gold layer and turned to the surface plasmon wave [25]. This phenomenon can be detected by the minimum reflectance at the detector. From the curve in figure 5, we can see the minimum point at the response curve of sensor based on SPR for index background of 0 mmol/L (0 %) glucose concentration. The minimum point describes the resonance angle of the sensor based on surface plasmon resonance. This curve also serves as the baseline reference to observe the ability of sensor based on surface plasmon resonance for diabetes monitoring.

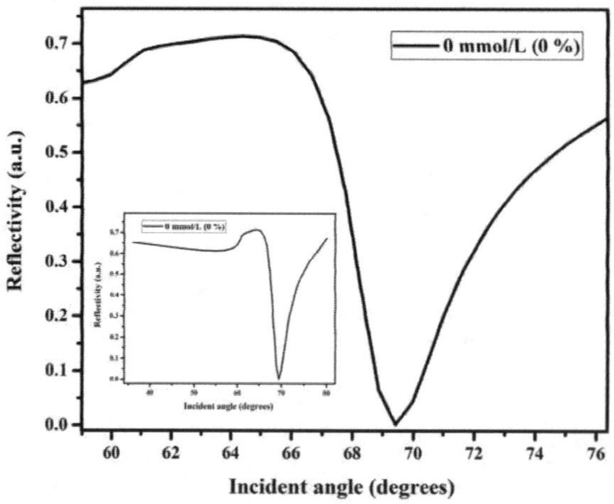

Figure 5. The response curve of sensor based on SPR in index background of 0 mmol/L (0 %) glucose concentration.

The increasing of analytic induces the analyte refractive index value itself to increased, hence the resonance angle is shifting to larger angles accordingly to the relation between analyte refractive index and SPR resonance angle in [24]. Figure 6.a shows that the resonance angle shift occurs for each glucose concentration change where 0 mmol/L (0 %) serve as the baseline. The ability of sensor based on SPR for diabetes monitoring has been proved by its ability to show shifting for very small glucose concentration change such as from the baseline point to 7 mmol/L (0.13 %), 12 mmol/L (0.22 %), 17 mmol/L (0.31 %), and 22 mmol/L (0.40 %). In figure 6.b, we analyze the sensitivity value of the sensor based on SPR for diabetes monitoring. The sensitivity of SPR based-sensor can be defined as the ratio of resonance angle shifts to the change of the analyte refractive index [25]. The sensitivity of SPR based-sensor determines its performance for diabetes monitoring application because in this simulation we used approximate glucose concentration to resembles glucose levels in human blood. The sensitivity of the sensor based on SPR in this study is 116.69 °/RIU. Moreover, with the value of 116.69 °/RIU, this sensor based on SPR provides fine performance in which these sensors able to detect such a low level of glucose concentration change.

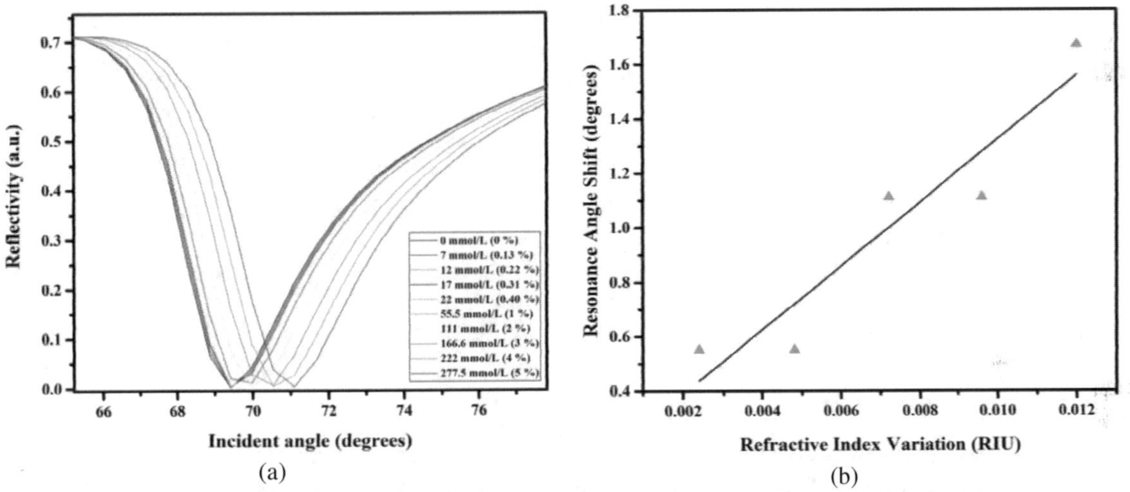

Figure 6. SPR based-sensor response curve using FDTD approach for (a) various glucose concentration and (b) resonance shift response to refractive index variation at 670 nm wavelength.

4. Conclusion

This study set out to simulate how effective sensor based on MRR and sensor based on SPR for diabetes monitoring. Both types of sensors are providing resonance shift for the variation of glucose concentration for healthy human and human with diabetes disease. In addition, the sensitivity of the sensor based on MRR and sensor based on surface plasmon resonance is 85.84 nm/RIU and 116.69 °/RIU respectively. Therefore, this study has shown that both sensors can be used for the purpose of diabetes monitoring.

Acknowledgments

We would like to thank Vice-Rector for Research, Business and Partnership, Universitas Pendidikan Indonesia for financially support this research.

References

[1] American Diabetes Association 2017 Classification and diagnosis of diabetes *Diabetes care* **40**(Supplement 1) S11-S24

[2] Tuomilehto J, Lindström J, Eriksson J G, Valle T T, Hämäläinen H, Ilanne-Parikka P ... and Salminen V 2001 Prevention of type 2 diabetes mellitus by changes in lifestyle among subjects

with impaired glucose tolerance *New England Journal of Medicine* **344**(18) 1343-1350
[3] Varma R, Bressler N M, Doan Q V, Gleeson M, Danese M, Bower J K ... and Turpcu A 2014 Prevalence of and risk factors for diabetic macular edema in the United States *JAMA ophthalmology* **132**(11) 1334-1340
[4] Anastas P and Eghbali N 2010 Green chemistry: principles and practice *Chemical Society Reviews* **39**(1) 301-312
[5] De Vos K, Bartolozzi I, Schacht E, Bienstman P and Baets R 2007 Silicon-on-Insulator microring resonator for sensitive and label-free biosensing *Optics express* **15**(12) 7610-7615
[6] Srivastava S K, Verma R and Gupta B D 2012 Surface plasmon resonance based fiber optic glucose biosensor *Third Asia Pacific Optical Sensors Conference* **8351** 83511Z
[7] Sadana A, Sadana N and Sadana R 2018 *A Fractal Analysis of Chemical Kinetics with Applications to Biological and Biosensor Interfaces* (Amsterdam: Elsevier)
[8] Van V 2016 *Optical Microring Resonators Theory, Techniques, and Applications* (Boca Raton: Taylor & Francis Group)
[9] Rabus D G 2007 *Integrated Ring Resonators* (Verlag Berlin Heidelberg: Springer)
[10] Chao C Y and Guo L J 2006 Design and optimization of microring resonators in biochemical sensing applications *Journal of lightwave technology* **24**(3) 1395-1402
[11] Mulyanti B, Hariyadi T, Menon P S, Shaari S, Haroon H, Ehsan A A, Majlis B Y, Hasanah L, Mahmudin D and Wiranto G 2014 Design and optimization of coupled microring resonators (MRRs) in Silicon-on-Insulator *Sains Malaysiana* **43**(2) 247–252
[12] Shahidinejad A 2017 A Literature Review on Microring Resonators *International Journal Series in Engineering Science (IJSES)* 14-22
[13] Mulyanti B, Hasanah L, Hariyadi T, Novitasari R, Pantjawati A B, Yuwono H and Khairurrijal K 2015 The Influence of Glucose Concentration to Resonant Wavelength Shift of Polymer-based Microring Resonators *Advanced Materials Research* **1112** 32-36
[14] Vollmer F, Braun D, Libchaber A, Khoshsima M, Teraoka I and Arnold S 2002 Protein detection by optical shift of a resonant microcavity *Applied physics letters* **80**(21) 4057-4059
[15] Chao C Y and Guo L J 2003 Biochemical sensors based on polymer microrings with sharp asymmetrical resonance *Applied Physics Letters* **83**(8) 1527-1529
[16] Amiri I S, Ariannejad M M, Kouhdaragh V, Seyedi S A and Yupapin P 2019 Microring resonator made by ion-exchange technique for detecting the CO2, H2O, and NaCl as cladding layer *Journal of King Saud University-Science* **31**(1) 27-32
[17] Miyazaki C M, Shimizu F M, Mejía-Salazar J R, Oliveira Jr O N and Ferreira M 2017 Surface plasmon resonance biosensor for enzymatic detection of small analytes *Nanotechnology* **28**(14) 145501
[18] Iwasaki Y, Horiuchi T and Niwa O 2001 Detection of electrochemical enzymatic reactions by surface plasmon resonance measurement *Analytical chemistry* **73**(7) 1595-1598
[19] Miyazaki C M, Shimizu F M and Ferreira M 2017 Surface plasmon resonance (SPR) for sensors and biosensors *Nanocharacterization Techniques* 183-200
[20] Md Zain A R 2009 *One-dimensional photonic crystal/photonic wire cavities based on silicon-on-insulator (SOI)* (Doctoral dissertation, UK: University of Glasgow).
[21] Yee K 1966 Numerical solution of initial boundary value problems involving Maxwell's equations in isotropic media *IEEE Transactions on antennas and propagation* **14**(3) 302-307
[22] Bogaerts W, De Heyn P, Van Vaerenbergh T, De Vos K, Kumar Selvaraja S, Claes T ... and Baets R 2012 Silicon microring resonators *Laser & Photonics Reviews* **6**(1) 47-73
[23] Schmidt S 2016 *Enhancing the performance of silicon photonic biosensors for clinical applications* (Doctoral dissertation, US: University of Washington)
[24] Kurihara K and Suzuki K 2002 Theoretical understanding of an absorption-based surface plasmon resonance sensor based on Kretchmann's theory *Analytical chemistry* **74**(3) 696-701
[25] Fouad S, Sabri N, Jamal Z A Z and Poopalan P 2017 Surface plasmon resonance sensor sensitivity enhancement using gold-dielectric material *Int. J. Nanoelectronics and Materials* **10** 149-158

An increase in crashworthiness capability using pyramid arrangement of the crush initiator

F Dionisius*, Y N Rohmat, M M Yusuf, S Suliono, E Haris and T Endramawan

Department of Mechanical Engineering, Politeknik Negeri Indramayu, Jalan Raya Lohbener Lama No. 08, Indramayu 45252, West Java Indonesia

*dionisiusfelix@gmail.com

Abstract. Collisions on vehicles could cause injury to passengers such as being thrown from a passenger room. To prevent this, a vehicle structure was not too strong which called by crashworthiness criteria through maximum impact force. This study was conducted to determine the smallest structural strength through maximum impact force in a square column with 0.8 mm in thickness. The column was given holes with 3 mm in diameter as crush initiators. The hole was arranged in pyramid pattern with different level and begun with 1 hole in the square column. The research used the quasi-static concept experimentally with 30 mm/min in movement speed of actuator. The machine used Universal Testing Machine (UTM) with 200 kN in capacities. The result showed that the pyramid arrangement with the highest level had the greatest maximum impact force 22.124 kN. Comparison of differences between the highest levels with the lowest levels reached 4.86%. While the difference between the highest level with a square column with 1 hole was 14.24%. So that the more holes arranged in a pyramid would reduce the maximum impact force during a collision.

1. Introduction

A vehicle must meet the element of safety in the face of an accident when driving. The vehicle has a structure that can protect passengers in the event of a collision. The vehicle structure is divided into two spaces, namely the crumple zone and safety zone. Crumple zone is an area that is intentionally destroyed [1]. While the safety zone is an area that must be strong in protecting passengers. Crumple zone can absorb energy and reduce injury to passengers which called by crashworthiness. The crumple zone is located on the front rail area of a vehicle. In general, the shape of the front rail profile is circular, square or rectangular hollow and others. Many researchers had carried out experiments on the front rail area with material modifications, profile forms [2] and placed triggers which called by crush initiator [3,4]. The modification produced crashworthiness criteria such as maximum impact force and specific energy absorption when experiencing axial loads both quasi static and quasi dynamic. Estrada et al [5] had studied structures with bi-tubular profiles to absorb energy during collisions. The profile with aluminum 6063-T5 material was given a hole as a crush initiator. The test used a finite element simulation method with quasi static loading. Pirmohammad and Esmaelli [6] carried out an optimization of the square and octagonal bi-tubal structures using NSGA-II and ANN. The structure was given circle, hexagon and ellipse holes as crush initiator. The maximum impact force and specific energy absorption were known by the elemental method analysis with LS-DYNA code. In addition, heat treatment in certain areas could also be a crush initiator. This method of treatment had been carried out by Peixinho et al [7] in using an aluminum square column. Tests were carried out with dynamic loads to obtain maximum energy and force absorption during collisions. Eyvazian et al [8] performed quasi static loading by using corrugated on circular tubes. Experimental and theoretical analysis was used to obtain energy absorption and maximum impact force when given axial loads.

Balaji and Annamalai [9] also conducted experiments and simulations on square hollow tubes with quasi static loading. The tube was given by a crush initiator in the form of a V-notch model and groove. Choiron et al. [10] modified the crash box using the initial fold. Finite element analysis was carried out to obtain a deformation pattern when given by dynamic axial loads. Using buckling initiator on a square tube was also carried out by Mashalkar and Tuljapure [11]. The tube was tested using finite element modelling with the help of Hyperworks and LS-DYNA. Loading used a quasi-dynamic method with variations in direction to get peak load and energy absorption.

Khalili et al [12] investigated thin-walled aluminum tubes when were given by dynamic loads. The test was conducted to determine the effect of energy transfer on trigger in the aluminum specimen. The square hollow structure was also investigated by Marzbanrad et al [13] with using the LS-DYNA numerical simulation method. The structure was given by a groove shaped hole with deep variation. This was done to obtain optimization with consideration of energy absorption and maximum impact force. Subramaniyah et al [14] had also conducted experiments on square tubes and circles with quasi static loading. Both tubes were given by holes as crush initiator to obtain maximum force and energy absorption. In addition, Mamalis et al [15] also carried out testing by using experimental and simulation methods on mild steel square tube. The tube was also given a crush initiator in the form of a hole to get a reduction of maximum impact force.

This paper discusses the use of crush initiator by modifying the pattern to form a pyramid arrangement. This arrangement was almost similar to the pattern from Dionisius et al [16] which used an inverted pyramid pattern. The objective of this paper was to determine the maximum collision force in quasi static loading. In addition, the smallest maximum impact force of some of these patterns is the best crashworthiness criteria in reducing passenger injuries. So that this research could be utilized by vehicle designers.

2. Methods

In achieving the desired results, there are several steps in this study as illustrated in Figure 1.

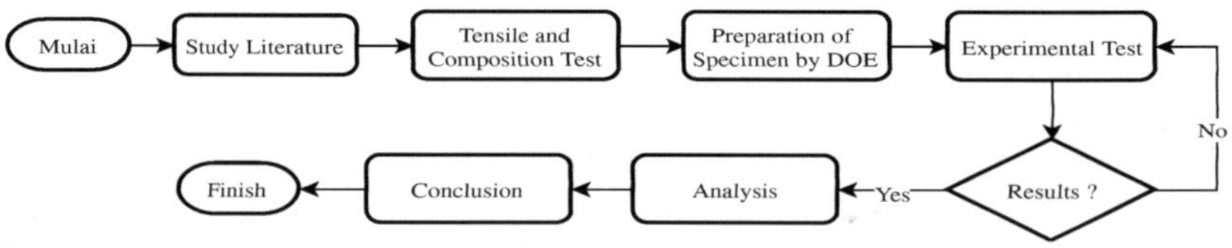

Figure 1. Flow chart of steps.

2.1. Concept of crashworthiness

Crashworthiness is a structure that can absorb energy and reduce injury to passengers during an accident. Crashworthiness has several criteria, one of which is the maximum impact force (F_{max}). The impact force is generated from the structure of the vehicle. If it is too large, it can cause passengers to leave the area. The use of structures with thin wall thickness could absorb energy and reduce collision forces as examined by Abramowicz [17] and Tarlochan et al [18]. In this case, the smallest maximum impact force was the desired crashworthiness criterion.

Crashworthiness testing used a Universal Testing Machine (UTM) with 200 kN in capacities. The actuator on the UTM moved in quasi static loading with constant speed of 30 mm/min or 0.5 mm/s as shown in Figure 2b. Testing of specimens was carried out until 100 mm in deformation. The UTM would produce a graph of force and displacement so that the maximum impact force for each specimen could be known. This test was done by 3 times for each design. The test was based on the concept in Figure 2a where the UTM actuator was an impactor.

IOP Conf. Series: Materials Science and Engineering **850** (2020) 012049 doi:10.1088/1757-899X/850/1/012049

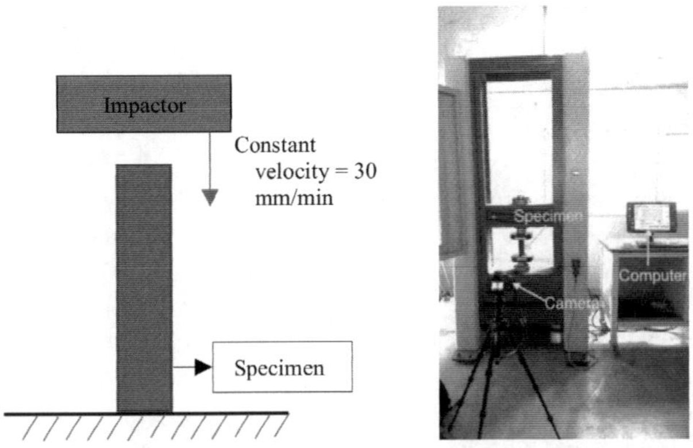

(a) (b)

Figure 2. Experimental of crashworthiness: a) Concept; b) By using UTM.

2.2. Tensile and composition test

Tensile testing was done to determine the characteristics of mechanical properties. The machine was Universal Testing Machine (UTM) which carried out in Mechanical Engineering Workshop, Politeknik Negeri Indramayu. The standard of this test used ASTM E8/E8M. The specimen was plate which taken from a square column with 0.8 mm in thickness as shown in Figure 3a. While the results could be seen in Figure 4 where was 192.80 MPa in yield strength and 267.5 MPa in ultimate tensile strength. These results were obtained through a tensile test as shown in Figure 3b.

Material composition testing used an Optical Emission Spectrometer machine with 20 elements in capacities. Material composition could be seen in Table 1. The type of material in this test was classified as low carbon steel because the number of carbon values was below 0.2%.

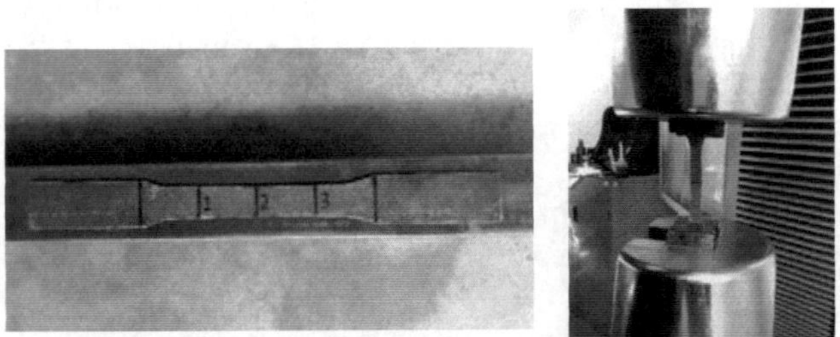

(a) (b)

Figure 3. Tensile test: a) Preparation of specimen; b) Experimental of tensile test.

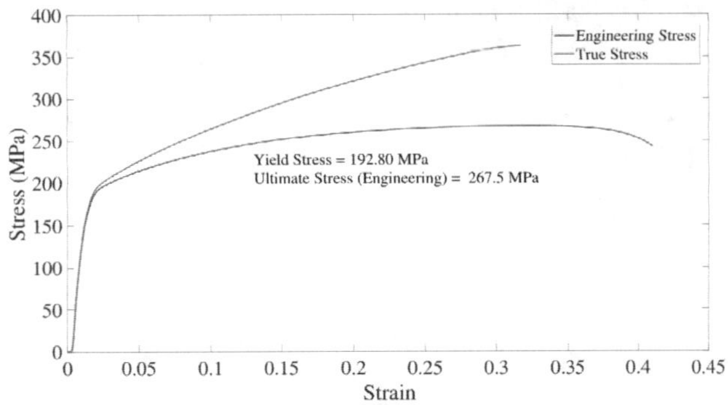

Figure 4. Graph of engineering and true stress.

Table 1. Composition of elements in material of specimen.

No.	Elements		Value (%)
1	Carbon	(C)	0,046
2	Silicon	(Si)	0,038
3	Sulfur	(S)	0,002
4	Phosphorus	(P)	0,016
5	Manganese	(Mn)	0,129
6	Nickel	(Ni)	0,01
7	Chromium	(Cr)	0,02
8	Copper	(Cu)	0,006
9	Wolfram / Tungsen	(W)	0,001
10	Titanium	(Ti)	0,002
11	Tin	(Sn)	0,001
12	Aluminium	(Al)	0,027
13	Antimony	(Sb)	0,001
14	Ferro / Iron	(Fe)	99,715

2.3. Design of Experimental (DOE)
The specimen was a square column with a thickness of 0.8 mm which had 36 x 36 (mm) in section size with 200 mm in length. The column was given by hole 3 mm in diameter as a crush initiator. The position of the hole was at 10 mm from the end of the specimen which called by NA. Then adding the same diameter hole was done by forming a pyramid arrangement. These additions were carried out in stages called first level to third level in design as shown in Figure 5.

3. Results and discussion
Crashworthiness testing in this paper was carried out to determine the maximum impact force in specimens with pyramid arrangement. The test produced progressive buckling up to 100 mm in deformation as shown in Figure 6. The total time of this deformation was 200 seconds. The test produced 4 folds where the value of the maximum force is at the first folding which called by folding mechanism. This could also be seen in
Figure **7** where there were 4 peak values for each specimen condition. Some of the peak values were the beginning of folding that occurred in each specimen during quasi-static loading.

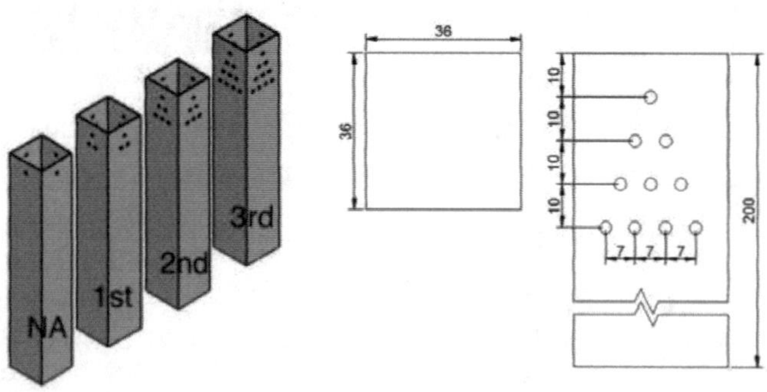

Figure 5. Design of experimental.

t = 0s t = 20s t = 100s t = 180s t = 200s

Figure 6. Progressive buckling.

Based on the graph in
Figure **7**, the maximum force value for each condition could be seen in **Error! Reference source not found.**. First level to third level design had a decrease in maximum collision force from 23.255 kN to 22.124 kN or 4.86%. While from the NA design to third level, the maximum impact force decreased by 14.24%. This showed that crush initiator with multilevel pyramid arrangement could reduce the maximum impact force during a collision.

The decrease in maximum impact force was caused by the location of different folds in each design. The difference could be seen in Figure 9. The first fold on the NA design was located in the area of 1 hole. While the level 1 design was in the area of 20 mm from the end of the specimen where there were 2 holes each side. The level 2 and level 3 designs also experienced initial fold in the area of 3 holes and 4 holes in parallel. This means that the location of the fold will depend on the area with the number of parallel holes. Therefore, the reduction of maximum impact force occurred in this study.

Based on the stress theory proposed by Khurmi and Gupta [19] that the force was the multiplication between stress and area during axial loading. Figure 10 shows that the area in each design is not the holes area. If the area is smaller, it will produce a small force and vice versa. So that the force produced when axial loads will be directly proportional to the area of each specimen.

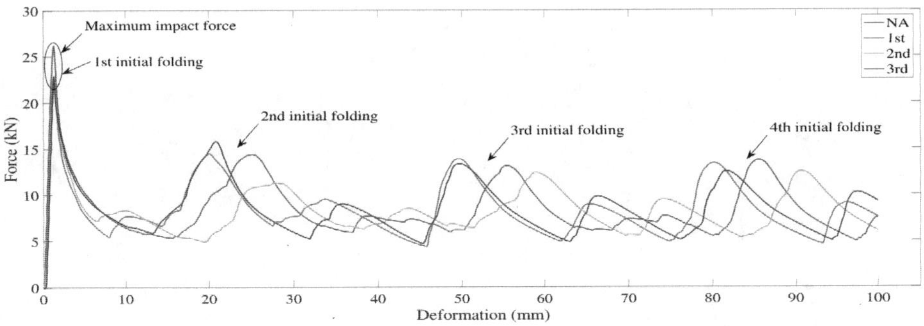

Figure 7. Graph of force and deformation during crashworthiness test.

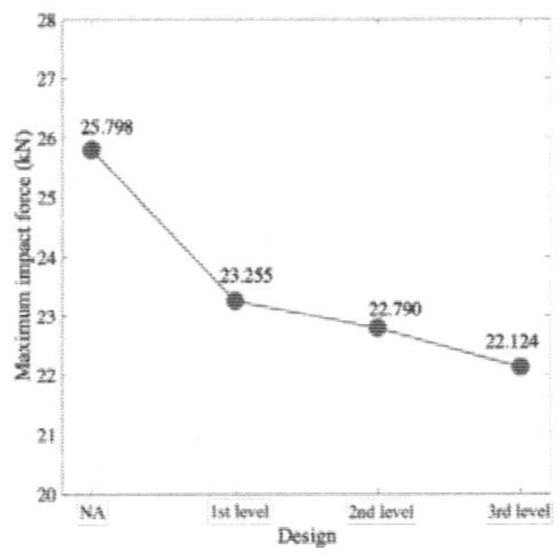

Figure 8. Maximum impact force of each condition.

(a) (b) (c) (d)

Figure 9. Initial buckling in design of: a) NA; b) 1st level; c) 2nd level; d) 3rd level.

4. Conclusion

Crashworthiness testing with quasi static loading had been done on thin-walled square column specimens. The specimen was given a crush initiator which began with 1 hole. Then the hole would be added to form a pyramid arrangement. This arrangement was identified as the first to the third level. The conclusion of the test was that the maximum impact force had decreased by 14.24% from the specimen with 1 hole to the third level of pyramid arrangement. The decrease also occurred in the first level of pyramid arrangement to third level that was 4.86%. So that the best crashworthiness criteria were in the level 3 pyramid arrangement which has the smallest maximum impact force.

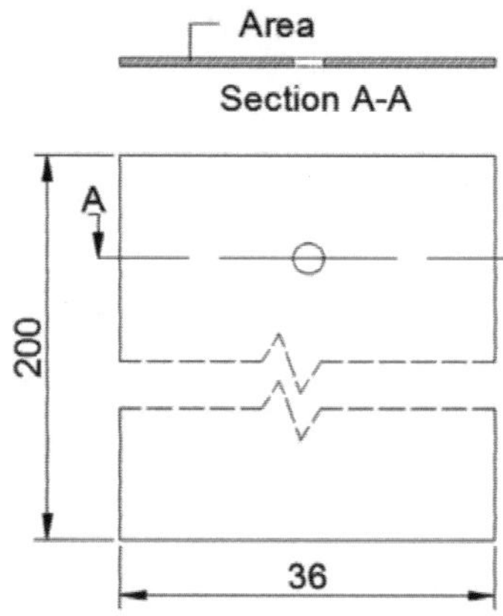

Figure 10. Area in NA design.

Acknowledgments

Thank you to the Centre of Research and Community Service of Politeknik Negeri Indramayu with PKM403/PL42/PM/2019 of contract numbers for funding assistance.

References

[1] Istiyanto J, Hakiman S, Sumarsono D A, Kiswanto G, Baskoro A S and Supriadi S 2014 Experiment and numerical study – Effects of crush initiators under quasi-static axial load of thin wall square tube *Appl. Mech. Mater.* **660** 628–32

[2] Hou S, Li Q, Long S, Yang X and Li W 2008 Multiobjective optimization of multi-cell sections for the crashworthiness design *Int. J. Impact Eng.* **35** 1355–67

[3] Malawat M, Istiyanto J and Sumarsono D A 2017 Effects of Wall Thickness and Crush Initiators

Position under Experimental Drop Test on Square Tubes *Appl. Mech. Mater.* **865** 612–8

[4] Malawat M, Sumarsono D A, Istiyanto J, Prayogo G and Dionisius F 2019 Theoretical Prediction Of Dynamic Axial Crushing On A Square Tube With Eight Holes Used As A Crush Initiator *Int. J. Technol.* **10** 1042–55

[5] Estrada Q, Szwedowicz D, Rodriguez-Mendez A, Elías-Espinosa M, Silva-Aceves J, Bedolla-Hernández J and Gómez-Vargas O A 2019 Effect of radial clearance and holes as crush initiators on the crashworthiness performance of bi-tubular profiles *Thin-Walled Struct.* **140** 43–59

[6] Pirmohammad S and Esmaeili-Marzdashti S 2019 Multi-objective crashworthiness optimization of square and octagonal bitubal structures including different hole shapes *Thin-Walled Struct.* **139** 126–38

[7] Peixinho N, Soares D, Vilarinho C, Pereira P and Dimas D 2012 Experimental Study of Impact Energy Absorption in Aluminium Square Tubes with Thermal Triggers *Mater. Res.* **15** 323–32

[8] Eyvazian A, Tran T N and Hamouda A M 2018 Experimental and theoretical studies on axially crushed corrugated metal tubes *Int. J. Non. Linear. Mech.* **101** 86–94

[9] Balaji G and Annamalai K 2017 An experimental and numerical scrutiny of crashworthiness variables for square column with V-notch and groove initiators under quasi-static loading *Cogent Eng.* **4** 1–20

[10] Choiron M A, Happy H K, Purnowidodo A and Rivai A 2019 Deformation Pattern and Energy Absorption Analysis on Initial Fold Crash Box by Oblique Crash Test *IOP Conference Series: Materials Science and Engineering* vol 494 pp 1–6

[11] B. S. Mashalkar and S. B. Tuljapure 2015 Effect of Buckling Initiator on Energy Absorption in Oblique loading of Bumper System *Int. J. Eng. Res. Technol.* **4** 1440–4

[12] Khalili P, Tarlochan F, Hamouda A M S and Al-Khalifa K 2015 Energy absorption capability of thin-walled aluminium tubes under crash loading *J. Mech. Eng. Sci.* **9** 1734–43

[13] Marzbanrad J, Ebrahimi-F M and Khosravi M 2014 Optimization of crush initiators on steel front rail of vehicle *Int. J. Automot. Eng.* **4** 749–57

[14] Subramaniyan S K, Kananasan A K, Yunus M R M, Mahzan S and Ghazali M I 2014 Crush characteristics and energy absorption of thin-walled tubes with through-hole crush initiators *Appl. Mech. Mater.* **606** 181–5

[15] Mamalis A G, Manolakos D E, Spentzas K N, Ioannidis M B, Koutroubakis S and Kostazos P K 2009 The effect of the implementation of circular holes as crush initiators to the crushing characteristics of mild steel square tubes: Experimental and numerical simulation *Int. J. Crashworthiness* **14** 489–501

[16] Dionisius F, Istiyanto J, Endramawan T and Andri 2018 Optimasi Desain Crashworthiness pada Tabung Persegi Berdinding Tipis Dengan Variasi Crush Initiator : Stripe-Parallelogram-Trigon (Design Optimization of Crashworthiness in Square Tube as Thin Walled Structure With Variations of Crush Initiator: Stripe-P *J. Teknol. Terap.* **4** 48–54

[17] Abramowicz W 2003 Thin-walled structures as impact energy absorbers *Thin-Walled Struct.* **41** 91–107

[18] Tarlochan F, Samer F, Hamouda A M S, Ramesh S and Khalid K 2013 Design of thin wall structures for energy absorption applications: Enhancement of crashworthiness due to axial and oblique impact forces *Thin-Walled Struct.* **71** 7–17

[19] Khurmi R S and Gupta J K 2005 *A Textbook of Machine Design* (New Delhi: Eurasia Publishing House (PVT.) LTD)

Rare earth metals utilization opportunities in the development of Indonesia defense industry technology

N F Jannah*, S Aritonang, S Sunardi, S N Adiprayoga and D D A Rajak

Indonesia Defense University, IPSC Bogor, West Java 16810, Indonesia

*jk.jannah@gmail.com

Abstract. Industrial technology developments that entered the era of 4.0, other than to utilize the development of technology which based on the internet of things, big data, artificial intelligence, cloud, etc., also need to be supported by advanced material technology capabilities, namely rare earth metals. In the defense industry, rare earth metals used in defense and security equipment such as military vehicle coating materials that are able to dodge or absorb waves of enemy radar detection, jet engine components of aircraft fighter, missile weapons systems, electronic components, subsea detectors, satellite devices and material for radar, etc. This research was conducted through literature review and using the SWOT analysis method on the opportunities for the use of rare-earth metals in Indonesia defense industry products. The data used are secondary data, which includes data on the distribution of rare earth metals in Indonesia and the specifications of the use of rare-earth metals in the defense industry. Based on the defense requirements for defense products and the percentage of rare earth metals content in the territory of Indonesia, one of the opportunities for the use of rare-earth metals for defense industry products is in the application of anti-radar layers.

1. Introduction

The defense industry is an industry that utilizes high technology and complex. Not only the operating system but since the stages of designing defense products to become finished products, defense industry products utilize a system that is mutually integrated and holistic. In this era of technological advancement 4.0, defense products are required to fulfill industrial demand in accordance with the requirements required by utilizing advanced technology and supported by the development of advanced materials.

In Indonesia, one of the advanced materials possessed is rare earth metal. Although its availability is not in the form of chunks of elements, it is scattered and bound in other complex compounds such as monazite and tin, where a simple enrichment process is needed to obtain it, but the potential of rare earth metals in Indonesia is worth taking into account. There have been many studies that provide an overview of the distribution of rare earth metals in Indonesia and explain the elemental enrichment processes that can be carried out. It is just that, for utilization in the field of defense is still very minimal. For this reason, this research was conducted based on literature review and secondary data on the number of rare earth metals in Indonesia as well as material requirements from defense products.

1.1. Defense industry

The end of the cold war is the starting point for technological development and the defense industry. It is also the beginning of the start of the competition and global market competitiveness. In the theory of revolution in military affairs (RMA), it is stated that the development of defense technology and the industry is caused by the presence of factors of liberalization and changing views on future warfare [1]. It is explained that war in the future is no longer a matter of taking up arms but a war that combines hard power and soft power.

So that the defense technology developed is not only combat equipment but a system that is mutually integrated and connected in one control to achieve certain objectives such as ground combat systems, air combat systems, and others.

The defense industry, according to law number 16 of 2012 concerning the defense industry, is a national industry consisting of state-owned and private enterprises which are tasked with fulfilling defense needs by producing defense and security equipment, as well as carrying out maintenance services in meeting strategic interests in defense and security in the Unitary State of the Republic of Indonesia [2]. In its development, the defense industry is targeted to have the ability to design weapons systems and defense equipment in accordance with Indonesia's strategic environment. Also, it is required to be able to realize the design into a prototype which will be developed and capable of producing or manufacturing. Then, it needs to be able to sell its products and compete in the global market, able to offer services related to maintenance, repair, and overhaul (MRO), able to provide spare parts from products manufactured, and be able to play a role in the integrated national and global defense logistics system. The Indonesia defense industry consists of several layers, namely tier 1 which is the main tool industry or lead integrator that produces the main tools for defense systems, tier 2 industries which are the main and strategic component industries as supporting industries, tier 3 industries namely non-tool components or industrial parts main defense system, and tier 4 which is the raw material industry (See Figure 1).

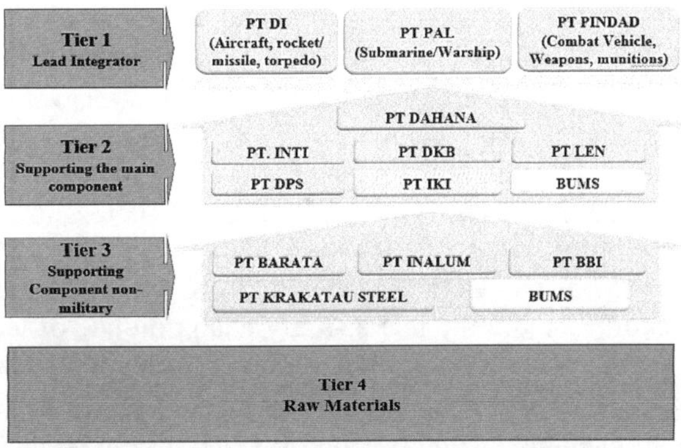

Figure 1. Defense industries elements.

1.2. Defense requirement: Air, sea, and land

The requirements of defense equipment for air, sea, and land, are based on the development of a generation of war followed by technological developments in defense. At this time, the generation of war has transitioned from the fourth generation to the fifth generation, which has the characteristics of using sophisticated weapons technology and changes in the context of the war. In addition, the fifth generation war, followed by industrial technology 4.0, has entered the realm of cyberspace where there is no longer a boundary between space and time in its implementation. This condition makes changes in the strategic environment and the types of threats that threaten state sovereignty. In addition, consideration of the requirements for air, sea, and land defense technology is also based on geographical structure, strategic environment, capabilities, and threats that always overshadow.

Knowing Indonesia's air defense power, according to AN Pratama in Kompas.com air magazine on October 8, 2018, the Indonesian Air Force has five mainstay aircraft covering the Sukhoi SU-30 used in carrying out attacks on terrain, Sukhoi SU-27, aircraft the T-50 Golden Eagle supersonic trainer, the EMB-314 Super Tucano anti-guerrilla trainer aircraft which serves to support reconnaissance, close air support, and rebellion crackdowns, as well as lightweight fighter jets for BAE Hawk 209 trainers [3]. Globally, the development of weapons technology in air defense has evolved into stealth technology and unmanned air vehicle (UAV).

The defense condition of Indonesian sea, when compared to the size of Indonesian sea, the number of fleets owned is still not able to secure the entire region. The Indonesian defense industry still opens wide opportunities for shipbuilding companies from abroad such as the Netherlands, Germany and South Korea due to the lack of capability in meeting the demand and technology standards used in the Indonesian fleet. Also, for operational support of the army, Indonesia still relies on Scorpion tanks, which are a lightweight tank class. If seen from the condition of Indonesia's land surface and its geographical conditions, the Indonesian army should be supported by playing battle tanks, assault helicopters, and counterinsurgency aircraft.

The references used in determining the requirements of defense equipment are the provisions of general forces standards, Indonesian military standards that are transmitted by the ministry of defense, global military standards, NATO Agreement (STANAG) standards, etc. What is emphasized in the requirements of air, sea, and land defense equipment is extreme demand in the use of advanced and renewable technology, advanced engineering, and advanced workforce. Also an extensive cooperation with the research community, in this case, is a research and development institution, as well as the role of academics and university. Moreover, the most important thing is to generate leading technologies and competencies of great importance to civil society. Determination of defense product requirements is determined by the Defense Industry Policy Committee, which is based on user needs, namely the navy, air force and army, which are supported by the role of defense technology research and development institutions and academics in universities.

1.3. The technology of industry 4.0

Industrial technology 4.0 is a form of technological transition revolution in stages from the first level to the fourth level. Where each level of revolution has its characteristics. During the 1.0 industrial revolution, the industry utilized steam-powered engine technology as a tool to produce the desired product. In the 2.0 industrial revolution, Humans began to develop industrial technology that utilizes mass production machines by relying on electricity and fuel use. Whereas in the 3.0 industrial revolution, technology began to be developed that utilizes information technology networks and automated machines. Until in industry 4.0, machines that were integrated with the internet were developed. The reasons for the industrial revolution and the increasingly changing use of technology are productivity, efficiency, effectiveness, and competitiveness. Therefore, humans are competing to start developing technologies that can facilitate their activities so they can increase production productivity. The industrial revolution 4.0 with machine technology that is mutually integrated is not just an engineering problem. However, every component involved in compiling the machine becomes an integrated system that takes part in industry 4.0. (See figure 2).

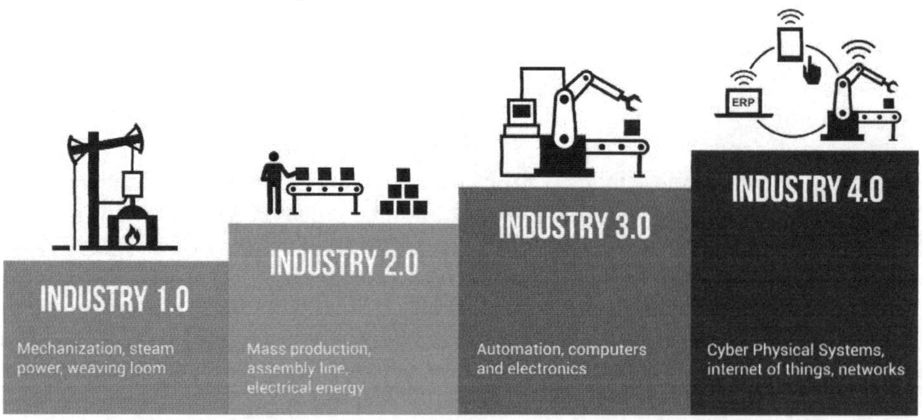

Figure 2. The industrial revolution [4].

IOP Conf. Series: Materials Science and Engineering **850** (2020) 012050 doi:10.1088/1757-899X/850/1/012050

Industry 4.0 is a unified system that communicates in real-time without time constraints by utilizing internet technology and cyber-physical systems to optimize the value of each process in the industry [5]. Meanwhile, according to Vasja Roblek et al., industrial technology 4.0 is a consequence of the use of internet technology connectivity to add value to functions and uses in both organizations and social environments [6]. The internet of things became the center of the industrial revolution. Industry 4.0 is related to the internet of things, industrial internet, and smart and cloud-based manufacturing. Also, it was stated that industry 4.0 is a strict limitation on the integration of humans and machinery in manufacturing [7]. The focus of technology development on industry 4.0 is shown in Figure 3.

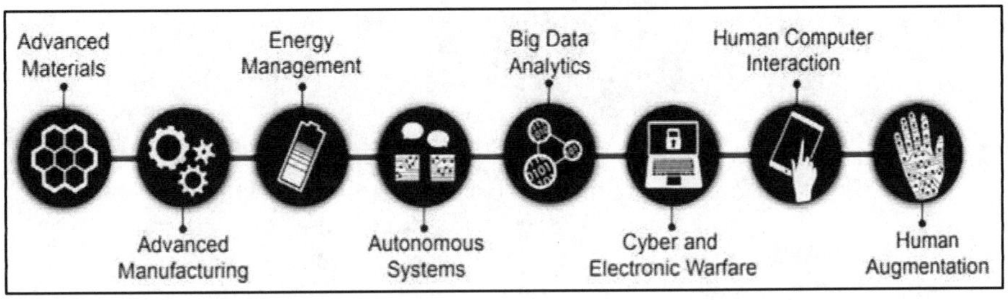

Figure 3. The focus of technology development [4].

1.4. Rare earth metals

Rare earth metal is an advanced materials group which has characteristics with a nanotechnology-based metamaterial structure to absorb or reflect electromagnetic waves, shape optimization, use in increasing computational power in small devices, and improved surfaces that are useful for coating technology. Some of the benefits contained in the use of rare-earth metals are, among others [8]:

- Rare earth metals used as advanced materials in the development of neo-magnets, namely magnets that have a magnetic field better than ordinary magnets so that the volume and weight of the magnetic mass can be optimized
- Manufacture of high strength low alloy (HSLA) steel, high carbon steel, super-alloy, and stainless steel in the metallurgical industry
- As a catalyst
- In the field of arbitration, earth metals are rarely used as control rods and radiation detector devices

Globally, rare earth metals mining commodity is calculated and relied upon to become a weapon to deal with trade wars like those carried out by China and the United States (US). In 2012, the Chinese government issued a policy to restrict the export of rare earth metals to the United States, which affected various fields such as the economy, politics, and high-tech products. The reasons given by the Chinese government regarding these restrictions are environmental problems and the presumption of rare earth metals as non-renewable natural resources [9]. The export restrictions carried out by China were then responded by the United States not as a significant obstacle because they have taken steps to limit the use of rare-earth metals from China and rearranged industrial supply chain mechanisms in the process of rare earth metal extraction and magnetic production [10]. In addition to China and the United States, there are five other countries which are listed as the most abundant rare earth producing countries in the world, namely Russia, Vietnam, Brazil, India, and Australia with a percentage as in Figure 4 [11].

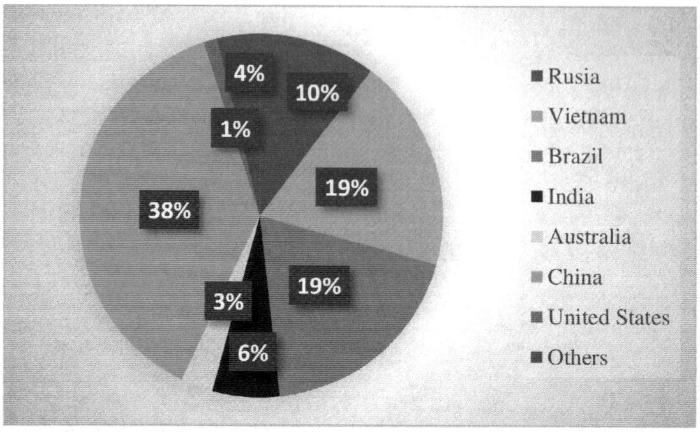

Figure 4. Countries with the most abundant rare earth content.

Rare earth metals, as the name implies, although it is rare but has a very high utilization value, so it is necessary to conduct a more in-depth study related to its use both for industrial development purposes in general and for the defense industry in particular. In Indonesia, rare earth metals can be found as follower metals in other minerals such as monazite, bauxite, zircon sand, alluvial gold and tin which can be found from large islands in Indonesia. In addition, in the Southeast Asian tin lane, there are world tin resources that pass through the Indonesian territory starting from Karimun Islands, Singkep, to the Bangka Belitung Islands. That are strategic potentials which can make a significant contribution to national development. The presence of rare earth metals in nature is not as a free element but as a complex compound bond. So, for its utilization, it is necessary to do the process of separating the bonds of these complex compounds. For example, in monazite minerals which are often found in Indonesia, they contain thorium in the form of oxides or thorium-phosphate salts [12]. In addition, other elements with relatively high concentrations, such as lanthanum (La) and Cerium (Ce) were found [13]. In addition, in the West Kalimantan region, analysis was carried out on a number of zircon sands namely 4,545.16 tons, 24.35 Kg of gold, and 520.07 tons of monazite obtained from a number of hypothetical resources in 402.67 Ha of a research area, giving information that there are 11.9 tons of rare earth metal [14].

2. Methods

This research was conducted by conducting literacy studies of literatures that discussed the use of rare earth metals and technological developments in raising the national defense industry. In general, the use of rare-earth metals used in battery technology. While its use in technology and the defense industry is still not widely discussed. For a discussion of technology and the defense industry that includes the development of industrial technology 4.0 is about the relationship between systems that are interconnected, implement interoperability functions, and are holistic.

In analyzing, the strategic planning evaluation method used is examined from parts of rare earth metal utilization, which include strengths, weaknesses, opportunities, and threats, or better known as SWOT analysis theory. The framework in this study is illustrated in Figure 5. Data analysis was carried out on secondary data from research results that supported each other.

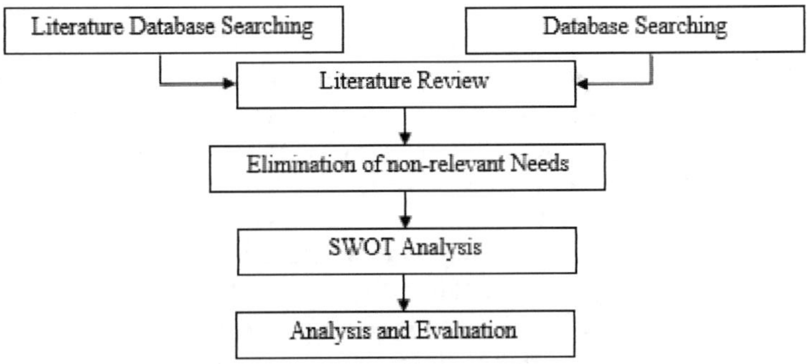

Figure 5. Research framework.

3. Results and discussion

In order to safeguard national sovereignty, the national defense needs to be strengthened and supported by the development and utilization of defense technology through the modernization of defense equipment that utilizes industrial technology development 4.0. The characteristics of defense technology are interoperability, where the whole system must be interconnected and holistic. The entire system, in addition to covering informatics, mechanical, cyber, and other systems, also includes the use of advanced materials. Indonesia is one country that has advanced material resources in the form of rare earth metals. Although its existence is not in the form of chunks and its quantity cannot be compared with China and the United States, but through a systematic, structured and sustainable processing and enrichment process can produce adequate results to support industries in Indonesia, including the defense industry. In law no. 16 of 2012 concerning the defense industry, things that need to be considered in meeting the needs of the national defense industry, namely defense and security equipment needed whether they can be made in Indonesia or must buy from other countries. It also concerns about defense industry involvement from every tier on 1 to 4 and the guarantee of trade returns, local content and/or offsets is at least 85% where the required fulfillment is at the beginning of 35% and a 5% increase every year.

The potential of natural resources in the form of rare earth metals that can be used in fulfilling the needs of raw materials is one of the potentials for technological development that will support the independence of the national defense industry. If based on the availability of rare earth metals in Indonesia, although they are still contained in complex compounds such as monazite, zircon sand, gold, alluvial tin, etc., the potential for their utilization can be described based on an analysis of strengths, weaknesses, opportunities, and threats it has. In addition to utilization in the field of electronic components, rare earth metals have also begun to be developed to become materials that can absorb radar waves. These materials can be used as a coating material in defense vehicles such as submarines, fighter aircraft, medium tanks, frigate ships, missile destroyers, and others. The development of rare earth metal as an anti-radar material in the X-band 8-12 GHz frequency range was carried out by BATAN which developed smart magnet material as a result of processing monazite sand into rare earth metals which were tested on the Patkamla Sadarin Navy Forces on Mutiara beach [15]. Radar wave absorbing material is important in developing defense capabilities as well as applied to Australia's defense capability standards, which conducts research on its application as smart material [16].

The characteristics of a material can absorb electromagnetic waves from the radar in two ways, namely by absorbing through a magnetic field and converting captured waves into heat energy [17]. Wave absorbent material must also have the ability to interact well between the electric field from the emission of radiation and the electrical components of electromagnetic radiation [18], and meet requirements such as layer thickness, mass of material that is light, has excellent durability, and has a frequency absorption range broad [19]. Wave absorbent material works by minimizing the radar cross-section (RCS), which is the object's ability to reflect the radar signal towards the receiver by performing a comparison between the backscatter per stationary (solid angle unit) of the target against the interrupted energy density [20]. In the field of defense, radar wave absorbent material has a very significant role in the development of stealth technology. These materials have the ability to prevent detection (radar, acoustics, infrared, etc.) by absorbing waves and suppressing the number of waves reflected from the metal structure found in the body of a combat vehicle [21].

Referring to the defense industry owned by developed countries such as the United States, rare earth metal needs in developing full stealth technology are 920 lbs or 418 Kg per 1 unit of F-35 fighter, 2359 Kg for 1 unit of destroyer belonging to the Arleigh United States Navy Burke DDG-51, and 4173 Kg for 1 unit of submarine with fast attack capability SSN-774 [22] (see figure 6). If based on the amount of defense budget held by the ministry of defense each year, it is known that the existing budget is not sufficient to develop rare earth metals as stealth technology in supporting defense forces. Therefore, the use of LTJ can be modified only as a mixture on the paint layer on military vehicles. It will reduce the amount of rare earth metal use and optimize the defense budget. Another solution that can be applied by the ministry of defense is to establish cooperation with local mining industries to enrich the elements so that the LTJ contained as waste material is not wasted.

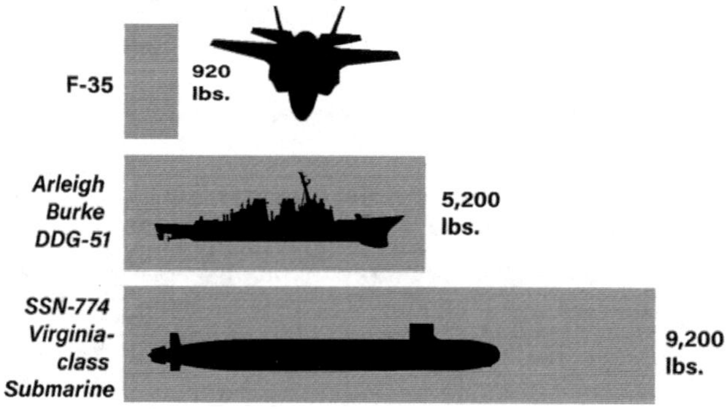

Figure 6. Rare earth ingredients.

If Indonesia can develop stealth technology in its defense products without having to depend on imported commodities from abroad, the Indonesian defense industry has been able to eliminate one of the obstacles to its independence, namely the embargo. The threat of an embargo on raw materials for rare earth metals as China did in the United States, not a frightening thing anymore. However, to carry out rare earth metal development in the Indonesian defense industry technology requires excellent and careful management of the availability of natural resources, human resources, and markets. Local mining industries, such as PT. Timah Tbk, as an example, can be developed to become a tier 4 defense industry that supports rare earth metal needs as a result of the enrichment of tin mining waste.

4. Conclusion

Based on the results of observations on the potential use of rare earth metals in the defense industry which is seen from its availability in nature and the amount of material utilized, the following points can be concluded:

- Even though it is not comparable to the amount of rare earth metal possessed by developed countries such as the United States and China, Indonesia has rare earth metal reserves which are sufficient to develop advanced materials in the defense industry.
- One of the opportunities for the use of rare-earth metals in the Indonesian defense industry is the development of stealth technology by utilizing anti-radar paint coatings.
- The potential for rare earth metal raw, material embargoes from abroad can be minimized by developing the 4th tier defense industry in Indonesia. For this reason, it is necessary to hold downstream industries to produce rare earth metals as part of the defense industry in Tier 4, namely raw materials.

- Weaknesses of the defense industry in developing rare earth metals in defense technology are caused by resource management that still needs to be improved, and there are limited defense budgets.

Meanwhile, there will be a lot of struggle to build up the industry on tier 4, which concern on raw materials, but it is worth to be developed because the benefit from its developing has a high-tech requirement completion.

References

[1] Rachmat A N 2016 Tantangan dan Peluang Perkembangan Teknologi Pertahanan Global Bagi Pembangunan Kekuatan Pertahanan Indonesia *Transformasi Global* **1**(2)
[2] Undang-undang No. 16 Tahun 2012 tentang Industri Pertahanan
[3] Pratama A N 2018 *Mengenal 5 Pesawat Andalan TNI AU (Kompas)* [Online] Retrieved from https://nasional.kompas.com/
[4] Shenoi R A, Bowker J A, Dzielendziak A S, Lidtke A K, Zhu G, Cheng F ... and Ross K 2015 *Global Marine Technology Trends 2030*
[5] Prasetyo H and Sutopo W 2018 Industri 4.0: Telaah Klasifikasi aspek dan arah perkembangan riset *J@ ti Undip: Jurnal Teknik Industri* **13**(1) 17-26
[6] Roblek V, Meško M and Krapež A 2016 A complex view of industry 4.0 *Sage Open* **6**(2) 2158244016653987
[7] Vaidya S, Ambad P and Bhosle S 2018 Industry 4.0–a glimpse *Procedia Manufacturing* **20** 233-238
[8] Suprapto S J 2009 Tinjauan tentang unsur tanah jarang *Buletin Sumber Daya Geologi* **4**(1) 36-47
[9] Tjarsono I and Suryadi B 2015 *Kebijakan Cina Membatasi Ekspor Mineral Logam Tanah Jarang (Ltj/rare Earth) Ke Amerika Serikat Tahun 2012* (Riau: Doctoral dissertation, Riau University).
[10] Green J A 2019 *what will the US defense industry do when China cuts off rare earth supplies?* [Online] Retrieved from https://www.defensenews.com/
[11] Ricardo E 2019 *Ini Data Cadangan Terbesar Rare Earth di 7 Negara* [Online] retrieved from https://www.cnbcindonesia.com/
[12] Herlani R, Muljono, Widiyati S and Mujari 2011 Mempelajari Pengaruh Logam Tanah Jarang Serium (Ce) dan Lantanum (La) pada Analisis Thorium dengan Metoda Pendar Sinar-X *Prosiding Seminar Penelitian dan Pengelolaan Perangkat Nuklir* ISSN **1410-8178**
[13] Fadliyah T 2017 *Peningkatan Kadar Ce Konsentrat $Ce(No_3)_4$ dengan Metode Ekstraksi menggunakan Ekstraktan Topo* (Makassar: Sekolah Menengah Analisis Kimia Makassar)
[14] Muliyana T and Putra E 2015 *Penyelidikan dan evaluasi potensi rare earth element (REE) dan mineral ikutan pada wilayah bekas tambang/tailing di Kecamatan Kendawangan* Kabupaten Ketapang, Provinsi Kalimantan Barat (Pusat Sumber Daya Geologi-Badan Geologi)
[15] BATAN 2019 *BATAN Kembangkan Teknologi Cat Anti Deteksi Radar pada Alutsista* [Online] retrieved from http://www.batan.go.id/
[16] Ibrahim M E, Scala C M, Wilson A R, Truong V T and Edwards D P 2009 *Advanced Applications of Smart Materials Research for the Enhancement of Australian Defence Capability* (No. DSTO-TN-0919) Defense Science And Technology Organization Victoria (Australia) Maritime Platforms DIV.
[17] Lee W J, Lee S E and Kim C G 2005 Tensile and electrical properties of polypyrrole/epoxy composites for radar absorbing materials *the 5th Korea-Japan Joint Symposium on Composite Materials* (pp. 25-26)
[18] Saville P, Huber T and Makeiff D 2005 *Fabrication of Organic Radar Absorbing Materials: A Report on the TIF Project* (No. DRDC-TR-2005-124) Defence Research And Development Atlantic Dartmouth (Canada).
[19] Hebeish A A, Elgamel M A, Abdelhady R A and Abdelaziz A 2008 Factors affecting the

performance of the radar absorbant textile materials of different types and structus *Progress In Electromagnetics Research* **3** 219-226

[20] Nicolaescu I 2006 Radar absorbing materials used for target camouflage *Journal of optoelectronics and advanced materials* **8**(1) 333

[21] Caffarena V R, Ogasawara T, Pinho M S and Capitaneo J L 2007 Synthesis and characterization of nanocrystalline Ba3Co0. 9Cu1. 1Fe24O41 powder and its application in the reduction of radar cross-section *Materials Science-Poland* **25**(3) 2007

[22] Grier P 2018 *Rare-Earth Uncertainty* [Online] Retrieved from http://www.airforcemag.com/

IOP Conf. Series: Materials Science and Engineering **850** (2020) 012051 doi:10.1088/1757-899X/850/1/012051

Electrospun of Poly (vinyl alcohol)/ Potassium hydroxide (PVA/KOH) nanofiber composites using the electrospinning method

M R Almafie[1], Z Nawawi[2], J Jauhari[3,*] and I Sriyanti[1,3]

[1] Department of Physics Education, Sriwijaya University, Palembang, Indonesia
[2] Department of Electrical Engineering, Sriwijaya University, Palembang, Indonesia
[3] Laboratory of Instrumentation and Nanotechnology Applications, Faculty of Computer Science, Sriwijaya University, Palembang, Indonesia

*jaidan_j@unsri.ac.id

Abstract. The Poly (vinyl alcohol)/Potassium hydroxide (PVA/KOH) nanofiber Composites for application as a supercapacitor electrolyte. Electrospinning method has been successfully used to synthesize composite of the PVA/KOH nanofiber. The PVA and KOH have been made with concentration are 5% w/w (PK1), 10% w/w (PK2) and 15% w/w (PK3), with a mass ratio of PVA/KOH is 10:1. The Physicochemical properties of PVA/KOH nanofibers with three various in the experiment were studied, including morphology, size, and chemical interaction. The microscope result shows that nanofiber of PK1, PK2 and PK3 have bead fiber and free-bead fiber, Where the PK1 is bead fiber and the PK2 and PK3 are free-bead fiber. The average diameter of PK1, PK2 and PK3 were 635, 826, 1021 nm, respectively. The FTIR results show that there is interaction between Poly (vinyl alcohol) and Potassium hydroxide (KOH) in the form of a spectrum and widening of transmittance are 3320 cm-1 and 3301cm-1 which identifies hydrogen bonds.

1. Introduction

Electrospinning was a spinning method used to produce microfiber to nanofiber [1-4]. The main components of electrospinning are the syringe, syringe pump, high voltage power supply, and collector [1,2,5]. The supporting components are charge-couple device (CCD) camera, monitor, humidity sensor [1,5]. The use of this apparatus starts with a voltage adjusted by high voltage power supply on a syringe that has a polymer solution. Polymer solutions will be attracted by the formation of a Taylor cone at the tip of the syringe and produce fiber on the collector [5,6]. The fiber formation was controlled through parameters such as viscosity, polymer, conductivity, voltage, a distance of the syringe tip to the collector, and flow rate [7]. Application of fiber produced through electrospinning methods for example tissue engineering [8], water filtration [9], air filtration [10], wound dressing [11,12], drug delivery [13,14], batteries [3] and supercapacitor [2,15].

The material used to produce fiber can come from various polymer syntheses through a combination of other materials, for example the synthesis of Poly (vinyl alcohol) (PVA) and Potassium hydroxide (KOH). PVA was chosen because it has high dielectric, semi-crystalline, non-toxic and low prices. But at high temperatures, there is instability in ionic conductivity [16]. This condition can be overcome through PVA synthesis with KOH, where KOH has a low melting point and high ionic conductivity

[17]. Comparison of ionic conductivity Lithium hydroxide (LiOH) 38,69 S.cm^2/mol and Sodium hydroxide (NaOH) 50,11 S.cm^2/mol lower than Potassium hydroxide (KOH) 75,51 S.cm^2/mol [18]. The synthesis of this second material is called PVA/KOH nanofiber composite which can be applied as an electrolytic capacitor. Electrolytes are ionized materials that become electrical conductors. KOH is a strong electrolyte because it is capable of perfectly ionizing into ion K and ion OH [19]. PVA-based electrolyte studies have been reported for example PVA/SC[20], PVA/LiCF$_3$SO$_3$ [21], PVA/Chitosan [22], However, PVA/KOH nanofiber composites have never been reported.

In this study we have synthesized PVA/KOH composite nanofiber using the electrospinning method. The morphology and diameter of the nanofiber were observed using a digital microscope and the chemical interactions between PVA and KOH were investigated using Fourier Transform Infrared Ray (FTIR).

2. Experimental methods

Synthesis of PVA/KOH nanofiber using PVA with MW 89.000 g/mol, KOH with MW 56 g/mol and distilled water (H_2O). Both materials are dissolved into distilled water respectively, where the PVA solution are set at concentrations at 5%, 10% and 15% (w/w). through stirring for 60 minutes at 60°C. While the concentration of KOH solution is set at 2.5% (w/w) without stirring. Then, both solutions are synthesized with a mass ratio of 10:1 and stirred again for 90 minutes at 80°C. After compositing each solution is labeled PK 1, PK 2, and PK 3 as intended in table 1.

Table 1. Formula sintesis PVA/KOH.

Label	Concentration PVA % (w/w)	Concentration KOH % (w/w)	Mass Ratio
PK 1	5	2,5	10:1
PK 2	10	2,5	10:1
PK 3	15	2,5	10:1

The solution is inserted into the syringe, then the electrospun uses electrospinning apparatus (Nachriebe 601) illustrated in the schematic Figure 1. Throughout electrospun, the 0.8 mm diameter syringe is placed in the humidity chamber (RH 60%). The solution flows due to being driven by the pump with speed 40 μl/h. Voltage 19 kV DC given so that the solution is pulled towards collector, it is required to establish a Taylor cone monitored using a charge-coupled device camera (CCD-Camera). The nanofiber is collected in a drum-shaped collector that has a speed of 200 rpm and the distance of the syringe tip to the collector is 10 cm.

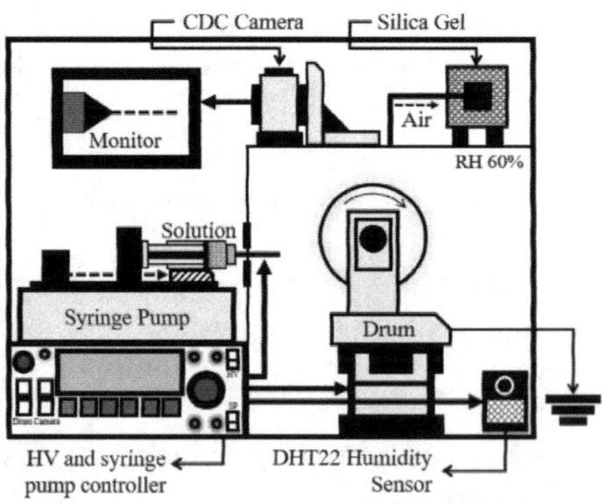

Figure 1. Schematic illustration of electrospinning apparatus.

The morphology of PK 1 nanofiber, PK 2 and PK 3 was observed using a digital microscope. The diameters of PK 1 nanofiber, PK 2 and PK 3 were measured using Image J 1.52a and the size distribution was analyzed using OriginPro 2018 ver. 9.5. The chemical interactions of PK1, PK2 and PK3 nanofiber were investigated using the Fourier Transform Infrared Ray (FTIR) spectrum range 500 – 4000 cm^{-1}.

3. Results and discussion

3.1. Poly (Vinyl alcohol)/ Potassium hydroxide image

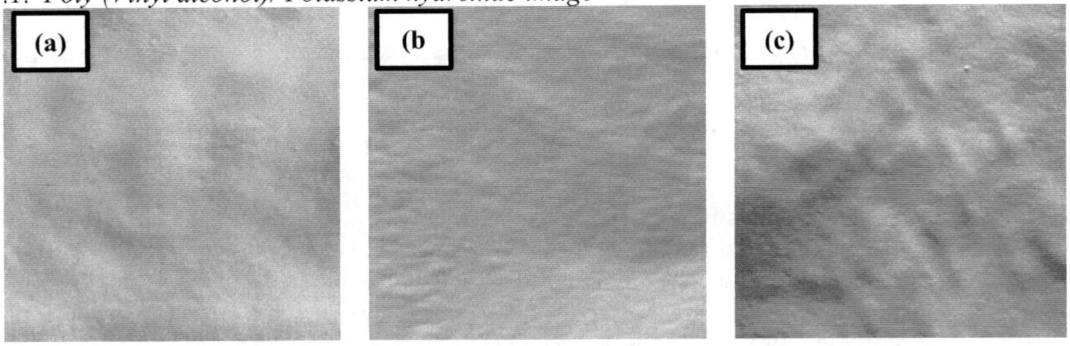

Figure 2. Nanofiber PVA/KOH (a) PK1, (b) PK 2, (c) PK 3.

The microscopic electrospun results using the electrospinning method are shown in figure 2. PK 1, PK 2 and PK 3 nanofiber has elastic, flexible, smooth white surface and is homogeneous. The formation of nanofiber is influenced by the balance between PVA/KOH solution and electrospinning process parameters. The mass ratio of PVA/KOH solution which is 10:1 is the maximum composition to produce the optimum nanofiber. Electrospun starts with synthesizing PVA and KOH. After composite, the solution is put into the syringe placed on the syringe pump. Then the solution is driven with a constant regulated speed. The DC voltage source is connected to the end of the syringe, acting as a positive electrode. Meanwhile, the negative electrode is the drum collector. These electrodes can be exchanged for poles. When the two electrodes are subjected to a very high electrical voltage, the solution which is initially totally neutral will be polarized due to the influence of the second potential difference of the electrode. If the syringe is connected to the positive pole of the DC voltage source, the surface of the solution will accumulate positive ions, while the negative ions gather in the middle such that the total electric field due to the polarization of the ions reaches zero [23]. Conversely, if the needle is connected to the negative pole of the DC voltage source, the negative ions accumulate on the surface of the solution, while the positive ions polarize in the middle of the solution. The charge in this polymer solution is spread throughout all parts of the solution, including on the surface of the solution. The accumulation of charges in solution causes the appearance of the Coulomb repulsive force between charges with the equation:

$$\vec{F_c} = k\frac{q_1 q_2}{r_{21}^2} \quad (1)$$

with Coulomb force and electric force due to the field between the two electrodes, namely:

$$F_{electric} = q\vec{E} \quad (2)$$

The influence of Coulomb force causes the polymer solution to be attracted to form a cone called Taylor Cone as shown in figure 3. When the resultant of these two forces exceeds the surface tension force of the polymer solution, which works in contrast to the Coulomb force and the electric force the charged polymer solution is accelerated by an electric field towards the collector, so that the polymer solution is attracted to and emits a very small size. As long as the polymer solution is attracted, the polymer undergoes elongation and the solvent evaporates due to the influence of the high voltage applied, so that the polymer formed at the surface of the collector is in the form of fiber in solid form [6,24].

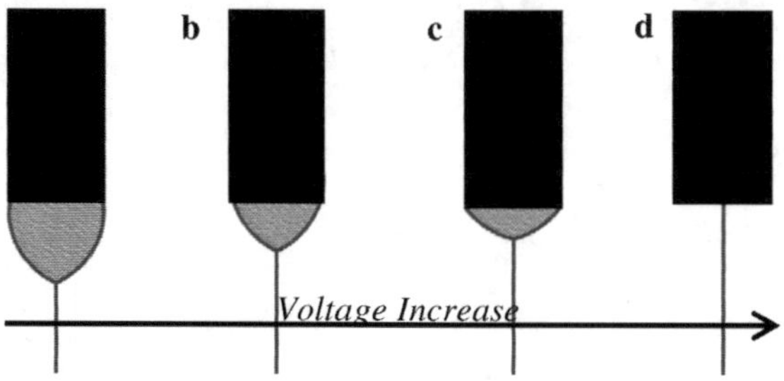

Figure 3. Illustration of a cone jet model if voltage is increased.

3.2. Morphology and diameter PVA/KOH nanofiber

The morphology and diameter distribution of PVA/KOH nanofiber is shown in Figure 4. PVA/KOH composites at low concentrations (5% (w/w) produce bead fiber, while at high concentrations (10% and 15% (w/w)) produce free-bead fiber. The morphology formed is influenced by viscosity which is a measure of the thickness of a fluid in inhibiting the fluid force, expressed in the concentration of the solution [25]. Viscosity also represents the extent to which the polymer chain attaches in solution. At low conditions the interaction between solvent molecules is more dominant than the interaction between polymer molecules and the interaction between polymer molecules and solvent molecules. Thus, the chain of bonds of polymer molecules is produced too little and evaporation of the air is faster as a result of bead fiber. In addition, solutions with low viscosity tend to form smaller size nanofiber while high viscosity solutions produce nanofiber of a larger size or free bead [26]. In this study, nanofiber diameter sizes increased from low concentrations (5% (w/w)) to high concentrations (15% (w/w)) such as PK 1 (635 nm), PK 2 (826 nm), and PK 3 (1021 nm).

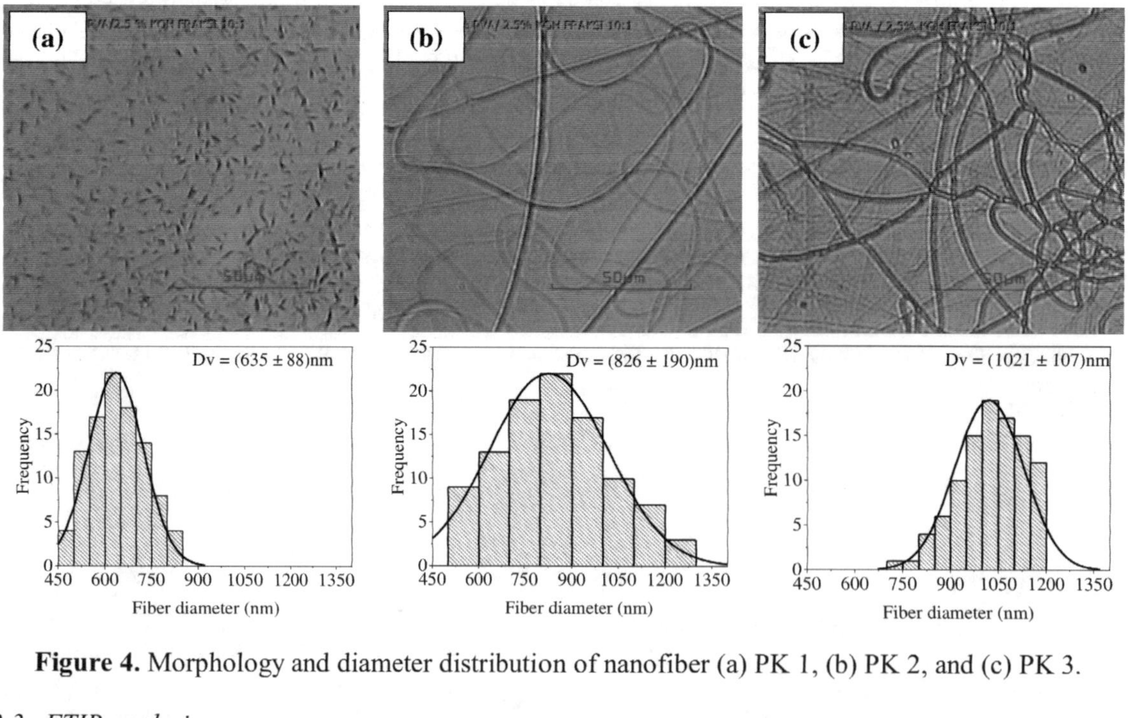

Figure 4. Morphology and diameter distribution of nanofiber (a) PK 1, (b) PK 2, and (c) PK 3.

3.3. FTIR analysis

FTIR is used to detect specific groups of polymers. The wave number used is in the range 500-4000 (cm^{-1}). Figure 5 showing the FTIR PVA spectrum, the peaks of PVA molecular transmittance are 3297 cm^{-1} a O-H stretching on the hydroxyl group, and peak 2942 and 2914 cm^{-1} identify asymmetrical and symmetrical CH2 stretching on alkyl groups [27–29]. Peak 1714 cm^{-1} identify absorption H_2O and peak 1427 cm^{-1} and 1244 cm^{-1} shown CH_2 beading while strengthening the bond between OH and CH [27–29]. While 1141 cm^{-1} and 1083 cm^{-1} each identifies a C−O stretching on ester groups, C=O stretching on aldehyde groups [27,29]. 916 cm^{-1} and 840 cm^{-1} each identifies CH_2 rocking and C–C rocking alkyl groups. Figure 6 show the KOH FTIR spectrum, the molecular transmittance peaks of KOH are 3500 cm^{-1} and 1405 cm^{-1} a OH strengthening in the hydroxyl group. Peak 1749 cm^{-1} and 1660 cm^{-1} identify absorption H_2O. While 885 cm^{-1} and 654 cm^{-1} identify the bond between K–O [19,30,31].

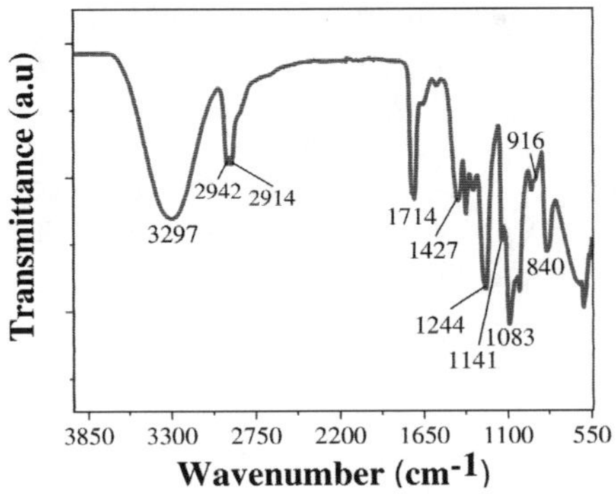

Figure 5. FTIR spectrum of Poly(vinyl alcohol) (PVA).

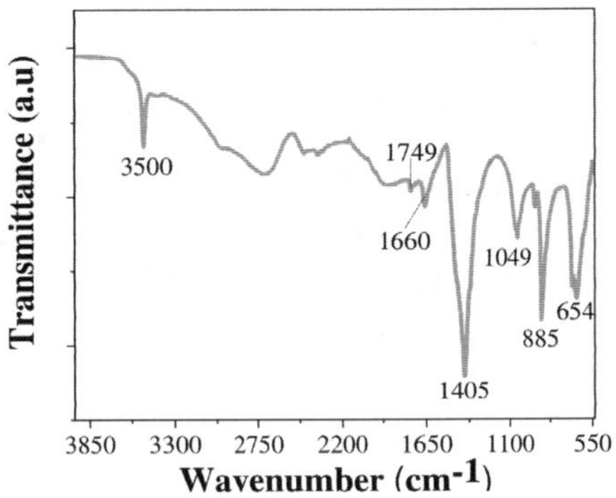

Figure 6. FTIR spectrum of Potassium hydroxide (KOH).

The results of PVA/KOH FTIR analysis of PK 1, PK 2, PK 3 nanofiber as in figure 7 show the change in intensity of PVA transmittance when combined with KOH. Addition of PVA in PVA/KOH concentrations also caused changes in peaks of PK 1, PK 2 and PK 3 nanofiber. First, in the range of 3800-3000 (cm^{-1}) there was an increase in the intensity of transmittance, where PK 3 nanofiber had a peak greater than PK 1 and PK 2. In this range, the three nanofiber have a tendency to peak similarity to the PVA. The increase in concentration causes an O–H shift in the hydroxyl group towards a larger number PK 1 (3301 cm^{-1}), PK 2 (3307 cm^{-1}) and PK 3 (3320 cm^{-1}).

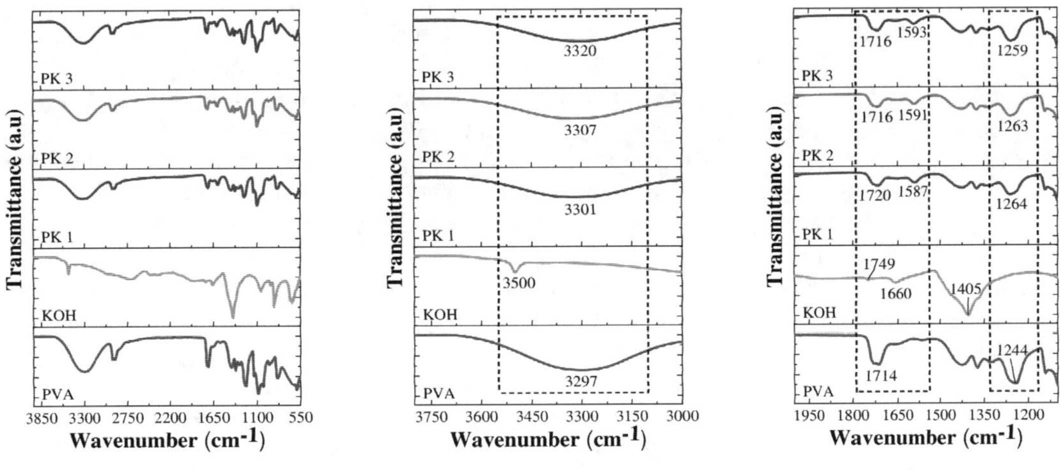

Figure 7. FTIR spectra of PVA/KOH PK 1, PK 2 and PK 3 nanofiber.

Second, the increase in PVA in PVA/KOH causes absorption by water to increase in the range 2000-1100 (cm^{-1}). When observed, PK 1 (1720 cm^{-1}), PK 2 (1716 cm^{-1}), PK 3 (1716 cm^{-1}) nanofiber weakened PVA transmittance (1714 cm^{-1}) which is very significant, this indicates that PVA is hydrophilic. Then, peak appearance PK 1 (1587 cm^{-1}), PK 2 (1591 cm^{-1}), PK 3(1593 cm^{-1}) indicates KOH interaction (1660 cm^{-1}) through absorption H_2O, where KOH is a strong base that is also hydrophilic. Third, an increase in PVA/KOH nanofiber makes a peak shift PK 1(1264 cm^{-1}), PK 2 (1263 cm^{-1}), PK 3 (1259 cm^{-1}) with increasingly sharp transmittance, this condition shows CH2 beading while strengthening the bond between OH and CH. Changes in peak FTIR in PK 1, PK 2 and PK 3 are seen as the result of PVA/KOH interactions.

4. Conclusion

PVA/KOH nanofiber has been successfully synthesized using the electrospinning method. PVA/KOH nanofiber is divided into PK 1, PK 2 and PK 3 characterized in the morphology, diameter and interaction of the PVA/KOH molecular chain. The three nanofiber have a morphology of elastic, flexible, smooth, white and homogeneous texture. The results of the microscope show that PK 1 is a fiber bead and PK 2 and PK 3 are free-bead fiber. The average diameter size of PK1, PK2 and PK3 is 635, 826, 1021 nm. FTIR results showed peaks between the hydroxyl OH group bonds, H_2O bonds by H_2O absorption, and CH_2 beading as well as strengthening the bond between OH and CH. Changes in peak FTIR in PK 1, PK 2 and PK 3 are seen as the result of PVA/KOH interactions.

Acknowledgments

This research was supported by Sriwijaya University, Republic of Indonesia under the University's Excellence Research (PUPT) Grant in the fiscal year 2019. Also, this research was supported by BPDPKS, the Republic of Indonesia under the Research and Development of Palm Oil in the fiscal year 2019.

References

[1] Ramakrishna S, Fujihara K, Teo W-E, Lim T-C and Ma Z 2005 *An Introduction to Electrospinning and Nanofibers* (World Scientific Publishing)

[2] Liu X, Marlow M N, Cooper S J, Song B, Chen X, Wu B and Brandon N P 2018 Flexible all-fiber electrospun supercapacitor *J. Power Sources* **384** 264–9

[3] Li L, Peng S, Lee J K Y, Ji D, Srinivasan M and Ramakrishna S 2017 Electrospun hollow nanofibers for advanced secondary batteries *Nano Energy* **39** 111–39

[4] Sriyanti I and Jauhari J 2019 Synthesis of polyvinyl acetate (PVAc) fibers using needleless electrospinning technique with straight wire electrode *J. Phys. Conf. Ser.* **1166**

[5] Yousefzadeh M 2016 Modeling and simulation of the electrospinning process *Electrospun Nanofibers* (Elsevier Ltd.) pp 277–301

[6] Khajavi R and Abbasipour M 2016 Controlling nanofiber morphology by the electrospinning process *Electrospun Nanofibers* (Elsevier Ltd.) pp 109–23

[7] Bhattarai R S, Bachu R D, Boddu S H S and Bhaduri S 2019 Biomedical applications of electrospun nanofibers: Drug and nanoparticle delivery *Pharmaceutics* **11**

[8] Ye K, Kuang H, You Z, Morsi Y and Mo X 2019 Electrospun Nanofibers for Tissue Engineering with Drug Loading and Release *Pharm.* **11**

[9] Tlili I and Alkanhal T A 2019 Nanotechnology for water purification: electrospun nanofibrous membrane in water and wastewater treatment *J. Water Reuse Desalin.*

[10] Zulfi A, Munir M M, Hapidin D A, Rajak A, Edikresnha D, Iskandar F and Khairurrijal K 2018 Air filtration media from electrospun waste high-impact polystyrene fiber membrane *Mater. Res. Express* **5** 35049

[11] Jauhari J, Wiranata S, Rahma A, Nawawi Z and Sriyanti I 2019 Polyvinylpyrrolidone/cellulose acetate nanofibers synthesized using electrospinning method and their characteristics *Mater. Res. Express* **6** 64002

[12] Sriyanti I, Edikresnha D, Rahma A, Munir M M, Rachmawati H and Khairurrijal K 2017 Correlation between Structures and Antioxidant Activities of Polyvinylpyrrolidone/ Garcinia mangostana L. Extract Composite Nanofiber Mats Prepared Using Electrospinnin *J. Nanomater.* **2017**

[13] Sriyanti I and Jauhari J 2019 Electrospun of poly(vinyl alcohol) nanofiber as carrier of Garcinia

mangostana L. pericarp extract *J. Phys. Conf. Ser.* **1170** 12056

[14] Sriyanti I, Edikresnha D, Rahma A, Munir M M, Rachmawati H and Khairurrijal K 2018 Mangosteen pericarp extract embedded in electrospun PVP nanofiber mats: Physicochemical properties and release mechanism of α-mangostin *Int. J. Nanomedicine* **13** 4927–41

[15] Chee W K, Lim H N, Andou Y, Zainal Z, Hamra A A B, Harrison I, Altarawneh M, Jiang Z T and Huang N M 2017 Functionalized graphene oxide-reinforced electrospun carbon nanofibers as ultrathin supercapacitor electrode *J. Energy Chem.* **26** 790–8

[16] Doble M and Kumar A 2005 CHAPTER 9 - Degradation of Polymers ed M Doble and A B T-B of I E Kumar (Burlington: Butterworth-Heinemann) pp 101–10

[17] Yan Y 2016 2 - Developments in fibers for technical nonwovens *Woodhead Publishing Series in Textiles* ed G B T-A in T N Kellie (Woodhead Publishing) pp 19–96

[18] Zhong C, Deng Y, Hu W, Qiao J, Zhang L and Zhang J 2015 A review of electrolyte materials and compositions for electrochemical supercapacitors *Chem. Soc. Rev.* **44** 7484–539

[19] Patterson B A, Call T G, Jardine J J, Origlia-Luster M L and Woolley E M 2001 Thermodynamics for ionization of water at temperatures from 278.15 K to 393.15 K and at the pressure 0.35 MPa: apparent molar volumes of aqueous KCl, KOH, and NaOH and apparent molar heat capacities of aqueous HCl, KCl, KOH, and NaOH *J. Chem. Thermodyn.* **33** 1237–62

[20] Lalia B S, Alkaabi M and Hashaikeh R 2015 Sulfated Cellulose/Polyvinyl Alcohol Composites as Proton Conducting Electrolyte for Capacitors *Energy Procedia* **75** 1869–74

[21] Malathi J, Kumaravadivel M, Brahmanandhan G M, Hema M, Baskaran R and Selvasekarapandian S 2010 Structural, thermal and electrical properties of PVA–LiCF3SO3 polymer electrolyte *J. Non. Cryst. Solids* **356** 2277–81

[22] Abraham A, Soloman P A and Rejini V O 2016 Preparation of Chitosan-Polyvinyl Alcohol Blends and Studies on Thermal and Mechanical Properties *Procedia Technol.* **24** 741–8

[23] Andrady A L 2008 *Science and technology of polymer nanofibers* (New Jersey: John Wiley and Sons)

[24] Seeram Ramakrishna K F and Ma W-E T T-C L & Z 2005 *An Introduction Electrospinning and Nanofibers* (Singapore: World Scientific Publishing)

[25] Nezarati R M, Eifert M B and Cosgriff-Hernandez E 2013 Effects of Humidity and Solution Viscosity on Electrospun Fiber Morphology *Tissue Eng. Part C Methods* **19** 810–9

[26] Salas C 2016 *Solution electrospinning of nanofibers* (Elsevier Ltd.)

[27] Jipa I M, Stoica A, Stroescu M, Dobre L M, Dobre T, Jinga S and Tardei C 2012 Potassium sorbate release from poly(vinyl alcohol)-bacterial cellulose films *Chem. Pap.* **66** 138–43

[28] Chatterjee B, Kulshrestha N and Gupta P N 2016 Nano composite solid polymer electrolytes based on biodegradable polymers starch and poly vinyl alcohol *Measurement* **82** 490–9

[29] Jia Z, Li Z, Li S, Li Y and Zhu R 2016 Adsorption performance and mechanism of methylene blue on chemically activated carbon spheres derived from hydrothermally-prepared poly(vinyl alcohol) microspheres *J. Mol. Liq.* **220** 56–62

[30] Mahmud Z S, Zaki N H M, Subban R H Y, Ali A M M and Yahya M Z A 2012 MG49-KOH-PC alkaline gel polymer electrolytes membrane for supercapacitors *CHUSER 2012 - 2012 IEEE Colloq. Humanit. Sci. Eng. Res.* 621–6

[31] Cebela M, Jankovic B, Hercigonja R, Lukic M, Dohcevic-Mitrovic Z, Milivojevic D and Matovic B 2017 Comprehensive characterization of BiFeO3 paowder synthesized by the hydrothermal procedure *Process. Appl. Ceram.* **10** 201–8

IOP Conf. Series: Materials Science and Engineering **850** (2020) 012052 doi:10.1088/1757-899X/850/1/012052

Analysis characteristic of cholesteryl acrylate conductivity with the addition of the PE-b-PEG doped ITO-nanoparticle in the development of liquid crystal technology

M Ismail*, A Afrizal, D I Syafei and M A Puspita

Department of Chemistry, Faculty of Mathematics and Natural Science, Universitas Negeri Jakarta, Jl. Rawamangun Muka, Jakarta, Indonesia

*mikanimochinus@gmail.com

Abstract. The study of liquid crystal material is very interesting and has the opportunity as a smart material such as cholesteryl acrylate. Liquid crystal cholesteryl acrylate can be made shynthesized by Acryloiloxy Butyloxy Benzoate (ABB) precursor and cholesterol. Cholesteryl acrylate was polymerized by the ultraviolet curing (UV-Curing) method to keep the crystal structure stable. During the polymerization process some ingredients are added which can increase and regulate the formation of liquid crystal structure phases. Cholesteryl acrylate can be doped by adding conductive oxide in the form of nano particle scale Indium Tin Oxide (18 nm). The optimum conductivity results were 5.70 x 10^{-7} S/m with the addition of ITO 10% w/w. The structure of the crystal phase formed can be directed by the addition of Polyethylene-block-Polyethylene Glycol during the UV-Curing process. The hydrophobic part of PE block will interact with hydrophobic liquid crystal molecules, while the hydrophilic part of the PEG block has a high affinity for the hydrophilic ITO electrode substrate. The added PE-b-PEG varies from 0.001 to 0.005 percent by weight of polymer cholesteryl acrylate composite. Conductance data is measured using LCR meters from a frequency of 50 Hertz to 5000 kHertz. The average conductance data for ITO is 0.09597838 S, cholesteryl acrylate is 5.20635 x 10^{-6} S, the polymer cholesteryl acrylate composite with doped ITO is 4.58284 x 10^{-6} S. When added PE-b-PEG 0,003 percent composite weight is obtained the highest average conductance data is 1.24234 x 10^{-5} S with an average conductivity value of 6.09 x 10^{-5} S/m. The varies composition of PE-b-PEG content is very influential to regulate the arrangement of crystals formed in electrical conductivity.

1. Introduction

Liquid crystal is a unique material and has very broad potential in various aspects of today's technology. A number of studies related to liquid crystalline material have been widely studied following the needs of the times. The researchers continue to strive to find new formulas in developing quality improvement materials such as liquid crystals. Today, liquid crystal material has been widely studied in the development of display technology [1]. Applications of liquid crystalline material can be used as biosensors such as temperature sensors, sensors that detect radiation-prone areas [2], and as an electro-optical material in a Liquid Crystal Display device [3]. This is because the properties of liquid crystals can polarize light, form thin, have an effective electric field response, and have low energy and voltage requirements [4]. In addition, liquid crystal material also has several advantages, namely when it wants to be synthesized, the material needed is easily obtained, and is more environmentally friendly because it comes from organic material [5].

Liquid crystalline cholesteryl acrylate material is a material that can be made by reacting it using Acryloiloxy Butyloxy Benzoate (ABB) and cholesterol precursors. The ABB precursor react with cholesteryl acrylate to form a collesteric structure [6]. Each layer in a colonic structure has a helical molecular structure on a regular basis. The layer in the collesteric structure also has a high level of stability. The value of the distance between the crystalline molecular liquid molecules which is twisted is called the pitch value in one full twist 360° [7]. The distance between the crystals can be inserted by the Indium Tin Oxide (ITO) nanoparticle dopant which serves to increase the conductivity value [8]. In addition to direct the arrangement of phases, Polyethylene-block-Polyethylene Glycol (PE-b-PEG) can be used which has hydrophobic and hydrophilic properties [9].

Further studies related to the liquid crystalline cholesteryl-acrylate when added PE-b-PEG have not been carried out. The results of the last study of liquid crystalline cholesterl acrylate were cholesteryl acrylate polymerized and coupled with ITO nanoparticles. The combination of nematic liquid crystals with a dopant of ITO nanoparticles can reduce the resistivity value from 7.73×10^{11} to 1.21×10^{11} Ω/m [10]. The ITO nanoparticle dopant used as a dopant from liquid crystalline cholesteryl acrylate gave the highest conductivity value of 5.70×10^{-7} S/m with an ITO concentration of 10% w/w [11]. The addition of PE-b-PEG arranged as a block copolymer is expected to be able to interact with liquid crystal molecules assembling helical arrangements to form a phase of colesteric structure [9]. Cholesteric phase is a mesophase structure condition from liquid crystalline cholesteryl acrylate which can conduct electric current [12].

The results obtained from this study are conductance data. Conductance data is very important as a reference to determine whether this material is a conductor, semiconductor, and insulator. Conductance is the ability of an object to deliver an electric current from a material and in SI units it has siemens (S) units. Where high conductant values indicate that the material is able to condition electric current well, a low conductance value indicates that the material cannot flow the charge perfectly [13]. Some metal conductive dopants that can modify the conductivity properties of collesteric liquid crystals are shown in Table 1.

Table 1. List of several conductive dopants in a polymeric liquid crystal [14].

Dopant	Particle Size	Density (kg/m3)
Carbonyl Iron	1.1 μm	7000
Indium Tin Oxide-ITO	3.38 μm	1200
Metal splinters of Aluminium	12 μm	2700

1.1. Electrical conductivity

In the intrinsic semiconductor, the electron concentration in the conduction energy band (n) equals the hole concentration in the valence energy band (p) so that with index i expressing "intrinsic", h is the Planck constant, m∗e and m∗h are mass respectively. Effective electrons and holes and Eg are energy gaps [15]. Based on the formula from Kittel, conductivity can be calculated by equation:

$$n_i = p_i = 2\left(\frac{2\pi k_B T}{h^2}\right)^{3/2} (m_e^* m_h^*)^{3/4} \, \mathrm{e}^{-E_g/(2k_B T)} \tag{1}$$

Electrical Conductivity (σ) in semiconductor material is strongly related to electron mobility (μh) and hole mobility (μh), the equation is:

$$\sigma = e(n\mu_e + p\mu_h) = en_i(\mu_e + \mu_h) \tag{2}$$

Equation (2) shows that conductivity depends on the concentration of the electric charge carrier. Each semiconductor material has a different gap energy. Silicon has an energy gap of 1.1 eV, germanium has a gap energy of 0.7 eV, and the dopant of ITO nanoparticles has an energy gap of 0.2 eV to 2.5 eV because it is a semiconductor material [16]. Calculations for gap energy can be formulated by Jorena with:

$$\mathrm{Eg} = ln\frac{\sigma\, o}{\sigma}\, 2\ \mathrm{kT} \tag{3}$$

Where σ_o has constant 2.173913043, σ as the conductivity value, k is the Boltzman constant value (8.62 x 10^{-5} eV), and T is the temperature value. In addition to that the electrical conductivity produced can also be formulated:

$$\sigma = G\ \frac{L}{A} \tag{4}$$

Where G is the conductance value (S), L is the thickness of the plate (1.96 x 10^{-3} m), and A is the cross-sectional area of the plate (4 x 10^{-4} m). Permitivity can also be measured through equations:

$$\varepsilon = \frac{C\ \times\ L}{A\ \varepsilon o} \tag{5}$$

Where C is the sample capacitance value (F), and ε_o is the relative permittivity value in vacuum which is equal to (8.85 x 10^{-12} C^2/Nm^2), L is the sample plate thickness and A is the cross-sectional area of the sample plate.

2. Methods

The steps carried out in this study were started by making Acryloiloxy Butyloxy Benzoate (ABB) precursors which were reacted with cholesterol to form a cholesteryl acrylate monomer. In order to maintain the structure of cholesteryl acrylate in order to remain stable photopolymerization process was carried out using the in Situ method by providing ultraviolet light radiation at a certain time. During the photopolymerization process a number of supporting materials were added which consisted of Indium Tin Oxide (ITO) and Polyethylene-block-Polyethylene Glycol (PE-b-PEG). The fixed variable is the cholesteryl acrylate monomer and ITO nanoparticle dopant, while the independent variable is a variation of PE-b-PEG which consists of a variation range of 0.001 to 0.005. Furthermore, the photopolymerization samples measured their electrosity using LCR Meter type Hioki 3532-50.

3. Results and discussion

The conductance value obtained can be measured using equation (4) and the permittivity value can be measured using equation (5). The frequency range used is starting from the smallest frequency of 5 Hz to the largest frequency of 5 MHz.

Based on Table 2, it can be seen that the cholesteryl acrylate is an organic material can be increased the conductivity value if it is given an ITO nanoparticle and PE-b-PEG material. ITO nanoparticles are conductive oxides which have a large conductivity of 0.47 S / m. The addition of PE-b-PEG can increase the value of conductivity because the helical twisted molecular structure forming the colesteric phase can form its own structure directed by PE-b-PEG. The addition of ITO nanoparticles dopants and PE-b-PEG material can increase the conductivity value. Therefore, the polymeric acrylate material that has been made in such a condition is classified as a semiconductor material. The smallest energy gap of the liquid crystalline polyester acrylate material is 0.135 eV derived from the addition of PE-b-PEG 0.003. The smaller the energy gap a material makes it easier for electrons to move from the valence band to the conduction band. A material if given an external electric field through an electrode, it can be seen the electrical phenomenon that appears (conductance and capacitance). Changing voltage frequency can cause a change in direction of electric dipole moment. If the material's dipole moment is more uniform, then this condition will reduce the external electric field.

Table 2. Conductance data, capacitance, conductivity, and permittivity of liquid crystalline cholesteryl acrylate.

Sample	Conductance (S)	Capacitance (F)	σ (S/m)	ε_o
Cholesteryl Acrylate	5.20635×10^{-6}	9.46×10^{-11}	2.55111×10^{-5}	52.35858236
ITO	0.095978	2.75×10^{-6}	0.470294062	1,520,613.148
PE-b-PEG	2.47611×10^{-5}	2.1×10^{-7}	0.000121329	116,079.5419
CA + ITO	4.58284×10^{-6}	5.37×10^{-11}	2.24559×10^{-5}	29.71871593
CA + ITO + PE-b-PEG 0,001	1.16643×10^{-5}	1.80×10^{-10}	5.71551×10^{-5}	99.8316056
CA + ITO + PE-b-PEG 0,002	1.18222×10^{-5}	1.03×10^{-10}	5.79286×10^{-5}	56.9674339
CA + ITO + PE-b-PEG 0,003	1.24234×10^{-5}	2.69×10^{-10}	6.08745×10^{-5}	148.9783775
CA + ITO + PE-b-PEG 0,004	8.06323×10^{-6}	7.65×10^{-11}	3.95098×10^{-5}	42.33352619
CA + ITO + PE-b-PEG 0,005	8.15005×10^{-6}	7.5×10^{-11}	3.99352×10^{-5}	41.50156322

The ability of the liquid crystalline cholesteryl acrylate material on delivering the largest electric current occurs when giving frequencies around 4 MHz to 5 MHz. Liquid crystalline cholesteryl acrylate material added with 0.003 PE-b-PEG material obtained greater conductance data compared to the addition of other PE-b-PEG materials. The average conductance data of cholesteryl acrylate dopant ITO nanoparticles under PE-b-PEG 0.003 is 1.242×10^{-5} S with conductivity values of 6.087×10^{-5} S / m. The optimum condition does not significantly influence the increase in the value of the electrical conductivity produced (Figure 1). The conductance value is influenced by the movement of the ion charge, the density of the charge, and the strength of the electric field produced.

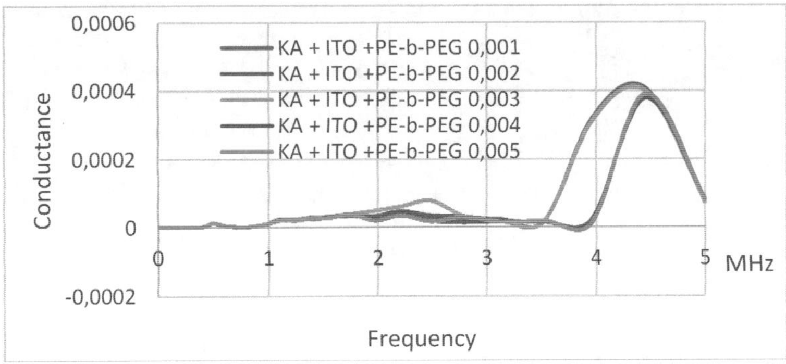

Figure 1. Relationship between conductance and frequency.

The addition of PE-b-PEG did not have a significant effect on the conductivity value of liquid crystalline cholesteryl acrylate. The biggest percentage is only giving effect as much as 58% when adding PE-b-PEG 0.003 (Figure 2). This is because PE-b-PEG only functions as a material that directs the formation of a collesteric structure in a polymeric cholesteryl acrylate.

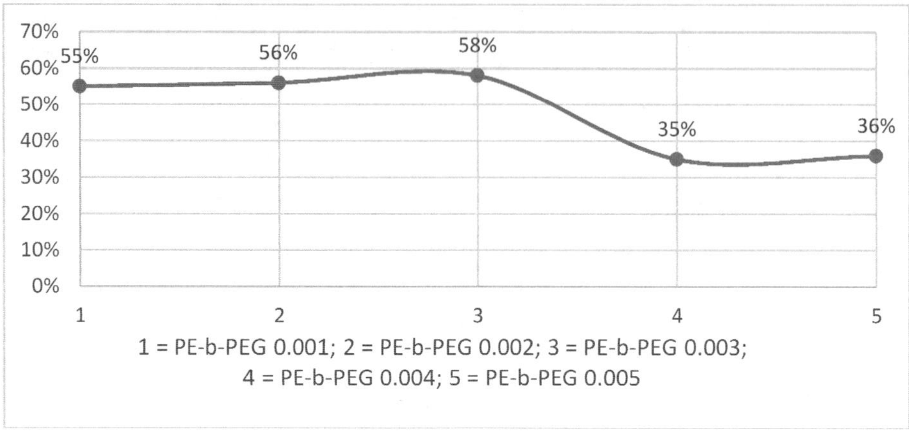

Figure 2. Percentage of influence of PE-b-PEG on conductivity.

The size of the amount of electric charge stored for an electric potential can be measured based on capacitance data (Figure 3). The highest average capacitance value is obtained when the addition of PE-b-PEG 0.003 which is equal to 2.69 x 10^{-10} F. In the frequency condition of 0.387 MHz there is a significant increase in capacitance, namely from 5.6616 x 10^{-12} to 2.2704 x 10^{-11}. The polymeric acrylate polymer dopant ITO nanoparticles at a frequency of 2.488 MHz has a large conductance with PE-b-PEG 0.003 compared to other PE-b-PEG. This explains that the liquid polyester crystalline acrylic can store a good electrical charge especially with the addition of PE-b-PEG 0.003 that at high frequency conditions there has been a change in the direction of the external electric field.

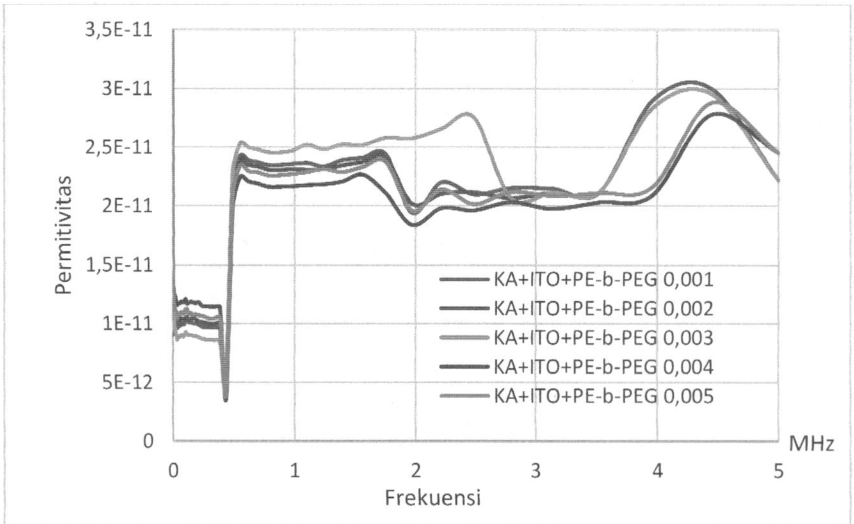

Figure 3. Relationship capacitance to frequency.

The energy gap of liquid crystals is very influential on the nature of liquid crystals. The smaller of energy gap, the electrons is easier from the valence band to move to the conduction band because the gap between spaces is narrow. Based on table data (3) the smallest energy gap is obtained for the composite sample of polymeric acrylate polymer dopant ITO nanoparticles with PE-b-PEG 0.003 of 0.135 eV. The narrow distance between gaps will facilitate electron mobility so that it can increase the conductivity value.

Table 3. The energy gap of the liquid crystalline cholesteryl acrylate in addition to the variation of PE-b-PEG.

Sample	**Energy Gap (eV)**
Cholesteryl Acrylate	0.146793
ITO	0.019795
KA + ITO + PE-b-PEG 0.001	0.136363275
KA + ITO + PE-b-PEG 0.002	0.136189472
KA + ITO + PE-b-PEG 0.003	0.135548091
KA + ITO + PE-b-PEG 0.004	0.141137292
KA + ITO + PE-b-PEG 0.005	0.140998807

4. Conclusion

Composite polymer liquid crystal polymer composites were successfully obtained by synthesis of ABB precursors and cholesterol. Next is the photolymerization to keep the structure of the structure stable, with ultraviolet radiation. During the UV-Curing process, the colloid acrylate monomer was mixed with ITO nanoparticles and PE-b-PEG materials with several variations. The addition of PE-b-PEG does not significantly increase the conductivity value, which only affects 58%. Based on the experimental results, the highest conductivity and capacitance data were obtained with the variation of PE-b-PEG of 0.003. In addition, the variation of PE-b-PEG 0.003 also has the lowest energy gap value which indicates that electron mobility can take place more smoothly.

Acknowledgement

This research was supported by Program Kreativitas Mahasiswa Menristekdikti number B/1856/B3.1/KM.02.01/2019.

References

[1] Bruce D W, Goodby J W, Sambles J R and Coles H J 2006 Introduction: New directions in liquid crystal science *Philos. Trans. R. Soc. A Math. Phys. Eng. Sci.* **364** 2567–71

[2] Zhang J, Su X, Yang D and Luan C 2018 Label-free liquid crystal biosensor for cecropin B detection *Talanta* **186** 60–4

[3] Liang X, Chen M, Guo S, Wang X, Zhang S, Zhang L and Yang H 2018 Programmable electro-optical performances in a dual-frequency liquid crystals / polymer composite system *Polymer (Guildf).* **149** 164–8

[4] Yang Z, Peng H, Wang W and Liu T 2010 Crystallization behavior of poly(ε-caprolactone)/layered double hydroxide nanocomposites *J. Appl. Polym. Sci.* **116** 2658–67

[5] Systems A M 2008 L Iquid Crystal Polymer : Enabling Next - Generation Conformal and Multilayer Electronics *Microw. J.* 83–115

[6] Umeir A F, Afrizal A and Riswoko A 2017 Uji Spektrofotometer Uv-Vis Pada Produk Blending Kristal Cair Kolesteril Akrilat Dengan Kristal Cair Tipe Ester Benzoat *JRSKT J. Ris. Sains dan Kim. Terap.* **3** 257–62

[7] Wang X Z Q F 2004 *Liquid Crystalline Polymers* (Singapore: World Scientific)

[8] Capozzi C J, Li Z, Samuels R J and Gerhardt R A 2008 Impedance spectroscopy and optical characterization of polymethyl methacrylate/indium tin oxide nanocomposites with three-dimensional Voronoi microstructures *J. Appl. Phys.* **104**

[9] Son I, Kim J H, Lee B, Kim C, Yoo J Y, Hyun K, Wu J P and Lee J H 2016 Vertical alignment of liquid crystals using an in situ self-assembled layer of an amphiphilic block copolymer *Macromol. Res.* **24** 235–9

[10] Liang B-J, Liu D-G, Shie W-Y, Tsai W-L, Hsu P-F and Louh R-F 2013 Effects of Nanoscaled Tin-Doped Indium Oxide on Liquid Crystals against Electrostatic Discharge *Crystals* **3** 530–53

IOP Conf. Series: Materials Science and Engineering **850** (2020) 012052 doi:10.1088/1757-899X/850/1/012052

[11] Afrizal; Eka Puji Astuti; Arif Rahman 2019 The Effect of Dopant Indium Tin Oxide (ITO) Nanoparticle Concentration on Characteristics and Conductivity Properties of Cholesteryl Acrylate Polymer-ITO Composite *Empower. Sci. Math. Glob. Compet.* 9–14
[12] Hikam M, Soegiyono B and Riswoko A 2012 XRD Pattern of Liquid Crystal Monomer Acrylate That Conjugated with Cholesterol and p-Hydroxyphenyl- 2-Methyl Butanoic **6** 1051–5
[13] Beiser A 1992 *Konsep Fisika Modern* (Jakarta: Erlangga)
[14] Trajkovska Petkoska A 2014 The role of conductive dopants in polymer cholesteric liquid crystals *Maced. J. Chem. Chem. Eng.* **33** 287
[15] Kittel C 2004 *Introduction to Solid State Physics* (John Wiley & Sons, Inc)
[16] Jorena 2009 Menentukan Energi Gap Semikonduktor Silikon Melalui Pengukuran Resistansi Bahan pada Suhu Beragam *J. Penelit. Sains* **12** 1–3

Vertical handoff model in next generation wireless networks

I Kustiawan[1,*], T Hariyadi[1] and D R Nurjannah[2]

[1] Department of Electrical Engineering Education, Universitas Pendidikan Indonesia, Jl. Dr. Setiabudhi No. 207, Bandung 40154, Indonesia
[2] Electronic Engineering Study Program, Politeknik TEDC Bandung, Jl. Pesantren No.2, Cimahi 40513, Indonesia

*iwan_kustiawan@upi.edu

Abstract. The purpose of this research is to propose a scheme for making handoff decisions effectively. We propose a handoff decision strategy through a fuzzy logic system to perform a seamless handoff. We combine four types of determinants, namely, RSS (received signal strength), data rate, network latency, and user preferences as input metrics to our fuzzy inference engine. RSS is pre-processed first by the Kalman filter to get rid of noise. The logic of the inference engine is realized on the User Equipment side as a faster reaction to network dynamics. Our scheme performance is evaluated by simulating mobility scenarios through the context of move-out and move-in scenarios. Simulation results show that our scheme outperforms other approaches in terms of reducing unnecessary handoff for all cases.

1. Introduction

Next generation wireless networks (NGWN) enable mobile users to explore heterogeneous networks using UE (user equipment) devices with multi-interface facilities when accessing real time and non-real time services. Multi-radio operation in a heterogeneous environment is now normal. 4th generation (broadband) mobile broadband technology has implemented multi-network integration such as WLAN and mobile telecommunications networks through the all-Internet protocol base for broadband mobile capacity [1,2].

Handoff decision is the process of choosing the best network access in a situation and deciding when to handoff or not at the right time. Traditional handoff or horizontal handoff involves the process between two adjacent cells on the same radio access technology. Certain link quality parameters such as RSS (received signal strength) or signal-to-noise ratio are commonly considered in handoff management. When the parameter in question is below a predetermined threshold, a handoff will be initiated. Due to differences in protocol structure and parameters involved in heterogeneous environments, horizontal handoff mechanisms do not serve well in the environment [3,4]. On the other hand, handoffs between different radio access technologies often occur when mobile users explore them. Specifically, handoffs that involve exchanges between different types of network access are called vertical handoffs [5].

As far as the handoff decision is concerned, traditional algorithms that use RSS as the main factor can cause not only ping-pong effects but also network load imbalances [6]. RSS is indeed practical but appears to be of limited use in the context of heterogeneous networks because RSS measured by different radio interfaces in general cannot be compared. However, RSS is still the most common metric that can be accepted by the device so we still consider RSS as a determining factor in triggering the handoff process. Also, in the case of tradeoffs among other metrics needed to make vertical handoff decisions, we consider other factors namely data rate and network latency as multi-criteria input in a fuzzy logic system to produce a single output value, which indicates whether vertical handoff is important immediately [7].

Our scheme focuses on communication simulations where IEEE 802.11 wireless networks are on cellular networks [8]. RSS provides information about the level of power received by the antenna. The power is reduced when the user moves away from the IEEE 802.11 access point (AP). Mobile users should switch to the cellular network before losing connection. Data rate is used to indicate the condition of network bandwidth; network latency is one of the QoS parameters that represent a round trip time delay that occurs in data communication on a network. Thus through this proposed design, we develop a scheme to determine when to carry out a moderate handoff decision.

2. Methods

Simulation methods are applied to find research solutions in our designs. This method directs us a road map to synthesize previous research work related to the design of research simulations and extend work to more specific areas such as identifying research questions, choosing a simulation approach, developing computational representations, elaborating basic rules of verification vs. experimentation, and evaluating research simulations [9].

Through this approach, we are motivated to look for visualization of small-scale or simple systems where the effect of certain controls on parameters is really expected to resemble the actual state or system. In this research, we use the Mamdani version of fuzzy logic as an inference engine to deduce handoff decisions [5]. Likewise, we evaluate and compare simulation results to verify the designed system performance. Four input metrics namely RSS, data rate, user preference and network latency are randomly combined research subjects. The fuzzy logic method and the filtering process are adapted to reasoning the output of the handoff decision making system (Figure 1).

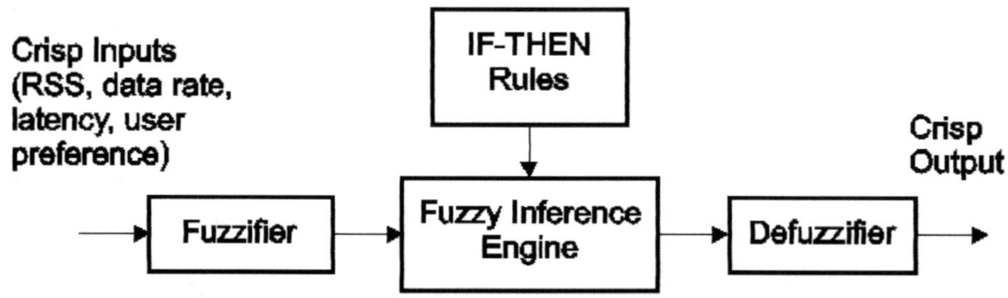

Figure 1. Block diagram of a fuzzy logic system.

The rule base contains IF-THEN rules that are required by the Fuzzy Inference System (FIS). We apply FIS Mamdani because it is intuitive, widely accepted and operates according to human instincts. The database defines the membership function of the fuzzy set. FIS generates fuzzified aggregate data, based on the fuzzy inference method. Defuzzifier converts fuzzified aggregate data into scalar values. The set of fuzzy output decisions is aggregated into a single fuzzy set and passed to the defuzzifier to be converted to the right quantity, the handoff factor, which determines whether a handoff is needed. We designed four fuzzy input variables and three fuzzy sets for each fuzzy variable, so that the maximum number of rules in our rule base is $3^4 = 81$. The sharp output from the defuzzifier block is used to select the network that is most suitable for the UE. The universe of discourse for handoff factor variables ranges from 0 to 10 as shown in Figure 2.

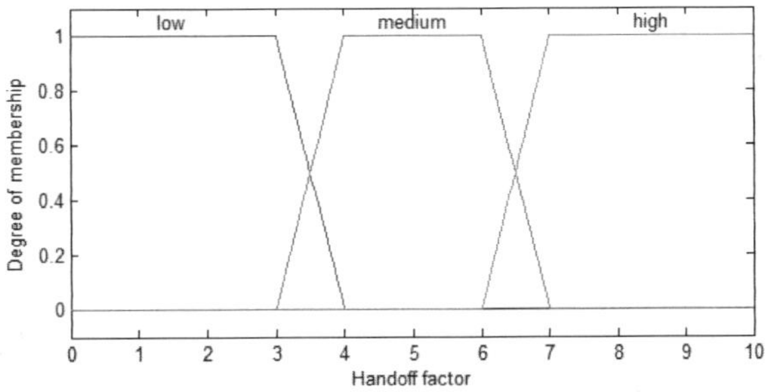

Figure 2. Membership function of the handoff factor.

The handoff process is diagramed in Figure 3. First, a UE constantly monitors RSSs of a serving point of attachment to the system, either an AP or a base station. Then, the Kalman filter will offset the effect of noise interference in which its Kalman-filtered RSSs will be merged with other three parameters to be processed by the FIS. After determining the move-out or move-in scenario, the output of the FIS will indicate whether the UE should initiate a vertical handoff or stay connected with the current network.

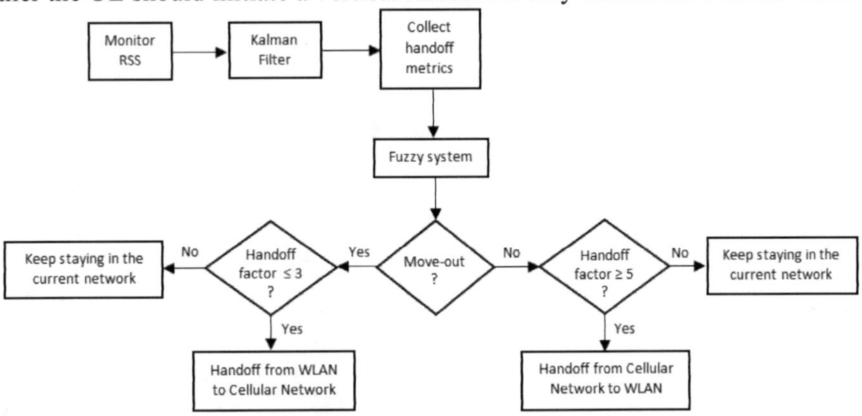

Figure 3. Workflow handoff process.

When the UE is departing from an AP, this is referred to as a *move-out* scenario, where the UE is expected to perform a vertical handoff to the cellular network soon. On the contrary, if the UE is associating to an AP from the cellular network, a *move-in* scenario happens, during which another vertical handoff shall occur shortly. Considering different mobility behaviours, handoff decision is made in the following lines:

- For the move-out scenario, if the deduced handoff factor is less than or equal to 3, then the UE selects the cellular network.
- For the move-in scenario, if the handoff factor is greater than or equal to 5, then the UE selects WLAN.
- If the handoff factor is between 3 and 5, then the UE stays connected with its current access network.

3. Results and discussion

Let us now investigate how the Kalman filter enacts the role of mitigating ping-pong effects. For simplicity, suppose that RSSs become a sole cause to trigger handoff, which is indicated with a control signal producing either 0 or 1. Accordingly, Figure 4 depicts handoff occurrences without and with using the Kalman filter in the move-out scenario. Each rising or falling edge of the signal represents a vertical handoff. From the figure it can be seen that, although RSS variations may cause frequent handoffs, such phenomena are greatly improved with the Kalman filter. Figure 4 also suggests the usefulness of the Kalman filter that prevents handoff from happening more frequently than necessary. While the Kalman filter is effective in dealing with RSSs, there are other practical determinants affecting handoff decision.

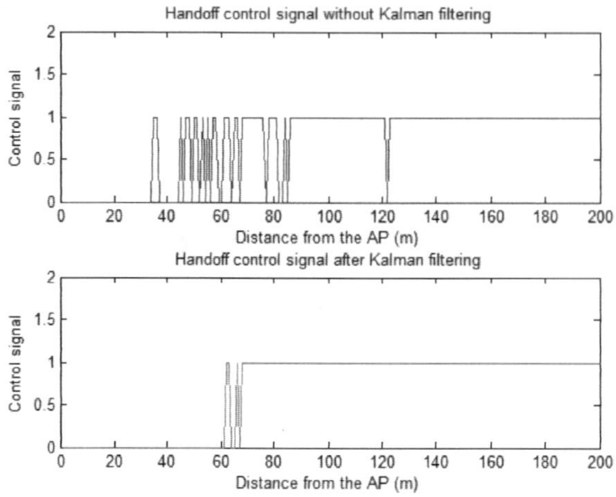

Figure 4. Handoff control signal without and with the Kalman filter.

Next we concentrate on how many handoffs are initiated during simulations. In the move-out scenario, the UE performs 36, 35, 29, and 21 vertical handoffs with the traditional fixed-RSS, Kalman-filtered RSS, Mamdani fuzzy logic, and our design, respectively (Figure 5). From Figure 5, it can be seen that the last three approaches reduce handoffs considerably in comparison to the traditional one, with relative reductions reaching 2.78%, 19.4%, and 41.67%, respectively. Relative reductions are assessed below:

$$\frac{(36-35)}{36} * 100 = 2.78\%$$
$$\frac{36-29}{36} * 100 = 19.4\%$$
$$\frac{36-21}{36} * 100 = 41.67\%$$

Concerning the move-in scenario, we have 33, 30, 30, and 18 handoffs resulting from the fixed-RSS, Kalman-filtered RSS, Mamdani fuzzy logic, and the proposed approach, respectively (Figure 6). With similar computation to that in Figure 6, we summarize that, when compared with the fixed-RSS approach, the last three schemes reduce handoffs by 9.09%, 9.09%, and 45.45%, respectively.

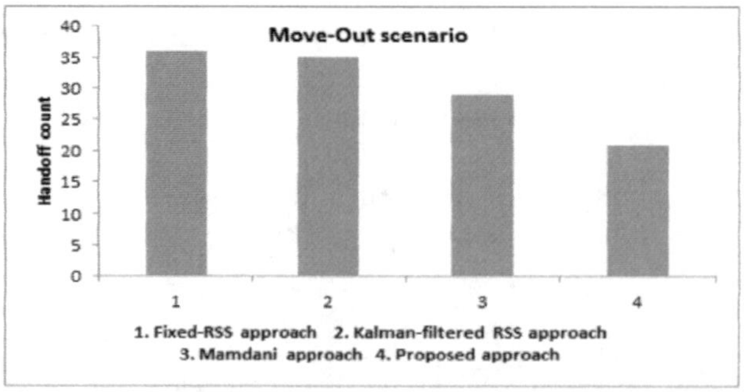

Figure 5. Handoff counts for different subject schemes in the move-out scenario.

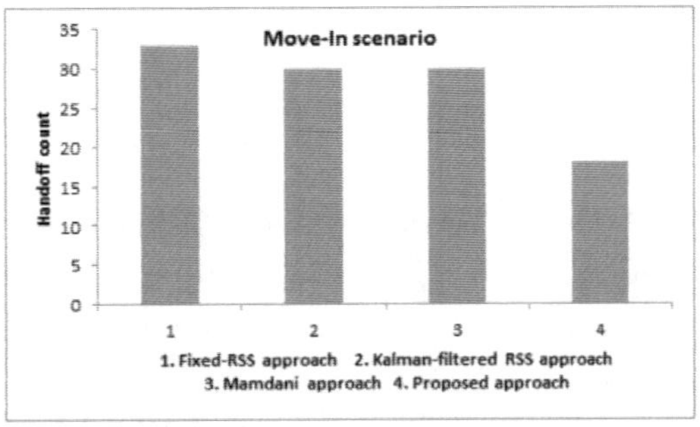

Figure 6. Handoff counts for different subject schemes in the move-in scenario.

4. Conclusion

Foregoing experiments show that the Kalman filter contributes to reducing signal fluctuations due to shadowing effects. Consequently, the Kalman-filtering RSS approach brings about fewer handoffs than the fixed-RSS scheme does. However, RSS as a single metric is not enough. Accordingly, the amalgamation of a Kalman filter and fuzzy system for vertical handoff decision making with multi input metrics is proposed in this study.

Combining a Kalman filter and a fuzzy system for smooth handoff in a heterogeneous networking environment, this paper devised a means to fuse four types of determinants for joint decision making. We developed a Mamdani fuzzy inference system that takes into account multi-attributes of different potency to resolve a moderate decision indicative of handoff initiation. Our approach was compared quantitatively with other schemes, showing the usefulness of our development in all scenarios.

References

[1] Jiang C, Zhang H, Ren Y, Han Z, Chen K C and Hanzo L 2016 Machine learning paradigms for next-generation wireless networks *IEEE Wireless Communications* **24**(2) 98-105

[2] Liu Y, Chen H H and Wang L 2016 Physical layer security for next generation wireless networks: Theories, technologies, and challenges *IEEE Communications Surveys & Tutorials* **19**(1) 347-76

[3] Çalhan A and Çeken C 2013 Case study on handoff strategies for wireless overlay networks *Comput. Stand. Interfaces* **35** 1 170–178

[4] Xenakis D, Passas N, Merakos L and Verikoukis C 2013 Mobility management for femtocells in LTE-advanced: Key aspects and survey of handover decision algorithms *IEEE Communications surveys & tutorials* **16**(1) 64-91

[5] Kustiawan I and Chi K H 2015 Handoff decision using a Kalman filter and fuzzy logic in heterogeneous wireless networks *IEEE Commun. Lett.* **19** 12 2258–2261

[6] Singhrova A and Prakash N 2012 Vertical handoff decision algorithm for improved quality of service in heterogeneous wireless networks *IET communications* **6**(2) 211-23

[7] Kunarak S and Suleesathira R 2010 Predictive RSS with fuzzy logic based vertical handoff algorithm in heterogeneous wireless networks *The 2010 International Conference on Advanced Technologies for Communications* 189-194 IEEE

[8] Kaltenberger F, Foukalas F, Holland O, Pietrzyk S, Thao S 2014 Vivier G. Spectrum overlay through aggregation of heterogeneous dispersed bands *2014 European conference on networks and communications (EuCNC)* 1-5 IEEE

[9] Yi J J, Lilja D J and Hawkins D M 2003 A statistically rigorous approach for improving simulation methodology *The Ninth International Symposium on High-Performance Computer Architecture* 281-291 IEEE

IOP Conf. Series: Materials Science and Engineering **850** (2020) 012054 doi:10.1088/1757-899X/850/1/012054

Analysis high gain waveguide slot antenna 2.3 GHz for Long Term Evolution-Time Division Duplex (LTE-TDD)

N F A Hakim[1,*], T Hariyadi[2], A B Pantjawati[2] and B Mulyanti[2]

[1] Electronics Engineering, Politeknik Gajah Tunggal, Tangerang, Banten, Indonesia
[2] Department of Electrical Engineering Education, Universitas Pendidikan Indonesia, Bandung, Indonesia

*nurulfahmiarief@gmail.com

Abstract. Waveguide slot antenna has many benefits in term of high gain, simple design and stable performance that can be applied to telecommunications systems. This paper will discuss about the design and analysis of waveguide slot antenna for LTE (Long Term Evolution) at a frequency of 2.3 to 2.4 GHz. The results obtained will be used for reference whether the antenna can be used for LTE-TDD frequencies. This research was conducted in two steps. The first step involved simulation of antenna design and optimization using 3D Electromagnetic Simulator software. Data simulation results showed that the working frequency of the antenna between 2.30 to. While in the second step, fabrication and measurement were carried out using a spectrum analyser and network analyser. The results obtained from measurements show that the antenna has a frequency between 2.28 to 2.41 GHz with bandwidth 130 MHz. The antenna has a directional radiation pattern with a gain at a frequency of 2.3 MHz of 15.68 dBi. From the simulation results and measurement, it can be concluded that the design of these antennas can be used for LTE-TDD applications.

1. Introduction

The demand for wireless communication services is constantly increasing, therefore the speed of data access should also increase. Multimedia Online Gaming (MMOG), Mobile TV, Web 2.0, and DSL contribute greatly to network traffic [1]. This is a challenge for the development of wireless communication systems providing wide frequency bands with better costs and performance.

The 4G technology, also known as Long Term Evolution (LTE), is designed to support wireless communication. LTE is a technology that provides many changes in wireless communication systems that are able to provide downlink data speeds of more than 100 megabits per second (Mbps), with low latency, and very efficient spectrum use. There are LTE cellular networks that use TDD (Time Division Duplex) and FDD (Frequency Division Duplex) technology. LTE-TDD uses the same frequency spectrum for uplink and downlink, while LTE-FDD for uplink and downlink uses a different frequency spectrum.

There has been a lot of development of LTE technology devices that aim to meet telecommunications needs [1-3]. One such device is an antenna, which acts as a transmitter and receiver. Research on the 2.3 GHz antenna frequency has been done a lot, but more is in the form of planar with low gain [2-5].

To get an antenna with a high gain, one of them is by making an antenna array. One antenna that is easy to create is an antenna waveguide slot. This antenna has advantages in simple design, wideband and high gain [6-8]. So that this type of antenna can be applied to 2.3 GHz LTE-TDD technology. This paper discusses the design of antenna waveguide slots that have a directional radiation pattern and gain greater than 14 dB at frequencies 2.3 to 2.4 GHz in accordance with LTE-TDD technology.

2. Antenna design

Design of the waveguide slot antenna is using monopole antenna feeder. This antenna has an impedance of 50 Ω to match it with the SMA connector. The antenna design and simulation is carried out using 3D Electromagnetic Simulator software. This software can display antenna performance parameters such as return loss, VSWR, impedance, gain, and radiation patterns. The design and dimensions of each parameter antenna is shown in Figure 1 and Table 1.

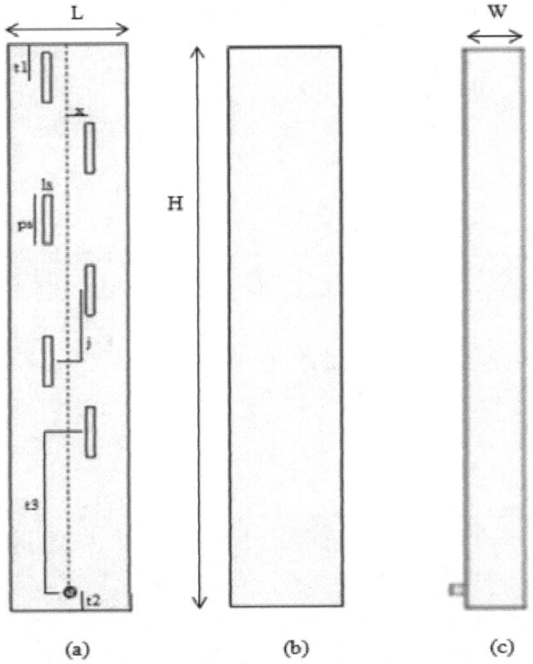

Figure 1. Waveguide slot antenna design (a) front (b) back side (c) right side.

Table 1. Dimension of the antenna.

Parameter	Size (mm)	Parameter	Size (mm)
t1	37.70	ps	55.20
t2	20	j	78.54
t3	158.33	x	20.43
Ls	7.85	W(width)	58.67
L (length)	113	H (height)	628

3. Simulation results

Return loss of the antenna is on the frequency 2.30 – 2.41 GHz. Figure 2 shows the return loss and bandwidth from the simulation results. Bandwidth from simulation is 108.89 MHz. The lowest return loss is at 2.33 GHz, which is -37.82 dB. Figure 3 shows VSWR simulation result. The antenna design has a directional radiation patterns. Figure 4 shows the radiation pattern of the antenna.

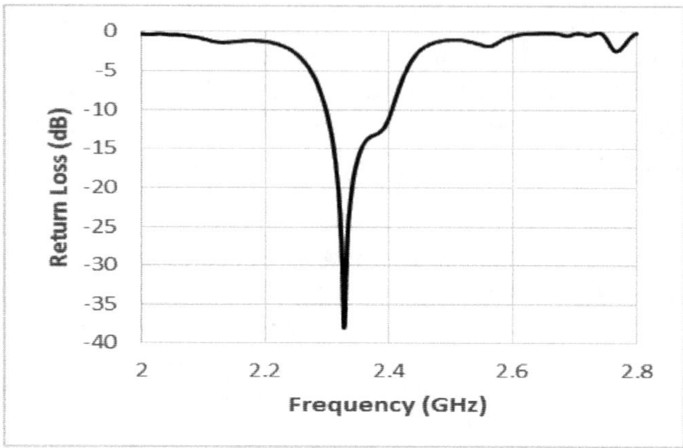

Figure 2. Return loss simulation result.

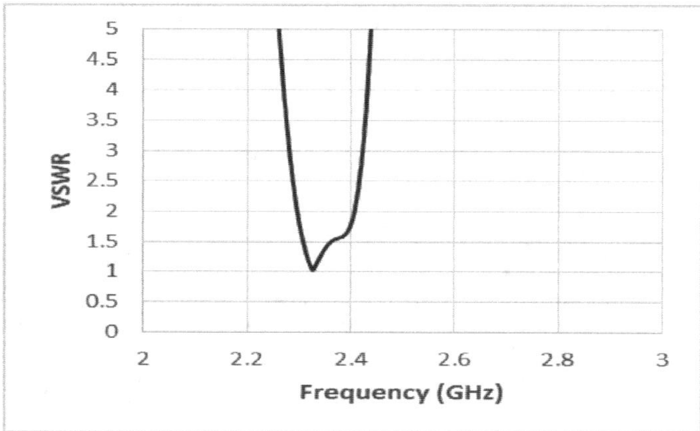

Figure 3. VSWR simulation result.

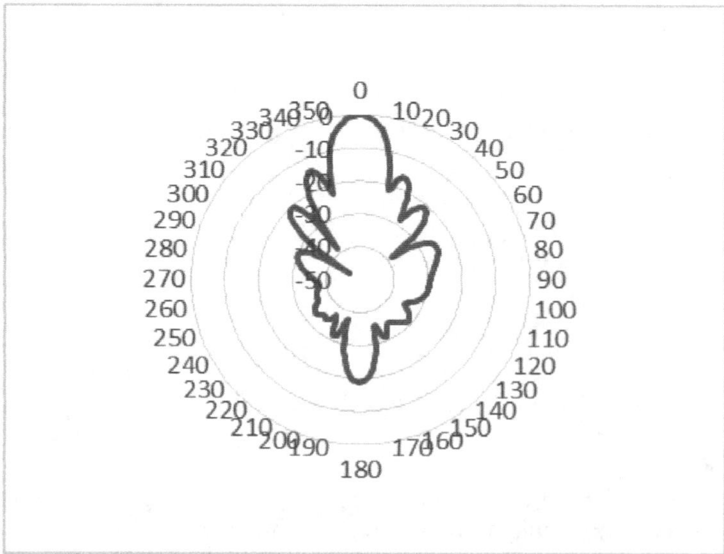

Figure 4. Radiation pattern waveguide slot antenna.

4. Fabrication results and analysis

After obtaining the simulation results, the next step is antenna realization. The material used for creating the antennas is a brass plate with a thickness 0.8 mm. The number of antenna slots is 6 pieces. Figure 5 shows the antenna realization. The purpose of antenna realization is to compare the results of simulations that have been carried out. The comparison of return loss simulation and fabrication is shown in Figure 6.

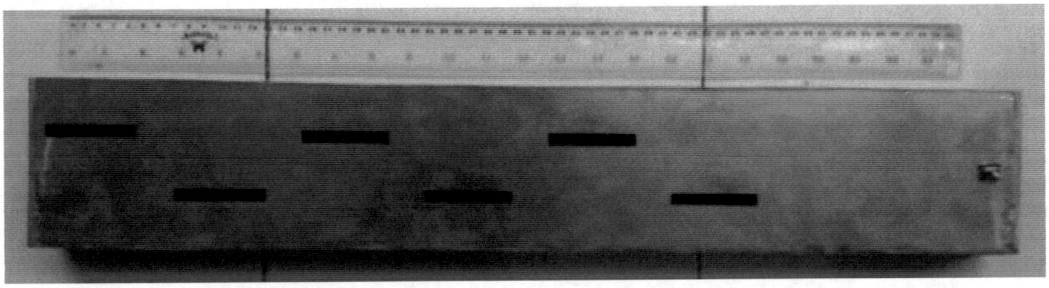

Figure 5. Waveguide slot antenna.

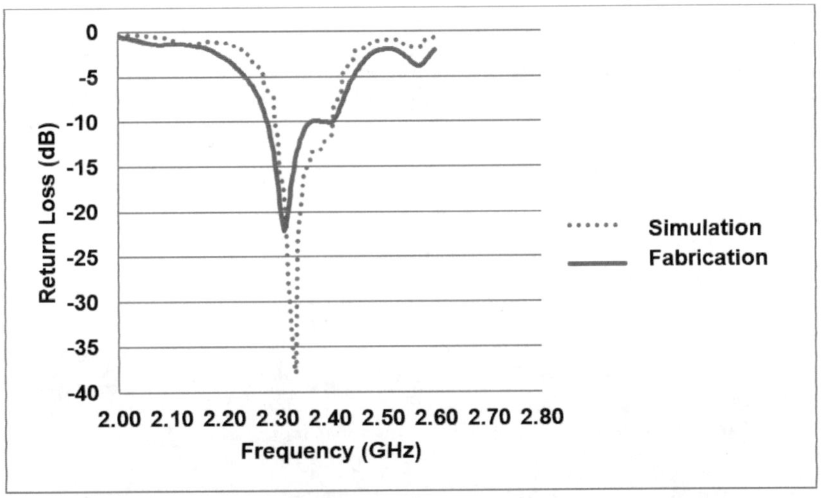

Figure 6. Comparison of return loss results between simulation and fabrication.

Based on Figure 6, it can be seen that there is a shift in the bandwidth and frequency of the antenna. For return loss the simulation results are obtained at -37,82 dB at a frequency of 2.33 GHz, while the fabrication results have a smaller value of -18.7 dB. Return loss of work frequency shifts to the left, but still in accordance with the desired specifications, which is <-10 dB in the frequency range 2.3 - 2.4 GHz. The bandwidth result from the simulation is 108.89 MHz, while the measurement is 130 MHz with a frequency range of 2.28 - 2.41 GHz. This difference can be caused by loss or bending on the connector cable.

Figure 7 showing a comparison of VSWR values from simulation and fabrication. The VSWR that obtained from the simulation is 1.03 at the frequency of 2.33 GHz, while the measurement results obtained 1.26 at the frequency of 2.33 GHz. The emergence of value differences is caused by several factors, one of them is interference of wave in the air and losses in the connector.

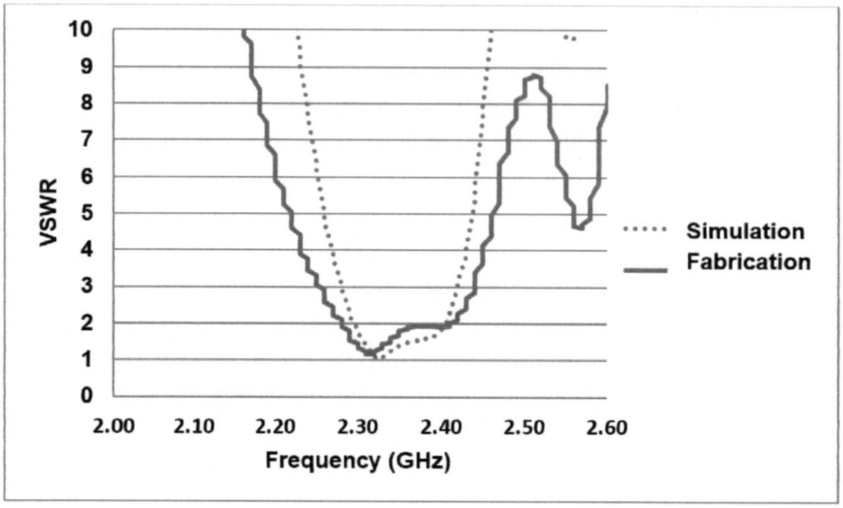

Figure 7. Comparison of VSWR results between simulation and fabrication.

In measuring the radiation pattern, the antenna is rotated every 10^0. The data obtained will be normalized to the maximum average value so that the antenna radiation pattern is formed. Comparison of the radiation pattern result between simulation and fabrication can be seen in Figure 8.

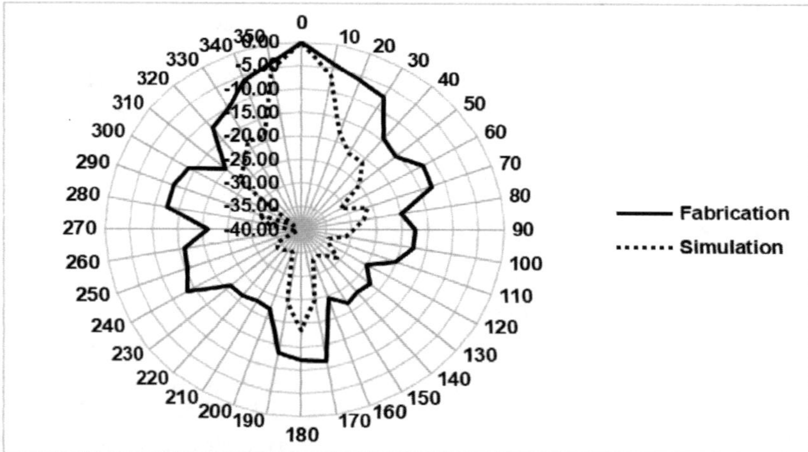

Figure 8. Radiation pattern results of simulation and fabrication.

Measurement of radiation pattern is carried out with a 10^0 angle elevation, this causes the results to be less precise. The cause of this radiation pattern difference is the existence of losses that interfere with the measurement process. In addition, the distance of antenna measurements that are too far away causes greater free space loss.

Measurement of gain is done by comparing the waveguide slot antenna with the horn antenna, which has a 12 dBi gain as a reference antenna. Table 2 is showing the gain values of each 10 MHz starting from 2.3 GHz to 2.4 GHz. This is done so that the antenna measurement becomes more precise. Graph of gain comparison from simulation results and fabrication of antenna waveguide slot can be seen in Figure 9.

Table 2. Comparison gain between simulation and fabrication.

Frequency (GHz)	Simulation (dBi)	Fabrication (dBi)
2.30	14.26	15.68
2.31	14.26	15.31
2.32	14.26	15.28
2.33	14.26	15.24
2.34	14.25	15.15
2.35	14.24	15.43
2.36	14.22	15.42
2.37	14.2	15.13
2.38	14.16	15.39
2.39	14.11	15.39
2.40	14.05	14.89

Formatting the title, authors and affiliations Figure 9 shows that the results of the gain obtained from the simulation and fabrication are slightly different. The fabricated gain is greater than the simulation. The gain in the simulation ranges from 14 dBi, while the fabrication ranges from 15 dBi. The cause of this difference is due to the existence of free space loss and loss on the connector cable. In addition, the presence of objects around the measurement area causes the measurement results obtained to be different from the simulation results.

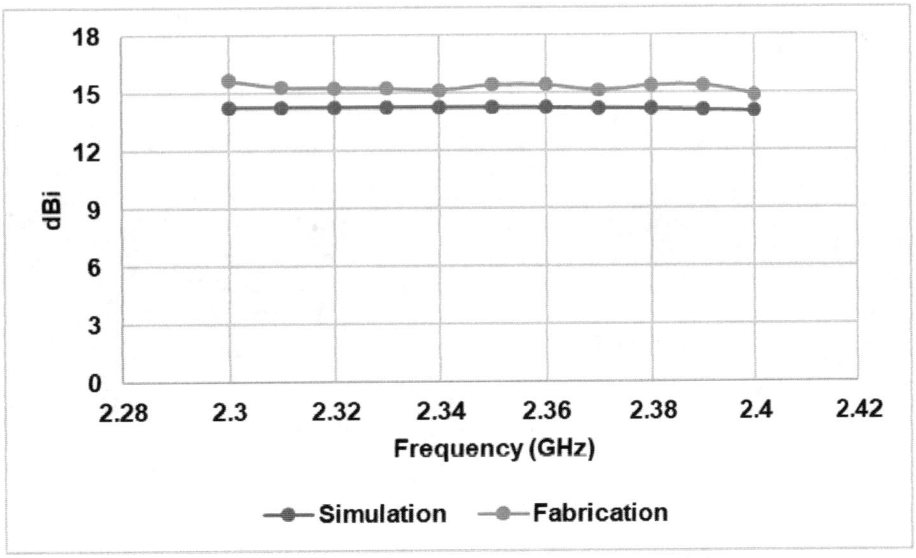

Figure 9. Comparison of gain values between simulation results and fabrication.

5. Conclusion

Waveguide slot antenna resulted in this research worked on the frequencies of 2.3 - 2.4 GHz with VSWR ≤ 2, therefore can be implemented on LTE (Long Term Evolution) technology. Antenna fabrication results from the simulation carried out have several differences, but still in accordance with the initial specifications. Antenna simulation return loss value <-10 dB in the frequency range of 2.30 GHz to 2.41 GHz or equivalent to VSWR ≤ 2. The measurement results obtained frequency range of

2.28 - 2.41 GHz with 130 MHz bandwidth. The average gain of an antenna is around 15 dBi with a directional radiation pattern.

References

[1] Chadchan S M and Akki C B 2010 3GPP LTE/SAE: An Overview *International Journal of Computer and Electrical Engineering* **2**(5) 806-14

[2] Hariyadi, T, Huda Y T and Mulyanti B 2010 A Small Ultra-Wideband Unidirectional Microstrip Antenna for Through-Wall Radar Application *Journal of Telecommunication, Electronic and Computer Engineering* **8**(1) 25-28

[3] Ratasuk R, Tolli D and Ghosh A 2010 Carrier Aggregation in LTE-Advanced *Vehicular Technology Conference* 1-5

[4] Chagharvarnd S, Hamid M R, Kamarudin M R and Khalily M 2014 Wideband Slot Antenna for 4G Applications *IEEE Asia-Pacific Conference on Applied Electromagnetics (APACE) 2014* 279–281

[5] Karimah S, Zulkifli F Y and Rahardjo E T 2014 Antenna design for femtocell LTE at frequency 2.3–2.4 GHz *2014 Makassar International Conference on Electrical Engineering and Informatics (MICEEI)* 50-53

[6] Ban Y-L, Chen J-H and Ying L-J 2012 Ultrawideband Antenna for LTE/GSM/UMTS Wireless USB Dongle Applications *Antennas and Wireless Propagation Letters* **11** 403-406

[7] Al-Husseini M, El-Hajj A and Kabalan Y K 2013 High-gain S-band Slotted Waveguide Antenna Arrays with Elliptical Slots and Low Sidelobe Levels *Progress in Electromagnetics Research Symposium Proceedings* 12-15

[8] Shi L, Zhao G and Sun H 2013 Research on high-precision waveguide slot array *2013 IEEE International Conference on Microwave Technology & Computational Electromagnetics* 39-42

IOP Conf. Series: Materials Science and Engineering **850** (2020) 012055 doi:10.1088/1757-899X/850/1/012055

Design of a compact UWB Wilkinson Power Divider using ring structured and tapered line matching transformer

R N Yuniar*, T Hariyadi, A Saripudin and B Mulyanti

Electrical Engineering Educational Department, Faculty of Technology and Vocational Education, Universitas Pendidikan Indonesia, Bandung, Indonesia

*regita.nurmalitay@student.upi.edu

Abstract. There is an increasing interest in the design of ultra-wideband (UWB) microwave components for UWB applications. One of the important components in the ultra-wideband application is power divider. In this paper, an ultra-wideband Wilkinson power divider (WPD) using ring structure and tapered line matching transformer is proposed. The aims of this research are to design an UWB WPD for feeding antenna UWB in use of detecting breast cancer. The method used in this research is simulation using finite integration based on CST STUDIO SUITE 2015 3D EM simulator software. This study shows the improvement of output responses due to the introduction of a tapered line and added isolation resistors. The bandwidth and input reflection coefficient are also improved by the ring structure. The proposed UWB WPD can be used for 1:2-way equal power division application in ultra-wideband range, such as UWB antenna which is operating in frequency 3-13 GHz. The simulation result shows that the return loss is better than -10 dB and the insertion loss is about 3.2 dB until 4.15 dB for the operating frequencies and impedance value of 50 ohms.

1. Introduction

Power divider is a passive component used in splitting or combining signals which plays an important role in the microwave applications [1]. Those applications such as microwave imaging, short-range indoor data transmission and positioning, healthcare like body area network, and through-the-wall radars [2]. The earliest and most widely used is Wilkinson power divider which can enable to reach the high isolation and matching between output ports. Nowadays, research works for design of ultra-wideband power divider and other UWB components are carried out extensively because of the advantages of UWB systems. UWB is radio technology which can be used in very low energy levels and operate from 3.1-10.6 GHz [3]. The advantages of the UWB systems are high data-rate and faster transmission capabilities, has low power spectral densities, that is about 41.3 dBm/MHz which is ensuring low interference with narrow-band and other wireless devices [4]. UWB also has advantage of high resolution, extremely low power consumption and low power emission [5].

Numerous works have been carried out to meet the requirements of UWB power divider such as wideband, high isolation, and compact size [6-11]. These works such as Wilkinson power divider [7], chain power divider [8], binary power divider [9], tapered lines power divider [10], defected ground structure [11], etc. Zou et al. has been investigated the performance of interconnected multi-way dividers and the effects the interconnecting lines have on the performance of the multi-way dividers [9]. The input return loss has the symmetric result about the centre frequency and found by the calculation. In

IOP Conf. Series: Materials Science and Engineering **850** (2020) 012055 doi:10.1088/1757-899X/850/1/012055

the design, an UWB power divider with wide frequency range from 3-13 GHz has been design [10]. The advantage of this design using tapered lines is a better performance and it requires shorter lengths and lower number of elements than the equivalent multi-section Wilkinson power divider has. Design in Michele et al. provides a frequency range spanning from 3.1 GHz to 10.6 GHz with a Chebyshev response using DGS structure [11]. The DGS technique is used to realize the high impedance lines. Even though the return loss is quite good in operating frequency range, but the losses in the high frequency still so high. In this paper, we proposed an UWB Wilkinson power divider designed using tapered line and interconnecting transmission line. The designing and simulation process carried out using EM simulator and microstrip technology.

2. Circuit design

The first step is to design a 2-output port power splitter/divider which should meet all of the specifications. The desired specifications of the UWB Wilkinson power divider are shown in Table 1.

Table 1. Parameter required for UWB WPD design.

Parameter	Value
Bandwidth	3 GHz-13 GHz
Insertion Loss	-3 dB ≤ IL ≤ -4 dB
Isolation	≤ 15 dB
Phase Unbalance	± 0.2°

The design of UWB WPD use Rogers RO4530B as the material of substrate with the dielectric constant of 3.66 and thickness of 1.524 mm. This material used because it has low material losses and good power handling. This material has electric tan δ value of 0.0037 and thermal conductivity is about 0.62 Wm/K.

The impedance is influenced by width of the transmission line that can be calculated using equation (3) [6].

$$Z_0 = \frac{120\pi}{\sqrt{\varepsilon_{eff}}[\frac{W}{H}+1.393+\frac{2}{3}\ln(\frac{W}{H}+1.444)} \tag{1}$$

If quarter-wavelength transmission lines with a characteristic impedance are added after the first stage, the input impedance at the input port of the first stage can be calculated by (2). The input port with this equation is perfectly matched at the centre frequency [9].

$$Z(f_0) = \frac{(Z_b Z_a{}^{-1})^2}{2} Z_0{}^2((Z_b Z_a{}^{-1})^2 \frac{Z_0}{2})^{-1} \tag{2}$$

Table 2 shows the parameters which is used in the design of an UWB power divider with ring structure and tapered line. Figure 1 shows the proposed design of an UWB power divider with the final dimension is 55 mm x 23.20 mm.

IOP Conf. Series: Materials Science and Engineering **850** (2020) 012055 doi:10.1088/1757-899X/850/1/012055

Table 2. Parameters used for UWB design.

Parameters	Description	Value (mm)
Tc	Conductor Thickness	0.035
Ts	Substrate's Thickness	1.524
L1	Length of the horizontal section of ring structure	2.82
Lw1	Feeder length	3
R	Radius	1
L2	Length of tapered	11
Wz0	line	3.17
W2	Width of 50Ω line	0.7
	Width of the ring	
W3	transmission line	3
L50	Width of tapered line	19
La	Length of 50Ω line	7.2
	Length of 50Ω line which is connected to output ports	

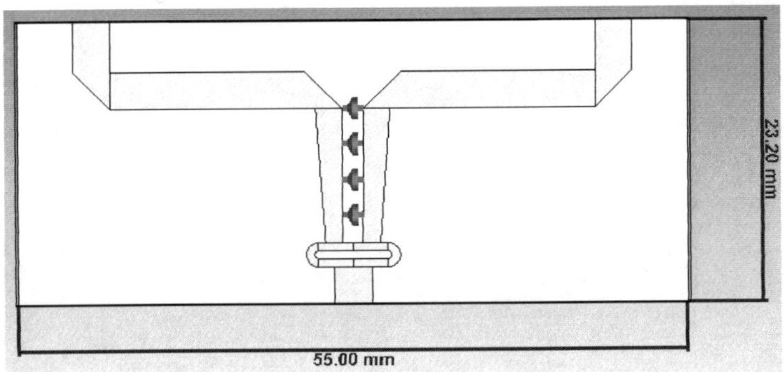

Figure 1. Proposed design of UWB Wilkinson Power Divider.

3. Results and discussion

From the simulation results, the values of return losses are smaller than -10 dB in the range of work frequencies, which are from 3 GHz to 13 GHz. Insertion losses that represent the amount of power sent to the output port are in the range of -3.2 dB to -4.15 dB. Return loss and insertion loss charts from the simulation results are shown in Figure 2.

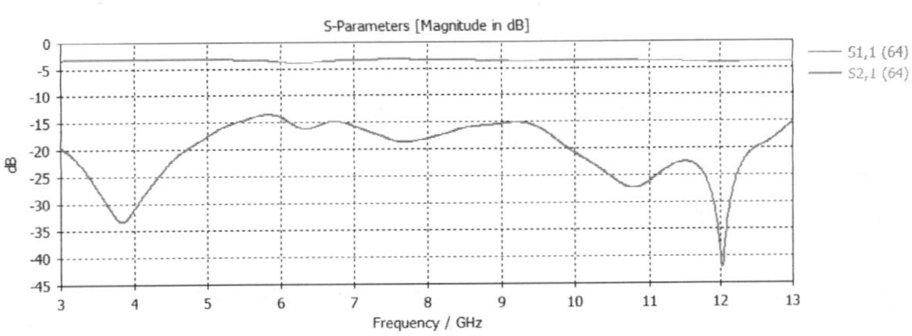

Figure 2. Return loss and insertion loss results of the proposed design.

Isolation between the output ports and return loss for each output ports are other parameters observed in the power divider. In this study, the design of the UWB power divider shows that the return loss output value has a value which is less than -10 dB at the operating frequencies with a maximum return loss of -33.5 dB. Figure 3 shows the results of return loss and insertion loss for each output port in the UWB power divider designed in this research. The return loss shows the performance of the designed UWB power divider.

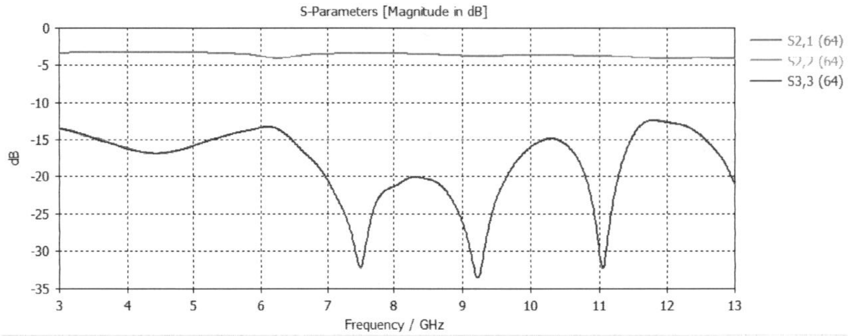

Figure 3. Return loss results for each output port.

The return loss from figure 3 has the value less than -10 dB from frequency less than 3 GHz until more than 13 GHz. So, the bandwidth is really wide. The maximum return loss value from 3GHz-13 GHz is -33 dB. Besides, the insertion loss value shows the loss is between -3 dB and -4.1 dB. It means that the losses are not too high.

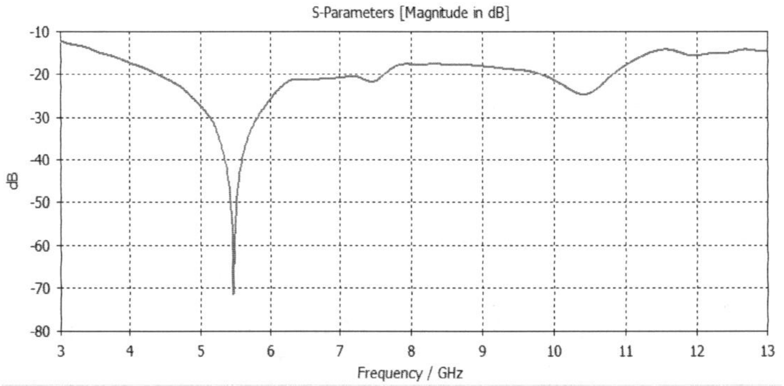

Figure 4. The results of the WPD UWB design isolation that has been made.

Figure 4 shows the isolation between the two output ports. From the graph, it can be seen that the isolation value is less than -15 dB in the range of operating frequencies of the device. Isolation can be repaired by adding an isolation resistor on the two ditched transmission lines. But this will have an effect on increasing the insertion loss value because of losses on resistive loads. A part of the total power sent will be dissipated to heat in each isolation resistor. Therefore, another way to improve isolation is to change the resistance values of each of the isolating resistors.

One of the important things in designing a component is the compatibility with the characteristic impedance of the transmission line, i.e. the impedance must be 50 Ω. Figure 5 shows the impedance value of the UWB power divider device designed to be in accordance with the impedance of the transmission line. The result shows that the input impedance is 50.035 Ω and output impedance is 49.98 Ω which has approached the characteristic impedance value of the transmission line. The impedance value in input or output sides reached with changing the width value of the transmission line.

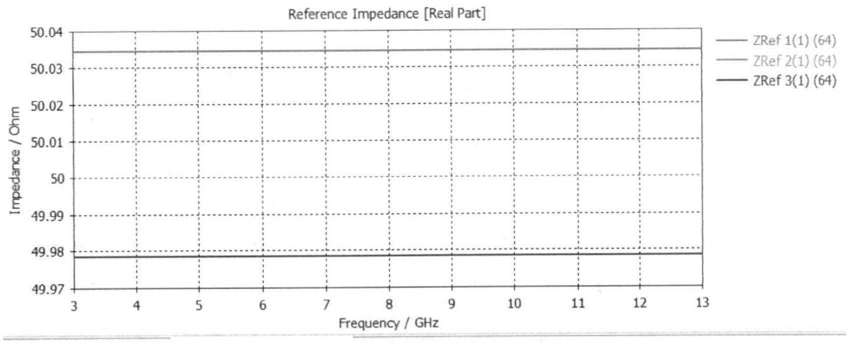

Figure 5. The impedance value of the proposed design of WPD UWB.

4. Conclusion

An UWB power divider using ring structure and tapered of transmission line has been realized. The developed UWB power divider has simulated return losses lower than -10 dB and isolation lower than -15 dB across the UWB frequency range, even higher, namely 3 - 13 GHz. The insertion loss varies from 3.2 dB up to 4.15 dB. The UWB power divider designed is suitable for amplifier, planar antenna feed network, and UWB radar antenna feed network.

References

[1] Hu S, Song K and Fan Y 2015 Ultra-wideband (UWB) eight-way ring-cavity power divider *International Journal of Microwave and Wireless Technologies* **7**(2) 115-120

[2] Hariyadi T, Huda Y T and Mulyanti B 2016 A small ultra-wideband unidirectional microstrip antenna for through-wall radar application *Journal of Telecommunication, Electronic and Computer Engineering (JTEC)* **8**(1) 25-28

[3] Federal Communication Commission 2002 *Ultra Wide-Band* (US: Federal Communication Commission)

[4] Tsang T K K and El-Gamal M N 2005 Ultra-wideband (UWB) Communication Systems: An Overview *IEEE-NEWCAS Conf.* 381-386

[5] Withington P, Fluhler H and Nag S 2003 Enhancing Homeland Security with Advanced UWB Sensors *IEEE Microwave Magazine* **4**(3) 51–58

[6] Pozar D M 2012 *Microwave Engineering* (New York: John Wiley & Sons)

[7] Wilkinson E J 1960 An Way Hybrid Power Divider *IEEE Trans. Microw. Theory Tech.* **8**(1) 116-118

[8] Antoniades M A and Eleftheriades G V 2005 A Broadband Series Power Divider Using Zero-Degree Metamaterial Phase-Shifting Lines *IEEE Microwave and Wireless Components Letters* **15**(11) 808-810

[9] Zhou J F, Morris K A and Lancaster M J 2007 General Design of Multiway Multisection Power Dividers by Interconnecting Two-Way Dividers *IEEE Trans. Microw. Theory Tech.* **55**(10) 2208-2215

[10] Mencia O, Pelaez P, Almoroz G and Alonso J I 2009 New Technique for the Design of Ultra Broadband Power Dividers based on tapered Lines *IEEE MTT-S Int. Microw. Symp. Digest* 493-496

[11] Squartecchia M, Cimoli B, Midili V, Johansen T K and Zhurbenko V 2017 Design of a Planar Ultra- Wideband Four-Way Power Divider/Combiner Using Defected Ground Structures *Proc. of the 47th Europ. Mic. Conf.* 9-12

IOP Conf. Series: Materials Science and Engineering **850** (2020) 012056 doi:10.1088/1757-899X/850/1/012056

Bit Error Rate performance analysis for Free Space Optic communication

M I Basudewa[1,*], Z H Bagaskara[1], S S A Damita[1], R F Putra[1] and D Ahmadi[2]

[1] Faculty of Electrical Engineering, Telkom University, Jl. Telekomunikasi Terusan Buah Batu Bandung 40257, Indonesia

[2] Faculty of Communication, Universitas Islam Bandung, Jl. Tamansari No. 1 Bandung 40116, Indonesia

*imansyah@student.telkomuniversity.ac.id

Abstract. This paper discusses the comparison of 4 modulation in Free Space Optic communication namely; Pulse Position Modulation (PPM), Pulse Amplitude Modulation (PAM), and On Off Keying Modulation (OOK). By Utilizing a light source with a length of 1550 nm, a simulation measured by Bit Error-Rate using Signal-to-Noise Ratio (SNR) in several distance. In this research, researchers used a quantitative approach by MATLAB simulation and using correlational studies to compare BER, SNR, distance and visibility. The results show that PPM is the most effective modulation in comparison with others.

1. Introduction

FSO technology utilize atmospheric channel as a propagating medium whose properties are random function of space and time [1]. There are many factors that cause attenuation in the optical signal like clouds, snow, haze, fog, rain, etc. In addition, link distance and pathloss are another factor that affect FSO communication. There are several channel models to estimate optical attenuation through fog using visibility as an input parameter [2]. Kim is one of the model that represent analysis of optical attenuation.

An implementation of FSO will be very helpful to a country due to information is an essential need for everyone to fulfil their needs. The research nowadays show that if the implementation of public information is running well, the effective and reliable governance will be realized. In addition, digital transformation is very necessary to improve the performance of government at the central and regional levels. So, the satisfaction people index on the city will be increased [3].

In a previous study that discussed BER, Power and Bandwidth performance by using several quadrature Phase Shift Keying (QPSK) modulation techniques, Binary Phase Shift Keying (BPSK), Differential Phase Shift Keying (DPSK), On-Off Keying (OOK) Return Zero (RZ) - Non Return Zero (NRZ), Pulse Amplitude Modulation (PAM) and Pulse Position Modulation (PPM).The research shows that L-PPM modulation technique is the best modulation to improve power efficiency in FSO [4]. In the other hands, M-PAM is usually used in communication system to increase bandwidth efficiency on FSO, where M is the data encoded as pulses that are sent [5].

In this research, a measurements will be carried out to determine the SNR variation and value of BER against distance with various modulation techniques. One of the challenge in implementing a FSO technology is attenuation that can limits the optical propagation path. Besides that, it affect signal quality and communication services. ITU-R model parameters is usually used to provides rainfall rate statistics that are closer to the value measured but surface rain gauge [6].

2. System models and channels

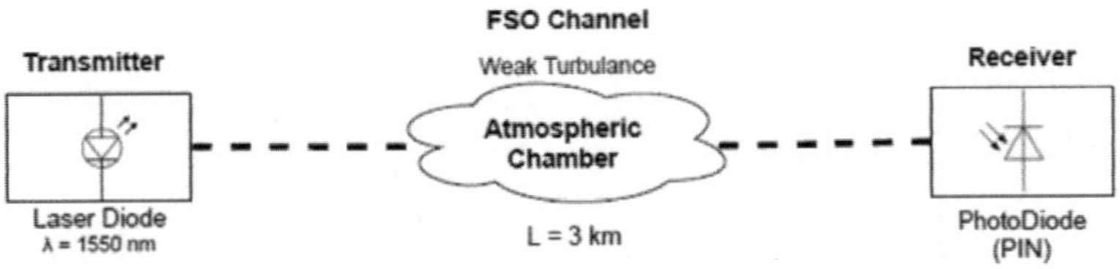

Figure 1. Free space optic channel.

Various factors that usually called attenuation will affect BER when sending data in FSO communication [7]. To find out the data transmission changes in Bit Error Rate (BER) value, it can be done by measuring the power level on the receiver side with the following equation:

$$Pr = Pt \cdot \tau t \cdot D^2/(\theta L^2) \cdot 10^{\left([-\gamma(\lambda)]^{(L/10)} \right)} \quad (1)$$

with Pr representing the efficiency of information delivery and Pt representing the efficiency of reception in optics, while D represents receiver diameter and θ is transmission angle divergence.

One of the factor that can affect FSO communication atmospheric attenuation is scattering factor. Scattering occurs when the transmission of light information collides with gas molecules or aerosols [8], so that the information signal will spread. In its application, FSO has a channel model one of them is the Kim model which has an equation:

$$\beta\alpha(\lambda) = \frac{3.91}{V} \cdot \left(\frac{\lambda}{550}\right)^{-\delta} \quad (2)$$

where λ is the wavelength for FSO and V is the visibility in (Km) where visibility is the visibility of an object.

The visibility category for Kim models is stated in the following table:

Table 1. Visibility of an object in Kim model.

δ	V
1.6	V > 50 km
1.3	6 km < V < 50 km
0.16	1 km < V < 6 km
0.34	0.5 km < V < 1 km
-0.5	V < 0.5 km

In this paper visibility is measured up to 6 km, the attenuation of syntax is caused by light turbulence caused by variations in the refractive index of temperature, pressure and particles in the atmosphere [9]. With the following equation:

$$[\sigma x]^2 = 1.23 \cdot C\_n^2 \, k^{(7/6)} L^{(11/6)} \quad (3)$$

Where σx is a variation of scintillation, C_n^2 is a refractive index structure parameter, k is 2/λ and L (km) is a distance from transmitter to receiver. Scintillation attenuation can be calculated by the following equation:

$$\gamma_{\text{scintillation}} = \sqrt{\sigma_x^2} \quad (4)$$

Table 2. Free space optic link parameter.

Parameter	Value
Wavelength	1550 nm
Link Distance	0 – 2 km
Power Transmitter	0.1 W
Diameter of Detector	10^{-5}
Beam Divergence	10^{-3}
Electron Charge	1.6×10^{-19}
Bandwidth	1 GHz
Responsivity	1.68
Dark Current	10^{-8}
Temperature	298K
Boltzman Constanta	1.38×10^{-23}
Noise Figure	1
Load Resistor	1000
Visibility	0 – 2 km

In this paper PPM, PAM, and OOK modulation will be compared. There is an advantage in PPM and OOK Modulation in the form of high efficiency of the power used but it has disadvantages in bandwidth. Whereas in PAM modulation, it has efficiency of bandwidth. But PAM modulation requires a lot of power [10]. This research also displays the observational data about the correlation between BER and link distance. Based on the testing that has been done, the SNR value will be the reference of the BER equation while the Photodetector PIN is used as the receiving device.

SNR (Signal Noise Ratio) is a comparison of the signal power with the interference in the channel. SNR can also be used as a benchmark of the quality of a communication system [11]. The SNR value is represented by the equation SNR:

$$(I\_p\text{^}2)/(2qB(Ip + ID) + 4KbTBFn/RL) \tag{5}$$

where Ip can be calculated from Ip = Pr R, B is the signal bandwidth, q is the unit of electron value, ID is the dark current value, T is representing the temperature value in the photodetector. Fn is a noise figure value which is equal to 1 and RL is the resistor load on photodetector.

BER (Bit Error Rate) is a unit of error sending data bits per total data bits sent. The BER has a unit ratio, the BER value depends on the SNR value and the use of the chosen modulation technique. In comparison, BER is more often used by researchers, technicians, industrial workers due to its simplicity [12]. PPM is a modulation where the position of the carrier signal pulses varies with the modulating signal. This modulation has the same signal amplitude and width. In this modulation, each pulse on the laser represents one or more bits of information based on their position to time. In PPM, several bits are sent at a time. The ppm block encoding is done by converting each value of "K" bits to L = 2 K [13].

$$PPM = \frac{1}{2} erfc(\frac{1}{2\sqrt{2}} \sqrt{SNR} \frac{L}{2} log2L) \tag{6}$$

PAM (Pulse Amplitude Modulation) is a modulation where the amplitude of the signal is varied based on the modulating signal. The pulse position and width of the modulated signal are fixed.

$$PAM = \frac{1}{2} erfc\left(\frac{\sqrt{SNRlog2M}}{2\sqrt{2}(M-1)}\right) \tag{7}$$

OOK-NRZ (On-Off Keying No Return Zero) is a type of modulation that is considered as on or sending data and if the value is 0, it's considered off or not sending data. In full wave OOK-NRZ, it has a value equal to 1. In addition, OOK-NRZ has greater bandwidth and greater power as well.

$$OOK - NRZ = \frac{1}{2} erfc\left(\frac{1}{2\sqrt{2}}\sqrt{SNR}\right) \tag{8}$$

OOK-RZ Modulation (On-Off Keying Return Zero) OOK-RZ (On-Off Keying Return Zero) is a type of modulation that only sends half wave. So it has less bandwidth than OOK-NRZ, but has lower power efficiency. The equation for OOK-RZ is:

$$\frac{1}{2} erfc\left(\frac{1}{2}\sqrt{SNR}\right) \tag{9}$$

3. Results and discussion

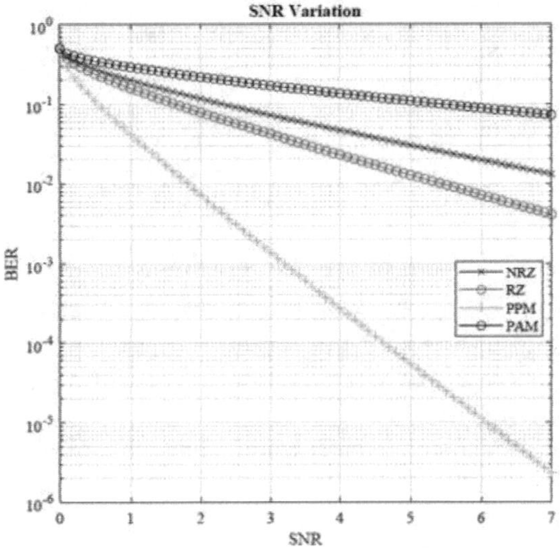

Figure 2. BER to SNR with various modulation.

In Figure 2 that includes Signal-to-Noise ratio and Bit Error Rate axis to identify the most efficient modulation that can be used in Free Space Optic Channel. The Figure Show that the most efficient modulation is the one who has the least value in BER and has the most value in SNR. PPM Modulation has the least BER value while PAM modulation has the biggest value of BER As the BER value increasing the quality of an information in a signal is decreasing.

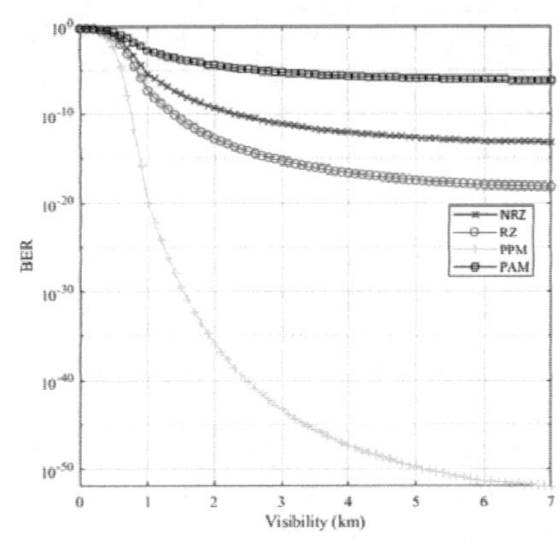

Figure 3. BER to visibility with various modulation.

In Figure 3 beside obtained the result of comparison in several modulation, there is a ratio BER to Visibility. In addition, PPM is also has the best performance according to the figure. Visibility is the distance how far the signal can be transmitted without interference or obstacles such as fog, rain droplets, aerosols, fog, haze and other particle.

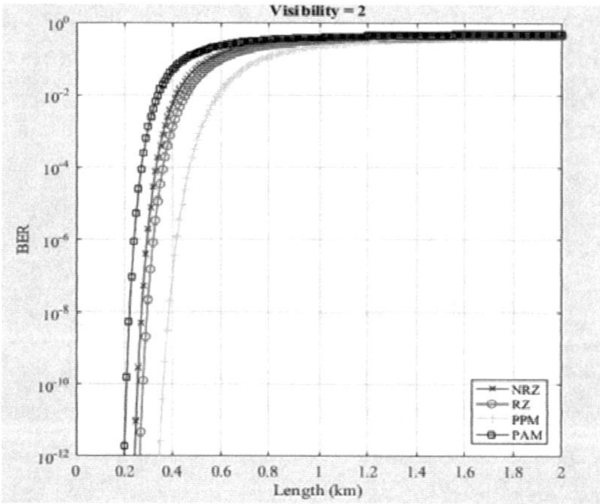

Figure 4. BER to length analysis with various modulation.

In Figure 4 beside obtained the result of comparison in several modulation, there is a ratio BER to Visibility. In addition, PPM is also has the best performance according to the figure because it can sent one or more bits information for each pulse of the laser. OOK-RZ modulation is the next due to its efficiency low power. OOK-NRZ is still more efficient than PAM modulation, even OOK-NRZ modulation consume more power and have greater bandwidth thank RZ-OOK modulation.

Table 3. Bit error rate in various modulation.

Distance	Value			
	PPM	OOK-RZ	OOK-NRZ	PAM
0.4 km	10^{-7}	10^{-3}	10^{-2}	10^{-1}
0.6 km	0.017	0.1	0.14	0.25
1 km	0.28	0.34	0.352	0.41
1.4 km	0.36	0.42	0.43	0.451
1.6 km	0.45	0.44	0.45	0.47

In Table 3 the result show that in several distances below 2 km due to optical communication standard. PPM modulation has the least BER value, then OOK-RZ Modulation, OOK-NRZ modulation and the last is PAM Modulation. The longer the link distance of FSO Channel the bigger the value of BER each modulation.

4. Conclusion

Analysis and simulation conducted in this paper show the four modulation experiments and the comparison of BER, SNR, distance link and visibility. It is proved that the most efficient modulation is PPM compared to OOK-RZ, OOK-NRZ, and PAM. The reason is because each pulse on the laser represents one or more bits of information based on their position to time. RZ-OOK modulation is more efficient than the NRZ-OOK modulation due lower power efficiency. The distance of this simulation used 1-2 Km due to the standard of optical communication free space optic implementation.

IOP Conf. Series: Materials Science and Engineering **850** (2020) 012056 doi:10.1088/1757-899X/850/1/012056

References

[1] Kaushal H and Kaddoum G 2016 Optical communication in space: challenges and mitigation techniques *IEEE communications surveys & tutorials* **19**(1) 57-96

[2] Mutaharrik M I, Syambas N R and Pamukti B 2019 Performance of On-Off Keying Modulation for Free Space Optics Communication *2019 International Conference on Electrical Engineering and Informatics (ICEEI)* 638-643

[3] Ahmadi D and Rachmiatie A 2019 Public Participation Model for Public Information Disclosure *Jurnal Komunikasi: Malaysian Journal of Communication* **35**(4)

[4] Gappmair W, Hranilovic S and Leitgeb E 2010 Performance of PPM on terrestrial FSO links with turbulence and pointing errors *IEEE Communications Letters* **14**(5) 468-470

[5] Song T and Kam P Y 2014 Efficient direct detection of M-PAM sequences with implicit CSI acquisition for the FSO system *2014 IEEE Globecom Workshops (GC Wkshps)* 475-480

[6] Ojo J S and Omotosho T V 2013 Comparison of 1-min rain rate derived from TRMM satellite data and raingauge data for microwave applications in Nigeria *J. Atmospheric Sol.-Terr. Phys.* **102** 17–25

[7] Ijaz M, Ghassemlooy Z, Perez J, Brazda V and Fiser O 2013 Enhancing the atmospheric visibility and fog attenuation using a controlled FSO channel *IEEE photonics technology letters* **25**(13) 1262-1265

[8] Zabidi S A, Al Khateeb W, Islam M R and Naji A W 2010 The effect of weather on free space optics communication (FSO) under tropical weather conditions and a proposed setup for measurement *International Conference on Computer and Communication Engineering (ICCCE'10)* 1-5

[9] Vit´asek J, L´atal J, Hejduk S, Bocheza J, Koudelka P, Skapa J, Siˇska Pˇ and Vaˇsinek V 2011 Atmospheric turbulences in free space optics channel *2011 34th International Conference on Telecommunications and Signal Processing (TSP)* 104–107

[10] Fatima K, Muhammad S S and Leitgeb E 2012 Adaptive coded modulation for FSO links *2012 8th International Symposium on Communication Systems, Networks Digital Signal Processing (CSNDSP)* 1-4

[11] Fiete R D and Tantalo T A 2001 Comparison of SNR image quality metrics for remote sensing systems *Optical Engineering* 40

[12] Gappmair W, Hranilovic S and Leitgeb E 2010 Performance of PPM on terrestrial FSO links with turbulence and pointing errors *IEEE Communications Letters* **14**(5) 468-470

[13] Elganimi T Y 2013 Performance comparison between OOK, PPM and pam modulation schemes for free space optical (FSO) communication systems: analytical study *International Journal of Computer Applications* **79**(11)

IOP Conf. Series: Materials Science and Engineering **850** (2020) 012057 doi:10.1088/1757-899X/850/1/012057

Design of array microstrip antenna with circular polarization for Broadband Tracking System Application

S Alam[1,*], I Surjati[2], L Sari[2], Y K Ningsih[2] and T Wibowo[3]

[1] Department of Electrical Engineering, Universitas Trisakti, DKI Jakarta, Indonesia
[2] Graduate Program of Electrical Engineering, Universitas Trisakti, DKI Jakarta, Indonesia
[3] Department of Electrical Engineering, Sekolah Tinggi Teknologi Indonesia, DKI Jakarta, Indonesia

*syah.alam@trisakti.ac.id

Abstract. This paper proposes the design and simulation of microstrip array antennas with 4x1 elements that have circular polarization for Broadband Antenna Tracking System (BATS) applications. The proposed antenna is designed to work at a frequency of 11 GHz by using FR-4 type substrate with a dielectric constant 4.3, thickness (h) 1.6 mm and loss tangent (tan α) 0.0265. Circular polarization with axial ratio $\leq$ 3 dB was obtained using the truncated corner technique while the 4x1 element array technique was used to increase the bandwidth of the proposed antenna. From the simulation results obtained return loss of -29.71 dB and VSWR 1.068 at the working frequency of 11 GHz. The bandwidth of the proposed antenna is 1663 MHz in the frequency range 10582 GHz - 12215 MHz. The axial ratio of the proposed antenna is $\leq$ 3 dB, which is 2.462 dB at the working frequency of 11 GHz. Based on the results obtained from the design and simulation process, the proposed antenna has worked in accordance with the specified working frequency and is suitable for the Broadband Tracking System application.

1. Introduction

Microwave radio communication systems have been widely used for long-distance and point-to-point communication. Based on regulations regulated by Cameron et al. [1] that the working frequency range for microwave radio is 10.7 - 11.7 GHz. One of the applications of microwave communication systems is Broadband Antenna Tracking System (BATS) [2]. BATS is used for radio communication systems in offshore oil drilling which serves to report the production of oil refineries on tankers in the ocean to platforms on land. Microwave radio communication systems require the accuracy of the transmitter antenna on the platform and the receiver on the tanker must always be in the line of sight position. For this reason, a microwave link radio system Broadband Antenna Tracking System (BATS) is used to control the azimuth of the antenna to keep it in line of sight. The movement of the antenna tracking depends on the received signal level on the receiving antenna on the tanker, if the reception is good, then the tracking is idle but if the reception is bad then tracking moves to find the best reception position. The type of antenna polarization used in BATS is vertical and horizontal. When the radio link uses horizontal polarization and the reception signal level decreases, the link will automatically use vertical polarization, and vice versa. This polarization transfer is generally found in the Indoor Unit (IDU) controller circuit as a protection switcher [3]. For this reason, an antenna that has circular polarization is needed to overcome the problem of overlapping between horizontal and vertical polarization so that the reception signal level is more stable and uninterrupted.

One type of antenna that can be optimized to produce circular polarization is the microstrip antenna [4]. One of the advantages of microstrip antennas is the low cost, compact design and suitable for high frequency so that it is effective for use in the field [5]. But microstrip antennas have several disadvantages including narrow bandwidth, low gain and low directivity [6]. Based on previous research, it has been described to produce circular polarization with axial ratio ≤ 3 dB on microstrip antennas can use several methods including truncated corner [7-8], proximity coupling [9-10] and fractal [11-12]. While increasing bandwidth and gain can be used array [13-14], log-periodic [15-16] and parasitic [17-18]. In the previous study by Varghese et al. [19], a microstrip antenna has been investigated with four-element cross fed array method for 11 GHz working frequency with a return loss of -24.138, VSWR 1.1326 and bandwidth of 210 MHz but polarization is not circular with axial ratio ≤ 3 dB. While the research conducted by Muludi and Aswoyo [20] has successfully designed a microstrip antenna with circular polarization using truncated corner and four-element array using a T junction which produces a return loss of -17,742 dB, VSWR of 1.298, bandwidth of 79.9 MHz and the axial ratio of 2.697 dB.

In this paper, the truncated corner technique is used to produce circular polarization with an axial ratio ≤ 3 dB. To increase gain and bandwidth of the proposed antenna, an array method with 4 elements is applied by arranging the antenna in parallel which is connected with the microstrip line. The focus of this study is to produce microstrip antennas with circular polarization with axial ratio ≤ 3 dB, return loss ≤ -10 dB and gain ≤ 5 dB which works at 11 GHz for Broadband Antenna Tracking System applications.

2. Design of antenna

2.1. Design of single patch of antenna

In this research, the initial design of a single patch of antenna uses a rectangular shape that works in the working frequency of 11 GHz. The proposed antenna is realized on a single layer substrate with relative permittivity (εr) of 4.3, substrate thickness (h) of 1.6 mm and loss tangent (tan δ) of 0.0265. The dimension of the rectangular microstrip antenna obtained using equations (1) until (4). The feed line used from the proposed antenna is microstrip line with an input impedance of 50 Ohms. The dimensions of the microstrip line are obtained using equations (5) and (6) [21].

$$W = \frac{C}{2f\sqrt{\frac{\varepsilon r+1}{2}}} \tag{1}$$

$$L\,eff = \frac{C}{2f\sqrt{\varepsilon_{reff}}} \tag{2}$$

$$\varepsilon r\ \text{eff} = \frac{\varepsilon r+1}{2} + \frac{\varepsilon r-1}{2}[1+12\frac{h}{W}]^{-\frac{1}{2}} \tag{3}$$

$$\Delta L = 0{,}412.h\frac{(\varepsilon_{reff}+0.3)\,(\frac{W}{h}+0.264)}{(\varepsilon_{reff}-0.258)\,(\frac{W}{h}+0.8)} \tag{4}$$

$$B = \frac{60\pi^2}{Z_0\sqrt{\varepsilon_{r\,eff}}}. \tag{5}$$

$$W_z = \frac{2h}{\pi}\left\{B-1-\ln(2B-1)+\frac{\varepsilon_r-1}{2\varepsilon_r}[\ln(B-1)+0.39-\frac{0.61}{\varepsilon_r}]\right\} \tag{6}$$

IOP Conf. Series: Materials Science and Engineering **850** (2020) 012057 doi:10.1088/1757-899X/850/1/012057

After calculating using equations (1) and (2), the dimensions of W is 7.6 mm and L is 6 mm, while the dimensions of the channel width (Wz) are 0.7 mm. The initial design of a single element microstrip antenna can be seen in Figure 1.

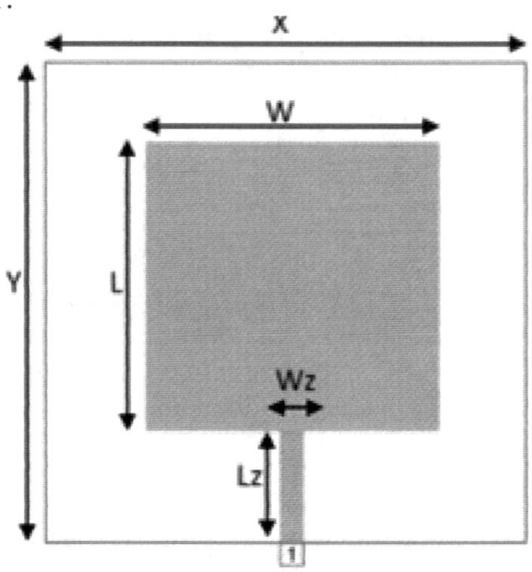

Figure 1. Design of microstrip antenna with rectangular patch.

Figure 1 shows that the dimensions of the substrate on a single element microstrip antenna are X = 10 mm and Y = 10 mm. Furthermore, a rectangular microstrip patch antenna has dimensions of Width (W) is 7.6 mm and Length (L) is 6 mm with dimensions of width (Wz) and length of microstrip line (Lz) is 0.7 mm and 3.5 mm.

2.2. Design of truncated corner microstrip antenna with array 4 element

To produce circular polarization on a single element microstrip antenna, a truncated corner method is required. The truncated corner method is carried out by cutting the edge of a single element patch antenna so that it occurs on the edge effect of the microstrip antenna. The biggest current flowing on the microstrip antenna is on the edge of the patch so that if the part is deformed it will have an impact of shiting the working frequency and produce circular polarization. The dimensions of the pieces on the edge of a single patch antenna element can be obtained using equation (7) while the cutting angle is 45 degrees with respect to the flat plane of the patch antenna [21].

$$\Delta L = \frac{1}{8} \times L \tag{7}$$

From equation (6) the dimensions of the pieces obtained on the patch edge of the microstrip antenna (ΔL) are 1.3 mm which are applied by cutting the upper and lower left edges of the single element antenna patch. In this paper, four patches linear array microstrip antenna technique is used for improving bandwidth and gain. The distance between patches antenna and its related length can be obtained using equation (7) and (8) below [22].

$$d = {}^{\lambda}/_{2} \tag{8}$$

$$d \text{ array} = d - L \tag{9}$$

The design of a single element antenna with the truncated corner technique and array with four elements can be seen in Figure 2.

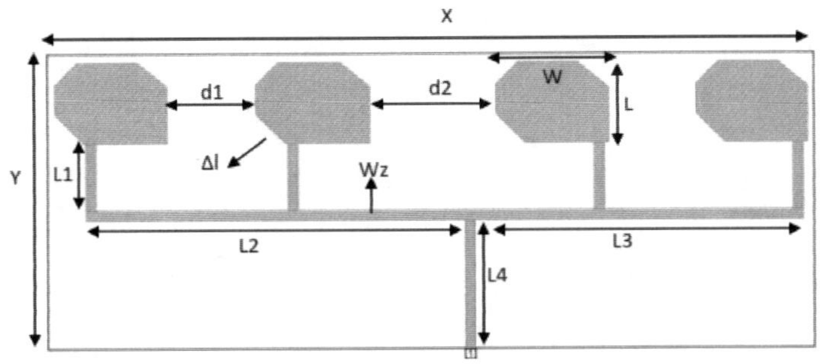

Figure 2. Design of microstrip antenna with truncated corner with array 4 element.

Figure 2 shows the design of the proposed antenna that has been developed using an array with four elements. Each patch antenna is separated by distance d1 and d2. This spacing arrangement between patches is done to prevent interference between the radiating elements. The overall dimension of the proposed antenna can be seen in table 1.

Table 1. The dimension of proposed antenna.

Parameter	Dimension (mm)	Parameter	Dimension (mm)
X	51.5	L2	25.4
Y	20	L3	22
W	7.8	L4	10
L	5.1	Wz	0.7
Δl	2.5	d1	5.9
L1	4.5	d2	8.4

Table 1 shows the overall dimensions of the proposed antenna. After optimization and trial error using AWR Microwave Office, the distance d1 5.9 mm and d2 8.4 mm. To produce the best axial ratio, the proposed antenna is given 3 slices on the edge of the patch.

3. Results and discussion

After the design process is complete, the next step is to do the simulation process using the AWR Microwave office. The simulated parameters are return loss, VSWR, axial ratio, gain and radiation pattern. Comparison of simulation results from the proposed antenna is shown in Figure 3 until Figure 8. The working frequency of the antenna observed in this paper is 11 GHz according to the frequency for the BATS application. Simulation is done by trial and error method using the AWR Microwave Office.

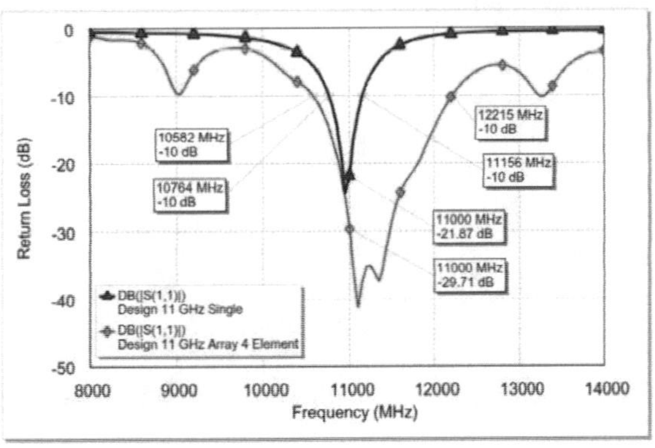

Figure 3. Comparison simulation result of return loss.

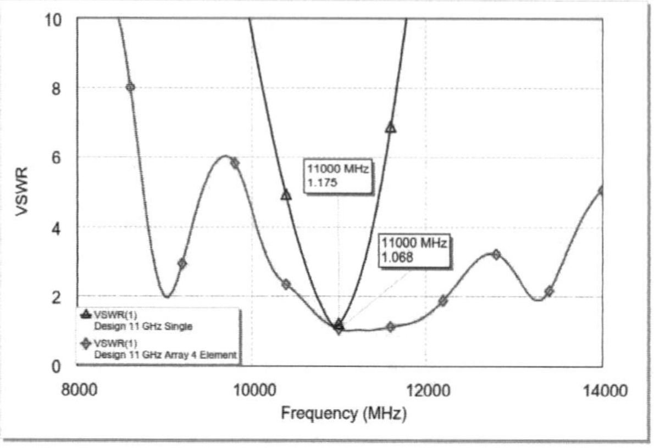

Figure 4. Comparison simulation result of VSWR.

Figure 3 and Figure 4 show a comparison of the simulation results of return loss and VSWR from the proposed antenna. The design of a single element antenna produces a return loss -21.87 dB, VSWR of 1.175 and bandwidth 392 MHz whereas after being optimized with an array of 4 elements the return loss increases until -29.71 dB and VSWR increase become 1.068. The bandwidth of a 4 element array antenna increases to 1663 MHz or 324 % compared to a single element antenna. Comparison of the simulation gain and axial ratio results is shown in Figure 5 and Figure 6.

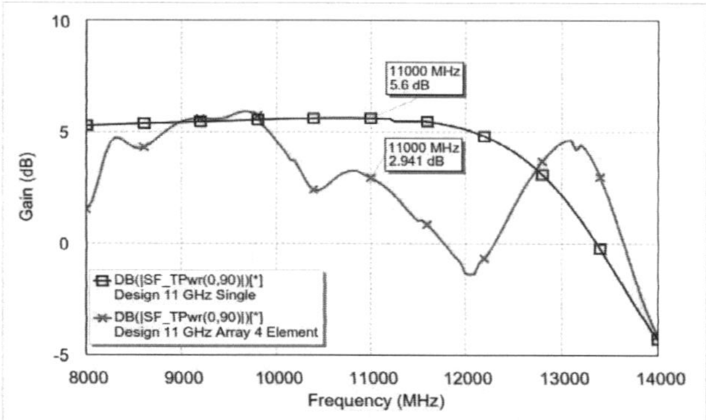

Figure 5. Comparison simulation result of gain.

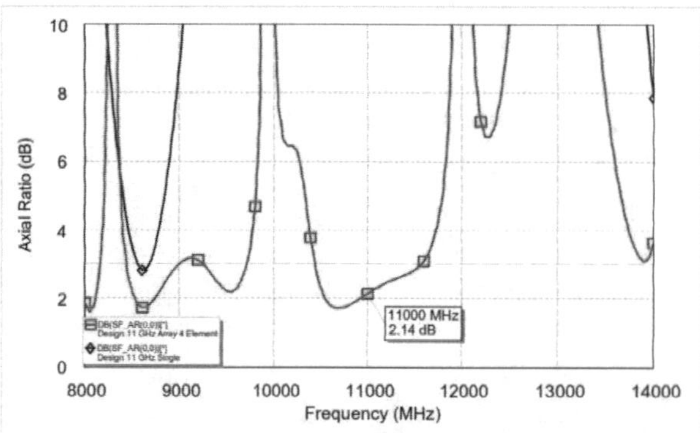

Figure 6. Comparison simulation result of axial ratio.

Figure 5 shows the simulation results of the gain of the proposed antenna. Single element antenna produces gain 5.6 dB while after optimizing the array it decreases becomes 2.941 dB at the working frequency of 11 GHz. The decrease in gain occurs because the bandwidth of the proposed antenna has increased significantly. Wide bandwidth produces smaller gain compared to a narrow bandwidth. This is shown when the antenna is designed with a single element, a narrower bandwidth with a higher gain is obtained. Figure 6 shows the simulation results of the axial ratio of the proposed antenna. The use of truncated corners in 4 element array antennas produces circular polarization with axial ratio 2.14 dB while on the antenna the single element axial ratio is still above 3 dB. Comparison of the overall simulation results from the proposed antenna can be seen in Table 2.

Table 2. Comparison of simulation result.

Condition	Parameter				
	Return Loss	**VSWR**	**Bandwidth**	**Gain**	**Axial Ratio**
Single Element	-21.87 dB	1.175	392 MHz	5.6 dB	-
Truncated Corner With Array 4 Element	-29.71 dB	1.068	1663 MHz	2.941 dB	2.941 dB

Table 2 shows the overall comparison of simulation results from the proposed antenna. Return loss and VSWR generated from antenna arrays with four elements increased by 35.94% and 9.10% compared to single element antennas. In addition, the bandwidth of the array antenna also increased by 324% compared to a single element but a decrease occurred for the gain of the antenna. The decrease in gain on the proposed antenna is due to the impact of increasing the bandwidth of the antenna significantly so that the antenna has a wider frequency range. The radiation pattern of the proposed antenna can be seen in Figure 7 and Figure 8. Figure 7 shows the radiation pattern of a single element antenna which has a narrow direction while Figure 8 shows a radiation pattern from an antenna array with four elements that have a wider directional angle. The red line is the electrical field (Ephi) and the black line is the magnetic field (Etheta) from the radiation pattern of the proposed antenna. From the whole simulation process, it can be concluded that the proposed antenna has produced circular polarization and has met the criteria for working at the 11 GHz working frequency for the BATS application with a minimum bandwidth of 1 GHz with a gain of 2 dB.

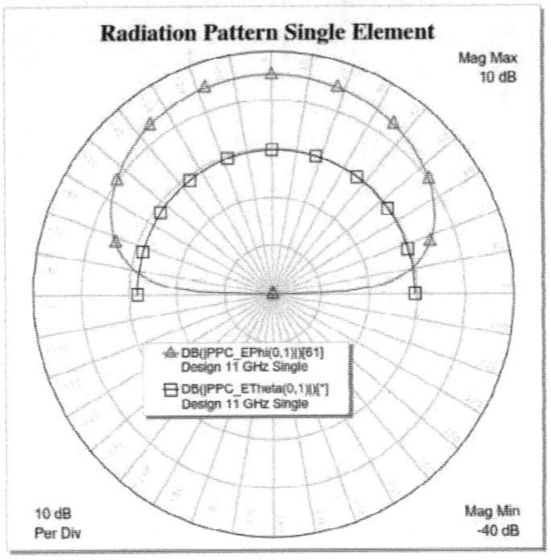

Figure 7. Radiation pattern of single element antenna.

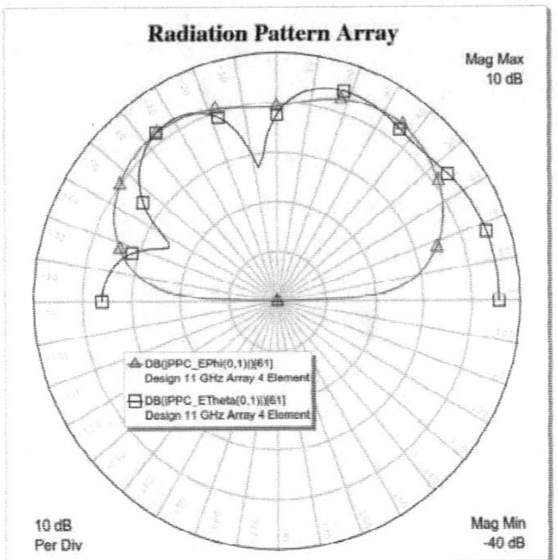

Figure 8. Radiation pattern of array antenna with 4 element.

4. Conclusion

In this paper, the microstrip antenna with rectangular patch developed using truncated corner with a four elements array has been successfully designed and investigated. From the simulation results obtained return loss -29.71 dB, VSWR 1.068 with bandwidth 1663 MHz. The gain generated from the proposed antenna is 2.94 dB while circular polarization is indicated by an axial ratio of 2.941 dB at the working frequency of 11 GHz. Return loss and VSWR generated from antenna arrays with four elements increased by 35.94% and 9.10% compared to single element antennas. In addition, the bandwidth of the array antenna also increased by 324% compared to a single element but a decrease occurred for the gain of the antenna. The decrease in gain on the proposed antenna is due to the impact of increasing the bandwidth of the antenna significantly so that the antenna has a wider frequency range. The future work of this research is to increase the gain by increasing the array to 16 elements and manufacturing and testing antennas in the laboratory.

References

[1] Cameron R J, Kudsia C M, and Mansour R R 2018 *Microwave filters for communication systems: fundamentals, design, and applications* (John Wiley & Sons)

[2] Hasyim A 2016 Planning and analysis of the reliability of line of sight microwave radio communication system on 12750–13250 MHz band *Buletin Pos dan Telekomunikasi* **14**(2) 147-160

[3] Grabner M 2019 Scintillation Characteristics from Experiment on 11 GHz Terrestrial Path *2019 13th European Conference on Antennas and Propagation (EuCAP)* (pp. 1-4) IEEE

[4] Alam S, Wijanto E, Harsono B, Samandatu F, Upa M and Surjati I 2018 Design Of Array and Circular Polarization Microstrip Antenna For LTE Communication *MATEC Web of Conferences* **218** 03006 EDP Sciences

[5] Shandal S, Mezaal Y S, Kadim M and Mosleh M 2018 New compact wideband microstrip antenna for wireless applications *Advanced Electromagnetics* **7**(4) 85-92

[6] Alam S, Wibisana I G N Y and Surjati I 2017 Miniaturization of array microstrip antenna using peripheral slits for wireless fidelity communication *2017 15th International Conference on Quality in Research (QiR): International Symposium on Electrical and Computer Engineering* (pp. 91-95) IEEE

[7] Anantha B, Merugu L and Somasekhar Rao P V D 2019 Polarization Reconfigurable Corner Truncated Square Microstrip Array Antenna *IETE Journal of Research* 1-8

[8] Bharathi A, Merugu L and Somasekhar Rao P V D 2018 Reconfigurable corner truncated square microstrip patch antennas for wireless communication applications *IETE Journal of Research* 1-14

[9] Kumar A, Ansari A Q, Kanaujia B K and Kishor J 2019 Dual Circular Polarization with Reduced Mutual Coupling Among Two Orthogonally Placed CPW-Fed Microstrip Antennas for Broadband Applications *Wireless Personal Communications* **107**(2) 759-770

[10] Odhekar A A, Kadam P A and Deshmukh A A 2018 Proximity Feed Broadband Circularly Polarized Circular Sectoral Microstrip Antenna *2018 Fourth International Conference on Computing Communication Control and Automation (ICCUBEA)* (pp. 1-6) IEEE

[11] Wei K, Zhu B and Tao M 2019 The Circular Polarization Diversity Antennas Achieved by A Fractal Defected Ground Structure *IEEE Access* **7** 92030-92036

[12] Pattnaik S, Sahu S, Dash S K K and Behera S S 2015 Rotated stacked dielectric resonator antenna with sierpinski fractal for circular polarization *2015 International Conference on Communications and Signal Processing (ICCSP)* (pp. 0361-0364) IEEE

[13] Kumar H and Kumar G 2018 Broadband monopulse microstrip antenna array for X-band monopulse tracking *IET Microwaves, Antennas and Propagation* **12**(13) 2109-2114

[14] Jing W, Wei W, Aimeng L and Meng G 2018 A Dual-band Dual-polarized Microstrip Antenna Array with Wideband and High Isolation for Ku-band *2018 IEEE International Symposium on Antennas and Propagation and USNC/URSI National Radio Science Meeting* (pp. 1011-

1012) IEEE

[15] Surjati I, Ningsih Y K, and Alam S 2017 Peripheral Slits Microstrip Antenna Using Log Periodic Technique for Digital Television Broadcasting *TELKOMNIKA (Telecommunication Computing Electronics and Control)* **15**(2) 628-635

[16] Yerlikaya M, Gültekin S S and Uzer D 2019 A Low Profile Wideband Log Periodic Microstrip Antenna Design for C-Band Applications *Advanced Electromagnetics* **8**(2) 48-52

[17] Surjati I, Alam S and Ningsih Y K 2018 Increasing Bandwidth of Log Periodic Array Microstrip Antenna Using Parasitic Air Gap for Digital Video Broadcasting Application *2018 4th International Conference on Wireless and Telematics (ICWT)* (pp. 1-5) IEEE

[18] Surjati I and Haidi J 2015 Increasing bandwidth triangular microstrip antenna using parasitic patch. In *2015 IEEE-APS Topical Conference on Antennas and Propagation in Wireless Communications (APWC)* (pp. 809-812) IEEE

[19] Varghese N M, Vincent S and Kumar O P 2016 Design and analysis of cross-fed rectangular array antenna; an X-band microstrip array antenna, operating at 11 GHz *2016 International Conference on Advances in Computing, Communications and Informatics (ICACCI)* (pp. 1261-1265) IEEE

[20] Muludi Z and Aswoyo B 2017 Truncated microstrip square patch array antenna 2× 2 elements with circular polarization for S-band microwave frequency *2017 International Electronics Symposium on Engineering Technology and Applications (IES-ETA)* (pp. 87-92) IEEE

[21] Wong K L 2004 *Compact and broadband microstrip antennas* (Vol. 168) (John Wiley and Sons)

[22] Visser H J 2006 *Array and phased array antenna basics* (John Wiley and Sons)

IOP Conf. Series: Materials Science and Engineering **850** (2020) 012058 doi:10.1088/1757-899X/850/1/012058

Design and realization of multi-section sum and diff 8 ways Wilkinson Power Divider at S-Band frequency for 3 dimensional radar application

S Herfiah*, T Hariyadi and A B Pantjawati

Department of Electrical Engineering Education, Universitas Pendidikan Indonesia, Jl. Dr. Setiabudhi 207, Bandung 40154, Indonesia

*shitaherfiah@student.upi.edu

Abstract. The Wilkinson Power Divider (WPD) is one of the most common components used in array antenna feeding networks. It is lossless if the output ports are matched, reciprocal, and largely isolated. Many designs and modifications such as multi-section, open circuited stubs, or tapered transmission were proposed to increase the bandwidth of the conventional WPD. In this paper, we will discuss the design of multi-section sum and diff Wilkinson Power Divider for 3-dimensional radar application using CST Studio Suite 2018. The WPD operates at S-Band Frequency, realized in microstrip using substrate Rogers R0-4003C (lossy) with thickness 1.524 mm and relative permittivity 3.55. The simulation and measurement results are met the desired specification, which are less than -20 dB return loss, more than -10 dB insertion loss, less than -20 dB port isolation, and 180° phase difference.

1. Introduction

Radar, stands for radio detection and ranging, is an object detection system that uses radio waves to determine an object parameter such as distance, height, direction, and speed [1]. The important part of the radar system is a transmitter (Tx) that connected to an antenna and sends electromagnetic waves towards the target, and a receiver (Rx) that connected to a receiving antenna and receives all the reflected waves from the radar target. Three-dimensional radar tracking is the development of two-dimensional radar tracking. Two-dimensional radar tracking only covers distance and angular direction, whereas three-dimensional radar tracking includes distance, angular direction, and height. Antennas are an important part of a radar system. Antennas used in radar systems usually have a better direction, where the main and null lobes can be adjusted in direction. This antenna can be realized by arranging some antenna elements into certain settings called arrays antenna. To supply power to the array antenna, a power divider is needed. Power dividers can be either a power divider or a power combiner. Power divider is one of the passive microwave components which can divide one input signal to two or more output signals [2].

The Wilkinson Power Divider (WPD) is a basically a three-port network that is lossless when all output ports are matched. Input power can be divided into two or more with the same amplitude, for two ways WPD using $\lambda/4$ impedance transformation [3]. In general, this WPD is considered a microstrip lines as shown in figure 1.

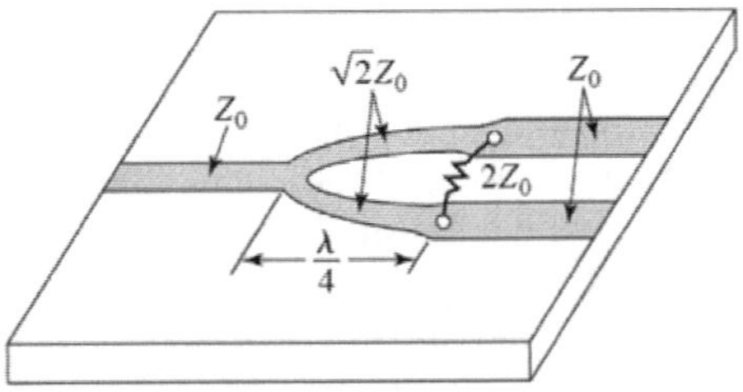

Figure 1. The simplest WPD.

Figure 1 is the simplest WPD consisting of two transmission line with a quarter wavelength channel. The WPD characteristic impedance (Z0) is 50 ohms and the resistor impedance is 2Z0 [3]. Cohn introduced WPD multi sections to widen the bandwidth of conventional WPD [4]. After that, there was a lot of research studies about WPD multi-section, some of the research results have disadvantages, namely increasing in insertion loss value and larger dimensions although it still provides a good isolation between the output ports and larger bandwidth [5]. Recently, many studies have addressed the design of the WPD multi-section [6-10]. In this paper, we will design a Wilkinson Power Divider 1x8 multi-section sum and diff working on the S-Band frequency, with frequency range 2.9–3.1 GHz. To realize the design, we first designed the 1x2 WPD then 1x4 WPD until it was developed up to 1x8 WPD. Sum and diff are part of the monopulse tracking system. Monopulse which is also known as simultaneous lobe comparison (simultaneous beams), is a technique for determining the angular location of a radar target or radiating source. This technique can eliminate errors caused by fluctuations in the target's echo amplitude and increase the data bit rate, because the angular information can be divided into two equal lobe angles [11-14].

2. Research methods

This research was carried out through four stages namely design, simulation, fabrication, and measurement. WPD is designed using CST Studio Suite 2018 software based on reference designs. In the simulation stage, optimization is done several times to get the desired specifications. After the optimal design is obtained, fabrication and measurement of return loss, insertion loss, port isolation and phase difference parameters are carried out. Each stage of the study was carried out through the flowchart as shown in figure 2.

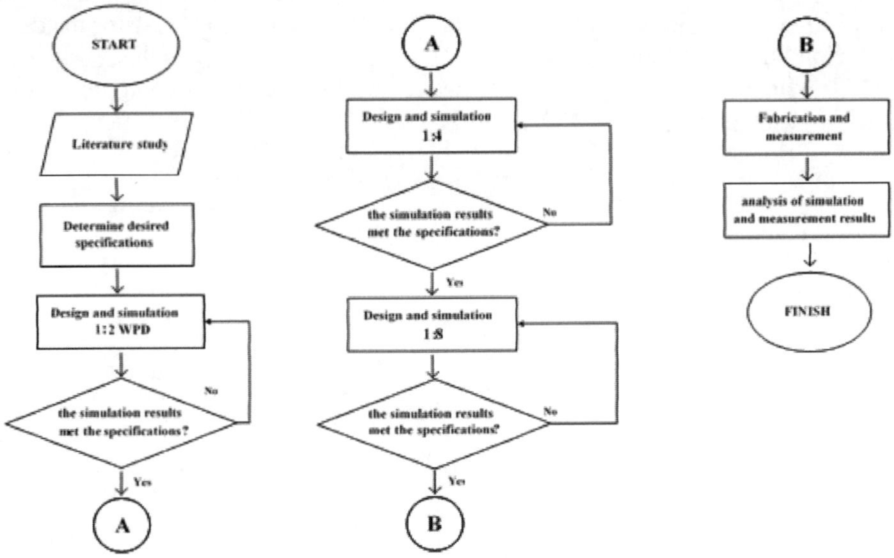

Figure 2. Research flowchart.

IOP Conf. Series: Materials Science and Engineering **850** (2020) 012058 doi:10.1088/1757-899X/850/1/012058

3. Results and discussion

This section describes the three stages of research, namely design, simulation and fabrication of the multisection sum and diff 1x8 Wilkinson power divider. The design stage is to determine the WPD specifications such as geometry shapes, operating frequency, return loss, insertion loss, and isolation port. The simulation stages are to find out the optimal result. After the optimal design is obtained, the fabrication and measurement are carried out.

3.1. WPD design

The WPD is designed using the Rogers RO-4003 C (lossy) as a substrate which has the specifications are listed in table 1, while the desired specifications are listed in table 2. To begin the design of WPD multisection, we first determine the characteristic impedance and resistor values for each section [4]. A theoretical calculation is done to find out the width and length of the line. Table 3 is a calculation of the dimensions of multi-section sum and diff WPD.

Table 1. Specification of substrate.

Substrate Type	Rogers RO-4003C (Lossy)
Relative permitivity (Ɛr)	3.55
Tangent Loss (tan δ)	0.0027
Thickness	1.524 mm

Table 2. Specification of WPD.

Parameter	Value
Return Loss Sum and Diff	≤ -15 dB
Insertion Loss Sum and Diff	≥-10 dB
Isolation Port	≤ -20 dB
Diff Phase	180°, ± 2.5°

3.2. Simulation results

Using the calculation results, simulation process is carried out using CST Studio Suite 2018 to met the desired specifications. The simulation is carried out several times by changing the dimensions and parameters of WPD so that optimal results are obtained. Calculated resistor value of both WPD are R_1=602,486 Ω, R_2 =92,356 Ω, and R_3=93,367 Ω. Because this resistor value is not available in the market, we use 100 Ω resistor instead of 92,456 Ω and 93,367 Ω, and 620 Ω resistor instead of 602,486 Ω during simulation and fabrication. The selected resistor value is slightly higher to increase isolation between ports [6]. Without using a resistor, high isolation value between the output ports will not provide [15].

Table 3. Calculation of parameter for multisections Sum and Diff WPD.

Dimension	Formula and Results
Impedance (Z) and resistor value (R) each sections	To get a value of impedance and resistor value use Paper Cohn Equation's [4], Impedance and resistor value are, Z_1= 54.25 Ω, R_1= 602.486 Ω Z_2= 70.71 Ω, R_2= 92. 356 Ω Z_3= 91.70 Ω, R_3= 93. 367 Ω

Table 3. Cont.

Dimension	Formula and Results
Widht of Transmission lines (W)	To get value of B use equation [2], $B=\frac{377\pi}{2Z_0\sqrt{\varepsilon_r}}$ And to get value of W use equation [2], $W=\frac{2}{\pi}[B-1-\ln(2B-1)+\frac{\varepsilon_r-1}{2\varepsilon_r}\{\ln(B-1)+0.39-\frac{0.61}{\varepsilon_r}\}]$ width for each section is, W_{Z3}= 1.066 mm; W_{Z2}= 1.859 mm W_{Z1}= 2.958 mm; W_{Z0}= 3.400 mm
ε_{eff}	To get the value ε_{reff} use equation below. In this case the value of $W_F/h \geq 1$ then $\varepsilon_{eff}=\frac{\varepsilon_r+1}{2}+\frac{\varepsilon_r-1}{2}(1+12(\frac{h}{w}))^{-0.5}$
Length of transmission lines (L)	To get the value length for each section use equation [2], $\lambda_0=\frac{c}{f}=\frac{3x10^8}{3x10^9}=100\text{mm}$ $\lambda_c=\frac{\lambda_0}{\sqrt{\varepsilon_{eff}}}$ $L=\frac{\lambda_c}{4}$ Length for each section is, L_{Z3}= 15.581 mm; L_{Z2}= 15.321 mm L_{Z1}= 15.064 mm; L_{Z0}= 14.992 mm For diff, L_{Z0} diff =44.976 mm
Metered Bend (X)	To get the value X for each section use [2], $D=w\sqrt{2}$ $X=D(0.52+0.65e^{(-1.35(\frac{w}{h}))}$ $A=(X-D/2)\sqrt{2}$ Metered bend for each section is, X_{Z3}= 1.164 mm; X_{Z2}= 1.696 mm X_{Z1}= 2.373 mm; X_{Z0}= 2.653 mm With value of A for each section is, A_{Z3}= 0.580 mm; A_{Z2}= 0.536 mm A_{Z1}= 0.393 mm; A_{Z0}= 0.352 mm

In designing WPD, we first designs WPD sum 2 ways, 4 ways, up to 8 ways. The final 3 sections WPD 8 ways shown in Figure 3, and it will be fabricated with the optimization dimensions listed in table 4. After designing the WPD sum, we then designed 3 section diff WPD 8 ways by adding $\lambda / 2$ to one of the Z_0 line lengths shown in figure 4.

Table 4. Final parameter of 3 section Sum and Diff WPD.

Parameter	Dimension Value
Length of Substrate	179 mm
Width of Substrat	56.5 mm
Impedance (Z) and Resistor Value (R)	Z_1= 54.25 Ω, R_1= 620 Ω Z_2= 70.71 Ω, R_2= 100 Ω Z_3= 91.70 Ω, R_3= 100 Ω
Width of Transmission Line (W)	Wz_3= 1.066 mm; Wz_2= 1.859 mm Wz_1= 2.550 mm; Wz_0= 3.400 mm
Length of Transmission Line (L)	Lz_3= 16.418 mm; Lz_2= 16.40 mm Lz_1= 15.728 mm; Lz_0= 54.62 mm Lz_0 *diff* = 45mm
Metered Bend (X)	Xz_3= 1.164 mm; Xz_2= 1.696 mm Xz_1= 2.274 mm; Xz_0= 3.500 mm With value of A for each section is, Az_3= 0.580 mm; Az_2= 0.539 mm Az_1= 0.416 mm; Az_0= 0.352 mm

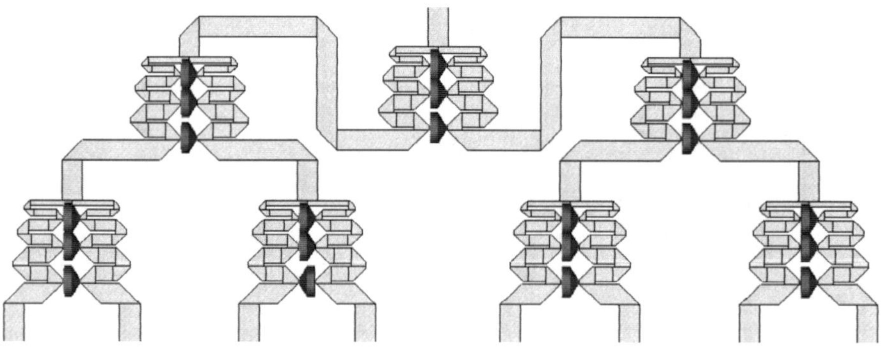

Figure 3. Model of 3 sections sum 8 ways WPD.

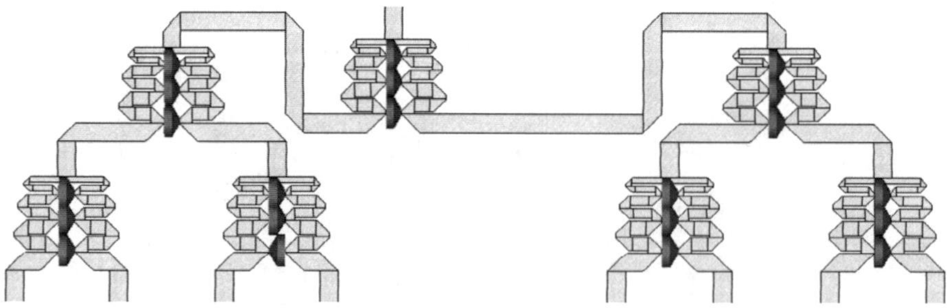

Figure 4. Model of 3 sections diff 8 ways WPD.

Figures 5 and 6 show the overall parameters of the simulation results for ports S11 through S99. The return loss has a value of less than -20 dB at a frequency of 3 GHz. Figures 9 and 10 show the isolation parameters of adjacent output ports such as S23, and S29 distant ports with values smaller than -20 dB. Figure 11 shows the different phase parameters in the WPD diff, where the results of S21, S31, S41, S51 are almost 180° different with S61, S71, S81, S91. Thus WPD are met the desired specifications, even though it is not exactly 180°, but when it is simulated with an antenna, the radiation pattern results are as expected.

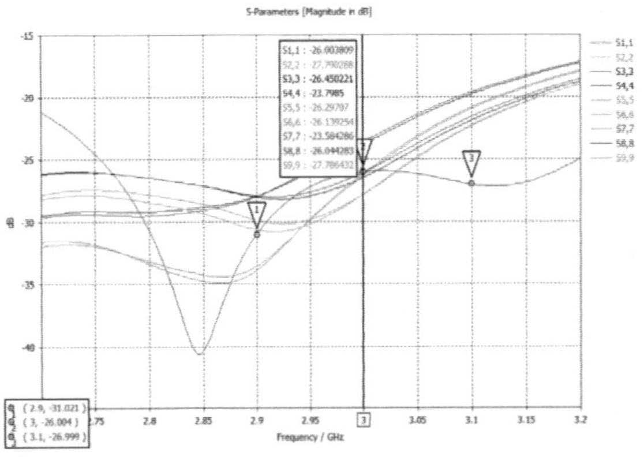

Figure 5. Return loss of sum 8 ways WPD.

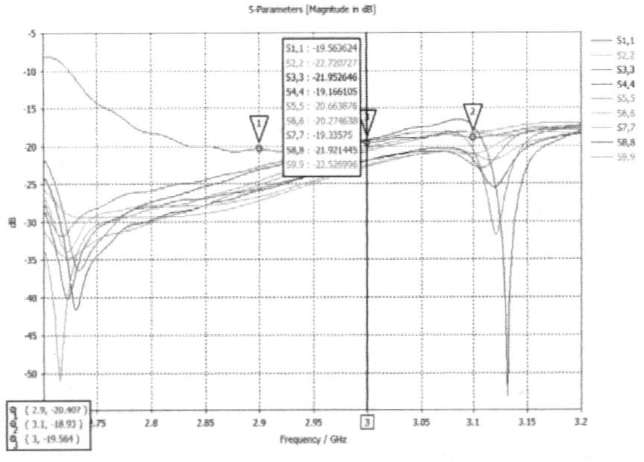

Figure 6. Return loss of diff 8 ways WPD.

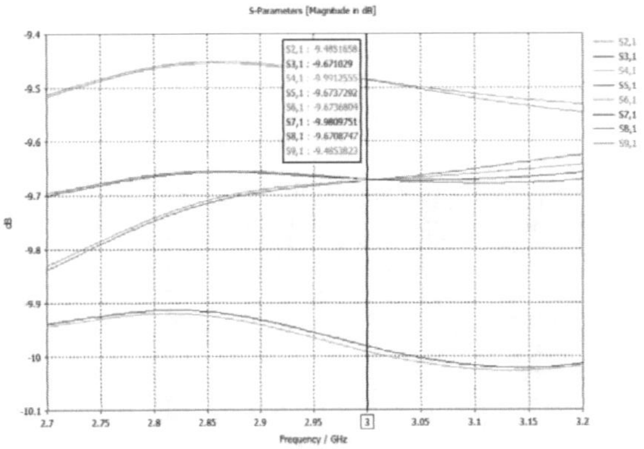

Figure 7. Insertion loss of sum 8 ways WPD.

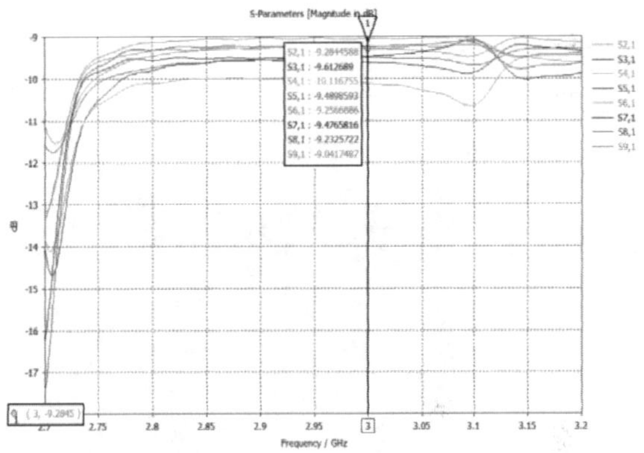

Figure 8. Insertion loss of diff 8 ways WPD.

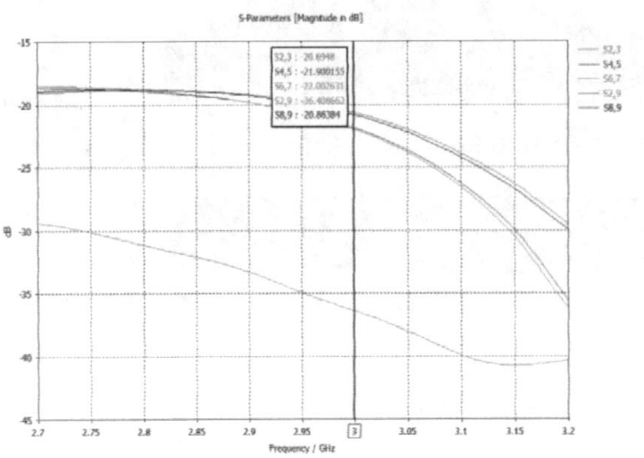

Figure 9. Port isolation of sum 8 ways WPD.

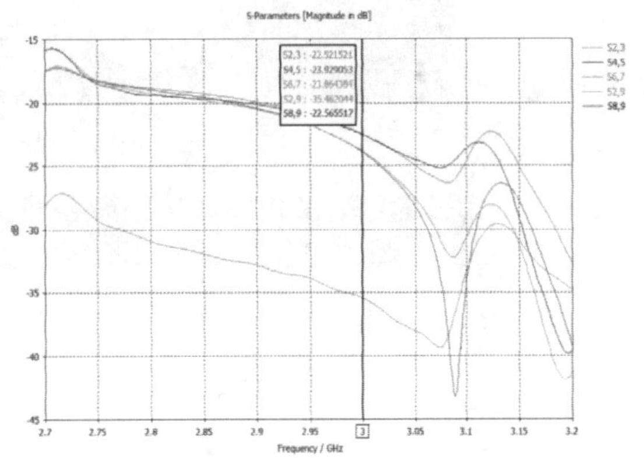

Figure 10. Port isolation of diff 8 ways WPD.

IOP Conf. Series: Materials Science and Engineering **850** (2020) 012058 doi:10.1088/1757-899X/850/1/012058

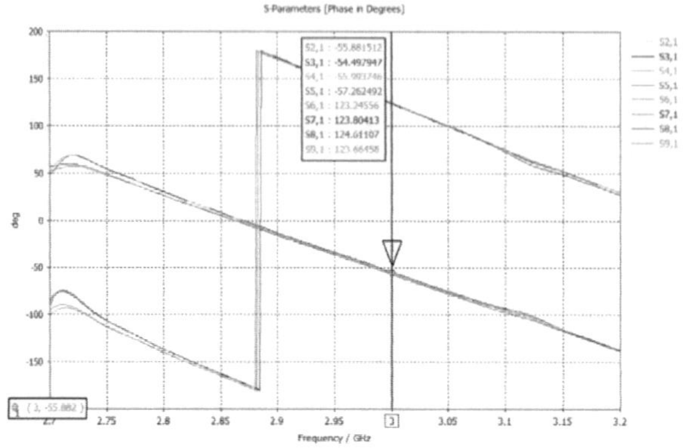

Figure 11. Difference phase of diff 8 ways WPD.

3.3. Realization

The WPD realization of the 3 section sum and diff 8 ways printed on the Rogers RO-4003 C (lossy) substrate, with the SMA connector and the resistor value used according to the simulation. Fabrication results of 3 section sum and diff 8 ways respectively as shown in figures 12 and 13.

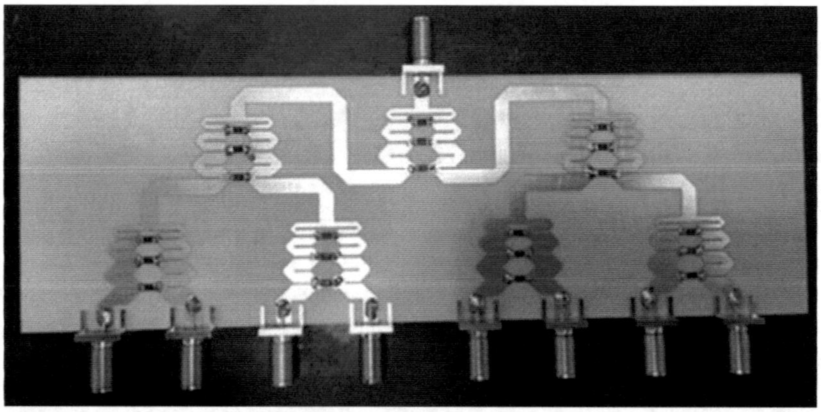

Figure 12. Fabrication of sum WPD.

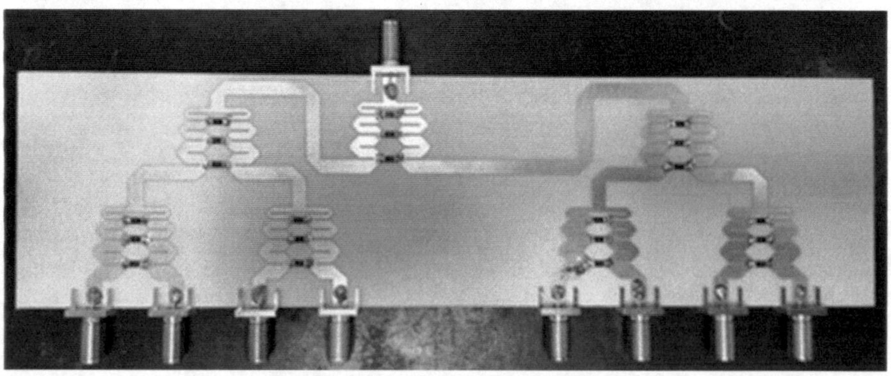

Figure 13. Fabrication of diff WPD.

The next step, the measurements results are compared with the simulation results. All parameters have been measured using the Vector Network Analyzer (VNA). Figures 14-19 show the comparison graph of simulation and measurement result. Figure 14 and 15 show the return loss parameters, which are less than -20 dB at a frequency of 2.9 GHz-3.1 GHz. Figures 16 and 17 show the port isolation parameters, which are smaller than -20 dB at a frequency of 2.9 GHz-3.1 GHz. Figure 18 and 19 show the insertion loss parameter. Ideally when the power divider is given input power of 0 dBm, then the eight output ports have a power of -9 dBm. the measurement and simulation results show that the output varies from -9 dB to -10 dBm for each output port at each sample frequency, meaning this still meets the desired specifications. Figure 19 shows 180° phase difference between S21 and S61 ports for the simulation results and 182° phase difference for measurement result.

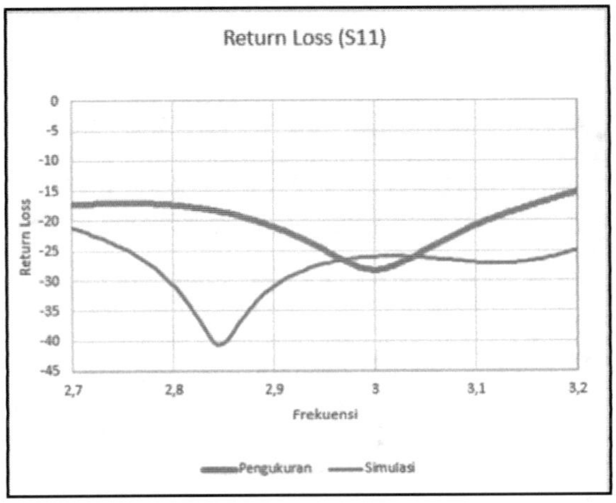

Figure 14. Graph comparison between return loss results of simulation and measurement Sum WPD.

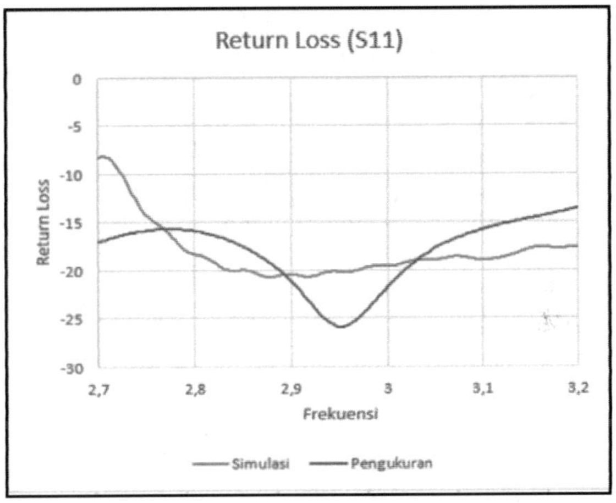

Figure 15. Graph comparison between return loss results of simulation and measurement Diff WPD.

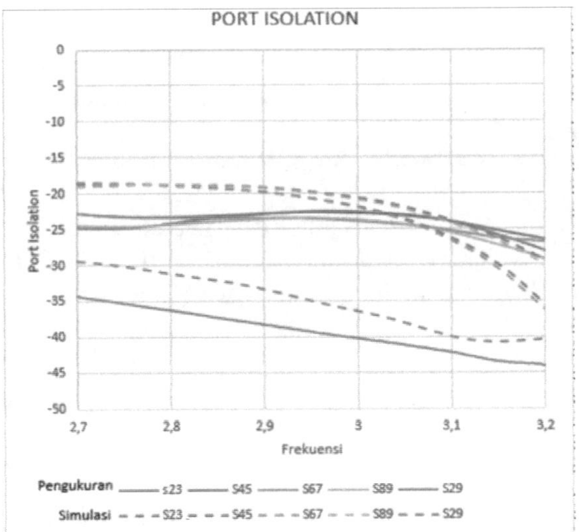

Figure 16. Graph comparison between port isolation results of simulation and measurement Sum WPD.

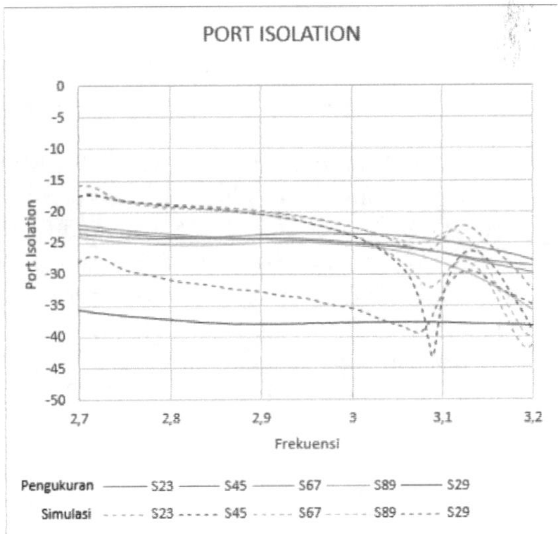

Figure 17. Graph comparison between port isolation results of simulation and measurement Diff WPD.

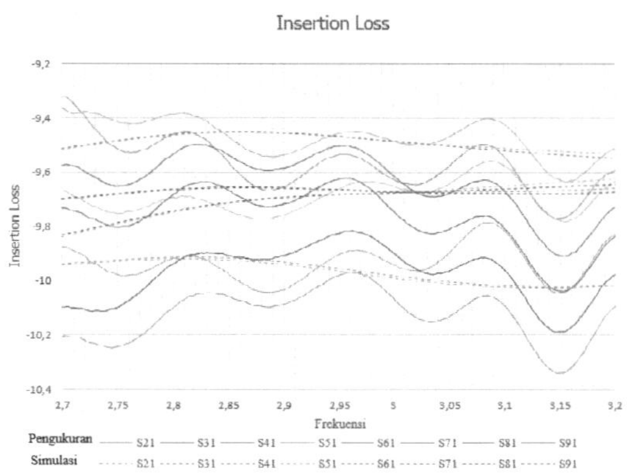

Figure 18. Graph comparison between insertion loss results of simulation and measurement Sum WPD.

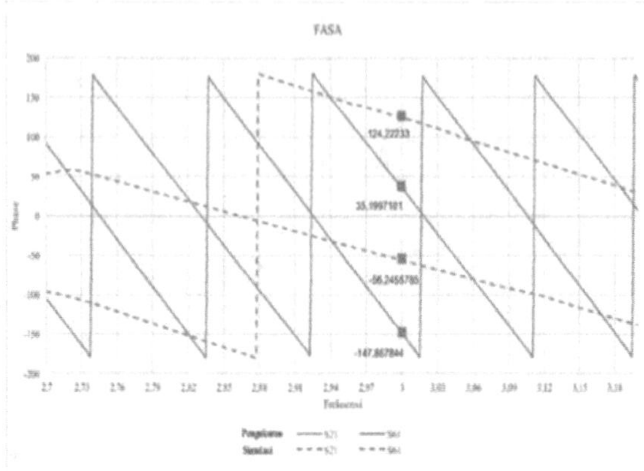

Figure 19. Graph comparison between insertion loss results of simulation and measurement Diff WPD.

4. Conclusion

The multisection sum and diff WPD for 3 dimensional radar applications has been designed and fabricated. The specification for the WPD at frequency 2.9 – 3.1 GHz has been achieved with the value of Return loss less than -15 dB, Insertion loss more than -10 dB, Port isolation less than -20 dB and Phase difference for diff WPD have acceptable and amplitude imbalance less than (± 2.5°).

References

[1] Oommen M, Lenin S, Ansari M S and Mishra S 2014 An analysis of digital signal processing in monopulse radar *International Journal of Advanced Research in Computer and Communications Engineering* **3**(3) 5665-5669

[2] Pozar D M 2011 *Microwave Engineering* (Hoboken: John Wiley and Sons, Inc.)

[3] Sowmiya M, Lotlikar A, Anagha E G and Rajesh A 2017 Design and analysis of Wilkinson power divider using 2D photonic crystal *Materials Science and Engineering Conference Series* **263** 5 052042

[4] Salimi P 2017 A Design Procedure for Multi - Section Micro - Strip Wilkinson Power Divider with Arbitrary Dividing Ratio **3** 48–54

[5] Honari M M, Mirzavand L, Mirzavand R Abdipour A and Mousavi P 2016 Theoretical design of broadband multisection Wilkinson power dividers with arbitrary power split ratio *IEEE Transactions on Components, Packaging and Manufacturing Technology* **6**(4) 605-612

[6] Kizilbey O, Bozdemir S and Yarman B S 2017 2–10 GHz multisection 2-way Wilkinson power divider with enhanced port match and isolation *2017 IEEE 18th Wireless and Microwave Technology Conference (WAMICON)* (pp. 1-4) IEEE

[7] Iqbal M, Camarchia V, Pirola M and Quaglia R 2016 A miniaturized wilkinson power divider for ultra wide-band operation *2016 21st International Conference on Microwave, Radar and Wireless Communications (MIKON)* (pp. 1-4) IEEE

[8] Okada Y, Kawai T and Enokihara A 2016 Compact lumped-element 5-way Wilkinson power divider with broadband operation *2016 46th European Microwave Conference (EuMC)* (pp. 202-205) IEEE

[9] Bao Z and Kong D 2014 The design of two-way N-section equal split Wilkinson power dividers and the loss analysis *Proceedings of 2014 3rd Asia-Pacific Conference on Antennas and Propagation* (pp. 134-136) IEEE

[10] Watkins G T 2014 A 1: 8 cascaded Wilkinson power divider *2014 44th European Microwave Conference* (pp. 219-222) IEEE

[11] Sherman S M and Barton D K 2011 *Monopulse principles and techniques* (Boston: Artech House)

[12] Zhou W, Xie J, Yuan Y and Li S 2014 Angle measurement accuracy analysis of sum-difference amplitude-comparison monopulse in onshore or shipborne ISAR *2014 19th International Conference on Digital Signal Processing* (pp. 531-535) IEEE

[13] Zheng P, Zhao G Q, Xu S H, Yang F and Sun H J 2016 Design of a W-band full-polarization monopulse Cassegrain antenna *IEEE Antennas and Wireless Propagation Letters* **16** 99-103

[14] Sheret T L, Allen B and Parini C G 2016 Monopulse sum and difference signals with compensation for a failed feed element *IET Microwaves, Antennas and Propagation* **10**(6) 645-650

[15] Wu Y, Liu Y and Xue Q 2010 An analytical approach for a novel coupled-line dual-band Wilkinson power divider *IEEE Transactions on Microwave Theory and Techniques* **59**(2) 286-294

Design and realization of 4 way Chebyshev Wilkinson power divider at 3 GHz frequency for 3 dimensional radar application

D R Khoirini*, T Hariyadi and A B Pantjawati

Department of Electrical Engineering Education, Universitas Pendidikan Indonesia, Jl. Dr. Setiabudhi 207, Bandung 40154, Indonesia

*dinarusyda@student.upi.edu

Abstract. In the radar system, detection of objects with high accuracy is needed. Therefore, a low sidelobe antenna is needed to avoid interference from other detected targets. One method to suppress sidelobe antenna is by using unequal power divider with Chebyshev polynomial excitation coefficients. In this paper, we will discuss about the design and fabrication of Chebyshev Wilkinson power divider for 3 dimensional radar application. This research was conducted through 4 stages, namely design, simulation, fabrication and measurement. This power divider is realized in the form of a microstrip using Rogers R0-4003C (lossy) substrate with a thickness of 1,524 mm and a dielectric constant of 3.55. The simulation and measurement results are met the desired specifications, the operating frequency is at 3 GHz with a bandwidth of 200 MHz, the return loss is less than -10 dB, the insertion loss value is in accordance with Chebyshev, the isolation value of the port is less than -10 dB, and the unbalance phase is smaller than 2.5°.

1. Introduction

Radar (Radio Detection and Ranging) is an object detection system that uses electromagnetic waves to determine the distance, altitude, direction and speed of objects. In radar technology, the object detected by a radar is called a target. The operating principle of radar is to measure the distance from the sensor to the target by measuring the travel time of electromagnetic waves propagating from the sensor, the target and back to the sensor [1-2]. Antenna is one of the important parts of the radar system. The antenna used in the radar system is an array antenna. To supply power to the array antenna, a power divider is needed. Power divider functions as a feeding network in array antennas that can match all antenna ports [3-6].

Wilkinson Power Divider (WPD) is basically a lossless 3 port network when all output ports are matched. Input power is divided into 2 or more outputs with the same amplitude (for equal power divider). Transformation of impedance $\lambda / 4$ is used for 1x2 WPD [7]. In general, WPD is considered a microstrip channel. The simplest WPD (1x2) consists of two transmission lines with a channel length of $\lambda / 4$, and by using resistor components which are set to increase the isolation value. In equal WPD, the characteristic impedance (Z0) is 50 Ω and the resistor impedance is 2Z0 [6-8].

WPD is classified as equal WPD and unequal WPD [9]. The difference between equal and unequal WPD is the power distribution. In equal WPD the power distribution is evenly distributed while unequal WPD is uneven [10]. One of the advantages of the unequal power divider is its ability to suppress the antenna sidelobe. The sidelobe emphasis is needed for applications that require high accuracy. On radar tracking, low sidelobe can prevent interference from other targets that are detected. One method to suppress sidelobe is by using Chebyshev tapered excitation on unequal power divider [11-14].

In this paper, a 4 way unequal WPD will be designed with Chebyshev tapered operating at a frequency of 3 GHz for 3 dimensional radar applications with 200 MHz bandwidth.

2. Research methods

WPD structure design is carried out through 4 stages, namely design, simulation, fabrication and measurement. The device design and simulation is done using CST Studio Suite software. The first step is designing and simulating a 1x2 WPD. After the simulation results meet the desired specifications, design and simulation is carried out for WPD 1x4. After the simulation results meet the final desired specifications, fabrication and measurement are carried out next. Stages of research can be seen in the flow chart in Figure 1.

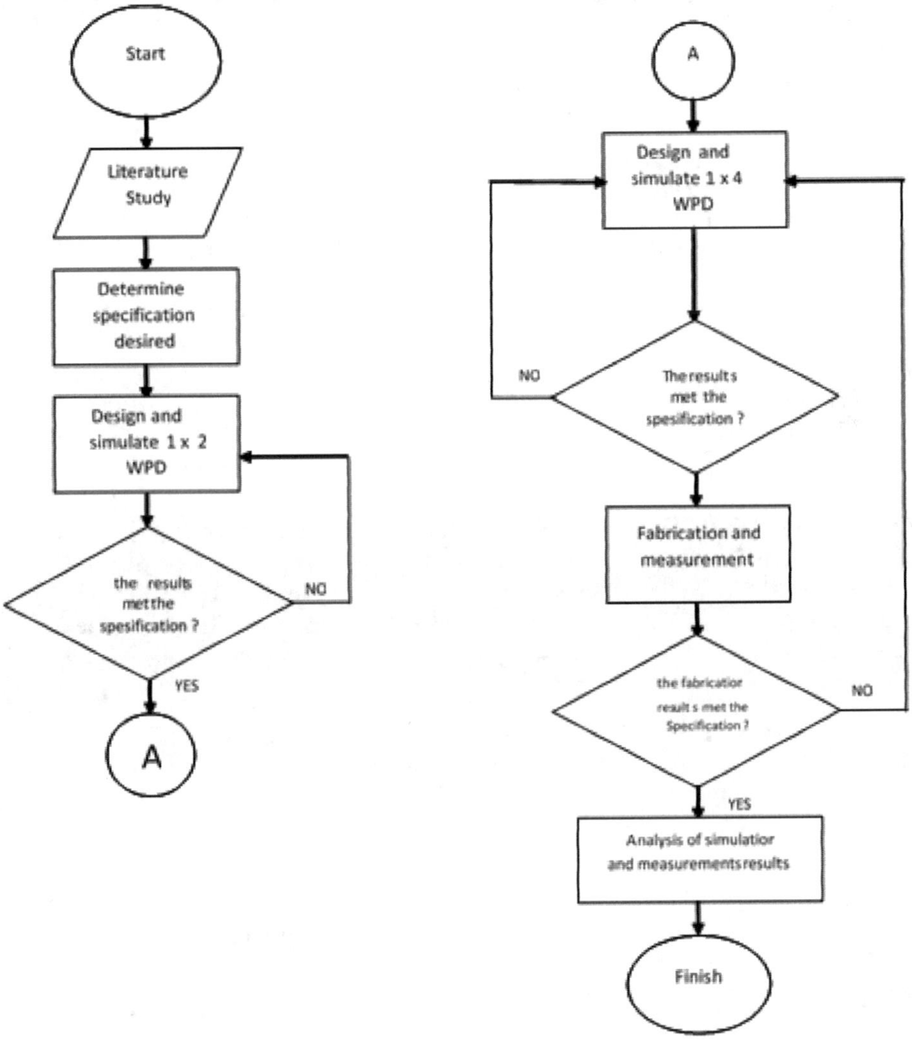

Figure 1. Research flow chart.

3. Results and discussion

This section explains the 4 Way Wilkinson power divider design and simulation results and the measurements result after fabrication.

3.1. Design

The design stage is carried out to determine the power divider specifications such as design form, dimensions, work frequency, return loss, insertion loss and isolation port. For substrate specifications can be seen in Table 1 and the WPD specifications in Table 2. To do the WPD design, it is necessary to determine the characteristic impedance and resistor value first [9]. Theoretical calculations are needed to find out the width and length of the transmission line. Calculation of WPD dimensions can be seen in table 3. WPD uses excitation coefficients based on the Chebyshev polynomial where the comparison of excitation coefficients is 0.3747: 1: 1: 0.3747 or the difference in the value of insertion loss 4.2: 0: 0: 4.2 (in dB).

Table 1. Substrate specifications.

Material	**Rogers RO-4003C *(Lossy)***
Dielectric constant (Ɛr)	3.55
Electric tand	0.0027
Substrate thickness	1.524 mm

Table 2. WPD specifications.

Parameter	**Value**
Return loss	≤ 10 dB
Isolation port	≤ 10 dB
Insertion loss difference	4.2 : 0 : 0 : 4.2
Phase unbalance	2.5°
Bandwidth	200 MHz

3.2. Simulation result

After the calculation is complete, the simulation process is then carried out using CST Studio Suite 2018. If the results of the WPD simulation obtained are not optimal, then the dimensions are changed. Optimization is carried out to get optimal results in accordance with the expected specifications. The result of the calculation of the resistor value is R1 = 100 =, R2 = 112,289 Ω, R3 = 112,289. Because the resistor with the value of the calculation results is not available on the market, then the closest rounding to the resistor value is available. In this simulation a 100 ohm resistor is used instead of a 112,289 ohm resistor. In designing WPD, we first designed 1x2 WPD then 1x4 WPD. So that we get 1x4 WPD shown in figure 2, which will then be realized with the optimization dimension listed in table 4.

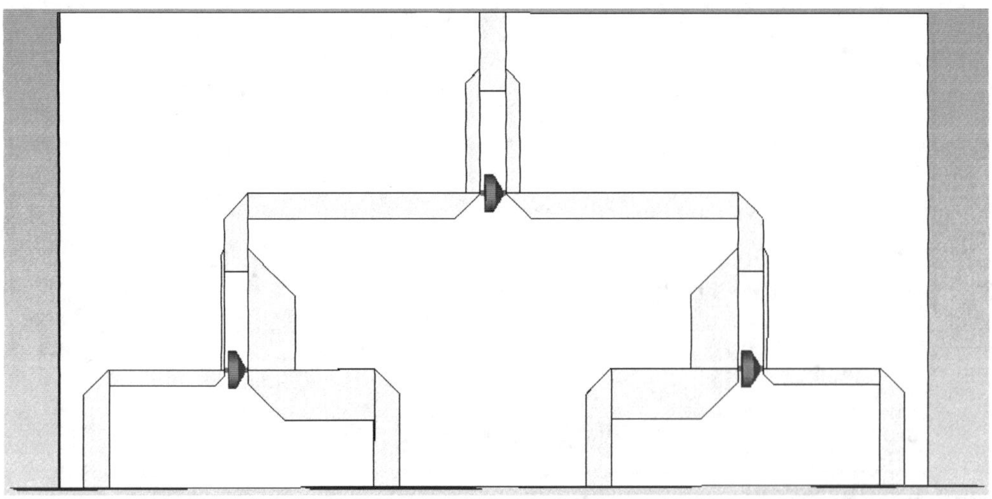

Figure 2. 1x4 way WPD model.

Table 3. WPD dimension calculation.

Dimension	Formula and Result
Impedance Explanation: Z = impedance K = comparison of output power port	$z_1 = z_0\sqrt{\frac{1+K^2}{K^3}}$, $Z_2 = K^2 Z_1$ $Z_3 = Z_0/K$, $Z_4 = Z_0 K$ $Z_{01} = \sqrt{2}Z_0$ [2] Z_1 = 112,408 Ω, Z_2 = 42.119 Ω Z_3 = 112.408 Ω, Z_4 = 30.606 Ω Z_0= 50, Z_{01} = 70.71
Resistor Explanation: R = resistor	$K^2 = P_3 / P_2$ $R = Z_0\left(K + \frac{1}{K}\right)$ [2] R_1 = 100 Ω R_2 = 112.289 Ω R_3 = 112.289 Ω
Transmission line width Explanation: W = transmission line width ε_r = relative permittivity	$B = \frac{377\pi}{2Z_0\sqrt{\varepsilon_r}}$ $W = \frac{2}{\pi}[B - 1 - \ln(2B - 1) + \frac{\varepsilon_r - 1}{2\varepsilon_r}\{\ln(B - 1) + 0.39 - \frac{0.61}{\varepsilon_r}\}]$ [2] W_{Z1} = 0.6275 mm W_{Z2} = 4.441 mm W_{Z3} = 1.385 mm W_{Z4} = 6.961 mm W_{Z0} = 3.4 mm W_{Z01} = 1.86 mm
Transmission line length Explanation: λ = wave length L = transmission line length	$\lambda_0 = \frac{c}{f} = \frac{3x10^8}{3x10^9} = 100mm$, $\lambda_c = \frac{\lambda_0}{\sqrt{\varepsilon_{eff}}}$ $\varepsilon_{eff} = \frac{\varepsilon_r + 1}{2} + \frac{\varepsilon_r - 1}{2}(1 + 12(\frac{h}{w}))^{-0.5}$, $\lambda/4 = \frac{\lambda_c}{4}$ [2] L_{Z1} = 15.733 mm, L_{Z2} = 14.837 mm L_{Z3} = 15.463 mm, L_{Z4} = 14.57 mm L_{Z01} = 15.322 mm
Mitered bends Explanation: X = mitered bends	$D = w\sqrt{2}$ $A = (X - D/2)\sqrt{2}$ $X = D(0.52 + 0.65e^{(-1.35(\frac{w}{h}))}$ [2] X_{Z1}= 0.792 mm X_{Z2}= 3.346 mm X_{Z3}= 1.392 mm X_{Z4}= 5.132 mm X_{Z0}= 2.654 mm X_{Z01}= 1.697 mm

Table 4. Dimensions of WPD optimization.

Parameter	Dimension Value
Substrate Length	111.838 mm
Substrate Width	60.462 mm
Transmission line width	Wz_1 = 0.5 mm, Wz_2 = 6 mm Wz_3 = 2.09 mm, Wz_4 = 6.5 mm Wz_0 = 3.23 mm, Wz_{01} = 1.781 mm
Transmission line length	Lz_1 = 14.57 mm, Lz_2 = 15.459 mm Lz_3 = 14.44 mm, Lz_4 = 16.37 mm Lz_{01} = 16mm
Mitered Bends	Xz_1= 1.03 mm, Xz_2= 6 mm Xz_3= 1.39 mm, Xz_4 = 4.79 mm Xz_{01} = 3.15 mm, Xz_{01} = 2 mm

Figure 3 shows the overall results for return loss from ports S11 to S55 at frequencies 2.9 - 3.1 GHz, which are smaller than -10 dB. Figure 4 shows the overall insertion loss results for ports S21 to S51 where the value meets the weighting of Chebyshev. Figure 5 shows the isolation value between ports both adjacent and far apart with values below -10 dB at 2.9 - 3.1 GHz. To find out the output port is in phase or not, it can be seen in the phase value of each output port. If the value of the phase is the same, it means it's in phase. It can be seen in figure 6 that the values of S21, S31, S41 and S51 are in phase with the phase unbalance smaller than 2.5 °.

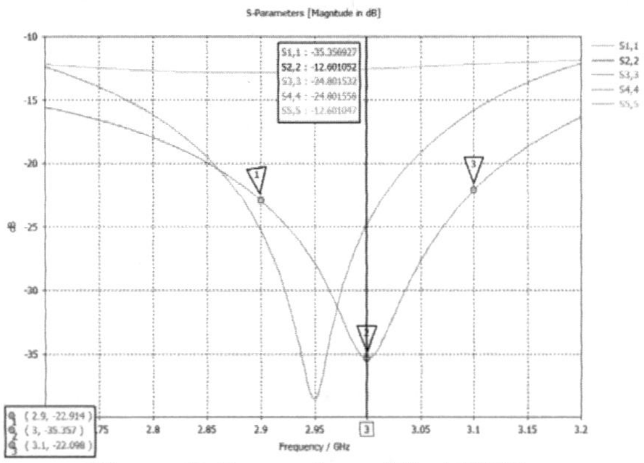

Figure 3. Return loss of 1x4 WPD.

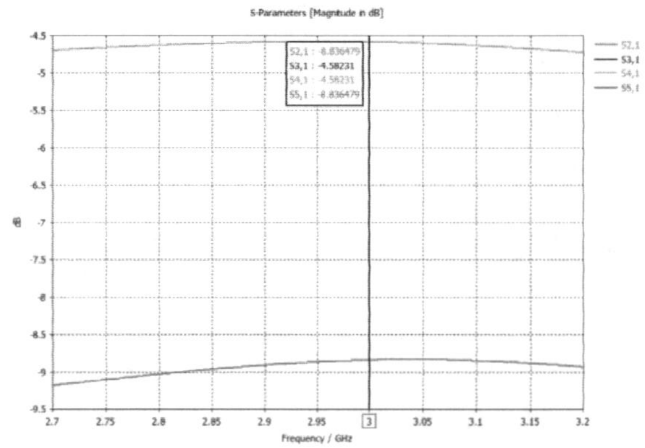

Figure 4. Insertion loss of 1x4 WPD.

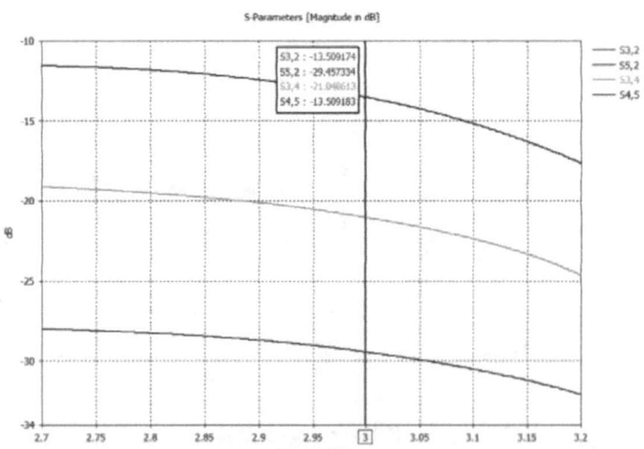

Figure 5. Isolation port of 1x4 WPD.

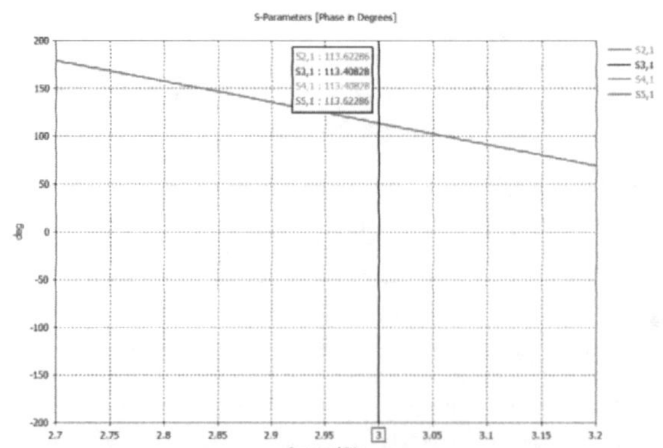

Figure 6. Phase of 1x4 WPD.

IOP Conf. Series: Materials Science and Engineering **850** (2020) 012059 doi:10.1088/1757-899X/850/1/012059

3.3. Fabrication result and measurement

The realization of the design from WPD 1x4 was printed on Rogers RO-4003 C (lossy) material, with SMA connector (SubMiniature version A) and 100 ohm resistor value. The results can be seen in figure 7. All parameters are measured using Vector Network Analyzer (VNA).

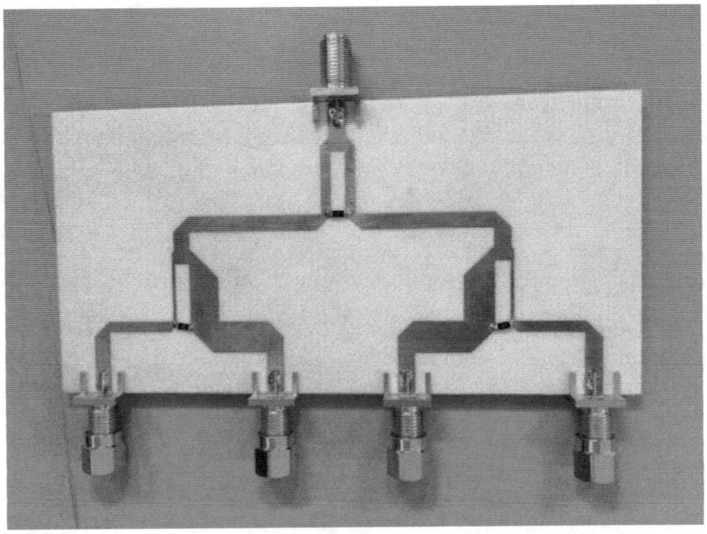

Figure 7. WPD fabrication result.

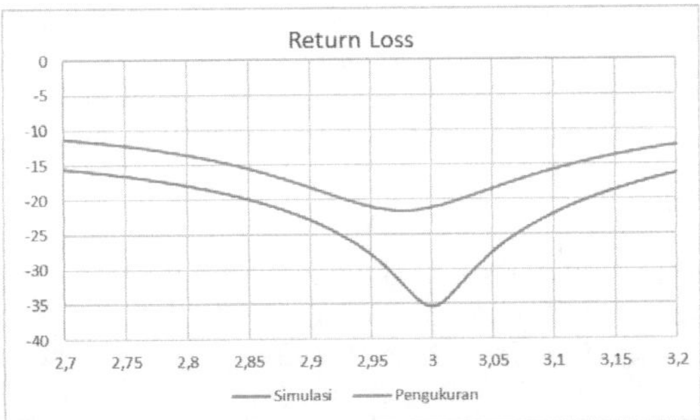

Figure 8. Comparison graph of return loss measurement and 1x4 WPD simulation.

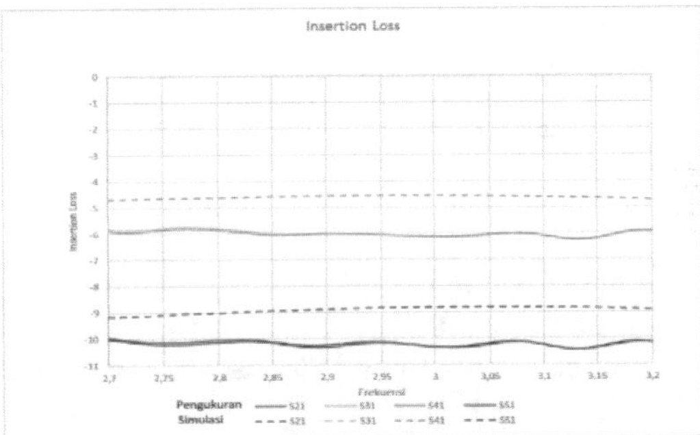

Figure 9. Comparison graph of insertion loss measurement and 1x4 WPD simulation.

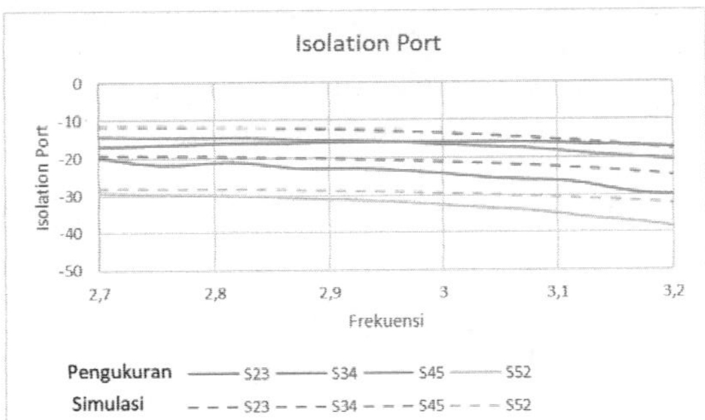

Figure 10. Comparison graph of isolation port measurement and 1x4 WPD simulation.

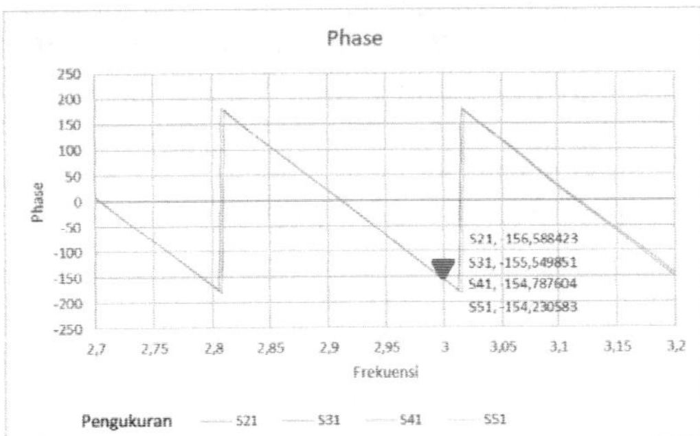

Figure 11. Phase measurement graph of 1x4 WPD.

IOP Conf. Series: Materials Science and Engineering **850** (2020) 012059 doi:10.1088/1757-899X/850/1/012059

From the measurements it was found that the results of each parameter met the specifications. Figure 8-10 show the simulations and measurements comparison charts. Figure 8 shows a return loss that less than -10 dB at 2.9 - 3.1 GHz. Figure 9 is an insertion loss that have met the specifications. Figure 10 is an isolation port that less than -10 dB at 2.9 - 3.1 GHz. Figure 11 shows the output port is in phase with a phase unbalance smaller than 2.5 °.

4. Conclusion

4 Way Chebyshev Wilkinson power divider at 3 GHz frequency with 200 MHz bandwidth for 3 dimensional radar applications has been designed and realized. Simulation results and fabrication results meet the desired specifications, where the return loss value is less than -10 dB, insertion loss values are in accordance with Chebyshev weighting, isolation ports are smaller than -10 dB and phase unbalance is smaller than 2.5 °.

References

[1] Bagad V S 2008 *Radar System* (Pune, India: Technical Publications Pune)
[2] Oommen M, Lenin S, Ansari M S and Mishra S 2014 An analysis of digital signal processing in monopulse radar *International Journal of Advanced Research in Computer and Communications Engineering* **3**(3) 5665-5669
[3] Pozar D 2011 *Microwave Enginnering Fourth Edition* (John Willey & Sons, Inc.)
[4] Ng C Y and Robertson I D 2002 A 140 GHz Wilkinson Power Divider/Combiner *7th IEEE High Frequency Postgraduate Student Colloquium*
[5] Manasseh S 2015 Evaluation of 1:2.1 Power Dividers for Radar Applications and Design. *i-manager's Journal on Circuits and Systems* **3**(2) 8-13
[6] Saranya M 2014 Microstrip Wilkinson Power Divider with Enhanced Isolation and Miniaturization for Defense Aplications *SSRG International Journal of Communication and Media Science* **1** 2
[7] Sowmiya M, Lotlikar A, Anagha E G and Rajesh A 2017 Design and analysis of Wilkinson power divider using 2D photonic crystal *Materials Science and Engineering Conference Series* **263** 5 052042
[8] Kumari R and Acharya V 2017 Design and Simulation of Two Way Power Divider in S Band *International Research Journal of Engineering and Technology (IRJET)*
[9] Watkins G T 2014 A 1:8 Cascaded Wilkinson Power Divider *2014 44th European Microwave Conference*
[10] Lee D, Do H, Jang Y, Kim H, Han S-M, Lim J, Choi K and Ahn D 2016 A New Design Method of the Unequal Wilkinson Power Divider Using an Arbitrarily Resistor Value *Microwave and Optical Technology Letters* **58**(10) 2450–2452
[11] Choe W and Jeong J 2016 N-Way Unequal Wilkinson Power Divider With Physical Output Port Separation *IEEE Microwave and Wireless Components Letters* **26**(4) 243–245
[12] Zulkifli F Y, Hidayat T, Basari and Rahardjo E T 2013 Sidelobe Level Suppression Using Unequal Four-Way Power divider for Proximity Coupled Microstrip Antenna *2013 Asia-Pacific Microwave Conference Proceedings (APMC)*
[13] Kasar Ö, Gözel M A and Kahriman M 2018 Broadband Wilkinson power divider based on chebyshev impedance transform method *International Advanced Researches and Engineering Journal* **2**(3) 299-303
[14] Cohn S 1968 A Class of Broadband Three-port TEM-mode Hybrids *IEEE Trans Microwave Theory Technic* **MTT-19** 2 110-116

IOP Conf. Series: Materials Science and Engineering **850** (2020) 012060 doi:10.1088/1757-899X/850/1/012060

Design of microstrip antenna *C-band* frequency for ground surveillance radar

H Y Salazar*, T Hariyadi and A B Pantjawati

Department of Electrical Engineering Education, Universitas Pendidikan Indonesia, Jl. Dr. Setiabudhi No.207, Bandung, Indonesia 40154

*hildayfs@gmail.com

Abstract. This paper presents designing of a conformal antenna using a number of microstrip patches in C band frequency range 4 GHz to 8 GHz. The type of microstrip C-band antenna was chosen because it is suitable for use on portable surveillance radar systems. Single element patches antenna with a carved slit is used to obtain a horizontal polarization. In this paper, an microstrip patch antenna C-band frequency for 3-dimensional radar application was designed using CST Studio Suite 2018. The antenna operates at frequency 5.6 GHz, with 40.55 x 22.65 mm dimension using substrate Rogers RO-4350B (lossy) with thickness 1. 524 mm and relative permittivity 3.66. The bandwidth obtained from the simulation is 200 MHz, the return loss values below -20 dB, VSWR less than 2, 5.6 dB gain, and has directional radiation pattern.

1. Introduction

Indonesia is a country that requires devices that can monitor land and waters throughout the country. One of the devices that can be used is Radar. Radar stands for Radio Detection and Ranging, this is a system for detecting objects that use Radio waves to determine the distance, altitude, direction, or speed of an object [1]. The radar system operates using radio wave transmission by detecting and translating echo signals reflected by the intended object. A portable surveillance radar system is needed because of the limited number of radars and the wide area that needs to be monitored [2].

In a wireless communication system, one component that plays an important role is the antenna. Rapid technological advances have brought antennas with light weight, affordable prices, and low profile while still being able to maintain high performance. In a radar system, it is also needed an antenna with low profile and lightweight characteristics, so that the current technology trend focuses a lot on the microstrip antenna design [3].

Radar Surveillance Ground is used as a detector of objects that move on land such as people, vehicles or aircraft that fly at a low altitude. This technology is designed using the Frequency Modulated Continuous Wave (FM-CW) technique, in which radar signals are transmitted continuously using two separate transmitting and receiving antennas. This technology requires a very low transmit power (≤ 10 Watt) so that it can be operated only by using a battery, but provides a wide monitoring area coverage [4,5].

The frequency range used for this surveillance radar is C-band. C-band is part of the microwave band of the electromagnetic spectrum for radio waves with frequencies between 4 to 8 GHz [6].

According to the definitions contained in the IEEE Standard Definitions of Terms for Antennas, the definition of an antenna is a device used to radiate or receive radio waves [7]. Microstrip antenna is a

IOP Conf. Series: Materials Science and Engineering **850** (2020) 012060 doi:10.1088/1757-899X/850/1/012060

metal conductor material attached to a ground plane. The microstrip antenna has several advantages, such as thin and small shape, has a light weight, easy to fabricate, easy to integrate with other electronic devices, and affordable price [8]. In general, microstrip antenna is shown in figure 1. Antenna polarization will greatly affect the performance of the radar. Antennas with horizontal polarization can reduce ground chaos rather than vertical polarization [2].

In this paper, a single microstrip antenna with horizontal polarization will be designed for ground surveillance radar. This antenna works on C-Band frequencies which are from the frequency range of 4-8 GHz and has an operating frequency of 5.6 GHz with a port impedance of 50 ohms. Horizontal polarization is obtained by making a slot on the antenna as shown in Figure 2.

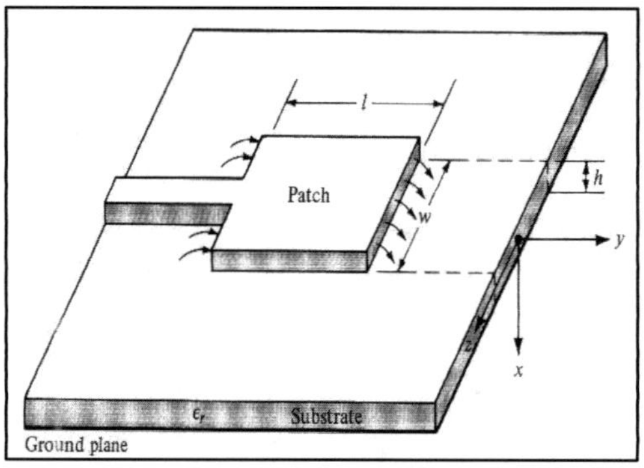

Figure 1. Microstrip antenna.

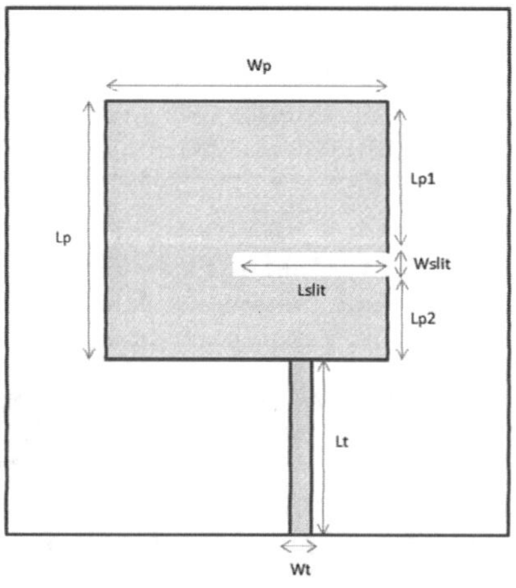

Figure 2. Microstrip antenna with a gap .

2. Researchs method

The research was conducted through three stages namely design, simulation and optimization to get optimal results by using CST Studio Suite software. The design is done based on the desired antenna specifications. After the design is completed, it is simulated and if the results are not appropriate to the desired specifications, the optimization process is carried out by changing the values of each dimension until it meets the specifications.

3. Results and Discussion

This section explains three steps of research, that is the design, simulation and optimization of the arrangement of C-Band frequency microstrip antennas. The design step is carried out to determine antenna specifications such as geometry, dimensions, work frequency, VSWR, radiation patterns, impedance, polarization and gain. The simulation step is done to find out the design results whether they are appropriate to the desired specifications, if not then it will proceed to the optimization step to get the desired results.

3.1. Design

Microstrip antenna is an antenna which is composed of three elements, that is : patch (radiating element), substrate, and ground. In designing this antenna, we use Rogers RO-4350 B (lossy) substrate which has the specifications listed in table 1. While the desired antenna specifications are listed in table 2.

Table 1. Substrate specifications.

Substrate type	Rogers RO-4350 B (*lossy*)
Relative dielectric constant (Ɛr)	3.66
Dielectric Loss tangent (tan δ)	0.0037
Substrate thick (h)	1.524 mm

Table 2. Antenna specifications.

Parameter	Value
Working Frequency (F0)	5.6 GHz
VSWR	< 2
Bandwidth	200 MHz
Impedance (Z_0)	50 Ω
Gain	> 5 dB
Radiation pattern	*Directional*
Polarization	*Horizontal*
Return Loss	<-10dB

3.1.1. Antenna type. The antenna form used was a modification of the journal references Paramayudha [9], Kuo F Y and Hwang [10]. There is a modified geometry shape as shown in figure 3. In this antenna design, there is a gap or slot to get horizontal polarization. Horizontal polarization occurs because of changes in the direction of surface current that flow on the patch. Before inserting a gap in a patch, the surface current flow vertically. With the gap in the patch, the surface will flow orizontally.

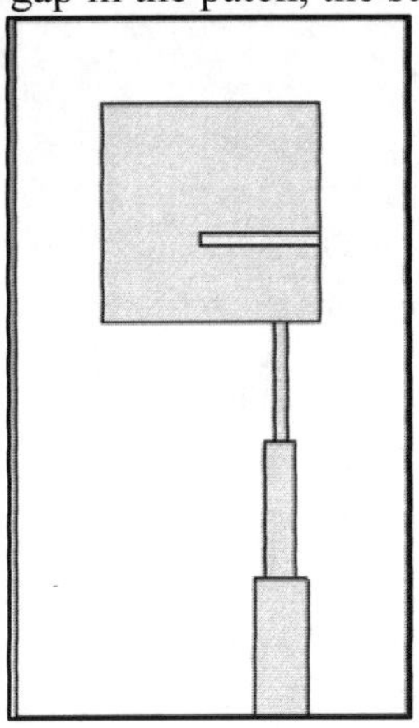

Figure 3. Modified antenna type.

3.1.2. Antenna design. For the initial design of the antenna dimensions, we use dimensional calculations for rectangular microstrip antennas as shown in table 3. This antenna uses the microstrip feeding rationing technique.

IOP Conf. Series: Materials Science and Engineering **850** (2020) 012060 doi:10.1088/1757-899X/850/1/012060

Table 3. Calculation of basic parameters for microstrip antennas.

No	Dimension	Formula and results
1.	Patch width	W = $\dfrac{c}{2f_0\sqrt{\dfrac{\varepsilon_r+1}{2}}}$ = 17.547 mm
2.	Patch Length	a. $\varepsilon_e = \dfrac{\varepsilon_r+1}{2} + \dfrac{\varepsilon_r-1}{2}\left(1+12\dfrac{h}{w}\right)^{-\frac{1}{2}}$ = 3.260 mm b. $\dfrac{\Delta L}{h} = 0.412\dfrac{(\varepsilon_e+0.3)\left[\dfrac{W}{h}+0.264\right]}{(\varepsilon_e-0.258)\left[\dfrac{W}{h}+0.813\right]}$ = 0.7114 mm c. $L = \dfrac{c}{2f_r\sqrt{\varepsilon_e}} - 2\Delta L$ = 14.835 mm
3.	a. Substrate width / ground b. Length of substrate/ground	a. Ws = 6h + W = (6 x 1.524) + 17.547 = 26.691 mm b. Ls = 6h + L = (6 x 1.524) + 14.835 = 23.979 mm
4.	a. Feed Channel Width b. Length of Feed Channels c. Width and length of matching impedance d. W50 length and width	a. For get B, value, use the equation $B = \dfrac{60\pi^2}{Z_0\sqrt{\varepsilon_r}}$ = 6.1844 and get the W_F value using the equation $W = \dfrac{2h}{\pi}\left\{B-1-ln(2B-1)+\dfrac{\varepsilon_r-1}{2\varepsilon_r}\left[ln(B-1)+0{,}39-\dfrac{0{,}61}{\varepsilon_r}\right]\right\}$ =3.33mm b. $L = \dfrac{1}{4}\lambda_d$. = 7 mm c. W= 1.813 mm L= 8 mm d. W50= 3.33 mm L50= 7.9 mm
5.	Relative Dielectric Constant (εreff)	To get the ε_{reff} value, use the equation below, In this case $W_F/h \geq 1$, $\varepsilon_{reff} = \dfrac{\varepsilon_r+1}{2} + \dfrac{\varepsilon_r-1}{2}\left(1+12\left(\dfrac{h}{w}\right)\right)^{\frac{1}{2}}$ = **2,72**

3.2. *Simulation*

After the calculation step, the simulation process is carried out using CST Studio Suite 2018. At this step, the optimization process is carried out, by changing the dimensions of the antenna, to obtain the desired specifications.

3.2.1. Initial design. The initial design list parameters are listed in table 4. Antenna simulation results using initial design parameters, are not met all the desired specifications. The operating frequency is 4.35 GHz instead of 5.6 GHz, the antenna impedance is 33.56 ohms instead of 50 ohms, polarization is not horizontal and gain is less than 5 dB. Return loss and VSWR values, -14.8 dB and 1.4 respectively, are already met the desired specifications.

Table 4. Initial design list parameters.

Name	Expression	Value	Description
pw	17.547	17.547	Patch width
feed_length	7	7	Feeder length
Lslit	5	5	Slit length
Lp1	6.1	6.1	Slit to patch bottom distance
Wslit	feed_width	6.1844	Slit width
feed_width	6.1844	6.1844	Feeder width
f_shit	2	2	Feeder shift
sub_width	1.524	1.524	Substrate thickness
lw	14.835	14.835	Patch length

Figure 4-9 shown respectively, the return loss, antenna impedance, VSWR, polarization, gain, and radiation pattern of initial design result.

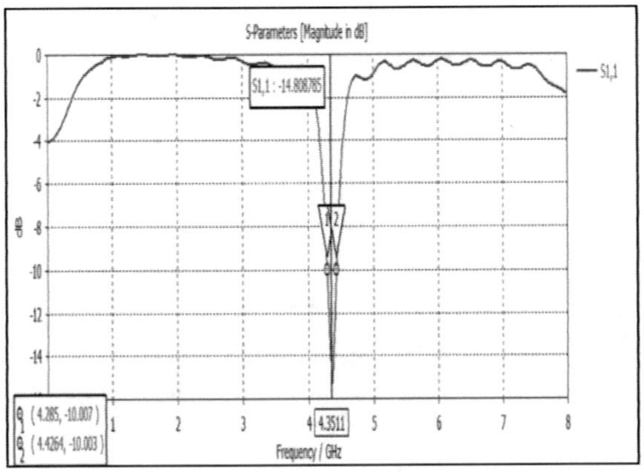

Figure 4. Initial design s-parameters.

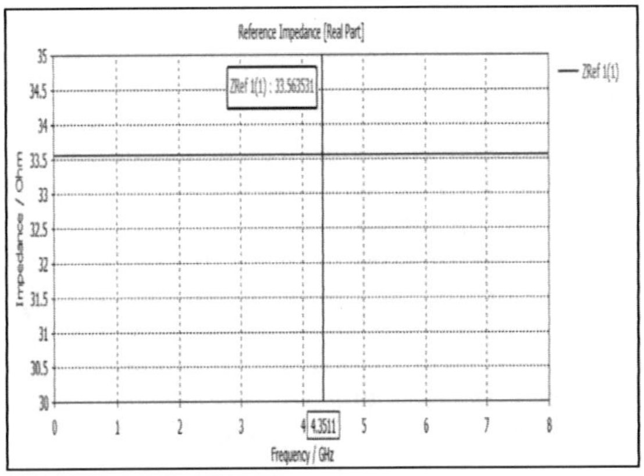

Figure 5. Impedance of initial design.

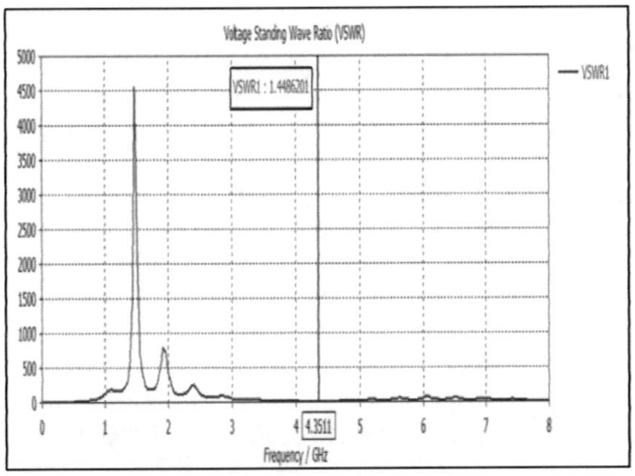

Figure 6. Initial design VSWR.

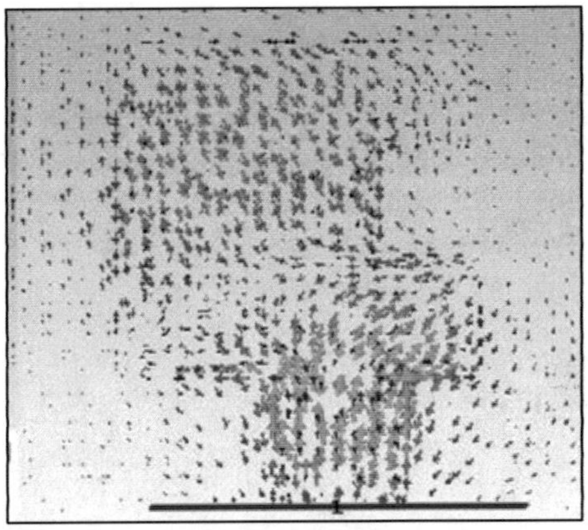

Figure 7. Initial design polarization.

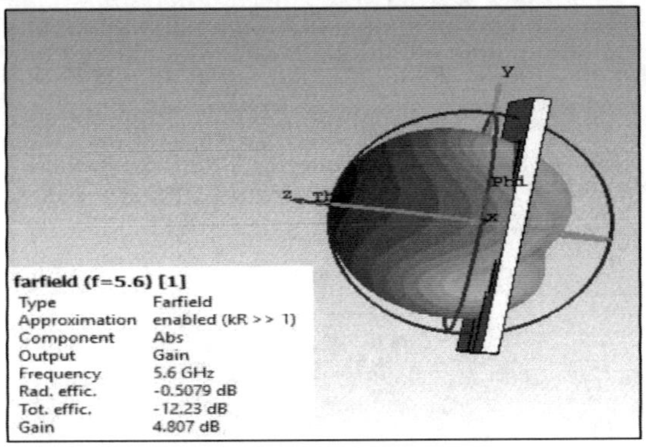

Figure 8. Initial design gain.

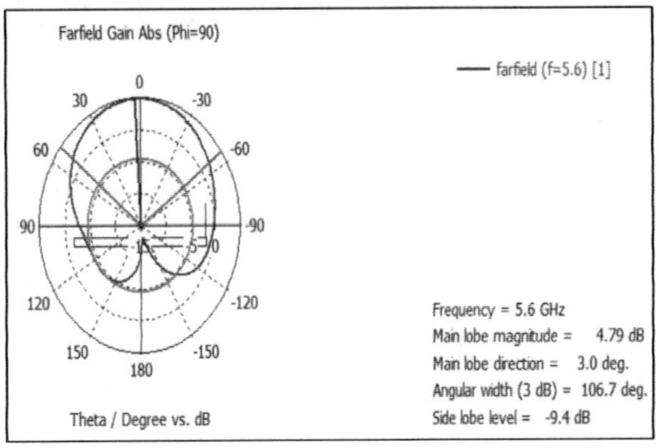

Figure 9. Initial design radiation pattern.

To get a design that fits the desired antenna specifications, optimization must be done by changing the antenna dimension values. Table 5 Shown the optimization results using parameters sweep.

Table 5. Optimization result using parameters sweep.

No.	Pw (mm)	Feed Width (mm)	Lp1	L slit	Feed Length (mm)	Frec. (GHz)	Imp (ohm)	Return Loss (dB)	Gain (dB)	Polari-zation	VSWR
1	17.547	6.18	6.1	5	7	4.3	33.5	-14.8	4	Vertical	1.4
2	16	6.18	6.1	5	7	4.8	33.5	-13.2	4.8	Vertical	1.4
3	16	5	6.1	5	7	4.72	38.9	-13.6	6.5	Vertical	1.5
4	9.75	0.66	3	5	5	7.12	105	-8.6	3.6	Vertical	2.1
5	12.65	0.66	3	5	25	7.64	106.7	-6.4	5.5	Vertical	2.4
6	12.65	0.66	3	7	25	5.68	106.8	-6.0	5.6	Horizontal	2.9
7	12.65	0.66	4.5	7	25	5.68	106.7	-5.8	5.5	Horizontal	3.05

The optimization using the sweep parameter, as in table 5, have not yielded satisfactory results. From the fifth experiment, only the gain meets the specifications. In the sixth experiment, we expanded the feed length and Lslit value to obtain horizontal polarization. The addition of LP1 in seventh experiment did not give better results yet. Next, we added a matching impedance by put in a new parameter, W50 (width of matching impedance) into the simulation as shown in table 6, and produces an antenna with the desired specifications.

Table 6. Optimization results using matching impedance.

No.	Pw (mm)	W50 (mm)	Feed Width (mm)	Lp1	lslit	Feed Length (mm)	Frec. (GHz)	Imp (ohm)	Return Loss (dB)	Gain (dB)	Polari-zation	VSWR
8	12.65	3.0375	0.75	4.5	7	25	5.6	50.11	-37	6.08	Horizontal	1.02
9	12.65	3.04	0.75	4.5	7	25	5.6	50.08	-37.05	6.08	Horizontal	1.02
10	12.65	3.0429	0.75	4.5	7	25	5.6	50.05	-37.1	6.08	Horizontal	1.02
11	12.65	3.048	0.75	4.5	7	25	5.6	50.00	-37.18	6.08	Horizontal	1.02
12	12.65	3.048	0.75	4.5	7	20	5.6	50.00	-20	6.2	Horizontal	1.2
13	12.65	3.048	0.75	4.5	7	15	5.6	50.00	-22	6.16	Horizontal	1.17
14	12.65	3.048	0.75	4.5	7	12	5.6	50.00	-26.1	5.9	Horizontal	1.1
15	12.65	3.048	0.75	4.5	7	7	5.6	50.00	-26.8	5.6	Horizontal	1.09

As shown in table 6, the addition of matching impedance can match the antenna impedance to 50 ohms line impedance. By reducing the feed length value as in 15th experiment, the VSWR will be smaller. Reduction in feed length as shown in 15th experiment, giving the desired antenna specifications.

3.2.2. Final design. After optimization, the antenna meets the specified specifications. The final dimension is shown in table 7.

Table 7. Final design parameters list.

Name	Expression	Value	Description
sub_add	5	5	Substrate width addition
pw	12.65	12.65	Patch width
feed_length	7	7	Feeder length
Lslit	7	7	Slit length
Lp1	4.5	4.5	Slit to patch bottom distance
Wslit	feed_width	0.75	Slit width
feed_width	0.75	0.75	Feeder width
f_shit	4	4	Feeder shift
sub_width	1.524	1.524	Substrate thickness
w50	3.048	3.048	Matching impedance

The results obtained from the final optimization are as shown in figure 10-15:

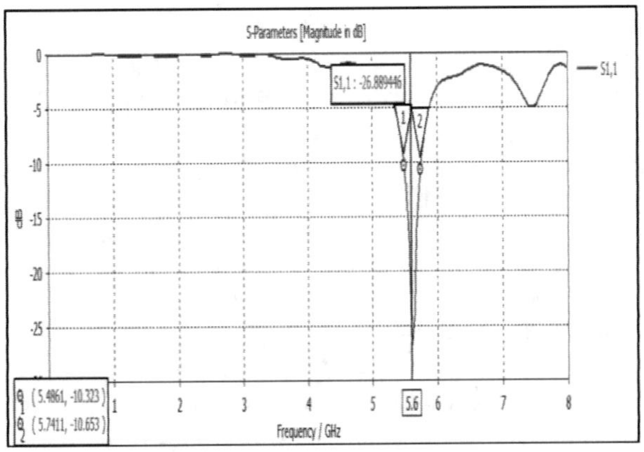

Figure 10. S-Parameter final design.

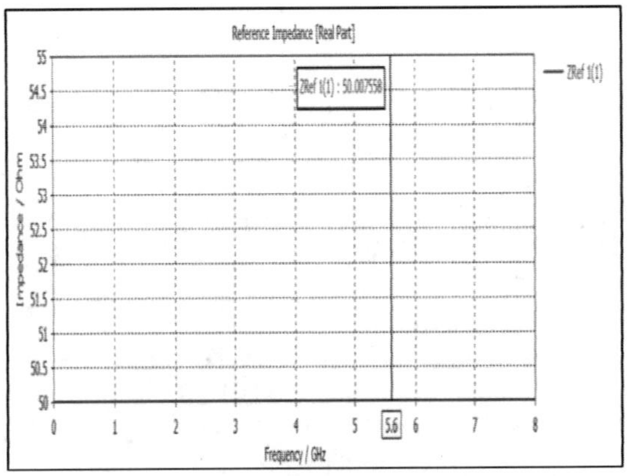

Figure 11. Final design impedance.

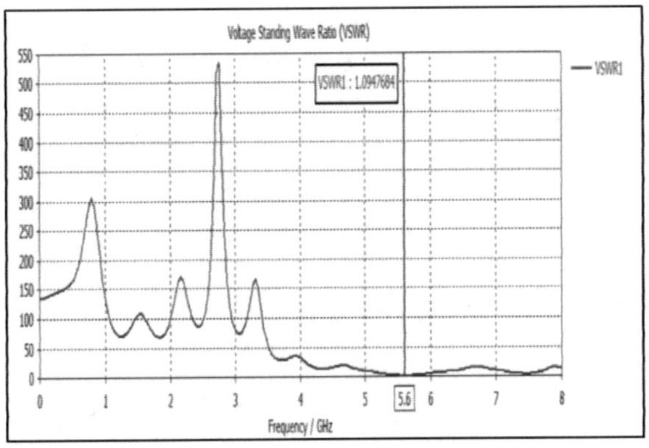

Figure 12. Final design VSWR.

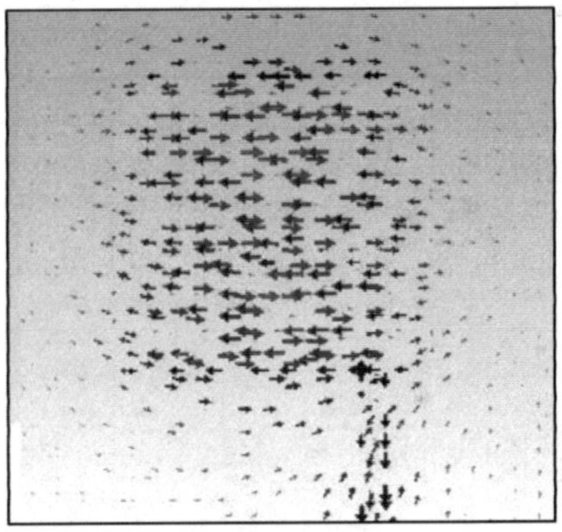

Figure 13. Final design polarization.

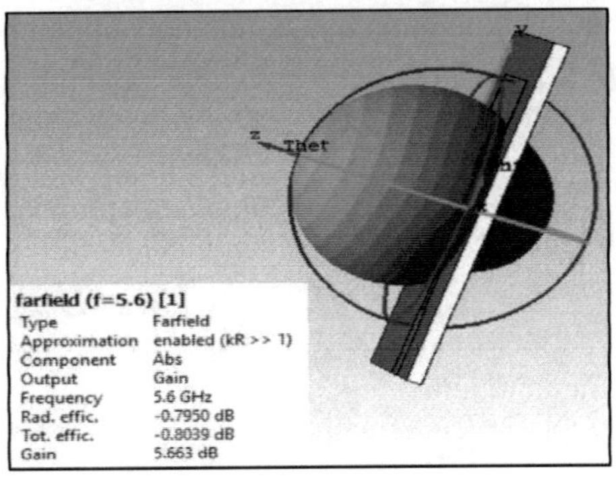

Figure 14. Final design gain.

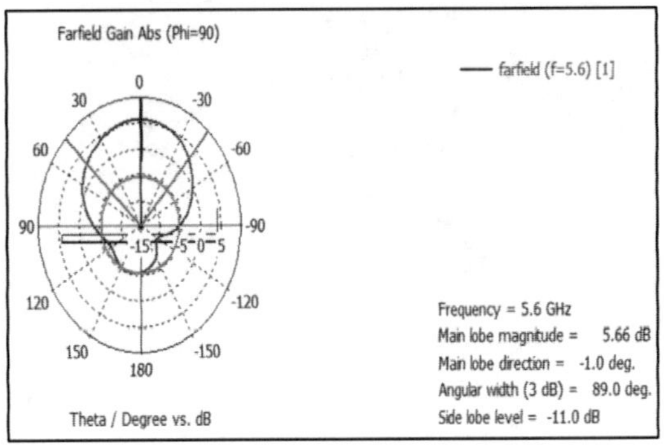

Figure 15. Final design radiation pattern.

4. Conclusion

A C-Band microstrip antenna for the Surveillance Ground radar has been designed. The desired antenna specifications have been reached, for the return loss value of -26.8dB, 1.09 VSWR, 50 ohm impedance, directional radiation pattern, horizontal polarization due to patch loopholes, 200 MHz bandwidth and gain of 5.6 dB. All parameters affect work frequency, so accuracy is needed in analyzing and determining dimensions.

References

[1] Molchanov P, Gupta S, Kim K and Pulli K 2015 Short-range FMCW monopulse radar for hand-gesture sensing *2015 IEEE Radar Conference (RadarCon)* (pp. 1491-1496) IEEE
[2] Briggs J N 2004 *Target detection by marine radar* **16** Iet.
[3] Beenamole K S 2009 Microstrip Antenna Designs for Radar Applications *DRDO Science Spectrum* 84-86
[4] Klugmann D 2016 FMCW radar in the digital age *2016 IEEE International Geoscience and Remote Sensing Symposium (IGARSS)* (pp. 7370-7373) IEEE
[5] Boskovic N, Jokanovic B, Oliveri F and Tarchi D 2015 High gain printed antenna array for FMCW radar at 17 GHz *2015 12th International Conference on Telecommunication in Modern Satellite, Cable and Broadcasting Services (TELSIKS)* (pp. 164-167) IEEE
[6] Samudrala S and Palle H C 2016 Rectangular microstrip patch antenna array for C-band scatterometer and MBI *2016 IEEE Annual India Conference (INDICON)* (pp. 1-4) IEEE
[7] Monebhurrun V 2018 Use of Standard Definitions of Terms for Antennas & Propagation: IEEE Std. 145 & IEEE Std. 211 *2018 IEEE Conference on Antenna Measurements & Applications (CAMA)* (pp. 1-2) IEEE
[8] Balanis C A 2016 *Antenna theory: analysis and design* (John wiley & sons)
[9] Paramayudha K, Amrullah Y S and Wahyu Y 2016 X-band antenna design for Indonesian ground surveillance man pack radar *2016 2nd International Conference of Industrial, Mechanical, Electrical, and Chemical Engineering (ICIMECE)* (pp. 1-3) IEEE
[10] Kuo F Y and Hwang R B 2014 High-isolation X-band marine radar antenna design *IEEE Transactions on Antennas and Propagation* **62**(5) 2331-2337

IOP Conf. Series: Materials Science and Engineering **850** (2020) 012061 doi:10.1088/1757-899X/850/1/012061

Analysis of the relevance between Vocational High School competencies and the needs of the industrial sector on Computer and Network Engineering Competencies

S Kamilah, A H S Budi* and W Purnama

Department of Electrical Engineering Education, Universitas Pendidikan Indonesia, Jalan Dr. Setiabudhi 207, Bandung 40154, Indonesia

*agusheri@aol.com

Abstract. The curriculum applied in vocational schools aims to synchronize graduate competencies with the demands of the industrial sector on workforce competencies. This study aims to analyze the relevance of the 2013 Vocational High School curriculum on Computer and Network Engineering Competencies and the needs of the workforce in the Telecommunications Industry Sector. This research was conducted using quantitative descriptive methods. The study population was the Vocational High School and the Telecommunications Industry in West Java Province and DKI Jakarta. The sample is determined by purposive sampling technique for the Telecommunications Industry and saturated sampling for Vocational High Schools. Participants in this study were 6 respondents from the Vocational High School and 21 respondents from the Telecommunications Industry. The research instrument used was a questionnaire. The analysis technique used is recall and precision. The results showed a high relevance between Basic Competence in the field of computer and network engineering and the need for workforce in the telecommunications industry sector by 83.84%.

1. Introduction

Vocational schools in implementing their education programs have a curriculum. The curriculum, in this case, is a set of plans and arrangements regarding the purpose, content, and material of the lesson as well as the methods used as guidelines for the implementation of learning activities to achieve certain educational goals. In the field of education, the curriculum is an important element in any form and model of education. Without a curriculum, it is difficult for educational planners to achieve their educational goals. The relevance of vocational schools to the workplace causes the curriculum implemented in schools to meet the minimum criteria demanded by the workplace. As in the Regulation of the Director-General of Primary and Secondary Education of the Ministry of Education and Culture number: 06 / D.D5 / KK / 2018 deciding on the third point that in each skill competency opened, vocational high school can specialize certain competencies (concentration of expertise) according to the demands the needs of the workforce are related to not ignoring the basic abilities of these skills [1]. Based on these regulations students are required to be able to master certain competencies according to the needs of the workforce.

Current labor issues state that the industry is still low in absorbing vocational graduates, one of which is due to minimal skills or expertise. To improve vocational graduates, the Head of Sub Directorate of the Directorate of Vocational Alignment and Industrial Cooperation of the Ministry of Education and

Culture pushed for the school level to synchronize the curriculum where content must always be updated. Because the demands of the workplace are always changing [2].

The gap that occurs between Vocational Schools and Industry, especially in the Nasional Vocational High School 4 Bandung, according to its graduates, is due to not yet being taught some of the basic competencies needed by industry. Not all basic competencies were taught due to practicum tools were not yet available and the teacher could not predict the technology needed by the industry that would develop in the future. Conformity between basic competencies in schools and what is needed by the industry will facilitate synergy between Vocational Schools and the needs of the industrial world. According to the minister of education and culture of the Republic of Indonesia Regulation Number 24 of 2016, the basic competencies in the 2013 curriculum must contain the ability and learning material for a subject in each education unit.

Based on the background, it is very necessary to review the relevance of vocational competencies with the competencies needed by the workplace. In this study, we focus on Computer and Network Engineering competencies skill.

2. Vocational education and curriculum

Vocational education is an effort to provide a stimulus in the form of learning experiences and interactions with the world outside of the students to help them develop themselves and their potential [3]. Curriculum is an important part of vocational education. The curriculum is a series of activities carried out by students in the school environment and under the responsibility of the school, in this case, is the vocational curriculum. As one expert explained, according to Saylor J. Gallen and William N. Alexander in his book "Curriculum Planning" states the curriculum is "The entire school effort to influence learning both take place in class, on the page or outside the school" [4]. The relevance of vocational schools to the workplace causes the curriculum implemented in schools to meet the criteria set by the workforce. To achieve industrial needs, revitalizing VHS is needed to encourage industries and Vocational Schools to carry out education by the needs of two parties namely school and industry [5].

The development of the curriculum in Indonesia that the curriculum has been replaced 11 times from the beginning of independence, this can be seen in figure 1.

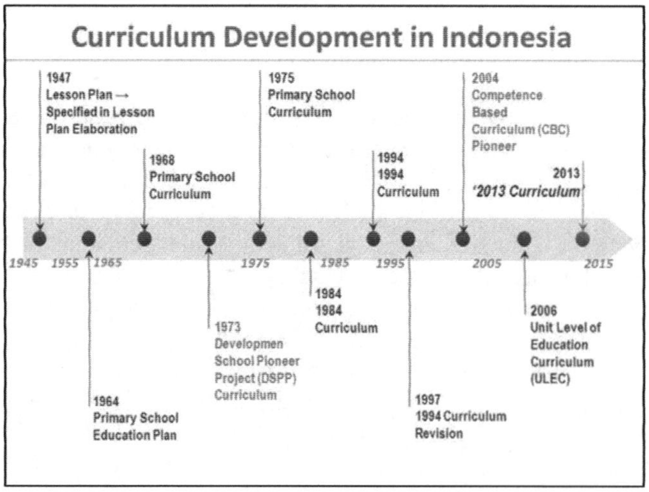

Figure 1. The development of the curriculum in Indonesia.

Based on figure 1, it can be seen from the development of curriculum usage since 1947 called the Lesson Plan to the revised 2013 curriculum. Since 2013 all schools including Vocational Schools are required to use the revised 2013 curriculum that is applied nationally.

IOP Conf. Series: Materials Science and Engineering **850** (2020) 012061 doi:10.1088/1757-899X/850/1/012061

2.1. Competence and basic competence
According to Government Regulation No. 32 of 2013 concerning national education standards, states "Competence is a set of attitudes, knowledge, and skills that must be owned, internalized, and mastered by students after learning a learning content, completing a program, or completing a particular education unit" [6]. The core competency in the 2013 curriculum is the level of ability to achieve graduate competence standards that must be possessed by students at each grade level. Basic competence in the 2013 curriculum contains the ability and learning material for a subject in each education unit that refers to core competencies [7].

2.2. Theory of suitability
The suitability is the appropriate subject matter; harmony (about opinions, understandings, tones, color combinations and so on; compatibility). Thus, what is meant by conformity in this study is the alignment or compatibility of the basic competencies of the revised 2013 curriculum with the needs of the working sector of the industrial sector. The suitability of basic competencies will facilitate the synergy between vocational high schools and the needs of the industrial world. This is in line with one of the objectives of vocational education, namely to prepare graduates to work directly in the industry [8].

2.3. Workplace and industry
According to law Number 3 of 2014 concerning Industry states "Industry is all forms of economic activities that process raw materials or utilize industrial resources to produce goods that have higher added value or benefits, including industrial services" [9]. The industry is an effort to process raw materials or semi-finished goods into finished goods that have added value to obtain profits [10]. The workplace is a complex environmental condition that has various interrelated aspects. Aspects of the workplace include aspects of the work environment and individual aspects. The workplace and the industrial world is a complex environment that has an important role in improving the economy.

2.4. KKNI
The Indonesian National Qualifications Framework (KKNI) is a competency qualification framework that can juxtapose, equalize and integrate between the field of education and the field of job training and work experience to provide recognition of work competencies by the work structure in various sectors [11]. Vocational graduates have the second level qualification in the KKNI as an operator. The basic competencies possessed by vocational high school graduates are adjusted to their respective competency skills. Figure 2 is the level of KKNI.

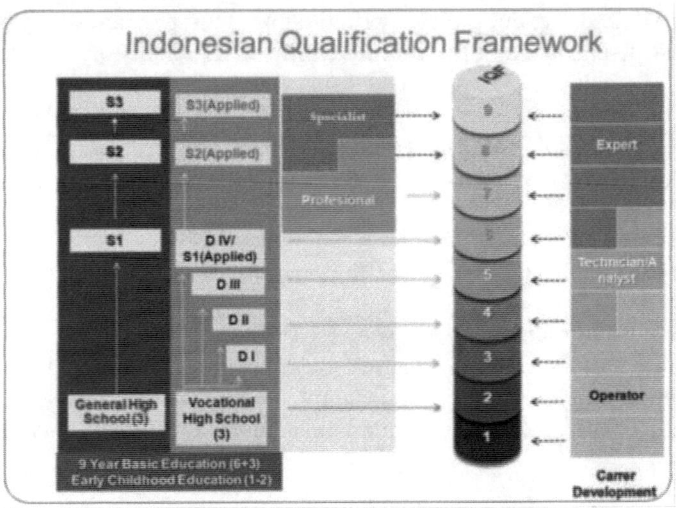

Figure 2. Level of KKNI.

2.5. SKKNI
The preparation of these competencies is based on the Indonesian National Work Competency Standards (SKKNI) which are adjusted to the industrial world or the business world for each skill competency. Based on the Minister of Manpower Regulation of the Republic of Indonesia Number 3 of 2016 concerning Indonesian National Work Competency Standards states that "Indonesian National Work Competency Standard is a formulation of work capabilities that includes aspects of knowledge, skills and/or expertise and work attitudes relevant to the implementation of tasks and Job requirements set in accordance with statutory provisions" [12].

3. Research methodology
In the study of analysis of relevancy between VHS competencies with industrial sector needs, especially in computer and network engineering competencies skill in 2013 curriculum, using descriptive research with a quantitative approach. Descriptive research is a method in examining the status of a group of humans, an object, a condition, a system of thought and the purpose of descriptive research helps provide an overview or description of a phenomenon under investigation [13]. Quantitative research is a study that requires the use of numbers in each stage of its research, starting from data collection, estimation of data and results of data [14].

3.1. Population and samples
The population in this study was Vocational High School and Telecommunication Industry. The sample is part of the number and characteristics of the population. To determine the sample to be used in the study, there are various sampling techniques used. This study uses purposive sampling and a saturated sampling technique. Purposive sampling is a sampling technique that sources data based on certain considerations [15]. Purposive sampling techniques are used to determine samples or participants in the Telecommunications Industry. Saturated sampling is a technique of determining the sample in which all members in the population are used as samples [15]. The technique used to determine the sample or participants at VHS 4 Bandung. Participants in this study were 6 respondents from Vocational High School and 21 respondents from PT. INTI, PT. LEN, PT. Dirgantara Indonesia (DI), PT. Telkom, PT. Aplikanusa Lintasarta, PT. Ghuma, PT. Gojek, PT. Nexwave, PT. Infomedia and PT. Indosat.

3.2. Research instruments
In this study, a closed and open questionnaire as instruments are used to collect data. The questionnaire uses a Guttman scale with yes or no answers. Industrial questionnaires are used to obtain data on the basic competencies needed by the industrial sector. The school questionnaire is used to obtain data on basic competencies in VHS 4 Bandung along with the basic competencies taught and not taught. The validity test of the instrument is using the reproducibility coefficient formula (kr) and the scalability coefficient (ks) [16]. The calculated result of kr value is 0.948 and ks value obtained after testing is 0.896, which shows that the instrument is suitable for use [16]. Test reliability using Kuder-Richardson 20 (KR 20) gives a result of 0.961 (the best result was> 0.8), that means the questionnaire used in the research was reliable [17].

3.3. Research procedure
This research was conducted through steps as follows:

3.3.1. Planning stage. At the planning stage, the researcher takes preparatory steps such as literature studies, determines the problems to be studied, develops problem formulation, chooses research methods, then determines the population and sample be studied, composes research instruments, performs validation and data collection techniques and analysis techniques data to be used.

3.3.2. Implementation phase. In the implementation phase, the researcher carried out implementation steps such as distributing questionnaires to the respondent then analyzing and processing the data with the chosen method.

3.3.3. Reporting stage. At the reporting stage, researchers compile a report by describing the findings that have been obtained according to the data found in the field and the results of data analysis that has been done, then provide suggestions for further research.

3.4. Data analysis
Data analysis is done by summing all the "Yes" answers by giving a value of 1 for each answer given by the respondent, adding up the percentage of basic competencies of each subject needed by the industry, and calculating the average value of basic competencies percentage for each subject needed by the industry using equation (1):

$$X = \frac{\sum X}{N} \tag{1}$$

Note:
X = The average value of the basic competencies of each subject needed by the industry
$\sum X$ = Amount of percentage of each basic competency needed by industry
N = Number of basic competency subjects in the Computer and Network Engineering competence skills (TKJ).

To find out the percentage value of the basic competencies needed in each productive study field is categorized in the following values [18]:

- Excellent (very relevant / appropriate) 76% -100%
- Good (relevant / appropriate) 56% - 75%
- Fair (less relevant / appropriate) 40% - 55%
- Poor (irrelevant / appropriate) <40%

Furthermore, looking for the value of suitability/relevance of the basic competencies in the expertise of Computer and Network Engineering expertise needed by the industrial world using equations (2) and (3) [19].

$$\text{Recall} = \frac{a}{a+c} \text{x } 100\ \% \tag{2}$$

$$\text{Precision} = \frac{a}{a+b} \text{x } 100\ \% \tag{3}$$

Note:
a (hits) = relevant / appropriate document
b (noise) = irrelevant / appropriate document
c (misses) = relevant / appropriate documents not found
d (reject) = irrelevant / appropriate document not found

4. Research results and discussion
The research result is shown in table 1.

IOP Conf. Series: Materials Science and Engineering **850** (2020) 012061 doi:10.1088/1757-899X/850/1/012061

Table 1. Number of basic competencies.

No.	Item	Basic competencies number
1.	Basic competencies TKJ program competency 2013 Curriculum Revision	135
2.	Basic competencies are taught at Vocational High School 4 Bandung	119
3.	Basic competencies are not taught at Vocational High School 4 Bandung	16
4.	Basic competencies are relevant and needed by the telecommunications industry	125
5.	Basic competencies are less relevant and less needed by the telecommunications industry	10
6.	Basic competencies needed by the telecommunications industry but not listed in the curriculum 2013 (R)	26

After knowing the basic competencies needed as shown in table 1, the calculation of suitability or relevance is as follows.

- Recall $= \frac{\text{a}}{\text{a}+c} x100\%$

  $= \frac{125}{125+26} x100\% = 77.64\ \%$

- Precision $= \frac{\text{a}}{\text{a}+b}\text{x } 100\ \%$

  $= \frac{125}{125+11}\text{x } 100\% = 92.60\ \%$

Based on the calculations above, it can be seen that the percentage of competency suitability of TKJ expertise with the needs of the workforce in the telecommunications industry sector has a recall value of 77.64% and a precision of 92.60%. To find out how much suitability or relevance, that the ideal condition of the effectiveness of an information retrieval system is when the recall ratio and precision are equal (1: 1) [20]. Therefore, the ratio of recall and precision will be calculated and multiplied by 100% to find out the percentage.

- Ratio $= \frac{\text{Recall}}{Precsion} = \frac{77.64}{92.60} =\ 0.8384$
- Relevance $= 0.8384\, x\ 100\% =\ 83.84\%$

Based on the calculation of the ratio, the value of suitability or relevance is 83.84% and it is said that basic competencies are very suitable with the needs of the workforce in the telecommunications industry sector.

- Basic competencies taught at TKJ program of Vocational High School 4 Bandung. There are 135 basic competencies in TKJ expertise competencies at VHS 4 Bandung, with details of 119 basic competencies which are taught basic competencies and 16 basic competencies are basic competencies which are not taught by teachers at VHS 4 Bandung. Competencies that are not taught include competence Testing computer performance, Managing repairs to application software installations, Installing various use resources on computer networks, Installing internet connections on workstations, Designing local networks (LANs), Installing local networks (LANs), Maintenance of local networks (LAN), Managing repair of local networks (LAN), Installing software programming languages, Creating interfaces (User Interfaces) on applications, Creating program codes for various control structures in application interfaces (User Interface), Formulating a simple package installer application, Using a fiber optic work tool, making a fiber optic connection, configuring a passive fiber optic network device, repairing a fiber optic network. Basic competencies that are not taught are due to several things including inadequate practice tools, besides that basic competencies are more emphasized for RPL expertise competencies rather than TKJ and insufficient time for the implementation of learning. basic competencies that is not taught is the basic competencies required by the Telecommunications Industry.

- Competencies needed by the Telecommunications Industry World on TKJ program. As technology develops, the needs of the industrial world, especially the telecommunications sector develop as well. Based on the research that has been carried out, the basic competencies required by the telecommunications industry is 151 basic competencies, of which 125 basic competencies are the basic competencies listed in the revised 2013 curriculum and 26 basic competencies are basic competencies added by the telecommunications industry. The most additional competency is in computer and network subjects where basic competencies related to fiber optic networks are needed. This is due to the high need for fiber optics for the modernization of telecommunications operator networks that previously used copper cables [21]. This further supports that VHS graduates are expected to be proficient in fiber optic related competencies.

The competencies needed by the industry as the era grows continues to increase following the development of science and technology (science and technology), local, national and global needs [22]. This shows the competencies needed must be relevant to current needs where the implication is that technical education providers must consider their existing curriculum, which must be embedded with the competencies needed [23]. This makes the need for creativity and innovation from the teacher to develop learning. The important thing about the implementation of the revised 2013 curriculum is the change in the mindset of teachers in carrying out their duties related to teaching. This curriculum is also expected to be able to overcome the problem of the gap between education and the workplace, then the long-term hope is to be able to form quality human resources [22]. Figure 3 is a circular diagram that maps the basic competencies needed by the telecommunications industry sector.

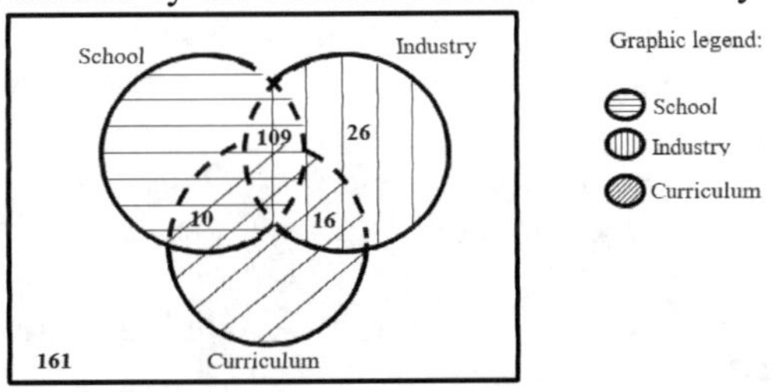

Figure 3. Circular diagram that maps the basic competencies needed by the telecommunications industry sector.

Note:

- The curriculum, with a value of 10 is the basic competencies taught but not needed by the workplace in the telecommunications industry sector.
- The curriculum, with a value of 16 is the basic competencies not taught and needed by the workplace in the telecommunications industry sector.
- The curriculum, with a value of 109 is the basic competencies taught and needed by the workplace in the telecommunications industry sector.
- Industry, with a value of 26 is the additional basic competencies from industry and needed by the workplace in the telecommunications industry sector.

5. Conclusion

Based on the results of the research, the relevance of basic competencies in the competence of computer and network engineering program and the needs of the telecommunications industry sector is 83.84% and suitable with industrial needs.

References

[1] Director General of Primary and Secondary Education of the Ministry of Education and Culture. Regulation No. 464 of 2018 concerning KI and KD subjects of national content (A), territorial content (B), basic areas of expertise (C1), basic skills programs (C2), expertise competency (C3).

[2] Leroux J A and Lafleur S 1995 Employability skills: The demands of the workplace *The Vocational Aspect of Education* **47**(2) 189-96

[3] Guile D and Griffiths T 2001 Learning through work experience *Journal of education and work* **14**(1)113-31

[4] Saylor J G and Alexander W M1974 *Planning curriculum for schools* Holt, Rinehart and Winston

[5] Murniati A R, Usman N and Azizah A 2016 Vocational School-Industry Partnership in Improving Graduate Competency *Jurnal Ilmiah Peuradeun* **4**(3) 269-80

[6] Indonesia Government Regulation Number 32 of 2013 *National Education Standards*

[7] Minister of Education and Culture. Regulation Number 24 of 2016 concerning core competencies and basic curriculum competencies in 2013 in primary and secondary education.

[8] Hamid M A, Nurtanto M, Rahmat A, Mutolib A, Nurhaji S, Fawaid M and Rizal S U 2018 The Analysis of Learning Implementation Plan (LIP) in Vocational Subjects Based on 2013 Curriculum *International Conference on Issues in Social and Education Research (ICISER 2017)* **2018** Atlantis Press

[9] Republic of Indonesia Law No. 34 of 2018 concerning Industry

[10] Lummus R R, Krumwiede D W and Vokurka R J 2001 The relationship of logistics to supply chain management: developing a common industry definition *Industrial management & data systems*

[11] Zamtinah M, Soenarto M and Mardapi D 2017 Developing a Model of Recognition of Work Experience and Learning Outcomes Based on Indonesian National Qualifications Framework for Vocational High Schools Students *InInternational Conference on Technology and Vocational Teachers (ICTVT 2017)* **2017** Atlantis Press

[12] Minister of Manpower 2016 *Regulation Number 3 of 2016 concerning Indonesian National Work Competency Standards*

[13] Nassaji H 2015 *Qualitative and descriptive research: Data type versus data analysis*

[14] Ludwig R and Johnston J 2016 How to build a quantitative research project *Radiologic technology* **87**(6) 713-5

[15] Teddlie C and Yu F Mixed methods sampling: A typology with examples *Journal of mixed methods research* **1**(1) 77-100

[16] Gouze K R and Nadelman L 1980 Constancy of gender identity for self and others in children between the ages of three and seven *Child Development* 275-8

[17] Fredrikson M 1983 Reliability and validity of some specific fear questionnaires *Scandinavian journal of psychology* **24**(1) 331-4

[18] Arikunto S 2013 *Manajemen pengajaran secara manusiawi* (Jakarta: Rineka Cipta)

[19] Chu H 2003 *Information representation and retrieval in the digital age* (Information Today, Inc.)

[20] Pao M L 1989 *Concepts of Information Retrieval* (Englewood, Colorado: Libraries Unlimited)

[21] Chomycz B 2000 *Fiber optic installer's field manual* (McGraw-Hill, Inc.)

[22] Ghobakhloo M 2018 The future of manufacturing industry: a strategic roadmap toward Industry 4.0 *Journal of Manufacturing Technology Management*

[23] Budi A, Kaprawi N and Yunos J 2012 *Sustainable Competencies for Electronic Engineering Graduate from Industry Perspectives* **37**

AUTHOR INDEX

Institute of Physics
Dirac House, Temple Back
Bristol BS1 6BE UK

ISSN: 1757-8981
ISBN 978-1-7138-3613-1

9 781713 836131

CURRAN ASSOCIATES INC.
proceedings.com